W9-BZC-256

www.wadsworth.com

wadsworth.com is the World Wide Web site for Wadsworth Publishing Company and is your direct source to dozens of online resources.

At *wadsworth.com* you can find out about supplements, demonstration software, and student resources. You can also send e-mail to many of our authors and preview new publications and exciting new technologies.

wadsworth.com
Changing the way the world learns®

A Note from the Author

Writing is my joy, sociology my passion. I delight in putting words together in a way that makes people learn or laugh or both. Sociology is one way I can do just that. It represents our last, best hope for planet-training our race and finding ways for us to live together. I feel a special excitement at being present when sociology, at last, comes into focus as an idea whose time has come.

I grew up in small-town Vermont and New Hampshire. When I announced I wanted to be an auto-body mechanic, like my dad, my teacher told me I should go to college instead. When Malcolm X announced he wanted to be a lawyer, his teacher told him a colored boy should be something more like a carpenter. The difference in our experiences says something powerful about the idea of a level playing field. The inequalities among ethnic groups run deep.

I ventured into the outer world by way of Harvard, the USMC, U.C. Berkeley, and 12 years teaching at the University of Hawaii. Along the way, I married Sheila two months after our first date, and we created Aaron three years after that: two of my wisest acts. I resigned from teaching in 1980 and wrote full-time for seven years, until the call of the classroom became too loud to ignore. For me, teaching is like playing jazz. Even if you perform the same number over and over, it never comes out the same twice, and you don't know exactly what it'll sound like until you hear it. Teaching is like writing with your voice.

At last, I have matured enough to rediscover and appreciate my roots in Vermont each summer. Rather than a return to the past, it feels more like the next turn in a widening spiral. I can't wait to see what's around the next bend.

THE BASICS OF SOCIAL RESEARCH

Second Edition

Earl Babbie

Chapman University

WADSWORTH

™

THOMSON LEARNING   Australia•Canada•Mexico•Singapore•Spain•United Kingdom•United States

WADSWORTH
THOMSON LEARNING

Sociology Editor: Lin Marshall
Assistant Editor: Analie Barnett
Editorial Assistant: Reilly O'Neal
Marketing Manager: Matthew Wright
Project Manager, Editorial Production: Jerilyn Emori
Print/Media Buyers: April Vanderbilt, Barbara Stephan
Permissions Editor: Stephanie Keough-Hedges
Production Service: Greg Hubit Bookworks

Text Designer: Lois Stanfield
Copy Editor: Molly D. Roth
Illustrator: Lotus Art
Cover Designer: Ross Carron
Cover Image: The Image Bank/Chris Close
Compositor: G&S Typesetters, Inc.
Text and Cover Printer: Von Hoffmann Press, Inc.

COPYRIGHT © 2002 Wadsworth Group. Wadsworth is an imprint of the Wadsworth Group, a division of Thomson Learning, Inc. Thomson Learning™ is a trademark used herein under license.

ALL RIGHTS RESERVED. No part of this work covered by the copyright hereon may be reproduced or used in any form or by any means—graphic, electronic, or mechanical, including photocopying, recording, taping, Web distribution, or information storage and retrieval systems—without the written permission of the publisher.

Printed in the United States of America
1 2 3 4 5 6 7 05 04 03 02

For permission to use material from this text, contact us by
Web: http://www.thomsonrights.com
Fax: 1-800-730-2215
Phone: 1-800-730-2214

ExamView™ and *ExamView Pro*™ are registered trademarks of FSCreations, Inc. Windows is a registered trademark of the Microsoft Corporation used herein under license. Macintosh and Power Macintosh are registered trademarks of Apple Computer, Inc. Used herein under license.

COPYRIGHT 2002 Thomson Learning, Inc. All Rights Reserved. Thomson Learning *Web Tutor*™ is a trademark of Thomson Learning, Inc.

Library of Congress Cataloging-in-Publication Data
Babbie, Earl R.
   The basics of social research / Earl Babbie.—2nd ed.
      p. cm.
   Includes bibliographical references and index.
   ISBN 0-534-51904-0
   1. Social sciences—Research.   2. Social sciences—Methodology.   I. Title.
H62.B18 2002
301'.07'2—dc21   2001017978

Wadsworth/Thomson Learning
10 Davis Drive
Belmont, CA 94002-3098
USA

For more information about our products, contact us:
Thomson Learning Academic Resource Center
1-800-423-0563
http://www.wadsworth.com

International Headquarters
Thomson Learning
International Division
290 Harbor Drive, 2nd Floor
Stamford, CT 06902-7477
USA

UK/Europe/Middle East/South Africa
Thomson Learning
Berkshire House
168-173 High Holborn
London WC1V 7AA
United Kingdom

Asia
Thomson Learning
60 Albert Street, #15-01
Albert Complex
Singapore 189969

Canada
Nelson Thomson Learning
1120 Birchmount Road
Toronto, Ontario M1K 5G4
Canada

Dedication

Evelyn Fay Babbie

Accompanying This Textbook

Study Guide for *The Basics of Social Research*, Second Edition, by Theodore C. Wagenaar and Earl Babbie

This practical book of activities is designed to help reinforce and extend your understanding of the material in *The Basics of Social Research,* Second Edition. Each chapter contains objectives, a summary, key terms, multiple-choice review questions, discussion questions, and exercises that engage you in actual research and involve the analysis of data from the 1973 through 1998 General Social Surveys.

SPSS *Companion for Research Methods*, Second Edition, by Robert Griffith Turner, Jr.

This concise, user-friendly guide introduces students to basic navigation in the *SPSS Student Version 10.0 CD-ROM,* and features data analysis exercises correlated chapter-by-chapter with *The Basics of Social Research,* Second Edition that include how to enter data; create, save, and retrieve files; produce and interpret data summaries; and more.

Experiencing *Social Research: Using MicroCase*® *ExplorIt* by David Ayers

Providing MicroCase-centered illustrations, summaries, and computerized exercises keyed to *The Basics of Social Research,* Second Edition, this workbook provides students with the opportunity to apply key methodological concepts and skills as they collect, build, and analyze data files that feature real sociological data.

To order a copy of these texts, contact your bookstore.

Also by Earl Babbie and available from Wadsworth

Social Research for Consumers

Survey Research Methods, Second Edition

Research Methods for Social Work (with Allen Rubin), Third Edition

The Sociological Spirit: Critical Essays in a Critical Science, Second Edition

Research Methods for Criminal Justice and Criminology, Third Edition (with Michael Maxfield)

CONTENTS IN BRIEF

CONTENTS

Chapter 5
Conceptualization, Operationalization, and Measurement   113

Chapter 6
Indexes, Scales, and Typologies   145

PREFACE

The book in your hands has been about three decades in the making. It began in the classroom, when I was asked to teach a seminar in survey research beginning in 1968. Frustrated with the lack of good textbooks on the topic, I began fantasizing something I called "A Survey Research Cookbook and Other Fables," which was published in 1973 with a more sober title: *Survey Research Methods*.

The book was an immediate success, except there were few courses limited to survey research. A number of instructors around the country asked if the same guy could write a more general methods book, and *The Practice of Social Research* appeared two years later, in 1975. The latter book has become a fixture in social research instruction, with the 9th edition published in 2001. The official Chinese edition was published in Beijing in 2000.

Over the life of *Practice,* successive revisions have been based in large part on suggestions, comments, requests, and (let's tell the truth) corrections from my colleagues around the country and, increasingly, around the world. In the most recent years, there has been a growing request for a shorter book with a more *applied* orientation.

From the beginning, social scientists have been interested in both pure and applied research. Some have primarily justified their efforts in terms of "knowledge for knowledge's sake," while others have focused on how their research could affect the quality of people's lives at a practical level. Over time, the emphasis on these two orientations has shifted back and forth.

Early U.S. sociologists such as Lester Ward and Jane Addams were strongly committed to social re-

form and saw their social scientific training as preparation for making life better for those around them.

Whereas the third quarter of the 20th century saw a greater emphasis on quantitative, pure research, the century ended with a renaissance of concern for applied sociological research (sometimes called *sociological practice*) and also a renewed interest in qualitative research. *The Basics of Social Research* was first published in 1999 in support of these trends. The second edition aims to increase and improve that support.

The book can also be seen as a response to changes in teaching methods and in student demographics. In addition to the emphasis on applied research, some alternative teaching formats have called for a shorter book, and student economics have argued for a paperback book. While standard methods courses have continued using *The Practice of Social Research,* I've been delighted to see that *Basics* seems to have satisfied a substantial market as well. The fine-tuning in this second edition is intended to help *Basics* serve this group even better than before.

CHANGES IN THE SECOND EDITION

The first prominent change to the book can be found in Chapter 3. The first edition devoted this chapter to a discussion of causation. However, a number of instructors asked that research ethics be moved out of the appendixes into a more prominent position, and I've responded by putting it in Chapter 3. The discussions of causation have been moved to Chapter 4, in a new section on "The Logic of Nomothetic Causation," following the introductory discussion of explanation as a purpose of research.

Other prominent changes in this edition of the book concern qualitative research methods. As such, Chapter 10 has been radically revised in order to present some of the major paradigms used in qualitative research. Further, I've added a chapter on qualitative data analysis (Chapter 13). Although I integrate qualitative and quantitative methods through the book, the methods of data analysis differ enough to merit separate treatment. For example, the computer programs available for the two approaches differ, and I've offered some illustrations of two popular programs: HyperResearch and NVivo. (Primers on several qualitative and quantitative programs can be found on the Web site for this book.)

Throughout the book, I have used a set of icons to draw students' attention to passages that relate to the use of various computer programs, including SPSS, MicroCase, InfoTrac, and the World Wide Web. My intention is to further integrate those resources into the book.

As in any book revision, extensive attention has been given to updating examples, reference material, statistical data, and research techniques (e.g., online surveys) as well as improving pedagogical methods. Many of these changes reflect suggestions by my fellow instructors. I've also added some specific features that I hope will assist both students and instructors.

For example, each chapter now begins with "An Opening Quandary" to pique the reader's interest. I've tried to identify a puzzle or seeming paradox related to each chapter. It is my intention that reading the chapter will clarify the quandary, with the concluding "A Quandary Revisited" ensuring that.

In view of the emphasis on applied social research, I've added several boxes called "Applying the Results" to point to how a particular logical discussion or research procedure has applications in the day-to-day world, often outside the research arena. For example, I point out how the issues of conceptualization and operationalization apply directly to the balloting confusion in Florida (and elsewhere) during the 2000 presidential election.

THE eBABBIE RESOURCE CENTER

Recent years have seen remarkable technological developments that impinge on both doing and teaching social research. Key among these is the World Wide Web. The first edition of this book had an appendix on "Research in CyberSpace," introducing students to the Web and listing a number of relevant resources. In this edition, I have moved that appendix to the Web itself. There are two reasons for this. First, the list of resources can be rapidly and repeatedly updated, without waiting three years for a revised textbook. Second, resource listings on the Web can be hotlinked, so students can access relevant Web sites with a click of the mouse rather than having to type in long and complex URLs.

Web links are only a small part of The eBabbie Resource Center, found at:

http://sociology.wadsworth.com/

In addition, you will find:

- Primers for the use of:
  SPSS 10.0
  MicroCase
  NVivo
  HyperResearch
- A statistics primer
- Online instructional materials
- Guides for budgeting research projects
- Flash Cards
- InfoTrac Exercises
- Student Practice Quizzes
- And more continuously evolving social resource aids!

INSTRUCTOR RESOURCES

Instructor's Manual with Test Bank

Margaret Jendrek has prepared another excellent instructor's manual to help instructors write examinations. In addition to the usual multiple-choice, true-false, and essay questions, the manual provides resources for planning lectures and gives suggested answers for some of the student problems in the study guide. Also included is a listing of print, film, and Internet resources for instructors, an appendix of the General Social Survey data, as well as a concise user guide for *InfoTrac College Edition* and tips for using *WebTutor.* Although students may not appreciate examinations as a general principle, I know that they benefit from the clarity Marty brings to that task.

ExamView for Windows and Macintosh

ExamView allows instructors to create, deliver, and customize tests and study guides (both print and online) in minutes with this easy-to-use assessment and tutorial system. *ExamView* offers both a Quick Test Wizard and an Online Test Wizard that guide you step by step through the process of creating tests. You can build tests of up to 250 questions using up to 12 question types from the *Instructor's Manual with Test Bank,* and with *ExamView's* complete word processing capabilities, you can enter an unlimited number of new questions or edit existing questions.

Readings in Social Research
by Diane Kholos Wysocki

The concepts and methodologies of social research come to life as students read the compelling articles in this unique collection. Wysocki includes an interdisciplinary range of social science readings that focus on important methods and concepts typically covered in the social research course. This reader is specifically designed to accompany the Babbie social research texts.

STUDENT RESOURCES

Study Guide

The student study guide accompanying this text is modeled after the study guide that Ted Wagenaar

and I prepared for *The Practice of Social Research.* The study guide continues to be a mainstay in my own teaching. Students tell me they use it heavily as a review of the text, and I count the exercises as half their grade in the course. I specify a certain number of points for each exercise depending on how hard it is and how much I want them to do it, and give a deadline for each exercise, typically right after we've covered the materials in class. Most exercises rate between 5 and 25 points.

Finally, I specify the total number of points that will rate an A on the exercises, the range of points representing a B, and so forth. From there on, it's up to the students. They can do whichever exercises they want and as many as they want, as long as they complete each by its deadline. Every exercise they submit gets them some fraction of the maximum points assigned to it. Though I end up with a fair amount of grading during the course, my experience is that those who do the exercises also do better on exams and papers. The study guide contains exercises for students who have access to SPSS, as well as plenty for those who don't.

SPSS Student Version 10.0 CD-ROM

This new CD-ROM includes SPSS software and preloaded data files to provide students with all the tools they need to use SPSS software at their own computers for a fraction of the cost of the commercial version. This SPSS student version features the same functionality as the SPSS commercial version with very few limitations. The first two preloaded GSS data files include 2500 cases and 62 variables (500 cases for each of these years: 1978, 1983, 1988, 1993, and 1998). The third and fourth files include 1500 cases (approximately 300 cases per year) and contain 50 variables.

SPSS Companion for Research Methods
by Robert Griffith Turner

This booklet is a great partner to the text and to the SPSS Student Version 10.0 CD-ROM. This concise, user-friendly guide is correlated chapter by chapter with the text and CD-ROM to help students learn basic navigation in SPSS. It includes chapter-specific exercises as well as information on how to enter data; create, save, and retrieve files; produce and interpret data summaries; and much more.

GSS Data Disk

Over the years, we have sought to provide up-to-date computer support for students and instructors. Because there are now many excellent programs for analyzing data, we have provided data to be used with them. With this edition, we have updated the data disk to include the most recent data. The first two files have 2500 cases and 62 variables from the General Social Survey, 500 cases for each of these years: 1978, 1983, 1988, 1993, 1998. The last two files have 1500 cases (approximately 300 cases per year) and include 50 variables. This data set has been used for many of the examples in the textbook as well as for select exercises in the student study guide.

A Simple Guide to SPSS for Windows, Versions 8.0, 9.0 & 10.0

by Lee A. Kirkpatrick and Brooke C. Feeney

Perfect for first-time users of SPSS, this concise, straightforward book teaches students what they need to know to perform such procedures as stem-and-leaf displays, *t*-tests, multiple regressions, scatterplots, and other basics.

MULTIMEDIA RESOURCES

Thomson Learning WebTutor™ 2.0

WebTutor is a content-rich, Internet-based teaching and learning tool correlated chapter by chapter to the text. This incredible online resource gives instructors and students a virtual environment rich with study, course-management, and communication tools. Instructors can provide virtual office

hours, post syllabi, set up threaded discussions, and track student progress. *WebTutor*'s content can be customized in a number of ways, from uploading images and other resources to adding Web links and creating practice materials. For students, *WebTutor* offers real-time access to many study tools, including flashcards (with audio), the *Newbury House* dictionary, practice quizzes, online tutorials, and Web links—all correlated to the text. *WebTutor* is available on both WebCT and Blackboard platforms.

InfoTrac College Edition

To supplement the readings listed at the end of each chapter, students will be able to access *InfoTrac College Edition,* an online library of full-text magazine and journal articles. To use the resource, students should locate *The Basics of Social Research,* Second Edition at the Wadsworth Sociology Web site (http://sociology.wadsworth.com/) and check the *InfoTrac College Edition* readings recommended for each chapter of the book. I have placed the recommendations on the Web site rather than in the book, since the readings available will be constantly changing and the recommended readings updated regularly. *The Basics of Social Research,* Second Edition Web site also contains several other useful resources, including online study quizzes for each chapter, links to sociology-related newsgroups on the Web, and lessons on surfing the Internet.

The Practice of Social Research Video

A lecture-launching video featuring myself includes six ten-minute segments introducing traditionally challenging concepts—operationalization, sampling, experimental design, formulation of theory, ethics, percentages/indexes, and variables.

ACKNOWLEDGMENTS

It would be impossible to acknowledge adequately all the people who have influenced this book. My earlier methods text, *Survey Research Methods,* was dedicated to Samuel Stouffer, Paul Lazarsfeld, and Charles Glock. I again acknowledge my debt to them.

I also repeat my thanks to those colleagues acknowledged for their comments during the writing of the first, second, and third editions of *Survey Research Methods.* The present book still reflects their contributions.

Many other colleagues helped me revise *The Basics of Social Research* and its predecessors. I particularly want to thank the instructors who reviewed the manuscript of this edition and made helpful suggestions:

Jeffrey A. Burr, University of Massachusetts–Boston

Douglas Forbes, University of Wisconsin–Marshfield

Albert Hunter, Northwestern University

Ross Koppel, University of Pennsylvania

Susan E. Marshall, University of Texas at Austin

William G. Staples, University of Kansas

Stephen F. Steele, Anne Arundel Community College

Yvonne Vissing, Salem State College

Over the years, I have become more and more impressed by the important role played by editors in books like this. Since 1973, I've worked with six sociology editors at Wadsworth, which has in-volved the kinds of adjustments you might need to make in six successive marriages. Happily, this edition of the book has greatly profited from my partnership with Lin Marshall. While Lin offers a publishing veteran's experience to the project, she also brings an enthusiasm for new technologies and pedagogies and creates exciting ways to take advantage of them.

In my experience, copy editors are the invisible heroes of publishing, and it has been my good fortune and pleasure to have worked with one of the very best, Molly Roth, for several years and books. Among her many gifts, Molly has the uncanny ability to hear what I am trying to say and find ways to help others hear it.

This edition of the book introduces you to a young sociologist you'll see and hear more of in the future: Sandrine Zerbib. Currently completing her doctorate at the University of California, Irvine, Sandrine is a first-rate methodologist and scholar, working both the qualitative and quantitative sides of the street. She is particularly sensitive to feminist perspectives, and her experiences as a woman add a new dimension to the sociomethodological concerns we share. Sandrine's efforts are most apparent in Chapters 10 and 13, but she has made contributions throughout the book. I look forward to a continuation of our partnership.

I have dedicated this book to my granddaughter, Evelyn Fay Babbie, born during the revision of the book and perfect in every regard. I hope that she will grow up in a world in which discrimination against women will finally seem a sadly quaint memory from a less enlightened past.

PROLOGUE

The Importance of Social Research

In many ways, the twentieth century hasn't been one of our better periods. Except for the relatively carefree twenties, we've moved from World War I to the Great Depression to World War II to the Cold War and its threat of thermonuclear holocaust to the tragedy of Vietnam. The thawing of the Cold War and the opening of Eastern Europe was a welcome relief, though it has in many ways heightened concern over the environmental destruction of our planet. And the thawing of the Cold War has hardly meant an end to war in general, as residents of Bosnia, Rwanda, and many other nations can attest. Americans now worry more than ever about the possibility of terrorism at home.

A case could be made that these are not the best of times. Many sage observers have written about the insecurity and malaise that characterize this century. All the same, the twentieth century has generated countless individual efforts and social movements aimed at creating humane social affairs, and most of those have arisen on college campuses. Perhaps you find these kinds of concerns and commitments in yourself.

As you look at the flow of events in the world, you can see the broad range of choices available if you want to make a significant contribution to future generations. Environmental problems are many and varied. Prejudice and discrimination are with us still. Millions die of hunger, and wars large and small circle the globe. There is, in short, no end to the ways you could demonstrate to yourself that your life matters, that you make a difference.

Given all the things you could choose from— things that really *matter*—why should you spend your time learning social research methods? I want to address this question at the start, because I'm going to suggest that you devote some of your time to such things as social theory, sampling, interviewing, experiments, computers, and so forth— things that can seem pretty distant from solving the world's pressing problems. Social science, though, is not only relevant to the major problems I've just listed—it also holds answers to them.

Many of the *big* problems we've faced and still face in this century have arisen out of our increasing technological abilities. The threat of nuclear terrorism is an example. Not unreasonably, we have tended to look to technology and technologists for solutions to those problems. Unfortunately, every technological solution so far has turned out to create new problems. At the beginning of this century, for example, many people worried about the danger of horse manure piling up in city streets. The invention of the automobile averted that problem. Now, no one worries about manure in the streets; we worry instead about a new and deadlier kind of pollutant in the air we breathe.

Similarly, in years past, we attempted to avoid nuclear attack by building better bombs and missiles of our own—so that no enemy would dare attack. But that only prompted our potential enemies to build ever bigger and more powerful weapons. Now, although the United States and Russia are exhibiting far less nuclear belligerence, similar contests elsewhere in the world could escalate. There is no technological end in sight for the insane nuclear weapons race.

The simple fact is that technology alone will never save us. It will never make the world work. You and I are the only ones who can do that. *The only real solutions lie in the ways we organize and run our social affairs.* This becomes evident when you consider all the social problems that persist today despite the clear presence of viable, technological solutions.

Overpopulation, for example, is a pressing problem in the world today. The number of people currently living on earth severely taxes our planet's life support systems, and this number rapidly increases year after year. If you study the matter you'll find that we already possess all the technological developments needed to stem population growth. It is technologically possible and feasible for us to stop population growth on the planet at whatever limit we want. Yet, overpopulation worsens each year.

Clearly, the solution to overpopulation is social. The causes of population growth lie in the forms, values, and customs that make up organized social life, and that is where the solutions are hidden. Those causes include beliefs about what it takes to be a "real woman" or a "real man," the perceived importance of perpetuating a family name, cultural tradition, and so forth. Ultimately, only social science can save us from overpopulation.

Or consider the problem of hunger on the planet. Some 13 to 15 million people die as a consequence of hunger each year. That amounts to 28 people a minute, every minute of every day, with 21 of them children. We all agree that this condition is deplorable; all would prefer it otherwise. But we tolerate this level of starvation in the belief that it is currently inevitable. We hope that perhaps one day someone will invent a method of producing food that will defeat starvation once and for all.

When we actually study the issue of starvation in the world, however, we can learn some astounding facts. First, the earth currently produces *more than enough food* to feed everyone. Moreover, this level of production does not even take into account farm programs that pay farmers not to plant and produce all the food they could.

Second, there are carefully planned and tested methods for ending starvation. In fact, since World War II, more than 30 countries have actually faced and ended their own problems of starvation. Some did it through food distribution programs. Others

focused on land reform. Some collectivized; others developed agribusiness. Many applied the advances of the Green Revolution. Taken together, these proven solutions make it possible to eliminate starvation totally.

Why then haven't we ended hunger altogether on the planet? The answer, again, lies in the organization and operation of our social life. New developments in food production will not end starvation any more than earlier ones have. People will continue to starve until we can *command* our social affairs rather than be enslaved by them.

Possibly, the problems of overpopulation and hunger seem distant to you, occurring somewhere "over there," on the other side of the globe. To save space, I'll simply remind you of the conclusion, increasingly reached, that there is no "over there" anymore: There is only "over here" in today's world. And regardless of how you view world problems, there is undeniably no end to the social problems in your own back yard—possibly even in your front yard: crime, inflation, unemployment, homelessness, cheating in government and business, child abuse, prejudice and discrimination, pollution, drug abuse, increased taxes, and reduced public services.

We can't solve our social problems until we understand how they come about and how they persist. Social science research offers a way to examine and understand the operation of human social affairs. It provides points of view and technical procedures that uncover things that would otherwise escape our awareness. Often, as the cliché goes, things are not what they seem; social science research can make that clear. One example illustrates this fact.

Poverty is a persistent problem in the United States, and none of its intended solutions is more controversial than *welfare.* Although the program is intended to give the poor a helping hand while they reestablish their financial viability, many complain that it has the opposite effect.

Part of the public image of welfare in action was crystallized by Susan Sheehan (1976) in her book, *A Welfare Mother,* which describes the situation of a three-generation welfare family, suggesting that the welfare system trapped the poor rather than liberating them. Martin Anderson (1978:56) agreed with Sheehan's assessment and charged that the welfare system had established a caste system in America, "perhaps as much as one-tenth of this nation—a caste of people almost totally dependent on the state, with little hope or prospect of breaking free. Perhaps we should call them the Dependent Americans."

George Gilder (1990) has spoken for many who believe the poor are poor mainly because they refuse to work, saying the welfare system saps their incentive to take care of themselves. Ralph Segalman and David Marsland (1989) support the view that welfare has become an intergenerational way of life for the poor in welfare systems around the world. Children raised in welfare families, they assert, will likely live their adult lives on welfare.

> This conflict between the intent of welfare as a temporary aid (as so understood by most of the public) and welfare as a permanent right (as understood by the welfare bureaucracy and welfare state planners) has serious implications. The welfare state nations, by and large, have given up on the concept of client rehabilitation for self-sufficiency, an intent originally supported by most welfare state proponents. What was to have been a temporary condition has become a permanent cost on the welfare state. As a result, welfare discourages productivity and self-sufficiency and establishes a new mode of approved behaviour in the society—one of acceptance of dependency as the norm.
>
> (Segalman and Marsland 1989:6–7)

These negative views of the effects of the welfare system are widely shared by the general public, even among those basically sympathetic to the aims of the program. Greg Duncan at the University of Michigan's Survey Research Center points out that census data would seem to confirm the impression that a hard core of the poor have become trapped in their poverty. Speaking of the percentage of the population living in poverty at any given time, he says,

> Year-to-year changes in these fractions are typically less than 1 percent, and the Census survey's other measures show little change in the

characteristic of the poor from one year to the next. They have shown repeatedly that the individuals who are poor are more likely to be in families headed by a woman, by someone with low education, and by blacks.

> Evidence that one-eighth of the population was poor in two consecutive years, and that those poor shared similar characteristics, is consistent with an inference of absolutely no turnover in the poverty population. Moreover, the evidence seems to fit the stereotype that those families that are poor are likely to remain poor, and that there is a hard-core population of poor families for whom there is little hope of self-improvement.
>
> (Duncan 1984:2–3)

Duncan continues, however, to warn that such snapshots of the population can conceal changes taking place. Specifically, an unchanging percentage of the population living in poverty does not necessarily mean the *same* families are poor from year to year. Theoretically, it could be a totally different set of families each year.

To determine the real nature of poverty and welfare, the University of Michigan undertook a "Panel Study of Income Dynamics" in which they followed the economic fate of 5,000 families from 1969 to 1978, or ten years, the period supposedly typified by Sheehan's "welfare mother." At the beginning, the researchers found that in 1978, 8.1 percent of these families were receiving some welfare benefits and 3.5 percent depended on welfare for more than half their income. Moreover, these percentages did not differ drastically over the ten-year period. (Duncan 1984:75)

Looking beyond these surface data, however, the researchers found something you might not have expected. During the ten-year period, about one-fourth of the 5,000 families received welfare benefits at least once. However, only 8.7 percent of the families were ever dependent on welfare for more than half their income. *"Only a little over one-half of the individuals living in poverty in one year are found to be poor in the next, and considerably less than one-half of those who experience poverty remain persistently poor over many years"* (Duncan 1984:3; emphasis original).

Only 2 percent of the families received welfare each of the 10 years, and less than 1 percent were continuously dependent on welfare for the 10 years. Table P-1 summarizes these findings.

These data paint a much different picture of poverty than people commonly assume. In a summary of his findings, Duncan says:

> While nearly one-quarter of the population received income from welfare sources at least once in the decade, only about 2 percent of all the population could be characterized as dependent upon this income for extended periods of time. Many families receiving welfare benefits at any given time were in the early stages of recovering from an economic crisis caused by the death, departure, or disability of a husband, a recovery that often lifted them out of welfare when they found full-time employment, or remarried, or both. Furthermore, most of the children raised in welfare families did not themselves receive welfare benefits after they left home and formed their own households.
>
> (Duncan 1984:4–5)

Many of the things social scientists study—including all the social problems you've just read about—generate deep emotions and firm convictions in most people. This makes effective inquiry into the facts difficult at best; all too often, researchers manage only to confirm their initial prejudices. The special value of social science research methods is that they offer a way to address such issues with logical and observational rigor. They let us all pierce through our personal viewpoints and take a look at the world that lies beyond our

own perspectives. And it is that "world beyond" that holds the solutions to the social problems we face today.

At a time of increased depression and disillusionment, we are continually tempted to turn away from confronting social problems and retreat into the concerns of our own self-interest. Social science research offers an opportunity to take on those problems and discover the experience of making a difference after all. The choice is yours; I invite you to take on the challenge. Your instructor and I would like to share the excitement of social science with you.

Table P-1

Incidence of Short- and Long-Run Welfare Receipt and Dependence, 1969–78

	Percent of U.S. Population:	
	Receiving Any Welfare Income	Dependent on Welfare for More Than 50% of Family Income
Welfare in 1978	8.1%	3.5%
Welfare in 1 or more years, 1969–78	25.2	8.7
Welfare in 5 or more years, 1969–78	8.3	3.5
Welfare in all 10 years, 1969–78	2.0	0.7
"Persistent welfare" (welfare in 8 or more years), 1969–78	4.4	2.0

Source: Greg J. Duncan, Years of Poverty, Years of Plenty: The Changing Fortunes of American Workers and Families (Ann Arbor: University of Michigan, 1984), 75.

The Basics of Social Research

Bonnie Kamin /PhotoEdit

Part One

AN INTRODUCTION TO INQUIRY

S*cience* is a familiar word used by everyone. Yet images of science differ greatly. For some, science is mathematics; for others, it's white coats and laboratories. It's often confused with technology or equated with tough high school or college courses.

Science is, of course, none of these things per se. It's difficult, however, to specify exactly what science is. Scientists, in fact, disagree on the proper definition. For the purposes of this book, we'll look at science as a method of inquiry—a way of learning and knowing things about the world around us. Contrasted with other ways of learning and knowing about the world, science has some special characteristics. We'll examine these traits in this opening set of chapters.

Dr. Benjamin Spock, the renowned author and pediatrician, began his books on child care by assuring new parents that they already knew more about child care than they thought they did. I want to begin this book on a similar note. It will become clear to you before you've read very far that you already know a great deal about the practice of scientific social research. In fact, you've been conducting scientific research all your life. From that perspective, the purpose of this book is to help you sharpen skills you already have and perhaps to show you some tricks that may not have occurred to you.

Part 1 of this book is intended to lay the groundwork for the discussions that follow in the rest of the book—to examine the fundamental characteristics and issues that make science different from other ways of knowing things. In Chapter 1, we'll begin with a look at native human inquiry, the sort of thing you've been doing all your life. In the course of that examination, we'll see some of the ways people go astray in trying to understand the world around them, and I'll summarize the primary characteristics of scientific inquiry that guard against those errors.

Chapter 2 deals with social scientific paradigms and theories, as well as the links between theory and research. We'll look at some of the theoretical paradigms that shape the nature of inquiry, largely determining what scientists look for and how they interpret what they see.

While most of this book concerns the art and science of doing social research, Chapter 3 introduces some of the political and ethical considerations that affect social research. We'll see the ethical norms that social researchers follow when they design and implement research. We'll also see how social contexts affect social research.

Overall, Part 1 constructs a backdrop against which to view the more specific aspects of research design and execution. By the time you complete Part 1, you should be ready to look at some of the more concrete aspects of social research.

1 HUMAN INQUIRY AND SCIENCE

Bonnie Kamin/PhotoEdit

What You'll Learn in This Chapter

We'll examine the way people learn about their world and the mistakes they make along the way. We'll also begin to see what makes science different from other ways of knowing.

In this chapter . . .

INTRODUCTION

This book is about knowing things—not so much *what* we know as *how* we know it. Let's start by examining a few things you probably know already.

You know the world is round. You probably also know it's cold on the dark side of the moon, and you know people speak Chinese in China. You

---

AN OPENING **QUANDARY**

The decision to have a baby is a deeply personal one over which many couples agonize. No one is in charge of who and how many will have babies in the United States in any given year. While you must get a license to marry or go fishing, you do not need a license to have a baby. Many couples delay pregnancy, some pregnancies happen by accident, and some pregnancies are planned. Given all these uncertainties and idiosyncrasies, how can baby food and diaper manufacturers know how much to produce from year to year? By the end of this chapter, you should be able to answer this question.

---

know that vitamin C prevents colds and that unprotected sex can result in AIDS.

How do you know? If you think for a minute, you'll see you know these things because somebody told them to you, and you believed what you were told. You may have read in *National Geographic* that people speak Chinese in China, and that made sense to you, so you didn't question it. Perhaps your physics or astronomy instructor told you it was cold on the dark side of the moon, or maybe you read it on the NASA Web page.

Some of the things you know seem obvious to you. If I asked you how you know the world is round, you'd probably say, "Everybody knows that." There are a lot of things everybody knows. Of course, at one time, everyone "knew" the world was flat.

Most of what you know is a matter of agreement and belief. Little of it is based on personal experience and discovery. A big part of growing up in any society, in fact, is the process of learning to accept what everybody around you "knows" is so. If you don't know those same things, you can't really be a part of the group. If you were to question seriously that the world is really round, you'd quickly find yourself set apart from other people. You might be sent to live in a hospital with other people who ask questions like that.

Although it's important for you to see that most of what you know is a matter of believing what you've been told, I also want you to see that there's nothing wrong with you in that respect. That's simply the way human societies are structured. The basis of knowledge is agreement. Because you can't learn through personal experience and discovery alone all you need to know, things are set up so you can simply believe what others tell you. You know some things through tradition, others from "experts."

There are other ways of knowing things, however. In contrast to knowing things through agreement, you can know them through direct experience—through observation. If you dive into a glacial stream flowing through the Canadian Rockies, you don't need anyone to tell you it's cold. The first time you stepped on a thorn, you knew it hurt before anyone told you.

When your experience conflicts with what everyone else knows, though, there's a good chance you'll surrender your experience in favor of the agreement. For example, imagine you've come to a party at my house. It's a high-class affair, and the drinks and food are excellent. In particular, you're taken by one of the appetizers I bring around on a tray: a breaded, deep-fried tidbit that's especially zesty. You have a couple—they're so delicious! You have more. Soon you're subtly moving around the room to be wherever I am when I arrive with a tray of these nibblies.

Finally, you can contain yourself no longer. "What are they?" you ask. I let you in on the secret: "You've been eating breaded, deep-fried worms!" Your response is dramatic: Your stomach rebels, and you promptly throw up all over the living room rug. What a terrible thing to serve guests!

The point of the story is that both of your feelings about the appetizer were quite real. Your initial liking for them was certainly real, but so was the feeling you had when you found out what you'd been eating. It should be evident, however, that the feeling of disgust you had when you discovered you were eating worms was strictly a product of the agreements you have with those around you that worms aren't fit to eat. That's an agreement you began the first time your parents found you sitting in a pile of dirt with half of a wriggling worm dangling from your lips. When they pried your mouth open and reached down your throat for the other half of the worm, you learned that worms are not acceptable food in our society.

Aside from these agreements, what's wrong with worms? They're probably high in protein and low in calories. Bite-sized and easily packaged, they're a distributor's dream. They are also a delicacy for some people who live in societies that lack our agreement that worms are disgusting. Some people might love the worms but be turned off by the deep-fried breading.

Here's a question to consider: "Are worms *really* good or *really* bad to eat?" And here's a more interesting question: "*How could you know* which was really so?" This book is about answering the second kind of question.

LOOKING FOR REALITY

Reality is a tricky business. You've probably long suspected that some of the things you "know" may not be true, but how can you really know what's real? People have grappled with this question for thousands of years.

One answer that has arisen out of that grappling is science, which offers an approach to both agreement reality and experiential reality. Scientists have certain criteria that must be met before they'll accept the reality of something they haven't personally experienced. In general, an assertion must have both *logical* and *empirical* support: It must make sense, and it must not contradict actual observation. Why do earthbound scientists accept the assertion that it's cold on the dark side of the moon? First, it makes sense, because the surface heat of the moon comes from the sun's rays. Second, the scientific measurements made on the moon's dark side confirm the expectation. So, scientists accept the reality of things they don't personally experience—they accept an agreement reality—but they have special standards for doing so.

More to the point of this book, however, science offers a special approach to the discovery of reality through personal experience, that is, to the business of inquiry. *Epistemology* is the science of knowing; *methodology* (a subfield of epistemology) might be called the science of finding out. This book is an examination and presentation of social science methodology, or how social scientists find out about human social life. You'll see that some of the methods coincide with the traditional image of science but others have been specially geared to sociological concerns.

In the rest of this chapter, we'll look at inquiry as an activity. We'll begin by examining inquiry as a natural human activity, something you and I have engaged in every day of our lives. Next, we'll look at some kinds of errors we make in normal inquiry, and we'll conclude by examining what makes science different. We'll see some of the ways science guards against common human errors in inquiry.

Ordinary Human Inquiry

Practically all people exhibit a desire to predict their future circumstances. We seem quite willing, moreover, to undertake this task using *causal* and *probabilistic* reasoning. First, we generally recognize that future circumstances are somehow caused or conditioned by present ones. We learn that getting an education will affect how much money we earn later in life and that swimming beyond the reef may bring an unhappy encounter with a shark. As students we learn that studying hard will result in better grades. Second, people also learn that such patterns of cause and effect are *probabilistic* in nature: The effects occur more often when the causes occur than when the causes are absent—but not always. Thus, students learn that studying hard produces good grades in most instances, but not every time. We recognize the danger of swimming beyond the reef without believing that every such swim will be fatal.

As we'll see throughout the book, science makes these concepts of causality and probability more explicit and provides techniques for dealing with them more rigorously than does casual human inquiry. It sharpens the skills we already have by making us more conscious, rigorous, and explicit in our inquiries.

In looking at ordinary human inquiry, we need to distinguish between prediction and understanding. Often, we can make predictions without understanding—perhaps you can predict rain when your trick knee aches. And often, even if we don't understand why, we're willing to act on the basis of a demonstrated predictive ability. The racetrack buff who finds that the third-ranked horse in the third race of the day always wins will probably keep betting without knowing, or caring, why it works out that way.

Whatever the primitive drives or instincts that motivate human beings, satisfying them depends heavily on the ability to predict future circumstances. However, the attempt to predict is often placed in a context of knowledge and understanding. If we can understand why things are related to one another, why certain regular patterns occur, we can predict even better than if we simply observe and remember those patterns. Thus, human inquiry aims at answering both "what" and "why" questions, and we pursue these goals by observing and figuring out.

As I suggested earlier our attempts to learn about the world are only partly linked to direct, personal inquiry or experience. Another, much larger, part comes from the agreed-upon knowledge that others give us. This agreement reality both assists and hinders our attempts to find out for ourselves. To see how, consider two important sources of our secondhand knowledge—tradition and authority.

Tradition

Each of us inherits a culture made up, in part, of firmly accepted knowledge about the workings of the world. We may learn from others that eating too much candy will decay our teeth, that the circumference of a circle is approximately twenty-two–sevenths of its diameter, or that masturbation will blind us. We may test a few of these "truths" on our own, but we simply accept the great majority of them, the things that "everybody knows."

Tradition, in this sense of the term, offers some clear advantages to human inquiry. By accepting what everybody knows, we are spared the overwhelming task of starting from scratch in our search for regularities and understanding. Knowledge is cumulative, and an inherited body of information and understanding is the jumping-off point for the development of more knowledge. We often speak of "standing on the shoulders of giants," that is, of previous generations.

At the same time, tradition may be detrimental to human inquiry. If we seek a fresh understanding of something everybody already understands and has always understood, we may be marked as fools for our efforts. More to the point, however, it rarely occurs to most of us to seek a different understanding of something we all "know" to be true.

Authority

Despite the power of tradition, new knowledge appears every day. Aside from our own personal inquiries, we benefit throughout life from new discoveries and understandings produced by others. Often, acceptance of these new acquisitions depends on the status of the discoverer. You're more likely to believe the epidemiologist who declares that the common cold can be transmitted through kissing, for example, than to believe your uncle Pete.

Like tradition, authority can both assist and hinder human inquiry. We do well to trust in the judgment of the person who has special training, expertise, and credentials in a given matter, especially in the face of controversy. At the same time, inquiry can be greatly hindered by the legitimate authority who errs within his or her own special province. Biologists, after all, do make mistakes in the field of biology. Biological knowledge changes over time.

Inquiry is also hindered when we depend on the authority of experts speaking outside their realm of expertise. For example, consider the political or religious leader with no biochemical expertise who declares that marijuana is a dangerous drug. The advertising industry plays heavily on this misuse of authority by, for example, having popular athletes discuss the nutritional value of breakfast cereals or movie actors evaluate the performance of automobiles.

Both tradition and authority, then, are double-edged swords in the search for knowledge about the world. Simply put, they provide us with a starting point for our own inquiry, but they can lead us to start at the wrong point and push us off in the wrong direction.

Errors in Inquiry and Some Solutions

Quite aside from the potential dangers of tradition and authority, we often stumble and fall when we set out to learn for ourselves. Let's look at some of the common errors we make in our casual inquiries and the ways science guards against those errors.

Inaccurate Observations Quite frequently, we make mistakes in our observations. For example, what was your methodology instructor wearing on the first day of class? If you have to guess, that's because most of our daily observations are casual and semiconscious. That's why we often disagree about what really happened.

In contrast to casual human inquiry, scientific observation is a conscious activity. Simply making observation more deliberate helps reduce error. If you had to guess what your instructor was wearing the first day of class, you'd probably make a mistake. If you had gone to the first class meeting with a conscious plan to observe and record what your instructor was wearing, however, you'd likely be more accurate.

In many cases, both simple and complex measurement devices help guard against inaccurate observations. Moreover, they add a degree of precision well beyond the capacity of the unassisted human senses. Suppose, for example, that you had taken color photographs of your instructor that day.

Overgeneralization When we look for patterns among the specific things we observe around us, we often assume that a few similar events are evidence of a general pattern. That is, we tend to overgeneralize on the basis of limited observations. This can misdirect or impede inquiry. (Think back to our now-broke racetrack buff.)

Imagine that you're a reporter covering an animal-rights demonstration. You have to turn in your story in just two hours. Rushing to the scene, you start interviewing people, asking them why they're demonstrating. If the first two demonstrators you interview give you essentially the same reason, you may simply assume that the other 3,000 are also there for that reason. Unfortunately, when your story appears, your editor gets scores of letters from protesters who were there for an entirely different reason.

Scientists guard against overgeneralization by seeking a sufficiently large sample of observations. The **replication** of inquiry provides another safeguard. Basically, this means repeating a study and checking to see if the same results are produced each time. Then, as a further test, the study may be repeated under slightly varied conditions.

Selective Observation One danger of overgeneralization is that it may lead to selective observation. Once you have concluded that a particular pattern exists and have developed a general understanding of why it does, you'll tend to focus on future events and situations that fit the pattern, and you'll ignore those that don't. Racial and ethnic prejudices depend heavily on selective observation for their persistence.

Sometimes a research design will specify in advance the number and kind of observations to be made, as a basis for reaching a conclusion. If you and I wanted to learn whether women were more likely than men to support freedom to choose an abortion, we'd commit ourselves to making a specified number of observations on that question in a research project. We might select a thousand people to be interviewed on the issue. Alternately, when making direct observations of an event, such as an animal-rights demonstration, social scientists make a special effort to find "deviant cases"—those who do not fit into the general pattern.

Illogical Reasoning There are other ways of handling observations that contradict our conclusions about the way things are in daily life. Surely one of the most remarkable creations of the human mind is "the exception that proves the rule." This idea doesn't make any sense at all. An exception can draw attention to a rule or to a supposed rule, but in no system of logic can it prove the rule it contradicts. Yet, we often use this pithy saying to brush away contradictions with a simple stroke of illogic.

What statisticians have called the *gambler's fallacy* is another illustration of illogic in day-to-day reasoning. A consistent run of either good or bad luck is presumed to foreshadow its opposite. An evening of bad luck at poker may kindle the belief that a winning hand is just around the corner; many a poker player has stayed in a game much too long because of that mistaken belief. Conversely, an extended period of good weather may lead you to worry that it is certain to rain on the weekend picnic.

Although all of us sometimes fall into embarrassingly illogical reasoning in daily life, scientists avoid this pitfall by using systems of logic consciously and explicitly. Chapter 2 will examine the logic of science in more depth. For now, it's enough to note that logical reasoning is a conscious activity for scientists, who always have their colleagues around to keep them honest.

These, then, are a few of the ways we go astray in our attempts to know and understand the world and some of the ways that science protects inquiry from these pitfalls. Accurately observing and understanding reality is not an obvious or trivial matter. Indeed, it's more complicated than I've suggested.

What's Really Real?

Philosophers sometimes use the phrase "naive realism" to describe the way most of us operate in our daily lives. When you sit at a table to write, you probably don't spend a lot of time thinking about whether the table is really made up of atoms, which in turn are mostly empty space. When you step into the street and see a city bus hurtling down on you, it's not the best time to reflect on methods for testing whether the bus really exists. We all live with a view that what's real is pretty obvious—and that view usually gets us through the day.

Even so, I hope you can see that the nature of "reality" is perhaps more complex than we tend to assume. As a philosophical backdrop for the discussions to follow, let's look at what are sometimes called premodern, modern, and postmodern views of reality (W. Anderson 1990).

The Premodern View   This view of reality has guided most of human history. Our early ancestors assumed that they saw things as they really were. In fact, this assumption was so fundamental that they didn't even see it as an assumption. No cavemom said to her cavekid, "Our tribe makes an assumption that evil spirits reside in the Old Twisted Tree." No, she said, "STAY OUT OF THAT TREE OR YOU'LL TURN INTO A TOAD!"

As humans evolved and became aware of their diversity, they came to recognize that others did not always share their views of things. Thus, they may have discovered that another tribe didn't believe the tree was wicked; in fact, the second tribe believed that the tree spirits were holy and beneficial. The discovery of this diversity led members of the first tribe to conclude that "some tribes I could name are pretty stupid." For them, the tree was still wicked, and they expected some misguided people to be moving to Toad City.

The Modern View   What philosophers call the *modern view* accepts such diversity as legitimate, a philosophical "different strokes for different folks." As a modern thinker you would say, "I regard the spirits in the tree as evil, but I know others regard them as good. Neither of us is right or wrong. There are simply spirits in the tree. They are neither good nor evil, but different people have different ideas about them."

It's probably easy for you to adopt the modern view. Some might regard a dandelion as a beautiful flower while others see only an annoying weed. To the premoderns, a dandelion has to be either one or the other. If you think it is a weed, it is *really* a weed, though you may admit that some people have a warped sense of beauty. In the modern view, a dandelion is simply a dandelion. It is a plant with yellow petals and green leaves. The concepts

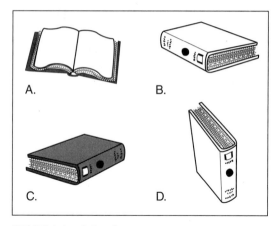

FIGURE 1-1   **A Book**

"beautiful flower" and "annoying weed" are subjective points of view imposed on the plant by different people. Neither is a quality of the plant itself, just as "good" and "evil" were concepts imposed on the spirits in the tree.

The Postmodern View   Philosophers also speak of a *postmodern view* of reality. In this view, neither the spirits nor the dandelion exists. All that's "real" are the images we get through our points of view. Put differently, there's nothing out there, it's all in here. As Gertrude Stein said of Oakland, "There's no there, there."

No matter how bizarre the postmodern view may seem to you on first reflection, it has a certain ironic inevitability. Take a moment to notice the book you're reading; notice specifically what it looks like. Since you're reading these words, it probably looks like Figure 1-1A.

But does Figure 1-1A represent the way your book "really" looks? Or does it merely represent what the book looks like from your current point of view? Surely, Figures 1-1B, C, and D are equally valid representations. But these views of the book are so different from one another. Which is the "reality"?

As this example illustrates, there is no answer to the question, "What does the book *really* look like?" All we can offer is the different ways it looks from different points of view. Thus, according to the postmodern view, there is no "book," only var-

FIGURE 1-2 **Wife's Point of View**

FIGURE 1-3 **Husband's Point of View**

ious images of it from different points of view. And all the different images are equally "true."

Now let's apply this logic to a social situation. Imagine a husband and wife arguing. When she looks over at her quarreling husband, Figure 1-2 is what the wife sees. Take a minute to imagine what you would think and feel if you were the woman in this drawing. How would you explain to an outsider or to your best friend what had happened? What solutions to the conflict would seem appropriate if you were this woman?

What the woman's husband sees is another matter altogether, as shown in Figure 1-3. Take a minute to imagine experiencing the situation from his point of view. What thoughts and feelings would you have? How would you tell your best friend what had happened? What solutions would seem appropriate for resolving the conflict?

Now, consider a third point of view. Suppose you're an outside observer watching this interaction between a wife and husband. What would it look like to you now? Unfortunately, we can't easily portray the third point of view without knowing something about the personal feelings, beliefs, past experiences, and so forth that you would bring to your task as "outside" observer. (Though I call you an *outside* observer, you are, of course, observing from *inside* your own mental system.)

To take an extreme example, if you were a confirmed male chauvinist, you'd probably see the fight pretty much the same way the husband saw it. On the other hand, if you were committed to the view that men are unreasonable bums, you'd see things the way the wife saw them.

But imagine instead that you see two unreasonable people quarreling irrationally with one another. Would you see them both as irresponsible jerks, equally responsible for the conflict? Or would you see them as two people facing a difficult human situation, each doing the best he or she can to resolve it?

Imagine feeling compassion for them and noticing how each of them attempts to end the hostility, even though the gravity of the problem keeps them fighting.

Notice how different these several views are. Which is a "true" picture of what is happening between the wife and the husband? You win the prize if you notice that your own point of view would again color your perception of what is happening here.

The postmodern view represents a critical dilemma for scientists. While their task is to observe and understand what is "really" happening, they are all human and, as such, have personal orientations that color what they observe and how they explain it. There is ultimately no way people can totally step outside their humanness to see and understand the world as it "really" is.

Whereas the modern view acknowledges the inevitability of human subjectivity, the postmodern view suggests that there is no "objective" reality to be observed in the first place. There are only our several subjective views.

We'll return to this discussion in Chapter 2 when we focus in on specific scientific paradigms. Ultimately, what you'll see is that (1) established scientific procedures sometimes allow you to deal effectively with this dilemma—that is, we can study people and help them through their difficulties without being able to view "reality" directly— and (2) the philosophical stances I've presented suggest a powerful range of possibilities for structuring our research.

Let's turn now to the foundations of the social scientific approaches to understanding. From there we can examine the specific research techniques social scientists use.

THE FOUNDATIONS OF SOCIAL SCIENCE

The two pillars of science are logic and observation. A scientific understanding of the world must (1) make sense and (2) correspond with what we observe. Both elements are essential to science and relate to three major aspects of the overall scientific enterprise: theory, data collection, and data analysis.

In the most general terms, scientific theory deals with logic, data collection with observation, and data analysis with patterns in what is observed and, where appropriate, the comparison of what is logically expected with what is actually observed. Though most of this textbook deals with data collection and data analysis—demonstrating how to conduct empirical research—recognize that social *science* involves all three elements. As such, Chapter 2 of this book concerns the theoretical context of research; Parts 2 and 3 focus on data collection; and Part 4 offers an introduction to the analysis of data. Figure 1-4 offers a schematic view of how the book addresses these three aspects of social science.

Let's turn now to some of the fundamental issues that distinguish social science from other ways of looking at social phenomena.

Theory, Not Philosophy or Belief

Social scientific theory has to do with what is, not with what *should* be. For many centuries, however, social theory has combined these two orientations. Social philosophers liberally mixed their observations of what happened around them, their speculations about why, and their ideas about how things ought to be. Although modern social scientists may do the same from time to time, realize that social *science* has to do with how things are and why.

This means that scientific **theory**—and science itself—cannot settle debates on value. Science cannot determine whether capitalism is better or worse than socialism except in terms of agreed-upon criteria. To determine scientifically whether capitalism or socialism most supported human dignity and freedom we would first have to agree on some measurable definitions of dignity and freedom. Our conclusions would depend totally on this agreement and would have no general meaning beyond it.

By the same token, if we could agree that suicide rates, say, or giving to charity were good measures of a religion's quality, then we could determine scientifically whether Buddhism or Christianity is the better religion. Again, our conclusion would be inextricably tied to the given criterion. As a practical matter, people seldom agree on criteria for determining issues of value, so science is seldom useful in settling such debates. In fact, questions like these are so much a matter of opinion and belief that scientific inquiry is often viewed as a threat to what is "already known."

We'll consider this issue in more detail in Chapter 12, when we look at evaluation research. As you'll see, social scientists have become increasingly involved in studying programs that reflect ideological points of view, such as affirmative action or welfare reform. One of the biggest problems

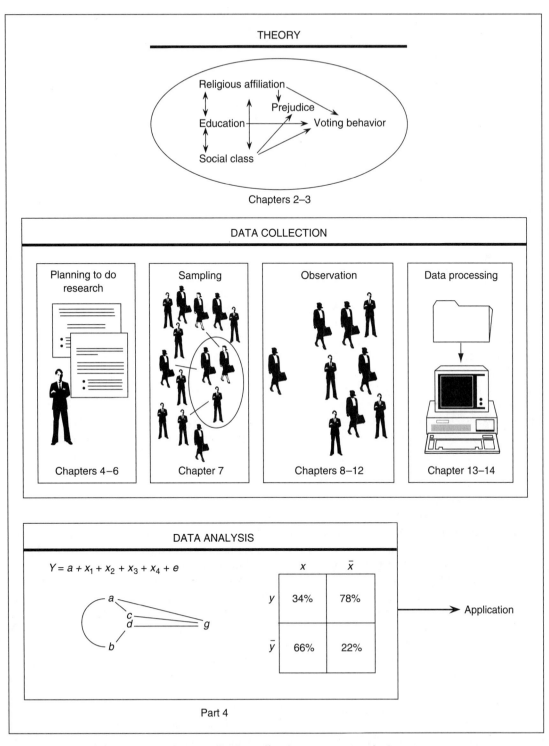

FIGURE 1-4 **Social Science = Theory + Data Collection + Data Analysis**

researchers face is getting people to agree on criteria of success and failure. Yet such criteria are essential if social scientific research is to tell us anything useful about matters of value. By analogy, a stopwatch can't tell us if one sprinter is better than another unless we first agree that speed is the critical criterion.

Social science, then, can help us know only what is and why. We can use it to determine what ought to be, but only when people agree on the criteria for deciding what's better than something else—an agreement that seldom occurs. With that understood, let's turn now to some of the fundamental bases on which social science allows us to develop theories about what is and why.

Social Regularities

In large part, social scientific theory aims to find patterns in social life. That aim, of course, applies to all science, but it is sometimes a barrier for people when they first approach social science.

Actually, the vast number of formal norms in society create a considerable degree of regularity. For example, only people who have reached a certain age can vote in elections. In the U.S. military, until recently only men could participate in combat. Such formal prescriptions, then, regulate, or regularize, social behavior.

Aside from formal prescriptions, we can observe other social norms that create more regularities. Republicans are more likely than Democrats to vote for Republican candidates. University professors tend to earn more money than do unskilled laborers. Men earn more than do women. The list of regularities could go on and on.

What about Exceptions?   The objection that there are always exceptions to any social regularity is also inappropriate. It doesn't matter that a particular woman earns more money than a particular man if men earn more than women overall. The pattern still exists. Social regularities represent probabilistic patterns; a general pattern need not be reflected in 100 percent of the observable cases.

This rule applies in physical science as well as social science. In genetics, for example, the mating of a blue-eyed person with a brown-eyed person will *probably* result in a brown-eyed child. The birth of a blue-eyed child does not challenge the observed regularity, however, since the geneticist states only that the brown-eyed offspring is more likely and, further, that brown-eyed offspring will be born in a certain percentage of the cases. The social scientist makes a similar, probabilistic prediction—that women overall are likely to earn less than are men. And the social scientist asks *why* this is the case.

Aggregates, Not Individuals

Social regularities do exist, then, and are worthy of theoretical and empirical study. As such, social scientists study primarily social patterns rather than individual ones. These patterns reflect the *aggregate* or collective actions and situations of many individuals. Although social scientists often study motivations and actions that affect individuals, they seldom study the individual per se. That is, they create theories about the nature of group, rather than individual, life.

Sometimes the collective regularities are amazing. Consider the birthrate, for example. People have babies for an incredibly wide range of personal reasons. Some do it because their parents want them to. Some think of it as a way of completing their womanhood or manhood. Others want to hold their marriages together. Still others have babies by accident.

If you have had a baby, you could probably tell a much more detailed, idiosyncratic story. Why did you have the baby when you did, rather than a year earlier or later? Maybe your house burned down and you had to delay a year before you could afford to have the baby. Maybe you felt that being a family person would demonstrate maturity, which would support a promotion at work.

Everyone who had a baby last year had a different set of reasons for doing so. Yet, despite this vast diversity, despite the idiosyncrasy of each individual's reasons, the overall birthrate in a society (the number of live births per 1,000 population) is remarkably consistent from year to year. See Table 1-1 for 20 years of birthrates for the United States.

TABLE 1-1   **Birthrates, United States: 1977–1996**

1977	15.1
1978	15.0
1979	15.6
1980	15.9
1981	15.8
1982	15.9
1983	15.6
1984	15.6
1985	15.8
1986	15.6
1987	15.7
1988	16.0
1989	16.4
1990	16.7
1991	16.3
1992	15.9
1993	15.5
1994	15.2
1995	14.8
1996	14.7

Source: Centers for Disease Control and Prevention, National Center for Health Statistics (1998), *Monthly Vital Statistics Report* 46 (11, Suppl.): 29.

TABLE 1-2   **Monthly Birthrates in the United States, 1998–1999**

	1998	*1999*
January	13.5	14.6
February	15.0	14.6
March	14.5	14.3
April	15.2	14.2
May	13.9	13.9
June	14.7	14.6
July	15.6	14.4
August	14.7	14.6
September	14.8	15.6
October	15.5	14.5
November	14.0	*
December	13.9	*

*Data not available
Source: Centers for Disease Control and Prevention, National Center for Health Statistics (2000), *Births, Marriages, Divorces, and Deaths: Provisional Data for October 1999* (Washington, DC: U.S. Government Printing Office): 2.

If the U.S. birthrate were 15.9, 35.6, 7.8, 28.9, and 16.2 in five successive years, demographers would begin dropping like flies. As you can see, however, social life is far more orderly than that. Moreover, this regularity occurs without society-wide regulation. As mentioned earlier, no one plans how many babies will be born or determines who will have them.

The regularity in birthrates even exists on a monthly basis, as Table 1-2 indicates. The greatest difference from month to month is 1.5 births per 1,000 population. Consider, though, that the 2000 birthrate for Niger was 54 per 1,000, while it was only 7 per 1,000 in China (Population Reference Bureau 2000). Clearly, the relative consistency of U.S. birthrates over time is not a function of human biology, but of social structure and culture.

Social scientific theories try to explain why aggregated patterns of behavior are so regular even when the individuals participating in them may change over time. It could be said that social scientists don't even seek to explain people per se. They try instead to understand the *systems* in which people operate, which in turn explain why people do what they do. The elements in such a system are not people but *variables.*

> The report from the Centers for Disease Control and Prevention on births, marriages, divorces, and deaths is available on the World Wide Web* at http://www .cdc.gov/nchs/data/nvs48_15.pdf

A Variable Language

Our most natural attempts at understanding are usually concrete and idiosyncratic. That's just the way we think.

Imagine that someone says to you, "Women ought to get back into the kitchen where they belong." You're likely to hear that comment in terms

*Each time the Internet icon appears, you'll be given helpful leads for searching the World Wide Web.

APPLYING THE RESULTS

Take a minute to reflect on some of the practical implications of the data you've just seen. The Opening Quandary for this chapter asked how baby food and diaper manufacturers could plan production from year to year. The consistency of U.S. birthrates suggests this is not the problem it might have seemed.

Who else might benefit from this kind of analysis? What about health care workers and educators? Can you think of anyone else?

What if we organized birthrates by region of the country, by ethnicity, by income level, and so forth? Clearly, these additional analyses could make the data even more useful. As you learn about the options available to social researchers, I think you'll gain an appreciation for the practical value that research can have for the whole society.

of what you know about the speaker. If it's your old uncle Harry who is also strongly opposed to daylight saving time, zip codes, and personal computers, you're likely to think his latest pronouncement simply fits into his rather dated point of view about things in general.

If, on the other hand, the statement issues forth from a politician who is trailing a female challenger and who has also begun making statements about women being emotionally unfit for public office and not understanding politics, you may hear his latest comment in the context of this political challenge.

In both examples, you're trying to understand the thoughts of a particular individual. In social science, researchers go beyond that level of understanding to seek insights into classes or types of individuals. Regarding the two examples just described, they might use terms such as *old-fashioned* or *bigot* to describe the kind of person who made

the comment. In other words, they try to place the individual in a set of similar individuals, according to a particular, defined concept.

By examining an individual in this way, social scientists can make sense out of more than one person. In understanding what makes the bigoted politician think the way he does, they'll also learn about other people who are "like him." In other words, they have not been studying bigots as much as *bigotry*.

Bigotry here is spoken of as a **variable** because it varies. Some people are more bigoted than others. Social scientists are interested in understanding the system of variables that causes bigotry to be high in one instance and low in another.

The idea of a system composed of variables may seem rather strange, so let's look at an analogy. The subject of a physician's attention is the patient. If the patient is ill, the physician's purpose is to help that patient get well. By contrast, a medical researcher's subject matter is different: the variables that cause a disease, for example. The medical researcher may study the physician's patient, but for the researcher that patient is relevant only as a carrier of the disease.

Of course, medical researchers care about real people, but in the actual research, patients are directly relevant only for what they reveal about the disease under study. In fact, when researchers can study a disease meaningfully without involving actual patients, they do so.

Social research involves the study of variables and the **attributes** that compose them. Social scientific theories are written in a language of variables, and people get involved only as "carriers" of those variables. Here's a closer look at what social scientists mean by variables and attributes.

Attributes or values are characteristics or qualities that describe an object—in this case, a person. Examples include *female, Asian, alienated, conservative, dishonest, intelligent,* and *farmer.* Anything you might say to describe yourself or someone else involves an attribute.

Variables, on the other hand, are logical groupings of attributes. Thus, for example, *male* and *female* are attributes, and *sex* or *gender* are the vari-

Some Common Social Concepts	
Female	Age
Upper class	African American
Young	Occupation
Social class	Gender
Race/ethnicity	Plumber

↓

Variable	Attributes
Age	Young, middle-aged, old
Gender	Female, male
Occupation	Plumber, lawyer, data-entry clerk . . .
Race/ethnicity	African American, Asian, Caucasian, Latino . . .
Social class	Upper, middle, lower . . .

FIGURE 1-5 **Variables and Attributes**

ables composed of these two attributes. The variable *occupation* is composed of attributes such as *farmer, professor,* and *truck driver. Social class* is a variable composed of a set of attributes such as *upper class, middle class,* and *lower class.* Sometimes it helps to think of attributes as the categories that make up a variable. (See Figure 1-5 for a schematic review of what social scientists mean by variables and attributes.)

The relationship between attributes and variables lies at the heart of both description and explanation in science. For example, we might describe a college class in terms of the variable *gender* by reporting the observed frequencies of the attributes *male* and *female:* "The class is 60 percent men and 40 percent women." An unemployment rate can be thought of as a description of the variable *employment status* of a labor force in terms of the attributes *employed* and *unemployed.* Even the report of family income for a city is a summary of attributes composing that variable: $3,124; $10,980; $35,000; and so forth.

The relationship between attributes and vari-

ables is more complicated in the case of explanation and gets to the heart of the variable language of scientific theory. Here's a simple example, involving two variables, *education* and *prejudice.* For the sake of simplicity, let's assume that the variable *education* has only two attributes: *educated* and *uneducated.* (Chapter 5 will address the issue of how such things are defined and measured.) Similarly, let's give the variable *prejudice* two attributes: *prejudiced* and *unprejudiced.*

Now let's suppose that 90 percent of the uneducated are prejudiced, and the other 10 percent are unprejudiced. And let's suppose that 30 percent of the educated people are prejudiced, and the other 70 percent are unprejudiced. This is illustrated graphically in Figure 1-6A.

Figure 1-6A illustrates a relationship or association between the variables *education* and *prejudice.* This relationship can be seen in terms of the pairings of attributes on the two variables. There are two predominant pairings: (1) those who are educated and unprejudiced and (2) those who are uneducated and prejudiced. Here are two other useful ways of viewing that relationship.

First, let's suppose that we play a game in which we bet on your ability to guess whether a person is prejudiced or unprejudiced. I'll pick the people one at a time (not telling you which ones I've picked), and you have to guess whether each person is prejudiced. We'll do it for all 20 people in Figure 1-6A. Your best strategy in this case would be to guess *prejudiced* each time, since 12 out of the 20 are categorized that way. Thus, you'll get 12 right and 8 wrong, for a net success of 4.

Now let's suppose that when I pick a person from the figure, I have to tell you whether the person is educated or uneducated. Your best strategy now would be to guess *prejudiced* for each uneducated person and *unprejudiced* for each educated person. If you follow that strategy, you'll get 16 right and 4 wrong. Your improvement in guessing prejudice by knowing education illustrates what it means to say that variables are related.

Second, by contrast, let's consider how the 20 people would be distributed if education and prejudice were unrelated to each other. This is illustrated

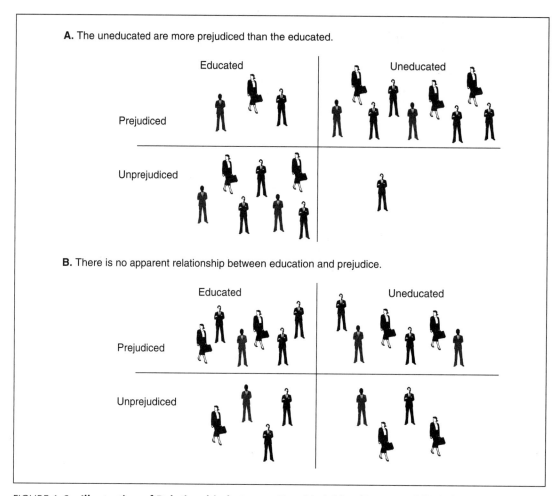

A. The uneducated are more prejudiced than the educated.

B. There is no apparent relationship between education and prejudice.

FIGURE 1-6   **Illustration of Relationship between Two Variables (Two Possibilities)**

in Figure 1-6B. Notice that half the people are educated, and half are uneducated. Also notice that 12 of the 20 (60 percent) are prejudiced. If 6 of the 10 people in each group were prejudiced, we would conclude that the two variables were unrelated to each other. Knowing a person's education would not be of any value to you in guessing whether that person was prejudiced.

We'll be looking at the nature of relationships among variables in some depth in Part 4 of this book. In particular, we'll see some of the ways relationships can be discovered and interpreted in research analysis. A general understanding of relationships now, however, will help you appreciate the logic of social scientific theories.

Theories describe the relationships we might logically expect among variables. Often, the expectation involves the idea of *causation.* A person's attributes on one variable are expected to cause, predispose, or encourage a particular attribute on another variable. In our example, it appeared that there is something about being educated that leads people to be less prejudiced than if they are uneducated.

As I'll discuss in more detail later in the book, *education* and *prejudice* in this example would be regarded as **independent** and **dependent variables,** respectively. *Prejudice* depends on something, hence it is called the dependent variable, which depends on an independent variable, in this

TABLE 1-3   **Education and Support for Segregation**

	Educational Level of Respondents			
	Less than Graduate	HS Graduate	Some College	College Graduate
Agree strongly	10%	7%	6%	1%
Agree slightly	8	5	5	4
Disagree slightly	19	26	18	10
Disagree strongly	62	62	70	85
100% =	(98)	(189)	(190)	(193)

case *education*. Although the educational levels of the people being studied vary, that variation is independent of prejudice.

Notice, at the same time, that educational variations can be found to depend on something else —such as the educational level of our subjects' parents. People whose parents have a lot of education are more likely to get a lot of education than those whose parents have little education. In this relationship, the subject's education is the dependent variable, the parents' education the independent variable. We can say the independent variable is the cause, the dependent variable the effect.

Returning to our first example, the discussion of Figure 1-6 has involved the interpretation of data. We looked at the distribution of the 20 people in terms of the two variables. In constructing a social scientific theory, we would derive an expectation regarding the relationship between the two variables from what we know about each. We know, for example, that education exposes people to a wide range of cultural variation and to diverse points of view—in short, it broadens their perspectives. Prejudice, on the other hand, represents a narrower perspective. Logically, then, we would expect that education and prejudice would be somewhat incompatible. We might therefore arrive at an expectation that increasing education would reduce prejudice, an expectation that would be supported by later observations.

Whereas Figure 1-6 illustrates two possibilities —that education reduces prejudice or that it has no effect—you might be interested in knowing what's actually the case. As one measure of prejudice, the 1996 General Social Survey asked a na-

tional sample of adults in the United States how they felt about the opinion, "White people have a right to keep Blacks out of their neighborhoods if they want to and Blacks should respect that right." Only 6 percent of the sample agreed strongly with the statement, with another 5 percent agreeing slightly. The majority—71 percent—strongly disagreed.

Table 1-3 presents an analysis of those data, grouping respondents according to their levels of educational attainment. The easiest way to read this table is to focus on the last line of percentages: those disagreeing strongly with the statement. Strong opposition to segregation increases steadily from 62 percent among those who hadn't completed high school to 85 percent among college graduates. This clearly supports the view that education reduces prejudice.

Notice that the theory has to do with the two variables *education* and *prejudice,* not with people per se. People are the carriers of those two variables, so the relationship between the variables can only be seen when we observe people. Ultimately, however, the theory uses a language of variables. It describes the associations we might logically expect to exist between particular attributes of different variables.

SOME DIALECTICS OF SOCIAL RESEARCH

There is no one way to do social research. (If there were, this would be a much shorter book.) In fact, much of the power and potential of social research lies in the many valid approaches it comprises.

Four broad and interrelated distinctions underlie these approaches. Though these distinctions can be seen as competing choices, a good social researcher thoroughly learns each of these orientations. This is what I mean by the "dialectics" of social research: There is a fruitful tension between the complementary concepts I'm about to describe.

Idiographic and Nomothetic Explanation

All of us go through life explaining things. We do it every day. You explain why you did poorly or well on an exam, why your favorite team is winning or losing, why you may be having trouble getting dates you enjoy. In our everyday explanations, we engage in two distinct forms of causal reasoning, though we do not ordinarily distinguish them.

Sometimes we attempt to explain a single situation exhaustively. Thus, for example, you may have done poorly on an exam because (1) you had forgotten there was an exam that day, (2) it was in your worst subject, (3) a traffic jam made you late for class, (4) your roommate had kept you up the night before the exam with loud music, (5) the police kept you until dawn demanding to know what you had done with your roommate's stereo—and with your roommate for that matter—and (6) a wild band of coyotes ate your textbook. Given all these circumstances, it is no wonder that you did poorly.

This type of causal reasoning is called an **idiographic** explanation. *Idio-* in this context means unique, separate, peculiar, or distinct, as in the word *idiosyncracy.* When we have completed an idiographic explanation, we feel that we fully understand the causes of what happened in this particular instance. At the same time, the scope of our explanation is limited to the case at hand. While parts of the idiographic explanation might apply to other situations, our intention is to explain one case fully.

Now consider a different kind of explanation. (1) Every time you study with a group, you do better on the exam than if you study alone. (2) Your favorite team does better at home than on the road.

(3) Athletes get more dates than do members of the biology club. This type of explanation—labeled **nomothetic**—seeks to explain a class of situations or events rather than a single one. Moreover, it seeks to explain "economically," using only one or just a few explanatory factors. Finally, it settles for a partial rather than a full explanation.

In each of these examples, you might qualify your causal statements with such words or phrases as "on the whole," "usually," or "all else being equal." Thus, you usually do better on exams when you've studied in a group, but not always. Similarly, your team has won some games on the road and lost some at home. And the gorgeous head of the biology club may get lots of dates, while the defensive lineman Pigpen-the-Terminator may spend a lot of Saturday nights alone punching heavy farm equipment. Such exceptions are acceptable within a broader range of overall explanation. As we noted earlier, patterns are real and important even when they are not perfect.

Both the idiographic and the nomothetic approaches to understanding can be useful to you in your daily life. The nomothetic patterns you discover might offer a good guide for planning your study habits, but the idiographic explanation is more convincing to your parole officer.

By the same token, both idiographic and nomothetic reasoning are powerful tools for social research. The researcher who seeks an exhaustive understanding of the inner workings of a particular juvenile gang or the corporate leadership of a particular multinational conglomerate engages in idiographic research: She or he tries to understand that particular group as fully as possible.

The researcher David Wellman (1995), for example, undertook an in-depth analysis of class consciousness among members of Local 10 of San Francisco's International Longshoremen's and Warehousemen's Union (ILWU). Wellman recognized that this particular union did not typify the U.S. labor movement. While he was interested in gaining insights into labor unionism generally and into the nature of capitalism, his immediate goal was to fully understand the history of Local 10.

Often, however, researchers aim at a more generalized understanding across a class of events, even though the level of understanding is inevitably more superficial. For example, those who seek to uncover the chief factors leading to juvenile delinquency are pursuing a nomothetic inquiry. They might discover, for example, that children from broken homes were more likely to be delinquent than were those from intact families. This explanation would extend well beyond any single child, but it would do so at the expense of a complete explanation.

In contrast to the Wellman study of Local 10, Susan Tiano (1994) sought to understand the overall impact of Third-World industrialization on the status of women. Does their movement into the industrial labor force signify liberation or oppression? Her survey of women factory workers in Mexico illustrates the nomothetic approach to understanding of any one child's delinquency.

As you can see, social scientists can access two distinct kinds of explanation. Just as the physicists sometimes treat light as a particle and other times as a wave, social scientists can search for relatively superficial universals today and probe the narrowly particular tomorrow. Both are good science, both are rewarding, and both can be fun.

Inductive and Deductive Theory

Like idiographic and nomothetic forms of explanation, inductive and deductive thinking both play a role in our daily lives. They, too, represent an important variation in social research.

There are two routes to the conclusion that you do better on exams if you study with others. On the one hand, you might find yourself puzzling, halfway through your college career, why you do so well on exams sometimes but poorly at other times. You might list all the exams you've taken, noting how well you did on each. Then you might try to recall any circumstances shared by all the good exams and all the poor ones. Did you do better on multiple-choice exams or essay exams? Morning exams or afternoon exams? Exams in the natural sciences, the humanities, or the social sciences? Times when you studied alone or . . . SHA-ZAM! It occurs to you that you have almost always done best on exams when you studied with others. This mode of inquiry is known as **induction.**

Inductive reasoning moves from the particular to the general, from a set of specific observations to the discovery of a pattern that represents some degree of order among all the given events. Notice, incidentally, that your discovery doesn't necessarily tell you *why* the pattern exists—just that it does.

Here's a very different way you might have arrived at the same conclusion about studying for exams. Imagine approaching your first set of exams in college. You wonder about the best ways to study—how much to review, how much to focus on class notes. You learn that some students prepare by rewriting their notes in an orderly fashion. Then you consider whether to study at a measured pace or pull an all-nighter just before the exam. Among these musings, you might ask whether you should get together with other students in the class or just study on your own. You could evaluate the pros and cons of both options.

Studying with others might not be as efficient, because a lot of time might be spent on things you already understand. On the other hand, you can understand something better when you've explained it to someone else. And other students might understand parts of the course you haven't gotten yet. Several minds can reveal perspectives that might have escaped you. Also, your commitment to study with others makes it more likely that you'll study rather than watch the special *Brady Bunch* retrospective.

In this fashion, you might add up the pros and cons and conclude, logically, that you'd benefit from studying with others. It seems reasonable to you, the way it seems reasonable that you'll do better if you study rather than not. Sometimes we say things like this are true "in theory." To complete the process, we test whether they're true in practice. For a complete test, you might study alone for half your exams and study with others for the other exams. This procedure would test your logical reasoning.

This second mode of inquiry, **deduction,** moves from the general to the specific. It moves from (1) a pattern that might be logically or theoretically expected to (2) observations that test whether the expected pattern actually occurs. Notice that deduction begins with "why" and moves to "whether," while induction moves in the opposite direction.

As you'll see later in this book, these two very different approaches are both valid avenues for science. Moreover, you'll see how they work together to provide ever more powerful and complete understandings.

Notice, by the way, that the deductive/inductive distinction is not necessarily linked to the nomothetic and idiographic modes. For example, idiographically and deductively, you might prepare for a particular date by taking into account everything you know about the person you're dating, trying to anticipate logically how you can prepare—what kinds of garb, behavior, hairstyle, oral hygiene, and so forth are likely to produce a successful date. Or, idiographically and inductively, you might try to figure out what it was exactly that caused your date to call 911. A nomothetic, deductive approach arises when you coach others on your "rules of dating," when you wisely explain why their dates will be impressed to hear them expound on the dangers of satanic messages concealed in rock and roll lyrics. When you later review your life and wonder why you didn't date more musicians, you might engage in nomothetic induction. Thus there are four possible approaches, which are used as much in life as in research.

We'll return to induction and deduction later in the book. At this point, let's turn to a third broad distinction that generates rich variations in social research.

Quantitative and Qualitative Data

The distinction between quantitative and qualitative data in social research is essentially the distinction between numerical and nonnumerical data. When we say someone is intelligent, we've made a qualitative assertion. A corresponding assertion about someone less fortunately endowed would be that he or she is "unintelligent." When psychologists and others measure intelligence by IQ scores, they are attempting to quantify such qualitative assessments. For example, the psychologist might say that a person has an IQ of 120.

Every observation is qualitative at the outset, whether it be your experience of someone's intelligence, the location of a pointer on a measuring scale, or a check mark entered in a questionnaire. None of these things is inherently numerical or quantitative, but sometimes it is useful to convert them to a numerical form. (Chapter 14 deals specifically with the quantification of data.)

Quantification often makes our observations more explicit. It can also make aggregating and summarizing data easier. Further, it opens up the possibility of statistical analyses, ranging from simple averages to complex formulas and mathematical models. Thus a social researcher might ask whether you tend to date people older or younger than yourself. A quantitative answer to this seems easily attained. The researcher asks how old each of your dates has been and calculates an average. Case closed.

Or is it? While "age" here represents the number of years people have been alive, sometimes people use the term differently; perhaps for some people "age" really means "maturity." Though your dates may tend to be a little older than you, they may act more immaturely and thus represent the same "age." Or someone might see "age" as how young or old your dates look or maybe the degree of variation in their life experiences, their worldliness. These latter meanings would be lost in the quantitative calculation of average age. Qualitative data are richer in meaning and detail than are quantified data. This is implicit in the cliché, "He is older than his years." The poetic meaning of this expression would be lost in attempts to specify how much older.

This richness of meaning is partly a function of ambiguity. If the expression meant something to you when you read it, that particular meaning arises from your own experiences, from people you have known who might fit the description of being "older than their years" or perhaps the times you have heard others use that expression. Two things are certain: (1) You and I probably don't mean exactly the same thing, and (2) you don't know exactly what I mean, and vice versa.

I have a young friend, Ray Zhang, who was responsible for communications at the 1989 freedom demonstrations in Tiananmen Square, Beijing. Following the Army clampdown, Ray fled south and was arrested and then released with orders to return to Beijing. Instead, he escaped from China and made his way to Paris. Eventually, he came to the United States, where he resumed the graduate studies he had been forced to abandon. I have seen him deal with the difficulties of getting enrolled in school without any transcripts from China, studying in a foreign language, meeting his financial needs—all on his own, thousands of miles from his family. Ray still speaks of one day returning to China to build a system of democracy.

Ray strikes me as someone "older than his years." You'd probably agree. This qualitative description, while it fleshes out the meaning of the phrase, still does not equip us to say *how much older* or even to compare two people in these terms without the risk of disagreeing as to which one is more "worldly."

It might be possible to quantify this concept, however. For example, we might establish a list of life experiences that would contribute to what we mean by *worldliness:*

Getting married
Getting divorced
Having a parent die
Seeing a murder committed
Being arrested
Being exiled
Being fired from a job
Running away with the circus

We might quantify people's worldliness as the number of such experiences they've had: the more such experiences, the more worldly we'd say they were. If we thought of some experiences as more powerful than others, we could give those experiences more points. Once we had made our list and point system, scoring people and comparing their worldliness would be pretty straightforward. We would have no difficulty agreeing on who had more points than whom.

To quantify a concept like worldliness, we need to be explicit about what we mean. By focusing specifically on what we'll include in our measurement of the concept, however, we also exclude any other meanings. Inevitably, then, we face a trade-off: Any explicated, quantitative measure will be

A QUANDARY **REVISITED**

This chapter opened with a question regarding uncontrolled variations in society—specifically, birthrates. We noted that there is no apparent control over who will or will not have a baby during a given year. Indeed, many babies occur by accident. For the most part, different people have the babies from one year to the next, and each baby is the result of idiosyncratic, deeply personal reasons.

As the data introduced in this chapter indicate, however, aggregate social life operates differently from individual experiences of living in society. While it is difficult to predict whether a specific person or couple will decide to have a child at a given time, there is a greater regularity at the level of groups, organizations, and societies, produced by social structure, culture, and other forces that individuals may or may not be aware of. Reflect, for example, on the impact of a housing industry that provides too few residences to accommodate large families, in contrast to one where accommodation is the norm. While that single factor would not absolutely determine the childbearing choices of a particular person or couple, it would have a predictable, overall effect across the whole society. And *social* researchers are chiefly interested in describing and understanding social patterns, not individual behaviors. This book will share with you some of the logic and tools social researchers use in that quest.

more superficial than the corresponding qualitative description.

What a dilemma! Which approach should we choose? Which is more appropriate to social research?

The good news is that we don't need to choose. In fact, we shouldn't. Both qualitative and quantitative methods are useful and legitimate in social research; we need both. Some research situations and topics are most amenable to qualitative examination, others to quantification.

However, because these two approaches call for different skills and procedures, you may feel more comfortable with and become more adept in one mode than the other. You'll be a stronger researcher, however, to the extent that you can learn both approaches. At the very least, you should recognize the legitimacy of both.

Finally, you may have noticed that the qualitative approach seems more aligned with idiographic explanations, while nomothetic explanations are more easily achieved through quantification. Though this is true, these relationships are not absolute. Moreover, both approaches present considerable "gray area." Recognizing the distinction between qualitative and quantitative research doesn't mean that you must identify your research activities with one to the exclusion of the other. A complete understanding of a topic often requires both techniques.

Pure and Applied Research

Social researchers have two distinct motivations: understanding and application. On the one hand, they are fascinated by the nature of human social life and are driven to explain it, to make sense out of apparent chaos. *Pure* research in all scientific fields is sometimes justified in terms of gaining "knowledge for its own sake."

At the same time, perhaps inspired by their subject matter, social researchers are committed to having what they learn make a difference, to see their knowledge of society put into action. Sometimes they focus on making things better. When I study prejudice, for example, I'd like what I discover to result in a more tolerant society. This is no different from the AIDS researcher trying to defeat that disease.

In applied social science, however, research is put into practice in many mundane ways as well. Experiments and surveys, for example, can be used in product marketing. In-depth interviewing techniques can be especially useful in social work encounters. Chapter 12 of this book deals with evaluation research, by which social researchers determine the effectiveness of social interventions.

As with the other dialectics just discussed, some social scientists are more inclined toward pure research, others toward application. Ultimately, both orientations are valid and vital elements in social research as a whole.

Now that you have learned about the foundations of social research, I hope you can see how vibrant and exciting such research is. All we need is an open mind and a sense of adventure—and a good grounding in the basics of social research.

Main Points

- The subject of this book is how we find out about social reality.

- Inquiry is a natural human activity. Much of ordinary human inquiry seeks to explain events and predict future events.

- When we understand through direct experience, we make observations and seek patterns of regularities in what we observe.

- Much of what we know, we know by agreement rather than by experience. In particular, two important sources of agreed-upon knowledge are tradition and authority. However, these useful sources of knowledge can also lead us astray.

- Science seeks to protect against the mistakes we make in day-to-day inquiry.

- Whereas we often observe inaccurately, researchers seek to avoid such errors by making observation a careful and deliberate activity.

- We sometimes jump to general conclusions on the basis of only a few observations, so scientists seek to avoid overgeneralization by committing themselves to a sufficient number of observations and by replicating studies.

- In everyday life we sometimes reason illogically. Researchers seek to avoid illogical reasoning by being as careful and deliberate in their reasoning as in their observations. Moreover, the public nature of science means that others are always there to challenge faulty reasoning.

❑ Three views of "reality" are the premodern, modern, and postmodern views. In the postmodern view, there is no "objective" reality independent of our subjective experiences. Different philosophical views suggest a range of possibilities for scientific research.

❑ Social theory attempts to discuss and explain what is, not what should be. Theory should not be confused with philosophy or belief.

❑ Social science looks for regularities in social life.

❑ Social scientists are interested in explaining human aggregates, not individuals.

❑ Theories are written in the language of variables.

❑ A variable is a logical set of attributes. An attribute is a characteristic, such as *male* or *female*. *Gender,* for example, is a variable made up of these attributes.

❑ In causal explanation, the presumed cause is the independent variable, while the affected variable is the dependent variable.

❑ Whereas idiographic explanations seek to understand specific cases fully, nomothetic explanations seek a generalized understanding of many cases.

❑ Inductive theories reason from specific observations to general patterns. Deductive theories start from general statements and predict specific observations.

❑ Quantitative data are numerical; qualitative data are not. Both types of data are useful for different research purposes.

❑ Both pure and applied research are valid and vital parts of the social research enterprise.

Key Terms

replication	dependent variable
theory	idiographic
variable	nomothetic
attribute	induction
independent variable	deduction

Review Questions

1. How would the discussion of variables and attributes apply to physics, chemistry, and biology?

2. Identify a social problem that you feel ought to be addressed and solved. What are the variables represented in your description of the problem? Which of those variables would you monitor in determining whether the problem was solved (for example, the percentage of corporate presidents who are women)?

3. Suppose you were interested in studying the quality of life among elderly people. What quantitative and qualitative indicators might you examine?

4. How might social research be useful to such professionals as physicans, attorneys, business executives, police officers, and newspaper reporters?

Additional Readings

Babbie, Earl. 1994. *The Sociological Spirit.* Belmont, CA: Wadsworth. This book is a primer in sociological points of view and introduces you to many of the concepts commonly used in social research.

Babbie, Earl. 1998. *Observing Ourselves: Essays in Social Research.* Prospect Heights, IL: Waveland Press. This is a collection of essays that expand some of the philosophical issues you will see in the following chapters, including objectivity, paradigms, determinism, concepts, reality, causation, and values.

Becker, Howard S. 1997. *Tricks of the Trade: How to Think about Your Research While You're Doing It.* Chicago: University of Chicago. This very approachable book offers an excellent "feel" for the enterprise of social scientific research, whether qualitative or quantitative. It is filled with research anecdotes that show social inquiry to be a lively and challenging endeavor.

Cole, Stephen. 1992. *Making Science: Between Nature and Society.* Cambridge, MA: Harvard University Press. If you're interested in a deeper examination of science as a social enterprise, you may find this a fascinating analysis.

Gallup, George Jr., Burns Roper, Daniel Yankelovich, et al. 1990. "Polls that Made a Difference." *The Public Perspective,* May/June, pp. 17–21. Several public opinion researchers talk about social research polls that have had an important impact on everyday life.

Hoover, Kenneth R. 1992. *The Elements of Social Scientific Thinking.* New York: St. Martin's Press. Hoover presents an excellent overview of the key elements in social scientific analysis.

Steele, Stephen F., and Joyce Miller Iutcovich, eds. 1997. *Directions in Applied Sociology,* Arnold, MD: Society for Applied Sociology. This book contains the presidential addresses of eleven presidents of the Society for Applied Sociology and provides an excellent overview of the issues involved in the application of social science knowledge.

Multimedia Resources

The Wadsworth Sociology Resource Center: Virtual Society

http://sociology.wadsworth.com/

Visit the companion Web site for the second edition of *The Basics of Social Research* to access a wide range of student resources. Begin by clicking on the Student Resources section of the book's Web site to access the following study tools:

- eBabbie Resource Center
- Planning a Research Project
- Doing Data Analysis
- Statistics Review
- Flash Cards
- Internet Links and Exercises
- InfoTrac College Edition: Exercises
- Quizzes

Visit the **eBabbie Resource Center** for an overview of each chapter and helpful online tutorials. Find information on budgeting and step-by-step examples of model research projects at **Planning a Research Project.** Learn how to use quantitative and qualitative data analysis programs at **Doing Data Analysis,** and brush up on your

statistics at **Statistics Review.** You can also further your study by accessing **Internet Links and Exercises** related to chapter materials, **Flash Cards, Quizzes,** and many other learning tools.

InfoTrac College Edition

http://www.infotrac-college.com/
wadsworth/access.html

Access the latest news and journal articles with InfoTrac College Edition, an easy-to-use online database of reliable, full-length articles from hundreds of top academic journals. Conduct an electronic search using the following search terms:

Deductive reasoning
Idiographic
Inductive reasoning
Nomothetic
Postmodernism research
Qualitative
Quantitative

2 PARADIGMS, THEORY, AND RESEARCH

John-Claude Lejeune

What You'll Learn in This Chapter

Here we'll examine some of the theoretical points of view that structure social scientific inquiry. This lays the groundwork for understanding the specific research techniques discussed throughout the rest of the book.

In this chapter . . .

AN OPENING **QUANDARY**

Scholars such as George Herbert Mead make a powerful argument that social life is really a matter of interactions and their residue. You and I meet each other for the first time, feel each other out, and mutually create rules for dealing with each other. The next time we meet, we'll probably fall back on these rules, which tend to stay with us. Think about your first encounters with a new professor or making a new friend. Mead suggests that all the social patterns and structures that we experience are created in this fashion.

Other scholars, such as Karl Marx, argue that social life is fundamentally a struggle among individuals and among groups. According to Marx, society is a class struggle in which the "haves" and the "have-nots" are pitted against each other in an attempt to dominate others and to avoid being dominated. He claimed that, rather than being mutually created individuals, rules for behavior grow out of the economic structure of a society.

Which of these very different views of society is true? Or does the truth lie somewhere else?

INTRODUCTION

Some restaurants in the United States are fond of conducting political polls among their diners before an upcoming election. Some people take these polls very seriously because of their uncanny history of predicting winners. By the same token, some movie theaters have achieved similar success by offering popcorn in bags picturing either donkeys or elephants. Years ago, granaries in the Midwest offered farmers a chance to indicate their political preferences through the bags of grain they selected.

Such oddities are of some interest. They all have the same pattern over time, however: They work for a while, but then they fail. Moreover, we can't predict when or why they will fail.

These unusual polling techniques point to the shortcoming of "research findings" based only on the observation of patterns. Unless we can offer logical explanations for such patterns, the regularities we've observed may be mere flukes, chance occurrences. If you flip coins long enough, you'll get ten heads in a row. Scientists might adapt a street expression to describe this situation: "Patterns happen."

Logical explanations are what theories seek to provide. Theory functions three ways in research. First, it prevents our being taken in by flukes. If we can't explain why Ma's Diner has been so successful in predicting elections, we run the risk of supporting a fluke. If we know why it has happened, we can anticipate whether it will work in the future.

Second, theories make sense of observed patterns in ways that can suggest other possibilities. If we understand the reasons why broken homes produce more juvenile delinquency than do intact homes—lack of supervision, for example—we can take effective action, such as after-school youth programs.

Finally, theories can shape and direct research efforts, pointing toward likely discoveries through empirical observation. If you were looking for your lost keys on a dark street, you could whip your flashlight around randomly—or you could use your memory of where you had been to limit your search to more likely areas. Theory, by analogy, directs researchers' flashlights where they are most likely to observe interesting patterns of social life.

This is not to say that all social science research is tightly intertwined with social theory. Sometimes social scientists undertake investigations simply to discover the state of affairs, such as an evaluation of whether an innovative social program is work-

ing or a poll to determine which candidate is winning a political race. Similarly, descriptive ethnographies, such as anthropological accounts of preliterate societies, produce valuable information and insights in and of themselves. However, even studies such as these often go beyond pure description to ask *why?* Theory is directly relevant to "why" questions.

This chapter explores some specific ways theory and research work hand in hand during the adventure of inquiry into social life. We'll begin by looking at several fundamental frames of reference, called *paradigms,* that underlie social theories and inquiry.

SOME SOCIAL SCIENCE PARADIGMS

There is usually more than one way to make sense of things. In daily life, for example, liberals and conservatives often explain the same phenomenon—teenagers using guns at school, for example—quite differently. So might the parents and teenagers themselves. But underlying these different explanations, or theories, are **paradigms**—the fundamental models or frames of reference we use to organize our observations and reasoning.

Paradigms are often difficult to recognize as such because they are so implicit, assumed, taken for granted. They seem more like "the way things are" than like one possible point of view among many. Here's an illustration of what I mean.

Where do you stand on the issue of human rights? Do you feel that individual human beings are sacred? Are they "endowed by their creator with certain inalienable rights," as asserted by the U.S. Declaration of Independence? Are there some things that no government should do to its citizens?

Let's get more concrete. In wartime, civilians are sometimes used as human shields to protect military targets. Sometimes they are pressed into slave labor or even used as mobile blood banks for military hospitals. How about organized programs of rape and murder in support of "ethnic cleansing"?

Those of us who are horrified and incensed by

such practices will probably find it difficult to see our individualistic paradigm as only one possible point of view among many. However, the Western (and particularly U.S.) commitment to the sanctity of the individual is regarded as bizarre by many other cultures in today's world. Historically, it is decidedly a minority viewpoint.

While many Asian countries, for example, now subscribe to some "rights" that belong to individuals, those are balanced against the "rights" of families, organizations, and the society at large. Criticized for violating human rights, Asian leaders often point to high crime rates and social disorganization in Western societies as the cost of what they see as our radical "cult of the individual."

I won't try to change your point of view on individual human dignity, nor have I given up my own. It's useful, however, to recognize that our views and feelings in this matter are the result of the paradigm we have been socialized into; they are not an objective fact of nature. All of us operate within many such paradigms. For example, the traditional Western view of the actual world as an objective reality distinct from our individual experiences of it is a deeply ingrained paradigm.

When we recognize that we are operating within a paradigm, two benefits accrue. First, we are better able to understand the seemingly bizarre views and actions of others who are operating from a different paradigm. Second, at times we can profit from stepping outside our paradigm. Suddenly we can see new ways of seeing and explaining things. We can't do that as long as we mistake our paradigm for reality.

Paradigms play a fundamental role in science, just as they do in daily life. Thomas Kuhn (1970) drew attention to the role of paradigms in the history of the natural sciences. Major scientific paradigms have included such fundamental viewpoints as Copernicus's conception of the earth moving around the sun (instead of the reverse), Darwin's theory of evolution, Newtonian mechanics, and Einstein's relativity. Which scientific theories "make sense" depends on which paradigm scientists are maintaining.

While we sometimes think of science as developing gradually over time, marked by important discoveries and inventions, Kuhn says it was typi-

cal for one paradigm to become entrenched, resisting substantial change. Eventually, however, as the shortcomings of that paradigm became obvious, a new paradigm would emerge and supplant the old one. Thus, the view that the sun revolves around the earth was supplanted by the view that the earth revolves around the sun. Kuhn's classic book on this subject is titled, appropriately enough, *The Structure of Scientific Revolutions.*

Social scientists have developed several paradigms for understanding social behavior. The fate of supplanted paradigms in the social sciences, however, has differed from what Kuhn has observed in the natural sciences. Natural scientists generally believe that the succession of paradigms represents progress from false views to true ones. No modern astronomer believes that the sun revolves around the earth, for example.

In the social sciences, on the other hand, theoretical paradigms may gain or lose popularity, but they're seldom discarded. Social science paradigms represent a variety of views, each of which offers insights the others lack while ignoring aspects of social life that the others reveal.

Each of the paradigms we're about to examine offers a different way of looking at human social life. Each makes certain assumptions about the nature of social reality. Ultimately, paradigms cannot be true or false; as ways of looking, they can only be more or less useful. Rather than deciding which paradigm is true or false, try to find ways they might be useful to you. As we shall see, each can open up new understandings, suggest different kinds of theories, and inspire different kinds of research.

Macrotheory and Microtheory

Let's begin with a discussion that encompasses many of the paradigms to be discussed. Some theorists focus their attention on society at large or at least on large portions of it. Topics of study for such **macrotheory** include the struggle among economic classes in a society, international relations, and the interrelations among major institutions in society, such as government, religion, and family. Macrotheory deals with large, aggregate entities of society or even whole societies.

Some scholars have taken a more intimate view of social life. **Microtheory** deals with issues of social life at the level of individuals and small groups. Dating behavior, jury deliberations, and student-faculty interactions are apt subjects for a microtheoretical perspective. Such studies often come close to the realm of psychology, but whereas psychologists typically focus on what goes on inside humans, social scientists study what goes on among them.

The distinction between macro- and microtheory crosscuts the paradigms we'll examine next. While some of them, such as symbolic interactionism and ethnomethodology, often work best at the microlevel, others, such as the conflict paradigm, can be pursued at either the micro- or the macrolevel.

Early Positivism

When the French philosopher Auguste Comte (1798–1857) coined the term *sociologie* in 1822, he launched an intellectual adventure that is still unfolding today. Most important, Comte identified society as a phenomenon that can be studied scientifically. (Initially he wanted to label his enterprise "social physics," but that term was taken over by another scholar.)

Prior to Comte's time, society simply *was*. To the extent that people recognized different kinds of societies or changes in society over time, religious paradigms predominantly explained these differences. The state of social affairs was often seen as a reflection of God's will. Alternatively, people were challenged to create a "City of God" on earth to replace sin and godlessness.

Comte separated his inquiry from religion, replacing religious belief with scientific objectivity. His "positive philosophy" postulated three stages of history. A "theological stage" predominated throughout the world until about 1300. During the next five hundred years, a "metaphysical stage" replaced God with ideas such as "nature" and "natural law." Finally, Comte felt he was launching the third stage of history, in which science would replace religion and metaphysics; knowledge would be based on observations through the five senses rather than on belief. Again, Comte felt that society

could be studied and understood logically and rationally, that sociology could be as scientific as biology or physics.

Comte's view came to form the foundation for subsequent development of the social sciences. In his optimism for the future, he coined the term *positivism* to describe this scientific approach, in contrast to what he regarded as negative elements in the Enlightenment. Only in recent decades has the idea of positivism come under serious challenge, as we'll see later in this discussion.

> To explore this topic in greater depth on the Web, search for "Auguste Comte," "positivism," or "positivist paradigm." *

Conflict Paradigm

Karl Marx (1818–1883) suggested that social behavior could best be seen as the process of conflict: the attempt to dominate others and to avoid being dominated. Marx focused primarily on the struggle among economic classes. Specifically, he examined the way capitalism produced the oppression of workers by the owners of industry. Marx's interest in this topic did not end with analytical study: He was also ideologically committed to restructuring economic relations to end the oppression he observed.

The conflict paradigm is not limited to economic analyses. Georg Simmel (1858–1918) was particularly interested in small-scale conflict, in contrast to the class struggle that interested Marx. Simmel noted, for example, that conflicts among members of a tightly knit group tended to be more intense than those among people who did not share feelings of belonging and intimacy.

In a more recent application of the conflict paradigm, when Michel Chossudovsky's (1997) analysis of the International Monetary Fund (IMF) and World Bank suggested that these two international organizations were increasing global poverty rather than eradicating it, he directed his attention to the competing interests involved in the process.

---

*Each time the Internet icon appears, you'll be given helpful leads for searching the World Wide Web.

In theory, the chief interest being served should be the poor people of the world or perhaps the impoverished, Third-World nations. The researcher's inquiry, however, identified many other interested parties who benefited: the commercial lending institutions who made loans in conjunction with the IMF and World Bank and multinational corporations seeking cheap labor and markets for their goods, for example. Chossudovsky's analysis concluded that the interests of the banks and corporations tended to take precedence over those of the poor people, who were the intended beneficiaries. Moreover, he found many policies were weakening national economies in the Third World, as well as undermining democratic governments.

Whereas the conflict paradigm often focuses on class, gender, and ethnic struggles, it would be appropriate to apply it whenever different groups have competing interests. For example, it could be fruitfully applied to understanding relations among different departments in an organization, fraternity and sorority rush weeks, or student-faculty-administrative relations, to name just a few.

These examples should illustrate some of the ways you might view social life if you were taking your lead from the conflict paradigm. To explore the applicability of this paradigm, you might take a minute to skim through a daily newspaper or news magazine and identify events you could interpret in terms of individuals and groups attempting to dominate each other and avoid being dominated. The theoretical concepts and premises of the conflict paradigm might help you make sense out of these events.

 To explore this topic in greater depth on the Web, search for "conflict theory," "conflict paradigm," or "Karl Marx."

Symbolic Interactionism

Whereas Marx chiefly addressed macrotheoretical issues—large institutions and whole societies in their evolution through the course of history—Georg Simmel (1858–1918) was more interested in the ways individuals interacted with one another,

or the "micro" aspects of society. He began by examining dyads (groups of two people) and triads (groups of three), for example. Similarly, he wrote about "the web of group affiliations."

Simmel was one of the first European sociologists to influence the development of U.S. sociology. His focus on the nature of interactions particularly influenced George Herbert Mead (1863–1931), Charles Horton Cooley (1864–1929), and others who took up the cause and developed it into a powerful paradigm for research.

Cooley, for example, introduced the idea of the "primary group," those intimate associates with whom we share a sense of belonging, such as our family, friends, and so forth. Cooley also wrote of the "looking-glass self" we form by looking into the reactions of people around us. If everyone treats us as beautiful, for example, we conclude that we are. See how fundamentally this paradigm differs from the society-level concerns of Marx.

Similarly, Mead emphasized the importance of our human ability to "take the role of the other," imagining how others feel and how they might behave in certain circumstances. As we gain an idea of how people in general see things, we develop a sense of what Mead called the "generalized other." Mead also felt that most interactions revolved around the process of individuals reaching a common understanding through language and other symbolic systems, hence the term *symbolic interactionism.*

Here's one way you might apply this paradigm to an examination of your own life. The next time you meet someone new, watch how your knowledge of each other unfolds through the process of interaction. Notice also any attempts you make to manage the image you are creating in the other person's mind.

Clearly this paradigm can lend insights into the nature of interactions in ordinary social life, but it can also help us understand unusual forms of interaction, as in the following case. Emerson, Ferris, and Gardner (1998) set out to understand the nature of "stalking." Through interviews with numerous stalking victims, they came to identify different motivations among stalkers, stages in the development of a stalking scenario, how people can recognize if they are being stalked, and what they can do about it.

> To explore this topic in greater depth on the Web, search for "interactionist paradigm," "interactionism," "symbolic interactionism," "George Herbert Mead," "Herbert Blumer," or "Georg Simmel."

Ethnomethodology

While some social scientific paradigms emphasize the impact of social structure (such as norms, values, and control agents) on human behavior, other paradigms do not. Harold Garfinkel, a contemporary sociologist, takes the point of view that people are continually creating social structure through their actions and interactions—that they are, in fact, creating their realities. Thus, when you and your instructor meet to discuss your term paper, even though there are myriad expectations about how you should act, the conversation will somewhat differ from any of those that have occurred before, and how you both act will somewhat modify your future expectations. That is, discussing your term paper will impact your future interactions with other professors and students.

Given the tentativeness of reality in this view, Garfinkel suggests that people are continuously trying to make sense of the life they experience. In a way, he suggests that everyone is acting like a social scientist: hence the term *ethnomethodology*, or "methodology of the people."

How would you go about learning about people's expectations and how they make sense out of their world? One technique ethnomethodologists use is to break the rules, to violate people's expectations. If you try to talk to me about your term paper, but I keep talking about football, any expectations you had for my behavior might come out. We might also see how you make sense out of my behavior. ("Maybe he's using football as an analogy for understanding social systems theory.")

In another example of ethnomethodology, John Heritage and David Greatbatch (1992) examined the role of applause in British political speeches:

How did the speakers evoke applause, and what function did it serve (for example, to complete a topic)? Research within the ethnomethodological paradigm often focuses on communication.

There's no end to the opportunities you have for trying on the ethnomethodological paradigm. For instance, the next time you get on an elevator, don't face front watching the floor numbers whip by (that's the norm, or expected behavior). Instead, just stand quietly facing the rear of the elevator. See how others react to this behavior. Just as important, notice how you feel about it. If you do this experiment a few times, you should begin to develop a feel for the ethnomethodological paradigm.*

We'll return to ethnomethodology in Chapter 10, when we discuss field research. For now, let's turn to a very different paradigm.

> To explore this topic in greater depth on the Web, search for "ethnomethodology" or "Harold Garfinkel."

Structural Functionalism

Structural functionalism, sometimes also known as "social systems theory," grows out of a notion introduced by Comte and others: A social entity, such as an organization or a whole society, can be viewed as an organism. Like organisms, a social system is made up of parts, each of which contributes to the functioning of the whole.

By analogy, consider the human body. Each component—such as the heart, lungs, kidneys, skin, and brain—has a particular job to do. The body as a whole cannot survive unless each of these parts does its job, and none of the parts can survive except as a part of the whole body. Or consider an automobile, composed of tires, steering wheel, gas tank, spark plugs, and so forth. Each of the parts serves a function for the whole; taken together, that system can get us across town. None of

---

*I am grateful to my colleague, Bernard McGrane, for this experiment. Barney also has his students eat dinner with their hands, watch TV without turning it on, and engage in other strangely enlightening behavior (McGrane 1994).

the individual parts would be of much use to us by itself, however.

The view of society as a social system, then, looks for the "functions" served by its various components. We might consider a football team as a social system—one in which the quarterback, running backs, offensive linemen, and others have their own jobs to do for the team as a whole. Or, we could look at a symphony orchestra and examine the functions served by the conductor, the first violinist, and the other musicians.

Social scientists using the structural functional paradigm might note that the function of the police, for example, is to exercise social control—encouraging people to abide by the norms of society and bringing to justice those who do not. We could just as reasonably ask what functions criminals serve in society. Within the functionalist paradigm, we'd see that criminals serve as job security for the police. In a related observation, Emile Durkheim (1858–1917) suggested that crimes and their punishment provided an opportunity for the reaffirmation of a society's values. By catching and punishing a thief, we reaffirm our collective respect for private property.

To get a sense of the structural-functional paradigm, thumb through your college or university catalog and assemble a list of the administrators (such as president, deans, registrar, campus security, maintenance personnel). Figure out what each of them does. To what extent do these roles relate to the chief functions of your college or university, such as teaching or research? Suppose you were studying some other kind of organization. How many of the school administrators' functions would also be needed in, say, an insurance company?

In applying the functionalist paradigm to everyday life, people sometimes make the mistake of thinking that functionality, stability, and integration are necessarily good, or that the functionalist paradigm makes that assumption. However, when social researchers look for the "functions" served by poverty, racial discrimination, or the oppression of women, they are not justifying such things. Rather, they seek to understand the roles such things play in the larger society as a way of understanding why they persist and how they could be eliminated.

> To explore this topic in greater depth on the Web, search for "social systems theory," "functionalism," or "Talcott Parsons." Parsons was the chief architect of the "social systems" paradigm and a leading U.S. sociologist.

Feminist Paradigms

When Ralph Linton concluded his anthropological classic, The Study of Man (1937:490), speaking of "a store of knowledge that promises to give man a better life than any he has known," no one complained that he had left women out. Linton was using the linguistic conventions of his time; he implicitly included women in all his references to men. Or did he?

When feminists (of both genders) first began questioning the use of masculine nouns and pronouns whenever gender was ambiguous, their concerns were often viewed as petty. Many felt the issue was one of women having their feelings hurt, their egos bruised. But be honest: When you read Linton's words, what did you picture? An amorphous, genderless human being, a hermaphrodite at once male and female, or a male persona?

In a similar way, researchers looking at the social world from a feminist paradigm have called attention to aspects of social life that are not revealed by other paradigms. In fact, feminism has established important theoretical paradigms for social research. In part it has focused on gender differences and how they relate to the rest of social organization. These paradigms have drawn attention to the oppression of women in many societies, which has in turn shed light on oppression in general.

Feminist paradigms have also challenged the prevailing notions concerning consensus in society. Most descriptions of the predominant beliefs, values, and norms of a society are written by people representing only portions of society. In the

United States, for example, such analyses have typically been written by middle-class white men—not surprisingly, they have written about the beliefs, values, and norms they themselves share. Though George Herbert Mead spoke of the "generalized other" that each of us becomes aware of and can "take the role of," feminist paradigms question whether such a generalized other even exists.

Further, whereas Mead used the example of learning to play baseball to illustrate how we learn about the generalized other, Janet Lever's research suggests that understanding the experience of boys may tell us little about girls.

> Girls' play and games are very different. They are mostly spontaneous, imaginative, and free of structure or rules. Turn-taking activities like jump rope may be played without setting explicit goals. Girls have far less experience with interpersonal competition. The style of their competition is indirect, rather than face to face, individual rather than team affiliated. Leadership roles are either missing or randomly filled. — (LEVER 1986:86)

Social researchers' growing recognition of the intellectual differences between men and women led the psychologist Mary Field Belenky and her colleagues to speak of *Women's Ways of Knowing* (1986). In-depth interviews with 45 women led the researchers to distinguish five perspectives on knowing that challenge the view of inquiry as obvious and straightforward:

- *Silence:* Some women, especially early in life, feel themselves isolated from the world of knowledge, their lives largely determined by external authorities.
- *Received knowledge:* From this perspective, women feel themselves capable of taking in and holding knowledge originating with external authorities.
- *Subjective knowledge:* This perspective opens up the possibility of personal, subjective knowledge, including intuition.
- *Procedural knowledge:* Some women feel they have mastered the ways of gaining knowledge through objective procedures.

- *Constructed knowledge:* The authors describe this perspective as "a position in which women view all knowledge as contextual, experience themselves as creators of knowledge, and value both subjective and objective strategies for knowing."
— (BELENKY ET AL. 1986:15)

"Constructed knowledge" is particularly interesting in the context of our previous discussions. The positivistic paradigm of Comte would have a place neither for "subjective knowledge" nor for the idea that truth might vary according to its context. The ethnomethodological paradigm, on the other hand, would accommodate these ideas.

To try out feminist paradigms, you might want to look into the possibility of discrimination against women at your college or university. Are the top administrative positions held equally by men and women? How about secretarial and clerical positions? Are men's and women's sports supported equally? Read through the official history of your school; is it a history that includes men and women equally? (If you attend an all-male or all-female school, of course, some of these questions won't apply.)

> To explore this topic in greater depth on the Web, search for "feminist paradigm," "feminist sociology," "feminist theory," and don't miss http://www.cddc.vt.edu/feminism/

Rational Objectivity Reconsidered

We began with Comte's assertion that we can study society rationally and objectively. Since his time, the growth of science, the decline of superstition, and the rise of bureaucratic structures have put rationality more and more at the center of social life. As fundamental as rationality is to most of us, however, some contemporary scholars have raised questions about it.

For example, positivistic social scientists have sometimes erred in assuming that humans will always act rationally. I'm sure your own experience offers ample evidence to the contrary. Many modern economic models also assume that people will make rational choices in the economic sector: They will choose the highest-paying job, pay the lowest price, and so forth. This assumption, however, ignores the power of such matters as tradition, loyalty, and image that compete with reason in determining human behavior.

A more sophisticated positivism would assert that we can rationally understand even nonrational human behavior. Here's an example. In the famous "Asch Experiment" (Asch 1958), a group of subjects is presented with a set of lines on a screen and asked to identify the two lines of equal length.

Imagine yourself a subject in such an experiment. You're sitting in the front row of a classroom in a group of six subjects. A set of lines (see Figure 2-1) is projected on the wall in front of you. The experimenter asks you, one at a time, to identify the line to the right (A, B, or C) that matches the length of line X. The correct answer (B) is pretty obvious to you. To your surprise, you find that all the other subjects agree on a different answer!

The experimenter announces that all but one of the group has gotten the correct answer; that is, you've gotten it wrong. Then a new set of lines is presented, and you have the same experience. The obviously correct answer is wrong, and everyone but you seems to understand that.

As it turns out, of course, you're the only *real* subject in the experiment—all the others are working with the experimenter. The purpose is to see whether you would be swayed by public pressure and go along with the incorrect answer. In one-third of the initial experiments, Asch found that his subjects did just that.

Choosing an obviously wrong answer in a simple experiment is an example of nonrational behavior. But as Asch went on to show, experimenters can examine the circumstances that lead more or fewer subjects to go along with the incorrect answer. For example, in subsequent studies, Asch varied the size of one group and the number of "dissenters" who chose the "wrong" (that is, the correct) answer. Thus, it is possible to study nonrational behavior rationally and scientifically.

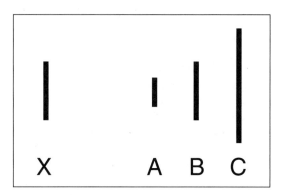

FIGURE 2-1 **The Asch Experiment**

More radically, we can question whether social life abides by rational principles at all. In the physical sciences, developments such as chaos theory, fuzzy logic, and complexity have suggested that we may need to rethink fundamentally the orderliness of physical events.

The contemporary challenge to positivism, however, goes beyond the question of whether people behave rationally. In part, the criticism of positivism challenges the idea that scientists can be as objective as the scientific ideal assumes. Most scientists would agree that personal feelings can and do influence the problems scientists choose to study, their choice of what to observe, and the conclusions they draw from their observations.

As with rationality, there is a more radical critique of objectivity. Whereas scientific objectivity has long stood as an unquestionable ideal, some contemporary researchers suggest that subjectivity might actually be preferred in some situations, as we glimpsed in the discussions of feminism and ethnomethodology. Let's take a moment to return to the dialectic of subjectivity and objectivity.

To begin with, all our experiences are inescapably subjective. There is no way out. We can see only through our own eyes, and anything peculiar to our eyes will shape what we see. We can hear things only the way our particular ears and brain transmit and interpret sound waves. You and I, to some extent, hear and see different realities. And both of us experience quite different physical "realities" than do bats, for example. In what to us is total darkness, a bat "sees" things such as flying insects by emitting a sound we humans can't hear. The reflection of the bat's sound creates a "sound picture" precise enough for the bat to home in on the moving insect and snatch it up. In a similar vein, scientists on the planet Xandu might develop theories of the physical world based on a sensory apparatus that we humans can't even imagine. Maybe they see X rays or hear colors.

Despite the inescapable subjectivity of our experience, we humans seem to be wired to seek an agreement on what is "really real," what is *objectively* so. Objectivity is a conceptual attempt to get beyond our individual views. It is ultimately a matter of communication, as you and I attempt to find a common ground in our subjective experiences. Whenever we succeed in our search, we say we are dealing with objective reality. This is the agreement reality discussed in Chapter 1.

While our subjectivity is individual, our search for objectivity is social. This is true in all aspects of life, not just in science. While you and I prefer different foods, we must agree to some extent on what is fit to eat and what is not, or else there could be no restaurants, no grocery stores, no food industry. The same argument could be made regarding every other form of consumption. There could be no movies or television, no sports.

Social scientists as well have found benefits in the concept of objective reality. As people seek to impose order on their experience of life, they find it useful to pursue this goal as a collective venture. What are the causes and cures of prejudice? Working together, social researchers have uncovered some answers that hold up to intersubjective scrutiny. Whatever your subjective experience of things, for example, you can discover for yourself that as education increases, prejudice tends to decrease. Because each of us can discover this independently, we say it is objectively true.

From the seventeenth century through the middle of the twentieth, the belief in an objective reality that people could see ever more clearly predominated in science. For the most part, it was held not simply as a useful paradigm but as The Truth. The term *positivism* generally represents the belief in a logically ordered, objective reality that we can come to know. This is the view challenged today by postmodernists and others.

Some say that the ideal of objectivity conceals as much as it reveals. As we saw earlier, much of what was regarded as scientific objectivity in years past was actually an agreement primarily among white, middle-class, European men. Experiences common to women, to ethnic minorities, or to the poor, for example, were not necessarily represented in that reality.

The early anthropologists are now criticized for often making modern, Westernized "sense" out of the beliefs and practices of nonliterate tribes around the world—sometimes portraying their subjects as superstitious savages. We often call orally transmitted beliefs about the distant past "creation myth," whereas we speak of our own beliefs as "history." Increasingly today, there is a demand to find the native logic by which various peoples make sense out of life.

Ultimately, we'll never know whether there is an objective reality that we experience subjectively or whether our concepts of an objective reality are illusory. So desperate is our need to know just what is going on, however, that both the positivists and the postmodernists are sometimes drawn into the belief that their view is real and true. There is a dual irony in this. On the one hand, the positivist's belief in the reality of the objective world must ultimately be based on faith; it cannot be proven by "objective" science, since that's precisely what's at issue. And the postmodernists, who say nothing is objectively so, do at least feel the absence of objective reality is *really* the way things are.

For social researchers, each approach brings special strengths, and each compensates for the weaknesses of the other. It's often most useful to "work both sides of the street," tapping into the rich variety of theoretical perspectives that can be brought to bear on the study of human social life.

The attempt to establish formal theories of society has been closely associated with the belief in a discoverable, objective reality. Even so, we'll see next that the issues involved in theory construction are of interest and use to all social researchers, from the positivists to the postmodernists—and all those in between.

TWO LOGICAL SYSTEMS REVISITED

In Chapter 1, I introduced deductive and inductive theory, with a promise that we would return to them later. It's later.

The Traditional Model of Science

Years of learning about "the scientific method," especially in the physical sciences, tends to create in students' minds a particular picture of how science operates. Although this traditional model of science tells only a part of the story, it's helpful to understand its logic.

There are three main elements in the traditional model of science, typically presented in the order in which they are implemented: theory, operationalization, and observation. Let's look at each in turn.

Theory At this point we're already well acquainted with the idea of theory. According to the traditional model of science, scientists begin with a theory, from which they derive hypotheses that they can test. So, for example, as social scientists we might have a theory about the causes of juvenile delinquency. Let's assume that we have arrived at the hypothesis that delinquency is inversely related to social class. That is, as social class goes up, delinquency goes down.

Operationalization To test any **hypothesis,** we must specify the meanings of all the variables involved in it: *social class* and *delinquency* in the present case. For example, *delinquency* might be specified as "being arrested for a crime," or "being convicted of a crime," and so forth. *Social class* might be specified as family income for this particular study.

Next, we need to specify how we'll measure the variables we have defined. **Operationalization** literally means the operations involved in measuring a variable. There are many ways we can pursue

this topic, each of which allows for different ways of measuring our variables.

For simplicity, let's assume we're planning to conduct a survey of high school students. We might operationalize delinquency in the form of the question: "Have you ever stolen anything? " Those who answer "yes" will be classified as delinquents in our study; those who say "no" will be classified as nondelinquents. Similarly, we might operationalize family income by asking respondents, "What was your family's income last year?" and providing them with a set of family income categories: under $10,000; $10,000–$24,999; $25,000–$49,999; and $50,000 and above.

At this point someone might object that "delinquency" can mean something more or different from having stolen something at one time or another, or that social class isn't necessarily exactly the same as family income. Some parents might think body piercing is a sign of delinquency even if their children don't steal, and to some "social class" might include an element of prestige or community standing as well as how much money a family has. For the researcher testing a hypothesis, however, the meaning of variables is exactly and only what the **operational definition** specifies.

In this respect, scientists are very much like Humpty Dumpty in Lewis Carroll's *Through the Looking Glass.* "When *I* use a word," Humpty Dumpty tells Alice, "it means just what I choose it to mean—neither more nor less."

"The question is," Alice replies, "whether you *can* make words mean so many different things. " To which Humpty Dumpty responds, "The question is, which is to be master—that's all."

Scientists have to be "masters" of their operational definitions for the sake of precision in observation, measurement, and communication. Otherwise, we would never know whether a study that contradicted ours did so only because it used a different set of procedures to measure one of the variables and thus changed the meaning of the hypothesis being tested. Of course, this also means that to evaluate a study's conclusions about juvenile delinquency and social class, or any other variables, we need to know how those variables were operationalized.

The way we have operationalized the variables in our imaginary study could be open to other problems, however. Perhaps some respondents will lie about having stolen anything, in which cases we'll misclassify them as nondelinquent. Some respondents will not know their family incomes and will give mistaken answers; others may be embarrassed and lie. We'll consider such issues in detail in Part 2.

Our operationalized hypothesis now is that the highest incidence of delinquents will be found among respondents who select the lowest family income category (under $10,000); a lower percentage of delinquents will be found in the $10,000–$24,999 category; still fewer delinquents will be found in the $25,000–$49,999 category; and the lowest percentage of delinquents will be found in the $50,000 and above category.

Observation The final step in the traditional model of science involves actual observation, looking at the world and making measurements of what is seen. Having developed theoretical clarity and expectations and having created a strategy for looking, all that remains is to look at the way things actually appear.

Let's suppose our survey produced the following data:

	Percentage delinquent
Under $10,000	20
$10,000–$24,999	15
$25,000–$49,999	10
$50,000 and above	25

Observations producing such data would confirm our hypothesis. But suppose our findings were as follows:

	Percentage delinquent
Under $10,000	15
$10,000–$24,999	15
$25,000–$49,999	15
$50,000 and above	15

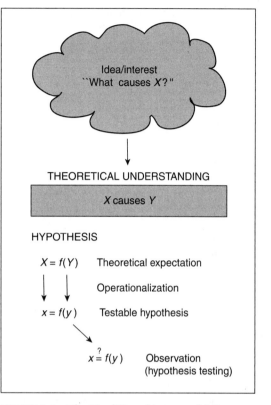

FIGURE 2-2 **The Traditional Image of Science**

These findings would disconfirm our hypothesis regarding family income and delinquency. *Disconfirmability* is an essential quality in any hypothesis. In other words, if there is no chance that our hypothesis will be disconfirmed, it hasn't said anything meaningful.

For example, the hypothesis that "juvenile delinquents" commit more crimes than do "nondelinquents" do cannot possibly be disconfirmed, because criminal behavior is intrinsic to the notion of delinquency. Even if we recognize that some young people commit crimes without being caught and labeled as delinquents, they couldn't threaten our hypothesis, since our observations would lead us to conclude they were law-abiding nondelinquents.

Figure 2-2 provides a schematic diagram of the traditional model of scientific inquiry. In it we see the researcher beginning with an interest in something or an idea about it. Next comes the development of a theoretical understanding. The theoretical considerations result in a hypothesis, or an expectation about the way things ought to be in the world if the theoretical expectations are correct. The notation $Y = f(X)$ is a conventional way of saying that Y (for example, delinquency) is a function of (is in some way caused by) X (for example, poverty). At that level, however, X and Y have general rather than specific meanings.

In the operationalization process, general concepts are translated into specific indicators and procedures. The lowercase x, for example, is a concrete indicator of capital X. Thus, while X is theoretical, x is something we could actually observe. If X stands for "poverty" in general, x might stand for "family income." If Y is the theoretical variable "juvenile delinquency," this could be measured as "self-reported crimes" on a survey.

This operationalization process results in the formation of a testable hypothesis: for example, increasing family income reduces self-reported theft. Observations aimed at finding out whether this is true are part of what is typically called hypothesis testing. (See the box "Hints for Stating Hypotheses" for more on this.)

SPSS
MC

The SPSS and MicroCase files on the disk available with this book allow you to skim data sets to learn the kinds of variables that might be operationalized in social research. Chapter 14, on quantitative analysis, will provide in-depth instruction on how to do this.*

Deduction and Induction Compared

The traditional model of science uses deductive logic (see Chapter 1). In this section, we're going to see how deductive logic fits into social scientific research and contrast it with inductive logic. W. I. B. Beveridge, a philosopher of science, describes these two systems of logic as follows:

---

*Each time the SPSS and MicroCase icons appear, they indicate that the topic under discussion could be pursued through the use of these software programs.

Logicians distinguish between inductive reasoning (from particular instances to general principles, from facts to theories) and deductive reasoning (from the general to the particular, applying a theory to a particular case). In induction one starts from observed data and develops a generalization which explains the relationships between the objects observed. On the other hand, in deductive reasoning one starts from some general law and applies it to a particular instance. — (BEVERIDGE 1950:113)

The classical illustration of deductive logic is the familiar syllogism "All men are mortal; Socrates is a man; therefore Socrates is mortal." This syllogism presents a theory and its operationalization. To prove it, you might then perform an empirical test of Socrates' mortality. That is essentially the approach discussed as the traditional model.

Using inductive logic, you might begin by noting that Socrates is mortal and by observing several other men as well. You might then note that all the observed men were mortals, thereby arriving at the tentative conclusion that all men are mortal.

Let's consider an actual research project as a vehicle for comparing the roles of deductive and inductive logic in theory and research.

A Case Illustration Years ago, Charles Glock, Benjamin Ringer, and I (1967) set out to discover what caused differing levels of church involvement among U.S. Episcopalians. Several theoretical or quasi-theoretical positions suggested possible answers. I'll focus on only one here—what we came to call the "Comfort Hypothesis."

In part, we took our lead from the Christian injunction to care for "the halt, the lame, and the blind" and those who are "weary and heavy laden." At the same time, ironically, we noted the Marxist assertion that religion is an "opiate for the masses." Given both, it made sense to expect the following, which was our hypothesis: "Parishioners whose life situations most deprive them of satisfaction and fulfillment in the secular society turn to the church for comfort and substitute rewards" (Glock et al. 1967:107–8).

Having framed this general hypothesis, we set about testing it. Were those deprived of satisfaction in the secular society in fact more religious than those who got more satisfaction from the secular society? To answer this, we needed to distinguish who was deprived. Our questionnaire included items that intended to indicate whether parishioners were relatively deprived or gratified in secular society.

To start, we reasoned that men enjoyed more status than do women in our generally male-dominated society. It followed that, if our hypothesis were correct, women should appear more religious than men. Once the survey data had been collected and analyzed, our expectation about gender and religion was clearly confirmed. On three separate measures of religious involvement—*ritual* (for example, church attendance), *organizational* (for example, belonging to church organizations), and *intellectual* (for example, reading church publications)—women were more religious than men. On our overall measure, women scored 50 percent higher than men.

In another test of the Comfort Hypothesis, we reasoned that in a youth-oriented society, old people would be more deprived of secular gratification than the young would be. Once again, the data confirmed our expectation. The oldest parishioners were more religious than were the middle-aged, who were more religious than were the young adults.

Social class—measured by education and income—afforded another test, which was successful. Those with low social status were more involved in the church than were those with high social status.

The hypothesis was even confirmed in a test that went against everyone's commonsense expectations. Despite church posters showing worshipful young families and bearing the slogan, "The Family That Prays Together Stays Together," the Comfort Hypothesis suggested that parishioners who were married and had children—the clear U.S. ideal at that time—would enjoy secular gratification in that regard. As a consequence, they should be *less* religious than those who lacked one or both family components. Thus, we hypothesized that parishioners who were both single and childless should be the most religious; those with either spouse or child should be somewhat less religious;

and those married with children—representing the ideal pictured on all those posters—should be least religious of all. That's exactly what we found!

Finally, the Comfort Hypothesis suggested that the various kinds of secular deprivation should be cumulative: Those with all the characteristics associated with deprivation should be the most religious; those with none should be the least. When we combined the four individual measures of deprivation into a composite measure (see Chapter 6 for methods of doing this), the theoretical expectation was exactly confirmed. Comparing the two extremes, we found that single, childless, old, lower-class female parishioners scored more than three times as high on the measure of church involvement than did young, married, upper-class fathers.

This research example clearly illustrates the logic of the deductive model. Beginning with general, theoretical expectations about the impact of social deprivation on church involvement, we derived concrete hypotheses linking specific measurable variables, such as age and church attendance. We then analyzed the actual empirical data to determine whether the deductive expectations were supported by empirical reality. Sounds good, right?

Alas, I've been fibbing a little bit just now. To tell the truth, although we began with an interest in discovering what caused variations in church involvement among Episcopalians, we didn't actually begin with a Comfort Hypothesis, or any other hypothesis for that matter. (In the interest of further honesty, Glock and Ringer initiated the study, and I joined it years after the data had been collected.)

A questionnaire was designed to collect information from parishioners that *might* shed some light on why some participated in the church more than others, but questionnaire construction was not guided by any precise, deductive theory. Once the data were collected, the task of explaining differences in religiosity began with an analysis of variables that have a wide impact on people's lives, including *gender, age, social class,* and *family status.* Each of these four variables was found to relate strongly to church involvement in the ways already described. Rather than being good news, this presented a dilemma.

HINTS FOR STATING HYPOTHESES

by Riley E. Dunlap
Department of Sociology, Washington State University

A hypothesis is the basic statement that is tested in research. Typically a hypothesis states a relationship between two variables. (Although it is possible to use more than two variables, you should stick to two for now.) Because a hypothesis makes a prediction about the relationship between the two variables, it must be testable so you can determine if the prediction is right or wrong when you examine the results obtained in your study. A hypothesis must be stated in an unambiguous manner to be clearly testable. What follows are suggestions for developing testable hypotheses.

Assume you have an interest in trying to predict some phenomenon such as "attitudes toward women's liberation," and that you can measure such attitudes on a continuum ranging from "opposed to women's liberation" to "neutral" to "supportive of women's liberation." Also assume that, lacking a theory, you'll rely on "hunches" to come up with variables that might be related to attitudes toward women's liberation.

In a sense, you can think of hypothesis construction as a case of filling in the blank: "_____ is related to attitudes toward women's liberation." Your job is to think of a variable that might plausibly be related to such attitudes, and then to word a hypothesis that states a relationship between the two variables (the one that fills in the "blank" and "attitudes toward women's liberation"). You need to do so in a precise manner so that you can determine clearly whether the hypothesis is supported or not when you examine the results (in this case, most likely the results of a survey).

The key is to word the hypothesis carefully so that the prediction it makes is quite clear to you as well as others. If you use age, note that saying "Age is related to attitudes toward women's liberation" does not say precisely how you think the two are related (in fact, the only way this hypothesis could be falsified is if you fail to find a statistically significant relationship of any type between age and attitudes toward women's liberation). In this case a couple of steps are necessary. You have two options:

1. "Age is related to attitudes toward women's liberation, with younger adults being more supportive than older adults." (Or, you could state the opposite, if you believed older people are likely to be more supportive.)
2. "Age is negatively related to support for women's liberation." Note here that I specify "support" for women's liberation (SWL) and then predict a negative relationship—that is, as age goes up, I predict that SWL will go down.

Glock recalls discussing his findings with colleagues over lunch at the Columbia faculty club. Once he had displayed the tables illustrating the impact of the variables and their cumulative effect, a colleague asked, "What does it all mean, Charlie?" Glock was at a loss. Why were those variables so strongly related to church involvement?

That question launched a process of reasoning about what the several variables had in common, aside from their impact on religiosity. (The composite index was originally labeled "Predisposition to Church Involvement.") Eventually we saw that each of the four variables also reflected differential status in the secular society, and then we had the thought that perhaps the issue of comfort was involved. Thus, the inductive process had moved from concrete observations to a general theoretical explanation.

A Graphic Contrast As the preceding case illustration shows, theory and research can usefully be done both inductively and deductively. Figure 2-3

In this hypothesis, note that both of the variables (*age,* the independent variable or likely "cause," and *SWL,* the dependent variable or likely "effect") range from low to high. This feature of the two variables is what allows you to use "negatively" (or "positively") to describe the relationship.

Notice what happens if you hypothesize a relationship between gender and SWL. Since *gender* is a nominal variable (as you'll learn in Chapter 5) it does not range from low to high —people are either male or female (the two attributes of the variable *gender*). Consequently, you must be careful in stating the hypothesis unambiguously:

1. "Gender is positively (or negatively) related to SWL" is not an adequate hypothesis, because it doesn't specify how you expect gender to be related to SWL—that is, whether you think men or women will be more supportive of women's liberation.
2. It is tempting to say something like "Women are positively related to SWL," but this really doesn't work because female is only an attribute, not a full variable (gender is the variable).
3. "Gender is related to SWL, with women being more supportive than men" would be my recommendation. Or, you could say, "with men being less supportive than women," which

makes the identical prediction. (Of course, you could also make the opposite prediction, that men are more supportive than women are, if you wished.)
4. Equally legitimate would be "Women are more likely to support women's liberation than are men. " (Note the need for the second "are," or you could be construed as hypothesizing that women support women's liberation more than they support men—not quite the same idea.)

The above examples hypothesized relationships between a "characteristic" (age or gender) and an "orientation" (attitudes toward women's liberation). Because the causal order is pretty clear (obviously age and gender come before attitudes, and are less alterable), we could state the hypotheses as I've done, and everyone would assume that we were stating causal hypotheses.

Finally, you may run across references to the **null hypothesis,** especially in statistics. Such a hypothesis predicts no relationship (technically, no statistically significant relationship) between the two variables, and it is always implicit in testing hypotheses. Basically, if you have hypothesized a positive (or negative) relationship, you are hoping that the results will allow you to reject the null hypothesis and verify your hypothesized relationship.

shows a graphic comparison of the deductive and inductive methods. In both cases, we are interested in the relationship between the number of hours spent studying for an exam and the grade earned on that exam. Using the deductive method, we would begin by examining the matter logically. Doing well on an exam reflects a student's ability to recall and manipulate information. Both of these abilities should be increased by exposure to the information before the exam. In this fashion, we would arrive at a hypothesis suggesting a *positive*

relationship between the number of hours spent studying and the grade earned on the exam. That is, we expect grades to increase as the hours of studying increase. If increased hours produced decreased grades, we would call it a *negative* relationship. The hypothesis is represented by the line in part 1(a) of Figure 2-3.

Our next step would be to make observations relevant to testing our hypothesis. The shaded area in part 1(b) of the figure represents perhaps hundreds of observations of different students, noting

APPLYING THE RESULTS

While many church leaders believe that the function of the churches is to shape members' behavior in the community, the Glock study suggests that church involvement primarily reflects a need for comfort by those who are denied gratification in the secular society. How might churches apply these research results?

On the one hand, churches might adjust their programs to the needs that were drawing their members to participation. They might study members' needs for gratification and develop more programs to satisfy them. On the other hand, churches could seek to remind members that the purpose of participation is to learn and practice proper behavior. Following that strategy would probably change participation patterns, attracting new participants in the church while driving away others.

how many hours they studied and what grades they got. Finally, in part 1(c), we compare the hypothesis and the observations. Because observations in the real world seldom if ever match our expectations perfectly, we must decide whether the match is close enough to confirm the hypothesis. Put differently, can we conclude that the hypothesis describes the general pattern that exists, granting some variations in real life?

Now let's address the same research question by using the inductive method. We would begin—as in part 2(a) of the figure—with a set of observations. Curious about the relationship between hours spent studying and grades earned, we might simply arrange to collect some relevant data. Then we'd look for a pattern that best represented or summarized our observations. In part 2(b) of the figure, the pattern is shown as a curved line running through the center of the curving mass of points.

The pattern found among the points in this case

suggests that with 1 to 15 hours of studying, each additional hour generally produces a higher grade on the exam. With 15 to about 25 hours, however, more study seems to slightly lower the grade. Studying more than 25 hours, on the other hand, results in a return to the initial pattern: More hours produce higher grades. Using the inductive method, then, we end up with a tentative conclusion about the pattern of the relationship between the two variables. The conclusion is tentative because the observations we have made cannot be taken as a test of the pattern—those observations are the source of the pattern we've created.

In actual practice, theory and research interact through a never ending alternation of deduction and induction. Walter Wallace (1971) has represented this process as a circle, which is presented in a modified form in Figure 2-4.

When Emile Durkheim ([1897] 1951) pored through table after table of official statistics on suicide rates in different areas, he was struck by the fact that Protestant countries consistently had higher suicide rates than Catholic ones. Why should that be the case? His initial observations led him to create a theory of religion, social integration, anomie, and suicide. His theoretical explanations led to further hypotheses and further observations.

In summary, the scientific norm of logical reasoning provides a two-way bridge between theory and research. Scientific inquiry in practice typically involves an alternation between deduction and induction. During the deductive phase, we reason *toward* observations; during the inductive phase, we reason *from* observations. Both deduction and induction are routes to the construction of social theories, and both logic and observation are essential.

Although both inductive and deductive methods are valid in scientific inquiry, individuals may feel more comfortable with one approach than the other. Consider this exchange in Sir Arthur Conan Doyle's *A Scandal in Bohemia,* as Sherlock Holmes answers Dr. Watson's inquiry (Doyle [1891] 1892:13):

"What do you imagine that it means?"

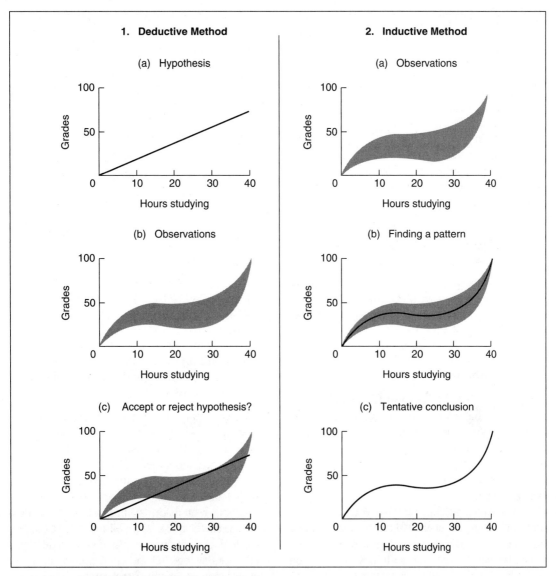

FIGURE 2-3 **Deductive and Inductive Methods**

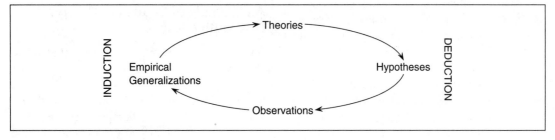

FIGURE 2-4 **The Wheel of Science**

"I have no data yet. It is a capital mistake to theorise before one has data. Insensibly one begins to twist facts to suit theories, instead of theories to suit facts."

Some social scientists would rally behind this inductive position, while others would take a deductive stance. Most, however, concede the legitimacy of both. With this understanding of the deductive and inductive links between theory and research, let's delve a little more deeply into how theories are constructed using these two different approaches.

DEDUCTIVE THEORY CONSTRUCTION

To see what is involved in deductive theory construction and hypothesis testing, let's imagine that you are going to construct a deductive theory. How would you go about it?

Getting Started

The first step in deductive theory construction is to pick a topic that interests you. It can be broad, such as "What's the structure of society? " or narrower, as in "Why do people support or oppose a woman's right to an abortion? " Whatever the topic, it should be something you're interested in understanding and explaining.

Once you've picked your topic, you then undertake an inventory of what is known or thought about it. In part, this means writing down your own observations and ideas about it. Beyond that, you need to learn what other scholars have said about it. You can do this by talking to other people and by reading what others have written about it. Appendix A provides guidelines for using the library—you'll probably spend a lot of time there.

Your preliminary research will probably uncover consistent patterns discovered by prior scholars. For example, religious and political variables will stand out as important determinants of attitudes about abortion. Findings such as these will be quite useful to you in creating your own theory.

Throughout this process, introspection is helpful. If you can look at your own personal processes—including reactions, fears, and prejudices you aren't especially proud of—you may be able to gain important insights into human behavior in general.

Constructing Your Theory

Although theory construction is not a lockstep affair, the following list of elements in theory construction should organize the activity for you.

1. Specify the topic.
2. Specify the range of phenomena your theory addresses. Will your theory apply to all of human social life, will it apply only to U.S. citizens, only to young people, or what?
3. Identify and specify your major concepts and variables.
4. Find out what is known (or what propositions have been demonstrated) about the relationships among those variables.
5. Reason logically from those propositions to the specific topic you are examining.

We've already discussed items (1) through (3), so let's focus now on (4) and (5). As you identify the relevant concepts and discover what has already been learned about them, you can begin to create a propositional structure that explains the topic under study. For the most part, social scientists have not created formal, propositional theories. Still, it is useful to look at a well-reasoned example. Let's look now at an example of how these building blocks fit together in actual deductive theory construction and empirical research.

An Example of Deductive Theory: Distributive Justice

A topic of central interest to scholars using the exchange paradigm (discussed earlier) is that of *distributive justice,* people's perception of whether they're being treated fairly by life, whether they're getting "their share." Guillermina Jasso describes the theory of distributive justice more formally, as follows:

> The theory provides a mathematical description of the process whereby individuals, reflecting on their holdings of the goods they value (such as beauty, intelligence, or wealth), compare themselves to others, experiencing a fundamental instantaneous magnitude of the justice evaluation (J), which captures their sense of being fairly or unfairly treated in the distributions of natural and social goods. — (JASSO 1988:11)

Notice that Jasso has assigned a letter to her key variable: J will stand for distributive justice. She does this to support her intention of stating her theory in mathematical formulas. Though theories are often expressed mathematically, we'll not delve too deeply into that practice here.

Jasso indicates that there are three kinds of postulates in her theory. "The first makes explicit the fundamental axiom which represents the substantive point of departure for the theory." She elaborates as follows:

> The theory begins with the received Axiom of Comparison, which formalizes the long-held view that a wide class of phenomena, including happiness, self-esteem, and the sense of distributive justice, may be understood as the product of a comparison process. — (JASSO 1988:11)

Thus, our sense of whether we are receiving a "fair" share of the good things of life comes from comparing ourselves with others. If this seems obvious to you, that's good. Remember, axioms are the taken-for-granted beginnings of theory.

Jasso continues to do the groundwork for her theory. First, she indicates that our sense of distributive justice is a function of "Actual Holding (A)" and "Comparison Holdings (C)" of some good. Let's consider money. My sense of justice in this regard is a function of how much I actually have, compared with how much others have. By specifying the two components of the comparison, Jasso can use them as variables in her theory.

Jasso then offers a "measurement rule" that further specifies how the two variables, A and C, will be conceptualized. This step is needed because some of the goods to be examined are concrete and commonly measured (such as money),

whereas others are less tangible (such as respect). The former kind, she says, will be measured conventionally, whereas the latter will be measured "by the individual's relative rank . . . within a specially selected comparison group." The theory will provide a formula for making that measurement (Jasso 1988:13).

Jasso continues in this fashion to introduce additional elements, weaving them into mathematical formulas for deriving predictions about the workings of distributive justice in a variety of social settings. Here is a sampling of where her theorizing takes her (1988:14–15).

- Other things [being] the same, a person will prefer to steal from a fellow group member rather than from an outsider.
- The preference to steal from a fellow group member is more pronounced in poor groups than in rich groups.
- In the case of theft, informants arise only in cross-group theft, in which case they are members of the thief's group.
- Persons who arrive a week late at summer camp or for freshman year of college are more likely to become friends of persons who play games of chance than of persons who play games of skill.
- A society becomes more vulnerable to deficit spending as its wealth increases.
- Societies in which population growth is welcomed must be societies in which the set of valued goods includes at least one quantity-good, such as wealth.

Jasso's theory leads to many other propositions, but this sampling should provide a good sense of where deductive theorizing can take you. To get a feeling for how she reasons her way to these propositions, let's look briefly at the logic involved in two of the propositions that relate to theft within and outside one's group.

- Other things [being] the same, a person will prefer to steal from a fellow group member rather than from an outsider.

Beginning with the assumption that thieves want to maximize their relative wealth, ask yourself whether that goal would be best served by stealing from those you compare yourself with or from outsiders. In each case, stealing will increase your Actual Holdings, but what about your Comparison Holdings?

A moment's thought should suggest that stealing from people in your comparison group will lower their holdings, further increasing your relative wealth. To simplify, imagine there are only two people in your comparison group: you and I. Suppose we each have $100. If you steal $50 from someone outside our group, you will have increased your relative wealth by 50 percent compared with me: $150 versus $100. But if you steal $50 from me, you will have increased your relative wealth 200 percent: $150 to my $50. Your goal is best served by stealing from within the comparison group.

- In the case of theft, informants arise only in cross-group theft, in which case they are members of the thief's group.

Can you see why it would make sense for informants (1) to arise only in the case of cross-group theft and (2) to come from the thief's comparison group? This proposition again depends on the fundamental assumption that everyone wants to increase his or her relative standing. Suppose you and I are in the same comparison group, but this time the group contains additional people. If you steal from someone else within our comparison group, my relative standing in the group does not change. Although your wealth has increased, the average wealth in the group remains the same (because someone else's wealth has decreased by the same amount). So my relative standing remains the same. I have no incentive to inform on you.

If you steal from someone outside our comparison group, your nefarious income increases the total wealth in our group, so my own wealth relative to that total is diminished. Since my relative wealth has suffered, I'm more likely to bring an end to your stealing.

This last deduction also begins to explain why informants are more likely to arrive from within the thief's comparison group. We've just seen how my relative standing was decreased by your theft.

How about other members of the other group? Each of them would actually profit from the theft, since you would have reduced the total with which they compare themselves. Hence, the theory of distributive justice predicts that informants arise from the thief's comparison group.

This brief and selective peek into Jasso's derivations should give you some sense of the enterprise of deductive theory. Realize, of course, that the theory guarantees none of the given predictions. The role of research is to test each of them empirically to determine whether what makes sense (logic) occurs in practice (observation).

There are two important elements in science, then: logical integrity and empirical verification. Both are essential to scientific inquiry and discovery. Logic alone is not enough, but on the other hand, the mere observation and collection of empirical facts does not provide understanding—the telephone directory, for example, is not a scientific conclusion. Observation, however, can be the springboard for the construction of a social scientific theory, as we shall now see in the case of inductive theory.

INDUCTIVE THEORY CONSTRUCTION

Quite often, social scientists begin constructing a theory through the inductive method by first observing aspects of social life and then seeking to discover patterns that may point to relatively universal principles. Barney Glaser and Anselm Strauss (1967) coined the term *grounded theory* in reference to this method.

Field research—the direct observation of events in progress—is frequently used to develop theories through observation (see Chapter 10). A long and rich anthropological tradition has seen this method used to good advantage.

Among contemporary social scientists, no one was more adept at seeing the patterns of human behavior through observation than Erving Goffman (1974:5):

A game such as chess generates a habitable universe for those who can follow it, a plane of being, a cast of characters with a seemingly unlimited number of different situations and acts through which to realize their natures and destinies. Yet much of this is reducible to a small set of interdependent rules and practices. If the meaningfulness of everyday activity is similarly dependent on a closed, finite set of rules, then explication of them would give one a powerful means of analyzing social life.

In a variety of research efforts, Goffman uncovered the rules of such diverse behaviors as living in a mental institution (1961) and managing the "spoiled identity" of disfiguration (1963). In each case, Goffman observed the phenomenon in depth and teased out the rules governing behavior. Goffman's research provides an excellent example of qualitative field research as a source of grounded theory.

Our earlier discussion of the Comfort Hypothesis and church involvement shows that qualitative field research is not the only method of observation appropriate to the development of inductive theory. Here's another detailed example to illustrate further the construction of inductive theory using quantitative methods.

An Example of Inductive Theory: Why Do People Smoke Marijuana?

During the 1960s and 1970s, marijuana use on U.S. college campuses was a subject of considerable discussion in the popular press. Some people were troubled by marijuana's popularity; others welcomed it. What interests us here is why some students smoked marijuana and others didn't. A survey of students at the University of Hawaii (Takeuchi 1974) provided the data to answer that question.

At the time of the study, countless explanations were being offered for drug use. People who opposed drug use, for example, often suggested that marijuana smokers were academic failures trying to avoid the rigors of college life. Those in favor of marijuana, on the other hand, often spoke of the search for new values: Marijuana smokers, they said, were people who had seen through the hypocrisy of middle-class values.

David Takeuchi's (1974) analysis of the data gathered from University of Hawaii students, however, did not support any of the explanations being offered. Those who reported smoking marijuana had essentially the same academic records as those who didn't smoke it, and both groups were equally involved in traditional "school spirit" activities. Both groups seemed to feel equally well integrated into campus life.

There were differences, however:

1. Women were less likely than men to smoke marijuana.
2. Asian students (a large proportion of the student body) were less likely than non-Asians to smoke marijuana.
3. Students living at home were less likely than those living in apartments to smoke marijuana.

As in the case of religiosity, the three variables independently affected the likelihood of a student's smoking marijuana. About 10 percent of the Asian women living at home had smoked marijuana, as contrasted with about 80 percent of the non-Asian men living in apartments. And, as in the religiosity study, the researchers discovered a powerful pattern of drug use before they had an explanation for that pattern.

In this instance, the explanation took a peculiar turn. Instead of explaining why some students smoked marijuana, the researchers explained why some didn't. Assuming that all students had some motivation for trying drugs, the researchers suggested that students differed in the degree of "social constraints" preventing them from following through on that motivation.

U.S. society is, on the whole, more permissive with men than with women when it comes to deviant behavior. Consider, for example, a group of men getting drunk and boisterous. We tend to dismiss such behavior with references to "camaraderie" and "having a good time," whereas a group of women behaving similarly would probably be regarded with great disapproval. We have an idiom, "Boys will be boys," but no comparable idiom for girls. The researchers reasoned, therefore, that women would have more to lose by smoking marijuana than men would. Being female, then, provided a constraint against smoking marijuana.

Students living at home had obvious constraints against smoking marijuana, compared with students living on their own. Quite aside from differences in opportunity, those living at home were seen as being more dependent on their parents—hence more vulnerable to additional punishment for breaking the law.

Finally, the Asian subculture in Hawaii has traditionally placed a higher premium on obedience to the law than have other subcultures, so Asian students would have more to lose if they were caught violating the law by smoking marijuana.

Overall, then, a "social constraints" theory was offered as the explanation for observed differences in the likelihood of smoking marijuana. The more constraints a student had, the less likely he or she would be to smoke marijuana. It bears repeating that the researchers had no thoughts about such a theory when their research began. The theory came from an examination of the data.

THE LINKS BETWEEN THEORY AND RESEARCH

Throughout this chapter, we have seen various aspects of the links between theory and research in social scientific inquiry. In the deductive model, research is used to test theories. In the inductive model, theories are developed from the analysis of research data. This section looks more closely into the ways theory and research are related in actual social scientific inquiry.

Whereas we have discussed two idealized logical models for linking theory and research, social scientific inquiries have developed a great many variations on these themes. Sometimes theoretical issues are introduced merely as a background for empirical analyses. Other studies cite selected empirical data to bolster theoretical arguments. In neither case is there really an interaction between theory and research for the purpose of developing new explanations. Some studies make no use of theory at all, aiming specifically, for example, at an ethnographic description of a particular social situation, such as an anthropological account of food and dress in a particular society.

As you read social research reports, however, you will very often find that the authors are conscious of the implications of their research for social theories and vice versa. Here are a few examples to illustrate this point.

When W. Lawrence Neuman (1998) set out to examine the problem of monopolies (the "trust problem") in U.S. history, he saw the relevance of theories about how social movements transform society ("state transformation"). He became convinced, however, that existing theories were inadequate for the task before him:

> State transformation theory links social movements to state policy formation processes by focussing on the role of cultural meaning in organized political struggles. Despite a resemblance among concepts and concerns, constructionist ideas found in the social problems, social movements, and symbolic politics literatures have not been incorporated into the theory. In this paper, I draw on these three literatures to enhance state transformation theory.
> — (NEUMAN 1998:315)

Having thus modified state transformation theory, Neuman had a theoretical tool that could guide his inquiry and analysis into the political maneuverings related to monopolies beginning in the 1880s and continuing until World War I. Thus, theory served as a resource for research and at the same time was modified by it.

In a somewhat similar study, Alemseghed Kebede and J. David Knottnerus (1998) set out to investigate the rise of Rastafarianism in the Caribbean. However, they felt that recent theories on social movements had become too positivistic in focusing on the mobilization of resources. Resource mobilization theory, they felt, downplays

> the motivation, perceptions, and behavior of movement participants . . . and concentrates instead on the whys and hows of mobilization. Typically theoretical and research problems include: How do emerging movement organizations seek to mobilize and routinize the flow of

resources and how does the existing political apparatus affect the organization of resources? — (1998:500)

To study Rastifarianism more appropriately, the researchers felt the need to include several concepts from contemporary social psychology. In particular, they sought models to use in dealing with problems of meaning and collective thought.

Frederika E. Schmitt and Patricia Yancey Martin (1999) were particularly interested in discovering what produced successful rape crisis centers and how such centers dealt with the organizational and political environments within which they operated. The researchers found theoretical constructs appropriate to their inquiry:

> This case study of unobtrusive mobilizing by [the] Southern California Rape Crisis Center uses archival, observational, and interview data to explore how a feminist organization worked to change police, schools, prosecutor[s], and some state and national organizations from 1974 to 1994. Mansbridge's concept of street theory and Katzenstein's concepts of unobtrusive mobilization and discursive politics guide the analysis.
> — (1999:364)

In summary, there is no simple recipe for conducting social science research. It is far more open-ended than the traditional view of science suggests. Ultimately, science depends on two categories of activity: logic and observation. As you'll see throughout this book, they can be fit together in many patterns.

THE IMPORTANCE OF THEORY IN THE "REAL WORLD"

At this point you may be saying, "Sure, theory and research are OK, but what do they have to do with the real world?" As we'll see later in this book, there are many practical applications of social research, from psychology to social reform. Think, for instance, how someone could make use of David Takeuchi's research on marijuana use.

But how does theory work in such applications?

A QUANDARY **REVISITED**

As we've seen, many different paradigms have been suggested for the study of society. The Opening Quandary asked which was true. It should have become apparent in this chapter that the answer is "None of the above." However, none of the paradigms is false, either.

By their nature, paradigms are neither true or false. They are merely different ways of looking and seeking explanations. Thus, they may be judged as useful or not useful in a particular situation but not true or false. Imagine that you and some friends are in a totally darkened room. Each of you has a flashlight. When you yourself turn on your flashlight, you create a partial picture of what's in the room, whereby some things are revealed, but others remain concealed. Now imagine your friends taking turns turning on their flashlights. Every person's flashlight presents a different picture of what's in the room, both revealing and concealing. Paradigms are like the flashlights in this gripping tale. Each offers a particular point of view that may or may not be useful in a given circumstance. None reveals the full picture, or the "truth."

In some minds, theoretical and practical matters are virtual opposites. Social scientists committed to the use of science know differently, however.

Lester Ward, the first president of the American Sociological Association, was committed to the application of social research in practice, or the use of that research toward specific ends. Ward (1906:5) distinguished pure and applied sociology as follows:

> Just as pure sociology aims to answer the questions What, Why, and How, so applied sociology aims to answer the question What for. The former deals with facts, causes, and principles, the latter with the object, end, or purpose.

No matter how practical and/or idealistic your aims, a theoretical understanding of the terrain may spell the difference between success and failure. As Ward saw it, "Reform may be defined as the desirable alteration of social structures. Any attempt to do this must be based on a full knowledge of the nature of such structures, otherwise its failure is certain" (1906:4).

Suppose you were concerned about poverty in the United States. The sociologist Herbert Gans (1971) suggests it is vital to understand the functions that poverty serves for people who are not poor. For example, the persistence of poverty means there will always be people willing to do the jobs no one else wants to do—and they'll work for very little money. The availability of cheap labor provides a great many affordable comforts for the nonpoor.

By the same token, poverty provides many job opportunities for social workers, unemployment office workers, police, and so forth. If poverty were to disappear, what would happen to social work colleges, for example?

I don't mean to suggest a conspiracy of people intent on keeping the poor in their place or that social workers secretly hope for poverty to persist. Nor do I want to suggest that the dark cloud of poverty has a silver lining. I merely want you to understand the point made by Ward, Gans, and many other sociologists: If you want to change society, you need to understand how it operates. As William White (1997) argued, "Theory helps create questions, shapes our research designs, helps us anticipate outcomes, helps us design interventions."

Main Points

- A paradigm is a fundamental model or scheme that organizes our view of something.

- Social scientists use a variety of paradigms to organize how they understand and inquire into social life.

- A distinction between types of theories that cuts across various paradigms is macrotheory (theories about large-scale features of society) versus microtheory (theories about smaller units or features of society).

- The positivistic paradigm assumes we can scientifically discover the rules governing social life.

- The conflict paradigm focuses on the attempt of one person or group to dominate others and to avoid being dominated.

- The symbolic interactionist paradigm examines how shared meanings and social patterns are developed in the course of social interactions.

- Ethnomethodology focuses on the ways people make sense out of life in the process

of living it, as though each were a researcher engaged in an inquiry.

- The structural functionalist (or social systems) paradigm seeks to discover what functions the many elements of society perform for the whole system—for example, the functions of mothers, labor unions, and radio talk shows.

- Feminist paradigms, in addition to drawing attention to the oppression of women in most societies, highlight how previous images of social reality have often come from and reinforced the experiences of men.

- Some contemporary theorists and researchers have challenged the long-standing belief in an objective reality that abides by rational rules. They point out that it is possible to agree on an "intersubjective" reality.

- In the traditional image of science, scientists proceed from theory to operationalization to observation. But this image is not an accurate picture of how scientific research is actually done.

- Social scientific theory and research are linked through two logical methods: Deduction involves the derivation of expectations or hypotheses from theories. Induction involves the development of generalizations from specific observations.

❑ Science is a process involving an alternation of deduction and induction.

❑ Guillermina Jasso's theory of distributive justice illustrates how formal reasoning can lead to a variety of theoretical expectations that can be tested by observation.

❑ David Takeuchi's study of factors influencing marijuana smoking among University of Hawaii students illustrates how collecting observations can lead to generalizations and an explanatory theory.

❑ In practice, there are many possible links between theory and research and many ways of going about social inquiry.

❑ Using theories to understand how society works is key to offering practical solutions to society's problems.

Key Terms

paradigms	operationalization
macrotheory	operational definition
microtheory	null hypothesis
hypothesis	

Review Questions

1. Consider the possible relationship between education and prejudice (mentioned in Chapter 1). How might that relationship be examined through (a) deductive and (b) inductive methods?

2. Select a social problem that concerns you, such as war, pollution, overpopulation, prejudice, or poverty, Then use one of the paradigms discussed in the chapter to address that problem. What would be the main variables involved in the study of that problem, including variables that may cause it or hold the key to its solution?

3. What, in your own words, is the difference between a paradigm and a theory?

4. You have been hired to evaluate how well a particular health maintenance organization (HMO) serves the needs of its clients. How might you implement this study using each of the following: (1) the interactionist paradigm, (2) the social systems or functionalist paradigm, (3) the conflict paradigm?

Additional Readings

Berger, Joseph, Morris Zelditch, Jr. , and Bo Anderson, eds. 1989. *Sociological Theories in Progress.* Newbury Park, CA: Sage. Several authors develop parts of a theory of social interaction, many of which focus on how we create expectations for each other's behavior.

Denzin, Norman K. , and Yvonna S. Lincoln. 1994. *Handbook of Qualitative Research.* Newbury Park, CA: Sage. Various authors discuss the process of qualitative research from the perspective of various paradigms, showing how they influence the nature of inquiry. The editors also critique positivism from a postmodern perspective.

DeVault, Marjorie L. 1999. *Liberating Method: Feminism and Social Research.* Philadelphia: Temple University Press. This book elaborates on some of the methods associated with the feminist paradigm and is committed to both rigorous inquiry and the use of social research to combat oppression.

Kuhn, Thomas. 1970. *The Structure of Scientific Revolutions.* Chicago: University of Chicago Press. In this exciting and innovative recasting of the nature of scientific development, Kuhn disputes the notion of gradual change and modification in science, arguing instead that established "paradigms" tend to persist until the weight of contradictory evidence brings their rejection and replacement by new paradigms. This short book is at once stimulating and informative.

Lofland, John, and Lyn H. Lofland. 1995. *Analyzing Social Settings: A Guide to Qualitative Observation and Analysis,* 3rd ed. Belmont, CA: Wadsworth. An excellent text on how to conduct qualitative inquiry with an eye toward discovering the rules of social life. Includes a critique of postmodernism.

McGrane, Bernard. 1994. *The Un-TV and 10 mph Car: Experiments in Personal Freedom and Everyday Life.* Fort Bragg, CA: The Small Press. Some excellent and imaginative examples of an ethnomethodological approach to society and to the craft of sociology. The book is useful for both students and faculty.

Reinharz, Shulamit. 1992. *Feminist Methods in Social Research.* New York: Oxford University Press. This book explores several social research techniques (such as interviewing, experiments, and content analysis) from a feminist perspective.

Ritzer, George. 1988. *Sociological Theory.* New York: Knopf. This is an excellent overview of the major theoretical traditions in sociology.

Sprague, Joey. 1997. "Holy Men and Big Guns: The Can[n]on in Social Theory." *Gender and Society* 11 (1): 88–107. This is an excellent analysis of the ways in which conventional social theory misses aspects of society that might be revealed in a feminist examination.

Turner, Jonathan H. , ed. 1989. *Theory Building in Sociology: Assessing Theoretical Cumulation.* Newbury Park, CA: Sage. This collection of essays on sociological theory construction focuses specifically on the question posed by Turner's introductory chapter, "Can Sociology Be a Cumulative Science?"

Turner, Stephen Park, and Jonathan H. Turner. 1990. *The Impossible Science: An Institutional Analysis of American Sociology.* Newbury Park, CA: Sage. Two authors bring two very different points of view to the history of U.S. sociologists' attempt to establish a science of society.

Multimedia Resources

The Wadsworth Sociology Resource Center: Virtual Society
http://sociology.wadsworth.com/

Visit the companion Web site for the second edition of *The Basics of Social Research* to access a wide range of student resources. Begin by clicking on the Student Resources section of the book's Web site to access the following study tools:

- eBabbie Resource Center
- Planning a Research Project
- Doing Data Analysis
- Statistics Review
- Flash Cards
- Internet Links and Exercises
- InfoTrac College Edition: Exercises
- Quizzes

Visit the **eBabbie Resource Center** for an overview of each chapter and helpful online tutorials. Find information on budgeting and step-by-step examples of model research projects at **Planning a Research Project.** Learn how to use quantitative and qualitative data analysis programs at **Doing Data Analysis,** and brush up on your statistics at **Statistics Review.** You can also further your study by accessing **Internet Links and Exercises** related to chapter materials, **Flash Cards, Quizzes,** and many other learning tools.

InfoTrac College Edition
http://www.infotrac-college.com/
wadsworth/access.html

Access the latest news and journal articles with InfoTrac College Edition, an easy-to-use online database of reliable, full-length articles from hundreds of top academic journals. Conduct an electronic search using the following search terms:

Conflict theory
Ethnomethodology
Feminism
Logical positivism
Macro-theory
Micro-theory
Social sciences functionalism
Sociological theory
Symbolic interactionism

3 THE ETHICS AND POLITICS OF SOCIAL RESEARCH

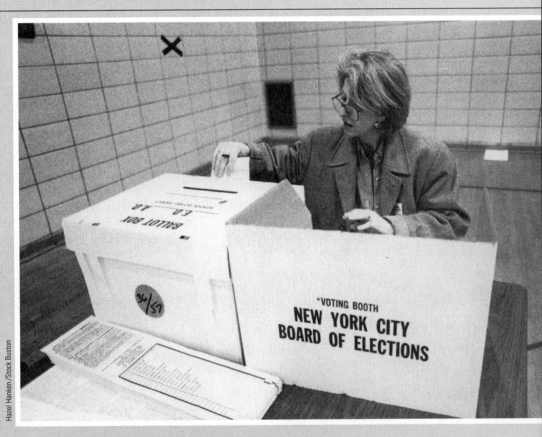

Hazel Hankin/Stock Buston

What You'll Learn in This Chapter

Social research takes place in a social context. Researchers must therefore take into account many ethical and political considerations alongside scientific ones in designing and executing their research. Often, however, clear-cut answers to thorny ethical and political issues are hard to come by.

In this chapter . . .

INTRODUCTION

My purpose in this book is to present a realistic and useful introduction to doing social research. For this introduction to be fully realistic, it must include four main constraints on research projects: scientific, administrative, ethical, and political.

Most of the book focuses on scientific and administrative constraints. We'll see that the logic of

AN OPENING **QUANDARY**

Whenever research ethics are discussed, the subject of the Nazi medical experiments of World War II often surfaces as the most hideous breach of ethical standards in history. Civilian and military prisoners in Nazi concentration camps were subjected to freezing; malaria; mustard gas; sulfanilamide; bone, muscle, and nerve transplantations; jaundice; sterilization; spotted fever; various poisons; and phosphorus burns, among other tortures. Many died, others were permanently maimed; all suffered tremendous physical and psychological pain. Some have argued, however, that the real breach of ethics did not lie in the suffering or the deaths per se. What could possibly be a worse ethical breach than that?

science suggests certain research procedures, but we'll also see that some scientifically "perfect" study designs are not administratively feasible, because they would be too expensive or take too long to execute. Throughout the book, therefore, we'll deal with workable compromises.

Before we get to the scientific and administrative constraints on research, it's useful to explore the two other important considerations in doing research in the real world—ethics and politics—which this chapter covers. Just as certain procedures are too impractical to use, others are either ethically prohibitive or politically difficult or impossible. Here's a story to illustrate what I mean.

Several years ago, I was invited to sit in on a planning session to design a study of legal education in California. The joint project was to be conducted by a university research center and the state bar association. The purpose of the project was to improve legal education by learning which aspects of the law school experience were related to success on the bar exam. Essentially, the plan was to prepare a questionnaire that would get detailed information about the law school experiences of individuals. People would be required to answer the questionnaire when they took the bar exam. By analyzing how people with different kinds of law school experiences did on the bar exam, we could find out what sorts of things worked and what didn't. The findings of the research could be made available to law schools, and ultimately legal education could be improved.

The exciting thing about collaborating with the bar association was that all the normally irritating logistical hassles would be handled. There would be no problem getting permission to administer questionnaires in conjunction with the exam, for example, and the problem of nonresponse could be eliminated altogether.

I left the meeting excited about the prospects for the study. When I told a colleague about it, I glowed about the absolute handling of the nonresponse problem. Her immediate comment turned everything around completely. "That's unethical. There's no law requiring the questionnaire, and participation in research has to be voluntary." The study wasn't done.

In retelling this story, it's obvious to me that requiring participation would have been inappropriate. You may have seen this even before I told you about my colleague's comment. I still feel a little embarrassed over the matter, but I have a specific purpose in telling this story about myself.

All of us consider ourselves ethical—not perfect perhaps, but as ethical as anyone else and perhaps more so than most. The problem in social research, as probably in life, is that ethical considerations are not always apparent to us. As a result, we often plunge into things without seeing ethical issues that may be apparent to others and may even be obvious to us when pointed out. When I reported back to the others in the planning group, for example, no one disagreed with the inappropriateness of requiring participation. Everyone was a bit embarrassed about not having seen it.

Any of us can immediately see that a study that requires small children to be tortured is unethical. I know you'd speak out immediately if I suggested that we interview people about their sex lives and then publish what they said in the local newspaper. But, as ethical as you are, you'll totally miss the ethical issues in some other situations—we all do.

The first half of this chapter deals with the ethics of social research. In part, it presents some of the broadly agreed-upon norms describing what's ethical in research and what's not. More important than simply knowing the guidelines, however, is becoming sensitized to the ethical component in research so that you'll look for it whenever you plan a study. Even when the ethical aspects of a situation are debatable, you should know that there's something to argue about.

Political considerations in research are also subtle, ambiguous, and arguable. Notice that the law school example involves politics as well as ethics. Although social researchers have an ethical norm that participation in research should be voluntary, this norm clearly grows out of U.S. political norms protecting civil liberties. In some nations, the proposed study would not have been considered unethical at all.

In the second half of this chapter, we'll look at social research projects that were crushed or nearly crushed by political considerations. As with ethical concerns, there is often no "correct" take on a given situation. People of goodwill disagree. I won't try to give you a party line about what is and is not politically acceptable. As with ethics, the point is to become sensitive to the political dimension of social research.

ETHICAL ISSUES IN SOCIAL RESEARCH

In most dictionaries and in common usage, *ethics* is typically associated with morality, and both deal with matters of right and wrong. But what is right and what wrong? What is the source of the distinction? For individuals the sources vary. They may be religions, political ideologies, or the pragmatic observation of what seems to work and what doesn't.

Webster's New World Dictionary is typical among dictionaries in defining *ethical* as "conforming to the standards of conduct of a given profession or group." Although this definition may frustrate those in search of moral absolutes, what we regard as morality and ethics in day-to-day life is a matter of agreement among members of a group. And, not surprisingly, different groups have agreed on different codes of conduct. Part of living successfully in a particular society is knowing what that society considers ethical and unethical. The same holds true for the social research community.

Anyone involved in social scientific research, then, needs to be aware of the general agreements shared by researchers about what is proper and improper in the conduct of scientific inquiry. This section summarizes some of the most important ethical agreements that prevail in social research.

Voluntary Participation

Often, though not always, social research represents an intrusion into people's lives. The interviewer's knock on the door or the arrival of a questionnaire in the mail signals the beginning of an activity that the respondent has not requested and that may require significant time and energy. Participation in a social experiment disrupts the subject's regular activities.

Social research, moreover, often requires that people reveal personal information about themselves—information that may be unknown to their friends and associates. And social research often requires that such information be revealed to strangers. Other professionals, such as physicians and lawyers, also ask for such information. Their requests may be justified, however, by their aims: They need the information in order to serve the personal interests of the respondent. Social researchers can seldom make this claim. Like medical scientists, they can only argue that the research effort may ultimately help all humanity.

A major tenet of medical research ethics is that experimental participation must be voluntary. The same norm applies to social research. No one should be forced to participate. This norm is far

easier to accept in theory than to apply in practice, however.

Again, medical research provides a useful parallel. Many experimental drugs are tested on prisoners. In the most rigorously ethical cases, the prisoners are told the nature and the possible dangers of the experiment, they are told that participation is completely voluntary, and they are further instructed that they can expect no special rewards —such as early parole—for participation. Even under these conditions, it's often clear that volunteers are motivated by the belief that they will personally benefit from their cooperation.

When the instructor in an introductory sociology class asks students to fill out a questionnaire that he or she hopes to analyze and publish, students should always be told that their participation in the survey is completely voluntary. Even so, most students will fear that nonparticipation will somehow affect their grade. The instructor should therefore be especially sensitive to such implications and make special provisions to eliminate them. For example, the instructor could insure anonymity by leaving the room while the questionnaires are being completed. Or, students could be asked to return the questionnaires by mail or to drop them in a box near the door just before the next course meeting.

This norm of voluntary participation, though, goes directly against several scientific concerns. In the most general terms, the scientific goal of generalizability is threatened if experimental subjects or survey respondents are all the kinds of people who willingly participate in such things. Because this orientation probably reflects more general personality traits, the results of the research might not be generalizable to all kinds of people. Most clearly, in the case of a descriptive survey, a researcher cannot generalize the sample survey findings to an entire population unless a substantial majority of the scientifically selected sample actually participates—the willing respondents and the somewhat unwilling.

As you'll see in Chapter 10, field research has its own ethical dilemmas in this regard. Very often, the researcher cannot even reveal that a study is being done, for fear that that revelation might significantly affect the social processes being studied. Clearly, the subjects of study in such cases are not given the opportunity to volunteer or refuse to participate.

Though the norm of voluntary participation is important, it is often impossible to follow. In cases where you feel ultimately justified in violating it, it is all the more important that you observe the other ethical norms of scientific research, such as bringing no harm to the people under study.

No Harm to the Participants

Social research should never injure the people being studied, regardless of whether they volunteer for the study. Perhaps the clearest instance of this norm in practice concerns the revealing of information that would embarrass subjects or endanger their home life, friendships, jobs, and so forth. We'll discuss this aspect of the norm more fully in a moment.

Because subjects can be harmed psychologically in the course of a social research study, the researcher must look for the subtlest dangers and guard against them. Quite often, research subjects are asked to reveal deviant behavior, attitudes they feel are unpopular, or personal characteristics that may seem demeaning, such as low income, the receipt of welfare payments, and the like. Revealing such information usually makes subjects feel at least uncomfortable.

Social research projects may also force participants to face aspects of themselves that they don't normally consider. This can happen even when the information is not revealed directly to the researcher. In retrospect, a certain past behavior may appear unjust or immoral. The project, then, can cause continuing personal agony for the subject. If the study concerns codes of ethical conduct, for example, the subject may begin questioning his or her own morality, and that personal concern may last long after the research has been completed and reported. For instance, probing questions can injure a fragile self-esteem.

It should be apparent from these observations that just about any research you might conduct runs the risk of injuring other people in some way. It isn't possible to insure against all these possible injuries, but some study designs make such injuries more likely than do others. If a particular research procedure seems likely to produce unpleasant effects for subjects—asking survey respondents to report deviant behavior, for example—the researcher should have the firmest of scientific grounds for doing it. If the research design is essential and also likely to be unpleasant for subjects, you'll find yourself in an ethical netherworld and may go through some personal agonizing. Although agonizing has little value in itself, it may be a healthy sign that you've become sensitive to the problem.

Increasingly, the ethical norms of voluntary participation and no harm to participants have become formalized in the concept of **informed consent.** This norm means that subjects must base their voluntary participation in research projects on a full understanding of the possible risks involved. In a medical experiment, for example, prospective subjects will be presented with a discussion of the experiment and all the possible risks to themselves. They will be required to sign a statement indicating that they are aware of the risks and that they choose to participate anyway. While the value of such a procedure is obvious when subjects will be injected with drugs designed to produce physical effects, for example, it's hardly appropriate when a participant observer rushes to the scene of urban rioting to study deviant behavior. While the researcher in this latter case is not excused from the norm of not bringing harm to those observed, gaining informed consent is not the means to achieving that end.

Although the fact often goes unrecognized, another possible source of harm to subjects lies in the analysis and reporting of data. Every now and then, research subjects read the books published about the studies they participated in. Reasonably sophisticated subjects can locate themselves in the various indexes and tables. Having done so, they may find themselves characterized—though not identified by name—as bigoted, unpatriotic, irreligious, and so forth. At the very least, such characterizations are likely to trouble them and threaten their self-images. Yet the whole purpose of the research project may be to explain why some people are prejudiced and others are not.

In one survey of churchwomen (Babbie 1967), ministers in a sample of churches were asked to distribute questionnaires to a specified sample of members, collect them, and return them to the research office. One of these ministers read through the questionnaires from his sample before returning them, and then he delivered a hellfire and brimstone sermon to his congregation, saying that many of them were atheists and were going to hell. Even though he could not identify the people who gave particular responses, it seems certain that many respondents were personally harmed by the action.

Like voluntary participation, avoiding harm to people is easy in theory but often difficult in practice. Sensitivity to the issue and experience with its applications, however, should improve the researcher's tact in delicate areas of research.

In recent years, social researchers have been gaining support for abiding by this norm. Federal and other funding agencies typically require an independent evaluation of the treatment of human subjects for research proposals, and most universities now have human-subject committees to serve this evaluative function. Although sometimes troublesome and inappropriately applied, such requirements not only guard against unethical research but can also reveal ethical issues overlooked by even the most scrupulous researchers.

Anonymity and Confidentiality

The clearest concern in the protection of the subjects' interests and well-being is the protection of their identity, especially in survey research. If revealing their survey responses would injure them in any way, adherence to this norm becomes all the more important. Two techniques—anonymity and confidentiality—assist researchers in this regard, although people often confuse the two.

Anonymity A research project guarantees **anonymity** when the researcher—not just the people who read about the research—cannot identify a given response with a given respondent. This implies that a typical interview survey respondent can never be considered anonymous, because an interviewer collects the information from an identifiable respondent. An example of anonymity is a mail survey in which no identification numbers are put on the questionnaires before their return to the research office.

As we'll see in Chapter 9 on survey research, assuring anonymity makes it difficult to keep track of who has or hasn't returned the questionnaires. Despite this problem, there are some situations in which you may be advised to pay the necessary price. In one study of drug use among university students, I decided that I specifically did not want to know the identity of respondents. I felt that honestly assuring anonymity would increase the likelihood and accuracy of responses. Also, I did not want to be in the position of being asked by authorities for the names of drug offenders. In the few instances in which respondents volunteered their names, such information was immediately obliterated on the questionnaires.

Confidentiality A research project guarantees **confidentiality** when the researcher can identify a given person's responses but essentially promises not to do so publicly. In an interview survey, for example, the researcher would be in a position to make public the income reported by a given respondent, but the respondent is assured that this will not be done.

Whenever a research project is confidential rather than anonymous, it is the researcher's responsibility to make that fact clear to the respondent. Moreover, researchers should never use the term *anonymous* to mean *confidential*.

With few exceptions (such as surveys of public figures who agree to have their responses published), the information respondents give must at least be kept confidential. This is not always an easy norm to follow, because, for example, the courts have not recognized social research data as the kind of "privileged communication" accepted in the case of priests and attorneys.

This unprotected guarantee of confidentiality produced a near disaster in 1991. Two years earlier, the Exxon *Valdez* supertanker had run aground near the port of Valdez in Alaska, spilling ten million gallons of oil into the bay. The economic and environmental damage was widely reported.

Less attention was given to the psychological and sociological damage suffered by residents of the area. There were anecdotal reports of increased alcoholism, family violence, and other secondary consequences of the disruptions caused by the oil spill. Eventually, 22 communities in Prince William Sound and the Gulf of Alaska sued Exxon for the economic, social, and psychological damages suffered by their residents.

For more information on this ecological disaster, see Picou, J. Steven, Duane A. Gill, and Maurie J. Cohen, eds. (1999). *The Exxon Valdez Disaster: Readings on a Modern Social Problem.* Dubuque, IA: Kendall-Hunt. This is described online at http://www.ssrc.msstate.edu/mafes/evos.html, along with additional Web site citations.*

To determine the amount of damage done, the communities commissioned a San Diego research firm to undertake a household survey asking residents very personal questions about increased problems in their families. The sample of residents were asked to reveal painful and embarrassing information, under the guarantee of absolute confidentiality. Ultimately, the results of the survey confirmed that a variety of personal and family problems had increased substantially following the oil spill.

When Exxon learned that survey data would be presented to document the suffering, they took an unusual step: They asked the court to subpoena

the survey questionnaires! The court granted the defendant's request and ordered the researchers to turn over the questionnaires—with all identifying information. It appeared that Exxon's intention was to call survey respondents to the stand and cross-examine them regarding answers they had given interviewers under the guarantee of confidentiality. Moreover, many of the respondents were Native Americans, whose cultural norms made such public revelations all the more painful.

Happily, the Exxon *Valdez* case was settled before the court decided whether it would force survey respondents to testify in open court. Unhappily, the potential for disaster remains.

The seriousness of this issue is not limited to established research firms. Rik Scarce was a graduate student at Washington State University when he undertook participant observation among animal-rights activists. In 1990, he published a book based on his research entitled *Ecowarriors: Understanding the Radical Environmental Movement.* In 1993, Scarce was called before a grand jury and asked to identify the activists he had studied. In keeping with the norm of confidentiality, the young researcher refused to answer the grand jury's questions and spent 159 days in the Spokane County jail.

You can use several techniques to guard against such dangers and ensure better performance on the guarantee of confidentiality. To begin, interviewers and others with access to respondent identifications should be trained in their ethical responsibilities. Beyond training, the most fundamental technique is to remove identifying information as soon as it's no longer necessary. In a survey, for example, all names and addresses should be removed from questionnaires and replaced by identification numbers. An identification file should be created that links numbers to names to permit the later correction of missing or contradictory information, but this file should not be available except for legitimate purposes.

Similarly, in an interview survey you may need to identify respondents initially so that you can recontact them to verify that the interview was conducted and perhaps to get information that was

---

*Each time the Internet icon appears, you'll be given helpful leads for searching the World Wide Web.

missing in the original interview. As soon as you've verified an interview and assured yourself that you don't need any further information from the respondent, however, you can safely remove all identifying information from the interview booklet. Often, interview booklets are printed so that the first page contains all the identifiers—it can be torn off once the respondent's identification is no longer needed. J. Steven Picou (1996a, 1996b) points out that even removing identifiers from data files does not always sufficiently protect respondent confidentiality, a lesson he learned during nearly a year in federal court. A careful examination of all the responses of a particular respondent sometimes allows others to deduce that person's identity. Imagine, for example, that someone said they were a former employee of a particular company. Knowing the person's gender, age, ethnicity, and other characteristics could make it possible for the company to identify that person.

Even if you intend to remove all identifying information, suppose you have not yet done so. What do you do when the police or a judge orders you to provide the responses given by your research subjects?

Deception

We've seen that the handling of subjects' identities is an important ethical consideration. Handling your own identity as a researcher can also be tricky. Sometimes it's useful and even necessary to identify yourself as a researcher to those you want to study. You'd have to be an experienced con artist to get people to participate in a laboratory experiment or complete a lengthy questionnaire without letting on that you were conducting research.

Even when you must conceal your research identity, you need to consider the following. Because deceiving people is unethical, deception within social research needs to be justified by compelling scientific or administrative concerns. Even then, the justification will be arguable.

Sometimes researchers admit that they're doing research but fudge about why they're doing it or for whom. Suppose you've been asked by a

---

APPLYING THE RESULTS

Upholding confidentiality is a real issue for practicing social researchers, even though they sometimes disagree about how to protect subjects. Harry O'Neill, the vice chair of the Roper Organization, for example, suggested that the best solution is to avoid altogether the ability to identify respondents with their responses:

> So how is this accomplished? Quite simply by not having any respondent-identifiable information available for the court to request. In my initial contact with a lawyer-client, I make it unmistakably clear that, once the survey is completed and validated, all respondent-identifiable information will be removed and destroyed immediately. Everything else connected with the survey—completed questionnaires, data tapes, methodology, names of interviewers and supervisors—of course will be made available. — (O'Neill 1992:4)

Board Chairman Burns Roper (1992:5) disagreed, saying that such procedures might raise questions about the validity of the research methods. Instead, Roper said that he felt he must be prepared to go to jail if necessary. (He noted that Vice Chair O'Neill promised to visit him in that event.)

---

public welfare agency to conduct a study of living standards among aid recipients. Even if the agency is looking for ways of improving conditions, the recipient-subjects are likely to fear a witch-hunt for "cheaters." They might be tempted, therefore, to give answers that make them seem more destitute than they really are. Unless they provide truthful answers, however, the study will not produce accurate data that will contribute to an effective improvement of living conditions. What do you do?

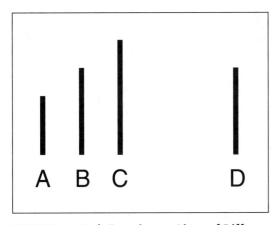

FIGURE 3-1   **Asch Experiment: Lines of Differing Lengths**

One solution would be to tell subjects that you're conducting the study as part of a university research program—concealing your affiliation with the welfare agency. Doing that improves the scientific quality of the study, but it raises a serious ethical issue.

Lying about research purposes is common in laboratory experiments. Although it's difficult to conceal that you're conducting research, it's usually simple—and sometimes appropriate—to conceal your purpose. Many experiments in social psychology, for example, test the extent to which subjects will abandon the evidence of their own observations in favor of the views expressed by others. Figure 3-1 shows the stimulus from the classic Asch experiment—frequently replicated by psychology classes—in which subjects are shown three lines of differing lengths (A, B, and C) and asked to compare them with a fourth line (D). Subjects are then asked, "Which of the first three lines is the same length as the fourth?"

 You can learn more about the Asch experiments by searching the Web for "Asch experiment" or "Solomon Asch."

You'd probably find it a fairly simple task to identify "B" as the correct answer. Your job would be complicated, however, by the fact that several other "subjects" sitting beside you all agree that A is the same length as D! In reality, of course, the others in the experiment are all confederates of the researcher, told to agree on the wrong answer. The purpose of the experiment is to see whether you'd give up your own judgment in favor of the group agreement. I think you can see that conformity is a useful phenomenon to study and understand, and it couldn't be studied experimentally without deceiving the subjects. We'll examine a similar situation in the discussion of a famous experiment by Stanley Milgram later in this chapter. The question is, how do we get around the ethical issue that deception is necessary for an experiment to work?

One appropriate solution researchers have found is to debrief subjects following an experiment. **Debriefing** entails interviews to discover any problems generated by the research experience so that those problems can be corrected. Even though subjects can't be told the true purpose of the study prior to their participation in it, there's usually no reason they can't know afterward. Telling them the truth afterward may make up for having to lie to them at the outset. This must be done with care, however, making sure the subjects aren't left with bad feelings or doubts about themselves based on their performance in the experiment. If this seems complicated, it's simply the price we pay for using other people's lives as the subject matter for our research.

As a social researcher, then, you have many ethical obligations to the subjects in your studies. The box entitled "Ethical Issues in Research on Human Sexuality" illustrates some of the ethical questions involved in a specific research area.

Analysis and Reporting

In addition to their ethical obligations to subjects, researchers have ethical obligations to their colleagues in the scientific community. These obligations concern the analysis of data and the way the results are reported.

In any rigorous study, the researcher should be more familiar than anyone else with the study's technical limitations and failures. Researchers have an obligation to make such shortcomings

ETHICAL ISSUES IN RESEARCH ON HUMAN SEXUALITY

by Kathleen McKinney
Department of Sociology, Illinois State University

When studying any form of human behavior, ethical concerns are paramount. This statement may be even more true for studies of human sexuality because of the topic's highly personal, salient, and perhaps threatening nature. Concern has been expressed by the public and by legislators about human sexuality research. Three commonly discussed ethical criteria have been related specifically to research in the area of human sexuality.

Informed Consent This criterion emphasizes the importance of both accurately informing your subject or respondent as to the nature of the research and obtaining his or her verbal or written consent to participate. Coercion is not to be used to force participation, and subjects may terminate their involvement in the research at any time. There are many possible violations of this standard. Misrepresentation or deception may be used when describing an embarrassing or personal topic of study because the researchers fear high rates of refusal or false data. Covert research, such as some observational studies, also violate the informed consent standard since subjects are unaware that they are being studied. Informed consent may create special problems with certain populations. For example, studies of the sexuality of children are limited by the concern that children may be cognitively and emotionally unable to give informed consent. Although there can be problems such as those discussed, most research is clearly voluntary, with informed consent from those participating.

Right to Privacy Given the highly personal nature of sexuality and society's tremendous concern with social control of sexuality, the right to privacy is a very important ethical concern for research in this area. Individuals may risk losing their jobs, having family difficulties, or being os-

known to their readers—even if admitting qualifications and mistakes makes them feel foolish.

Negative findings, for example, should be reported if they are at all related to the analysis. There is an unfortunate myth in scientific reporting that only positive discoveries are worth reporting (journal editors are sometimes guilty of believing this as well). In science, however, it's often as important to know that two variables are *not* related as to know that they are.

Similarly, researchers must avoid the temptation to save face by describing their findings as the product of a carefully preplanned analytical strategy when that is not the case. Many findings arrive unexpectedly—even though they may seem obvious in retrospect. So an interesting relationship was uncovered by accident—so what? Embroidering such situations with descriptions of fictitious hypotheses is dishonest. It also does a disservice to less experienced researchers by leading them into thinking that all scientific inquiry is rigorously preplanned and organized.

In general, science progresses through honesty and openness; ego defenses and deception retard it. Researchers can best serve their peers—and scientific discovery as a whole—by telling the truth about all the pitfalls and problems they've experienced in a particular line of inquiry. Perhaps they'll save others from the same problems.

Institutional Review Boards

The issue of research ethics in studies involving humans is now also governed by federal law. Any agency (such as a university or a hospital) wishing tracized by peers if certain facets of their sexual lives are revealed. This is especially true for individuals involved in sexual behavior categorized as deviant (such as transvestism). Violations of right to privacy occur when researchers identify members of certain groups they have studied, release or share an individual's data or responses, or covertly observe sexual behavior. In most cases, right to privacy is easily maintained by the researcher. In survey research, self-administered questionnaires can be anonymous and interviews can be kept confidential. In case and observational studies, the identity of the person or group studied can be disguised in any publications. In most research methods, analysis and reporting of data should be at the group or aggregate level.

Protection from Harm Harm may include emotional or psychological distress, as well as physical harm. Potential for harm varies by research method; it is more likely in experimental studies where the researcher manipulates or does something to the subject than in observational or survey research. Emotional distress, however, is a possibility in all studies of human sexuality. Respondents may be asked questions that elicit anxiety, dredge up unpleasant memories, or cause them to evaluate themselves critically. Researchers can reduce the potential for such distress during a study by using anonymous, self-administered questionnaires or well-trained interviewers and by wording sensitive questions carefully.

All three of these ethical criteria are quite subjective. Violations are sometimes justified by arguing that risks to subjects are outweighed by benefits to society. The issue here, of course, is who makes that critical decision. Usually, such decisions are made by the researcher and often a screening committee that deals with ethical concerns. Most creative researchers have been able to follow all three ethical guidelines and still do important research.

---

to receive federal research support must establish an "Institutional Review Board" (IRB), a panel of faculty (and possibly others) who review all research proposals involving human subjects so that they can guarantee that the subjects' rights and interests will be protected. While the law applies specifically to federally funded research, many universities apply the same standards and procedures to all research, including that funded by nonfederal sources and even research done at no cost, such as student projects.

The chief responsibility of an IRB is to ensure that the risks faced by human participants in research are minimal. In some cases, the IRB may ask the researcher to revise the study design; in other cases, the IRB may refuse to approve a study. Where some minimal risks are deemed unavoidable, researchers are required to prepare an "informed consent" form that describes those risks clearly. Subjects may participate in the study only after they have read the statement and signed it as an indication that they know the risks and voluntarily accept them.

Much of the impetus for establishing IRBs had to do with medical experimentation on humans, and many social research study designs are generally regarded as exempt from IRB review. An example is an anonymous survey sent to a large sample of respondents. The guideline to be followed by IRBs,

The "Protection of Human Subjects" law may be found online at http://ohrp.osophs.dhhs.gov/humansubjects/guidance/45cfr46.htm

as contained in the Federal Exemption Categories (45 CFR 46.101 [b]) exempts a variety of research situations:

(1) Research conducted in established or commonly accepted educational settings, involving normal educational practices, such as (i) research on regular and special education instructional strategies, or (ii) research on the effectiveness of or the comparison among instructional techniques, curricula, or classroom management methods.

(2) Research involving the use of educational tests (cognitive, diagnostic, aptitude, achievement), survey procedures, interview procedures or observation of public behavior, unless:

(i) information obtained is recorded in such a manner that human subjects can be identified, directly or through identifiers linked to the subjects; and (ii) any disclosure of the human subjects' responses outside the research could reasonably place the subjects at risk of criminal or civil liability or be damaging to the subjects' financial standing, employability, or reputation.

(3) Research involving the use of educational tests (cognitive, diagnostic, aptitude, achievement), survey procedures, interview procedures, or observation of public behavior that is not exempt under paragraph (b)(2) of this section, if:

(i) the human subjects are elected or appointed public officials or candidates for public office; or (ii) Federal statute(s) require(s) without exception that the confidentiality of the personally identifiable information will be maintained throughout the research and thereafter.

(4) Research involving the collection or study of existing data, documents, records, pathological specimens, or diagnostic specimens, if these sources are publicly available or if the information is recorded by the investigator in such a manner that subjects cannot be identified, directly or through identifiers linked to the subjects.

(5) Research and demonstration projects which are conducted by or subject to the approval of Department or Agency heads, and which are designed to study, evaluate, or otherwise examine:

(i) Public benefit or service programs; (ii) procedures for obtaining benefits or services under those programs; (iii) possible changes in or alternatives to those programs or procedures; or (iv) possible changes in methods or levels of payment for benefits or services under those programs.

(6) Taste and food quality evaluation and consumer acceptance studies, (i) if wholesome foods without additives are consumed or (ii) if a food is consumed that contains a food ingredient at or below the level and for a use found to be safe, or agricultural chemical or environmental contaminant at or below the level found to be safe, by the Food and Drug Administration or approved by the Environmental Protection Agency or the Food Safety and Inspection Service of the U.S. Department of Agriculture.

Paragraph (2) of the excerpt exempts much of the social research described in this book. Nonetheless, universities sometimes apply the law's provisions inappropriately. As chair of a university IRB, for example, I was once asked to review the letter of informed consent that was to be sent to medical insurance companies, requesting their agreement to participate in a survey that would ask which medical treatments were covered under their programs. Clearly the humans involved were not at risk in the sense anticipated by the law. In a case like that, the appropriate technique for gaining informed consent is to mail the questionnaire. If a company returns it, they've consented. If they don't, they haven't.

Other IRBs have suggested that researchers need to obtain permission before observing participants in public gatherings and events, before conducting surveys on the most mundane matters, and so forth. Christopher Shea (2000) has chronicled several such questionable applications of the law while supporting the ethical logic that originally prompted the law.

See Shea's article on IRBs at http://www
.linguafranca.com/print/0009/humans
.html

Professional Codes of Ethics

Ethical issues in social research are both important
and ambiguous. For this reason, most of the professional associations of social researchers have
created and published formal codes of conduct describing what is considered acceptable and unacceptable professional behavior. As one example,
Figure 3-2 presents the code of conduct of the
American Association for Public Opinion Research
(AAOPR), an interdisciplinary research association
in the social sciences. Most professional associations have such codes of ethics. See, for example,
the American Sociological Association, the American Psychological Association, the American Political Science Association, and so forth. You can find
many of these on the associations' Web sites.

TWO ETHICAL CONTROVERSIES

As you may already have guessed, the adoption
and publication of professional codes of conduct
have not totally resolved the issue of research
ethics. Social researchers still disagree on some
general principles, and those who agree in principle often debate specifics.

This section briefly describes two research projects that have provoked ethical controversy and
discussion. The first project studied homosexual
behavior in public restrooms, and the second examined obedience in a laboratory setting.

Trouble in the Tearoom

As a graduate student, Laud Humphreys became
interested in the study of homosexual behavior. He
developed a special interest in the casual and fleet-
ing same-sex acts engaged in by some male non-
homosexuals. In particular, his research interest
focused on homosexual acts between strangers

meeting in the public restrooms in parks, called
"tearooms" among homosexuals. The result was
the publication in 1970 of *Tearoom Trade.*

What particularly interested Humphreys about
the tearoom activity was that the participants
seemed otherwise to live conventional lives as
"family men" and accepted members of the community. They did nothing else that might qualify
them as homosexuals. Thus, it was important to
them that they remain anonymous in their tearoom
visits. How would you study something like that?

Humphreys decided to take advantage of the
social structure of the situation. Typically, the tearoom encounter involved three people: the two
men actually engaging in the sexual act and a look-
out, called the "watchqueen." Humphreys began
showing up at public restrooms, offering to serve
as watchqueen whenever it seemed appropriate.
Since the watchqueen's payoff was the chance to
watch the action, Humphreys was able to conduct
field observations as he would in a study of political rallies or jaywalking behavior at intersections.

To round out his understanding of the tearoom
trade, Humphreys needed to know something
more about the people who participated. Since the
men probably would not have been thrilled about
being interviewed, Humphreys developed a different solution. Whenever possible, he noted the license numbers of participants' cars and tracked
down their names and addresses through the police. Humphreys then visited the men at their
homes, disguising himself enough to avoid recognition, and announced that he was conducting a
survey. In that fashion, he collected the personal
information he couldn't get in the restrooms.

As you can imagine, Humphreys' research provoked considerable controversy both inside and
outside the social scientific community. Some critics charged Humphreys with a gross invasion of
privacy in the name of science. What men did in
public restrooms was their own business. Others
were mostly concerned about the deceit involved
—Humphreys had lied to the participants by
leading them to believe he was only a voyeur-
participant. Even people who felt that the tearoom participants were fair game for observation

FIGURE 3-2   **Code of Conduct of the American Association for Public Opinion Research**

CODE OF PROFESSIONAL ETHICS AND PRACTICES

We, the members of the American Association for Public Opinion Research, subscribe to the principles expressed in the following code.

Our goal is to support sound practice in the profession of public opinion research. (By public opinion research we mean studies in which the principal source of information about individual beliefs, preferences, and behavior is a report given by the individual himself or herself.)

We pledge ourselves to maintain high standards of scientific competence and integrity in our work, and in our relations both with our clients and with the general public. We further pledge ourselves to reject all tasks or assignments which would be inconsistent with the principles of this code.

THE CODE

I.   *Principles of Professional Practice in the Conduct of Our Work*

A. We shall exercise due care in gathering and processing data, taking all reasonable steps to assume the accuracy of results.

B. We shall exercise due care in the development of research designs and in the analysis of data.

1. We shall employ only research tools and methods of analysis which, in our professional judgment, are well suited to the research problem at hand.
2. We shall not select research tools and methods of analysis because of their special capacity to yield a desired conclusion.
3. We shall not knowingly make interpretations of research results, nor shall we tacitly permit interpretations, which are inconsistent with the data available.
4. We shall not knowingly imply that interpretations should be accorded greater confidence than the data actually warrant.

C. We shall describe our findings and methods accurately and in appropriate detail in all research reports.

II. *Principles of Professional Responsibility in Our Dealings with People*

A. The Public:

1. We shall cooperate with legally authorized representatives of the public by describing the methods used in our studies.
2. We shall maintain the right to approve the release of our findings whether or not ascribed to us. When misinterpretation appears, we shall publicly disclose what is required to correct it, notwithstanding our obligation for client confidentiality in all other respects.

B. Clients or Sponsors:

1. We shall hold confidential all information obtained about the client's general business affairs and about the findings of research conducted for the client, except when the dissemination of such information is expressly authorized.
2. We shall be mindful of the limitations of our techniques and facilities and shall accept only those research assignments which can be accomplished within these limitations.

C. The Profession:

1. We shall not cite our membership in the Association as evidence of professional competence, since the Association does not so certify any persons or organizations.
2. We recognize our responsibility to contribute to the science of public opinion research and to disseminate as freely as possible the ideas and findings which emerge from our research.

D. The Respondent:

1. We shall not lie to survey respondents or use practices and methods which abuse, coerce, or humiliate them.
2. We shall protect the anonymity of every respondent, unless the respondent waives such anonymity for specified uses. In addition, we shall hold as privileged and confidential all information which tends to identify the respondent.

Source: American Association for Public Opinion Research, *By-Laws* (May 1977). Used by permission.

because they used a public facility protested the follow-up survey. They felt it was unethical for Humphreys to trace the participants to their homes and to interview them under false pretenses.

Still others justified Humphreys' research. The topic, they said, was worth study. It couldn't be studied any other way, and they regarded the deceit as essentially harmless, noting that Humphreys was careful not to harm his subjects by disclosing their tearoom activities.

The tearoom trade controversy has never been resolved. It's still debated, and it probably always will be, since it stirs emotions and involves ethical issues people disagree about. What do you think? Was Humphreys ethical in doing what he did? Are there parts of the research that you believe were acceptable and other parts that were not?

 See the discussion by Joan Sieber online at http://go.to/tprizes for more on the political and ethical context of the "tearoom" research.

Observing Human Obedience

The second illustration differs from the first in many ways. Whereas Humphreys' study involved participant observation, the setting of this study was in the laboratory. Humphreys' study was sociological, this one psychological. And whereas Humphreys examined behavior considered by many to be a form of deviance, the researcher in this study examined obedience and conformity.

One of the more unsettling clichés to come out of World War II was the German soldier's common excuse for atrocities: "I was only following orders." From the point of view that gave rise to this comment, any behavior—no matter how reprehensible—could be justified if someone else could be assigned responsibility for it. If a superior officer ordered a soldier to kill a baby, the fact of the order supposedly exempted the soldier from personal responsibility for the action.

Although the military tribunals that tried the war crime cases did not accept this excuse, social researchers and others have recognized the extent to which this point of view pervades social life. People often seem willing to do things they know would be considered wrong by others if they can claim that some higher authority ordered them to do it. Such was the pattern of justification in the My Lai tragedy of Vietnam, when U.S. soldiers killed more than 300 unarmed civilians—some of them young children—simply because their village, My Lai, was believed to be a Viet Cong stronghold. This sort of justification appears less dramatically in day-to-day civilian life. Few would disagree that this reliance on authority exists, yet Stanley Milgram's study (1963, 1965) of the topic provoked considerable controversy.

To observe people's willingness to harm others when following orders, Milgram brought 40 adult men from many different walks of life into a laboratory setting designed to create the phenomenon under study. If you had been a subject in the experiment, you would have had something like the following experience.

You've been informed that you and another subject are about to participate in a learning experiment. Through a draw of lots, you're assigned the job of "teacher" and your fellow subject the job of "pupil." The "pupil" is led into another room and strapped into a chair; an electrode is attached to his wrist. As the teacher, you're seated in front of an impressive electrical control panel covered with dials, gauges, and switches. You notice that each switch has a label giving a different number of volts, ranging from 15 to 315. The switches have other labels, too, some with the ominous phrases "Extreme-Intensity Shock," "Danger—Severe Shock," and "XXX."

The experiment runs like this. You read a list of word pairs to the learner and then test his ability to match them up. Because you can't see him, a light on your control panel indicates his answer. Whenever the learner makes a mistake, you're instructed by the experimenter to throw one of the switches —beginning with the mildest—and administer a shock to your pupil. Through an open door between the two rooms, you hear your pupil's response to the shock. Then you read another list of word pairs and test him again.

As the experiment progresses, you administer ever more intense shocks, until your pupil screams for mercy and begs for the experiment to end. You're instructed to administer the next shock anyway. After a while, your pupil begins kicking the wall between the two rooms and continues to scream. The implacable experimenter tells you to give the next shock. Finally, you read a list and ask for the pupil's answer—but there is no reply whatever, only silence from the other room. The experimenter informs you that no answer is considered an error and instructs you to administer the next higher shock. This continues up to the "XXX" shock at the end of the series.

What do you suppose you really would have done when the pupil first began screaming? When he began kicking on the wall? Or when he became totally silent and gave no indication of life? You'd refuse to continue giving shocks, right? And surely the same would be true of most people.

So we might think—but Milgram found out otherwise. Of the first 40 adult men Milgram tested, nobody refused to continue administering the shocks until they heard the pupil begin kicking the wall between the two rooms. Of the 40, 5 did so then. Two-thirds of the subjects, 26 of the 40, continued doing as they were told through the entire series—up to and including the administration of the highest shock.

As you've probably guessed, the shocks were phony, and the "pupil" was a confederate of the experimenter. Only the "teacher" was a real subject in the experiment. As a subject, you wouldn't actually have been hurting another person, but you would have been led to think you were. The experiment was designed to test your willingness to follow orders to the point of presumably killing someone.

Milgram's experiments have been criticized both methodologically and ethically. On the ethical side, critics have particularly cited the effects of the experiment on the subjects. Many seem to have experienced personally about as much pain as they thought they were administering to someone else. They pleaded with the experimenter to let them stop giving the shocks. They became extremely upset and nervous. Some had uncontrollable seizures.

How do you feel about this research? Do you think the topic was important enough to justify such measures? Would debriefing the subjects be sufficient to ameliorate any possible harm? Can you think of other ways the researcher might have examined obedience?

 There is a wealth of discussion regarding the Milgram experiments on the Web. Search for "Milgram experiments," "human obedience experiments," or "Stanley Milgram."

THE POLITICS OF SOCIAL RESEARCH

As I indicated earlier, both ethics and politics hinge on ideological points of view. What is unacceptable from one point of view will be acceptable from another. Although political and ethical issues are often closely intertwined, I want to distinguish between them in two ways.

First, the ethics of social research deals mostly with the methods employed; political issues tend to center on the substance and use of research. Thus, for example, some critics raise ethical objections to the Milgram experiments, saying that the methods harmed the subjects. A political objection would be that obedience is not a suitable topic for study, either because (1) we should not tinker with people's willingness to follow orders from higher authority or (2), from the opposite political point of view, because the results of the research could be used to make people *more* obedient.

The second distinction between ethical and political aspects of social research is that there are no formal codes of accepted political conduct. Although some ethical norms have political aspects —for example, specific guidelines for not harming subjects clearly relate to our protection of civil liberties—no one has developed a set of political norms that all social researchers accept.

The only partial exception to the lack of political norms is the generally accepted view that a researcher's personal political orientation should not interfere with or unduly influence his or her scientific research. It would be considered improper for a researcher to use shoddy techniques or to distort or lie about his or her research as a way of furthering the researcher's political views. As you can imagine, however, studies are often enough attacked for allegedly violating this norm.

Objectivity and Ideology

In Chapter 1, I suggested that social research can never be totally objective, because researchers are human and therefore necessarily subjective. Science, as a collective enterprise, achieves the equivalent of objectivity through intersubjectivity. That is, different scientists, having different subjective views, can and should arrive at the same results when they employ accepted research techniques. Essentially, this will happen to the extent that each can set personal values and views aside for the duration of the research.

The classic statement on objectivity and neutrality in social science is Max Weber's lecture "Science as a Vocation" ([1925] 1946). In this talk, Weber coined the phrase *value-free sociology* and urged that sociology, like other sciences, needed to be unencumbered by personal values if it was to make a special contribution to society. Liberals and conservatives alike could recognize the "facts" of social science, regardless of how those facts were in accordance with their personal politics.

Most social researchers have agreed with this abstract ideal, but not all. Marxist and neo-Marxist scholars, for example, have argued that social science and social action cannot and should not be separated. Explanations of the status quo in society, they contend, shade subtly into defenses of that same status quo. Simple explanations of the social functions of, say, discrimination can easily become justifications for its continuance. By the same token, merely studying society and its ills without a commitment to making society more humane has been called irresponsible.

Quite aside from abstract disagreements about whether social science can or *should* be value-free, many have argued about whether particular research undertakings *are* value-free or whether they represent an intrusion of the researcher's own po-

litical values. Typically, researchers have denied such intrusion, and their denials have then been challenged. Let's look at some examples of the controversies that have raged and continue to rage over this issue.

Social Research and Race   Nowhere have social research and politics been more controversially intertwined than in the area of racial relations. Social researchers studied the topic for a long time, and often the products of the social research have found their way into practical politics. A few brief references should illustrate the point.

In 1896, when the U.S. Supreme Court established the principle of "separate but equal" as a means of reconciling the Fourteenth Amendment's guarantee of equality to African Americans with the norms of segregation, it neither asked for nor cited social research. Nonetheless, it is widely believed that the Court was influenced by the writings of William Graham Sumner, a leading social scientist of his era. Sumner was noted for his view that the mores and folkways of a society were relatively impervious to legislation and social planning. His view has often been paraphrased as "stateways do not make folkways." Thus the Court ruled that it could not accept the assumption that "social prejudices may be overcome by legislation" and denied the wisdom of "laws which conflict with the general sentiment of the community" (Blaunstein and Zangrando 1970:308). As many a politician has said, "You can't legislate morality."

When the doctrine of "separate but equal" was overturned in 1954 (*Brown* v. *Board of Education of Topeka*), the new Supreme Court decision was based in part on the conclusion that segregation had a detrimental effect on African-American children. In drawing that conclusion, the Court cited several sociological and psychological research reports (Blaunstein and Zangrando 1970).

For the most part, social researchers in this century have supported the cause of African-American equality in the United States, and their convictions often have been the impetus for their research. Moreover, they've hoped that their research will lead to social change. There is no doubt, for example, that Gunnar Myrdal's classic two-volume study (1944) of race relations in the United States had a significant impact on the topic of his research. Myrdal amassed a great deal of data to show that the position of African Americans directly contradicted U.S. values of social and political equality. Further, Myrdal did not attempt to hide his own point of view in the matter.

 You can pursue Myrdal's landmark research further online by searching for "Gunnar Myrdal" or "An American Dilemma."

Many social researchers have become directly involved in the civil rights movement, some more radically than others. Given the broad support for ideals of equality, research conclusions supporting the cause of equality draw little or no criticism. To recognize how solid the general social science position is in this matter, we need only examine a few research projects that have produced conclusions disagreeing with the predominant ideological position.

Most social researchers have—overtly, at least—supported the end of school segregation. Thus, an immediate and heated controversy was provoked in 1966 when James Coleman, a respected sociologist, published the results of a major national study of race and education. Contrary to general agreement, Coleman found little difference in academic performance between African-American students attending integrated schools and those attending segregated ones. Indeed, such obvious things as libraries, laboratory facilities, and high expenditures per student made little difference. Instead, Coleman reported that family and neighborhood factors had the most influence on academic achievement.

Coleman's findings were not well received by many of the social researchers who had been active in the civil rights movement. Some scholars criticized Coleman's work on methodological grounds, but many others objected hotly on the grounds that the findings would have segregationist political consequences. The controversy that raged around the Coleman report was reminiscent

of that provoked a year earlier by Daniel Moynihan (1965) in his critical analysis of the African-American family in the United States.

Another example of political controversy surrounding social research in connection with race concerns IQ scores. In 1969, Arthur Jensen, a Harvard psychologist, was asked to prepare an article for the *Harvard Educational Review* examining the data on racial differences in IQ test results (Jensen 1969). In the article, Jensen concluded that genetic differences between African Americans and whites accounted for the lower average IQ scores of African Americans. Jensen became so identified with that position that he appeared on college campuses across the country discussing it.

Jensen's research has been attacked on numerous methodological bases. Critics charged that much of the data on which Jensen's conclusion was based were inadequate and sloppy—there are many IQ tests, some worse than others. Similarly, it was argued that Jensen had not taken social-environmental factors sufficiently into account. Other social researchers raised still other methodological objections.

Beyond the scientific critique, however, Jensen was condemned by many as a racist. He was booed and his public presentations were drowned out by hostile crowds. Ironically, Jensen's reception by several university audiences was not significantly different from the reception received by abolitionists a century before, when the prevailing opinion favored leaving the institution of slavery intact.

 To examine a more recent version of the controversy surrounding race and achievement, search the Web for differing points of view concerning "The Bell Curve"—sparked by a book with that title by Richard J. Herrnstein and Charles Murray.

Many social researchers limited their objections to the Moynihan, Coleman, and Jensen research to scientific, methodological grounds. The political firestorms ignited by these studies, however, point out how ideology often shows up in matters of social research. Although the abstract model of science is divorced from ideology, the practice of science is not.

Project Camelot Among social researchers, *Camelot* is a household term in discussions of research and politics, frequently referenced with no further description. Irving Louis Horowitz (1967), a man who has criticized government agencies on occasion, said that Project Camelot "has had perhaps the worst public relations record of any agency or subagency of the U.S. government" (p. vi). What provoked such a stir?

On December 4, 1964, the Special Operations Research Office of American University sent an announcement to several social researchers about a proposed project concerning the topic of internal war within a nation. The announcement contained, in part, the following description:

> Project *Camelot* is a study whose objective is to determine the feasibility of developing a general social systems model which would make it possible to predict and influence politically significant aspects of social change in the developing nations of the world. Somewhat more specifically, its objectives are:
>
> *First,* to devise procedures for assessing the potential for internal war within national societies;
>
> *Second,* to identify with increased degrees of confidence those actions which a government might take to relieve conditions which are assessed as giving rise to a potential for internal war. — (HOROWITZ 1967:47)

Of course, few people are openly in favor of war, and most would support research aimed at ending or preventing it. By the summer of 1965, however, with the national debate on Vietnam gaining momentum, Camelot was being hotly argued in social science circles as a Department of Defense attempt to co-opt researchers into a counterinsurgency effort in Chile. Some claimed that the Defense Department intended to sponsor social research aimed at putting down political and potentially rev-

olutionary dissatisfaction in that volatile Latin-American nation. Whatever the motivations of the social researchers, it was feared that their research would be used to strengthen established regimes and thwart popular reformist and revolutionary movements in foreign countries.

Many social researchers who had agreed in principle to participate in the project soon felt they were facing a lesson learned decades before them by Robert Oppenheimer and some of the other scientists involved in the atomic bomb project—that scientific findings can be used for purposes that the scientists themselves oppose. Charges and countercharges were hurled around professional circles. Names were called, motives questioned. Old friendships ended. The Defense Department was roundly damned by all for attempting to subvert social research. Foreign relations with Latin America simultaneously chilled and heated up. Finally, under the cloud of growing criticism, Camelot was canceled and dismantled.

It's interesting to imagine what might have happened to Project Camelot had it been proposed to a steadfastly conservative and anticommunist social research community. I think there is no doubt that it would have been supported, executed, and completed without serious challenge or controversy. Certainly war per se was not the issue. There was no serious criticism when Samuel Stouffer organized the research branch in the army during World War II to conduct research aimed at supporting the war effort, making U.S. soldiers more effective fighters. Ultimately, science is neutral on the topics of war and peace, but scientists are not.

As a final example, consider what Hamnett et al. found out when they examined research ethics as seen from the standpoint of the countries that U.S. researchers sometimes study. Among other things, they pointed out the following:

> Governments in many parts of the Third World are increasingly making demands on cross-national researchers. . . . These range from restrictions on research not directly relevant to national development priorities to requirements for collaboration with host country institutions and scholars. The view that the exploitation

of any national resource, including social or cultural data, should be of benefit to that country provides one rationale for such requirements.

— (HAMNETT ET AL. 1984:6)

Politics with a Little "p"

While social research is often confounded by political ideologies, the "politics" of social research runs far deeper still. Social research in relation to contested social issues simply cannot remain antiseptically objective—particularly when differing ideologies are pitted against one another in a field of social science data.

The same is true when research is invoked in disputes between people with conflicting interests. For instance, social researchers who have served as "expert witnesses" in court would probably agree that the scientific ideal of a "search for truth" seems hopelessly naive in a trial or lawsuit. While expert witnesses technically do not represent either side in court, they are, nonetheless, engaged by only one side to appear, and their testimony tends to support the side of the party who pays for their time. This doesn't necessarily mean that these witnesses will lie on behalf of their patrons, but the contenders in a lawsuit are understandably more likely to pay for expert testimony that supports their case than for testimony that attacks it.

Thus, as an expert witness, you appear in court only because your presumably scientific and honest judgment happens to coincide with the interests of the party paying you to testify. Once you arrive in court and swear to tell the truth, the whole truth, and nothing but the truth, however, you find yourself in a world foreign to the ideals of objective contemplation. Suddenly the norms are those of winning and losing. As an expert witness, of course, all you have to lose is your respectability (and perhaps the chance to earn fees as an expert witness in the future). Still, such stakes are high enough to create discomfort for most social researchers.

I recall one case in federal court when I was testifying on behalf of some civil service workers who had had their cost-of-living allowance (COLA) cut

on the basis of research I thought was rather shoddy. I was engaged to conduct more "scientific" research that would demonstrate the injustice worked against the civil servants (Babbie 1982:232–43).

I took the stand, feeling pretty much like a respected professor and textbook author. In short order, however, I found I had moved from the academy to the hockey rink. Tests of statistical significance and sampling error were suddenly less relevant than a slap shot. At one point, an attorney from Washington lured me into casually agreeing that I was familiar with a certain professional journal. Unfortunately, the journal did not exist. I was mortified and suddenly found myself shifting domains. Without really thinking about it, I now was less committed to being a friendly Mr. Chips and more aligned with ninja-professor. I would not be fully satisfied until I, in turn, could mortify the attorney, which I succeeded in doing.

Even though the civil servants got their cost-of-living allowance back, I have to admit I was also concerned with how I looked in front of the courtroom assemblage. I tell you this anecdote to illustrate the personal "politics" of human interactions involving presumably scientific and objective research. We need to realize that as human beings social researchers are going to act like human beings, and we must take this into account in assessing their findings. This recognition does not invalidate their research or provide an excuse for rejecting findings we happen to dislike, but it does need to be taken into account.

Politics in Perspective

Although the ethical and the political dimensions of research are in principle distinct, they do intersect. Whenever politicians or the public feel that social research is violating ethical or moral standards, they'll be quick to respond with remedies of their own. Moreover, the standards they defend may not be those of the research community. And even when researchers support the goals of measures directed at the way research is done, the means specified by regulations or legislation can hamstring research.

There is a special concern among legislators for research on children. Although the social research norms discussed in this chapter would guard against bringing any physical or emotional harm to children, some of the restrictive legislation introduced from time to time borders on the actions of one particular Western city, which shall remain nameless. In response to concerns that a public school teacher had been playing New Age music in class and encouraging students to meditate, the city council passed legislation stating that no teacher could do anything that would "affect the minds of students"!

There are three main lessons that I hope you will take away from this discussion. First, science is not untouched by politics. The intrusion of politics and related ideologies is not unique to social research; the natural sciences have experienced and continue to experience similar situations. But social science, in particular, is a part of social life. Social researchers study things that matter to people, things they have firm, personal feelings about, and things that affect their lives. Moreover, researchers are human beings, and their feelings often show through in their professional lives. To think otherwise would be naive.

Second, science does proceed in the midst of political controversy and hostility. Even when researchers get angry and call each other names, or when the research community comes under attack from the outside, scientific inquiry persists. Studies are done, reports are published, and new things are learned. In short, ideological disputes do not bring science to a halt, but they do make it more challenging—and exciting.

Finally, an awareness of ideological considerations enriches the study and practice of social research methods. Many of the established characteristics of science, such as intersubjectivity, function to cancel out or hold in check our human shortcomings, especially those we are unaware of. Otherwise, we might look into the world and never see anything but a reflection of our personal biases and beliefs.

A QUANDARY **REVISITED**

The Nazi medical experiments were outrageous in many ways. Some of the experiments can only be described as ghoulish and sadistic. Often the scientific caliber of the experiments was shoddy. One could argue, however, that some people today suffer and even die from research. We often condone these risks by virtue of the benefits to humankind expected to follow from the research. Some of the Nazi doctors, no doubt, salved their own consciences with such reflections.

The Nazi medical experiments breached a fundamental ethical norm discussed in this chapter. This is reflected in the indictments of the Nuremberg Trials, which charged several medical personnel in the Nazi war machine with "plans and enterprises involving medical experiments *without the subjects' consent,* upon civilians and members of the armed forces of nations then at war with the German Reich [emphasis mine]" (*Trials of War Criminals* 1949–1953). Even if the most hideous experiments had not been conducted, and even accepting that there is always some risk when human research is undertaken, it is absolutely unacceptable to subject people to risks in research without their informed consent.

 For more on the ethics of Nazi experiments, see http://www.ushmm.org/ research/doctors/indiptx.htm

Main Points

- In addition to technical, scientific considerations, social research projects are likely to be shaped by administrative, ethical, and political considerations.

- What is ethical and unethical in research is ultimately a matter of what a community of people agree is right and wrong.

- Researchers agree that participation in research should normally be voluntary. This norm, however, can conflict with the scientific need for generalizability.

- Researchers agree that research should not harm those who participate in it, unless they willingly and knowingly accept the risks of harm by giving their informal consent.

- Whereas anonymity refers to the situation in which even the researcher cannot identify specific information with the individuals it describes, confidentiality refers to the situation in which the researcher promises to keep information about subjects private. The most straightforward way to ensure confidentiality is to destroy identifying information as soon as it's no longer needed.

- Many research designs involve a greater or less degree of deception of subjects. Because deceiving people violates common standards of ethical behavior, deception in research requires a strong justification—and even then the justification may be challenged.

- Social researchers have ethical obligations to the community of researchers as well as to subjects. These obligations include reporting results fully and accurately as well as disclosing errors, limitations, and other shortcomings in the research.

- Professional associations in several disciplines publish codes of ethics to guide researchers. These codes are necessary and helpful, but they do not resolve all ethical questions.

- Laud Humphrey's study of "tearoom" encounters and Stanley Milgram's study of obedience raise ethical issues that are debated to this day.

- Social research inevitably has a political and ideological dimension. Although science is neutral on political matters, scientists are not. Moreover, much social research inevitably involves the political beliefs of people outside the research community.

- Although most researchers agree that political orientation should not unduly influence research, in practice it can be very difficult to separate politics and ideology from the conduct of research. Some researchers maintain that research can and should be an instrument of social action and change. More subtly, a shared ideology can affect the way research is received by other researchers.

- Even though the norms of science cannot force individual researchers to give up their personal values, the intersubjective character of science provides a guard against "scientific" findings being the product of bias only.

Key Terms

informed consent
anonymity
confidentiality
debriefing

Review Questions

1. Consider the following real and hypothetical research situations. What is the ethical component in each example? How do you feel about it? Do you think the procedures described are ultimately acceptable or unacceptable? You might find it useful to discuss some of these situations with classmates.

   a. A psychology instructor asks students in an introductory psychology class to complete questionnaires that the instructor will analyze and use in preparing a journal article for publication.

   b. After a field study of deviant behavior during a riot, law enforcement officials demand that the researcher identify those people who were observed looting. Rather than risk arrest as an accomplice after the fact, the researcher complies.

c. After completing the final draft of a book reporting a research project, the researcher-author discovers that 25 of the 2,000 survey interviews were falsified by interviewers. To protect the bulk of the research, the author leaves out this information and publishes the book.

d. Researchers obtain a list of right-wing radicals they wish to study. They contact the radicals with the explanation that each has been selected "at random" from among the general population to take a sampling of "public opinion."

e. A college instructor who wants to test the effect of unfair berating administers an hour exam to both sections of a specific course. The overall performance of the two sections is essentially the same. The grades of one section are artificially lowered, however, and the instructor berates the students for performing so badly. The instructor then administers the same final exam to both sections and discovers that the performance of the unfairly berated section is worse. The hypothesis is confirmed, and the research report is published.

f. In a study of sexual behavior, the investigator wants to overcome subjects' reluctance to report what they might regard as shameful behavior. To get past their reluctance, subjects are asked, "Everyone masturbates now and then; about how much do you masturbate?"

g. A researcher studying dorm life on campus discovers that 60 percent of the residents regularly violate restrictions on alcohol consumption. Publication of this finding would probably create a furor in the campus community. Because no extensive analysis of alcohol use is planned, the researcher decides to keep this finding quiet.

h. To test the extent to which people may try to save face by expressing attitudes on matters they are wholly uninformed about, the researcher asks for their attitudes regarding a fictitious issue.

i. A research questionnaire is circulated among students as part of their university registration packet. Although students are not told they must complete the questionnaire, the hope is that they will believe they must—thus ensuring a higher completion rate.

j. A researcher pretends to join a radical political group in order to study it and is successfully accepted as a member of the inner planning circle. What should the researcher do if the group makes plans for the following?

   (1) A peaceful, though illegal, demonstration

   (2) The bombing of a public building during a time it is sure to be unoccupied

   (3) The assassination of a public official

2. Review the discussion of the Milgram experiment on obedience. How would you design a study to accomplish the same purpose while avoiding the ethical criticisms leveled at Milgram? Would your design be equally valid? Would it have the same effect?

3. Suppose a researcher who is personally in favor of small families—as a response to the problem of overpopulation—wants to conduct a survey to determine why some people want many children and others don't. What personal-involvement problems would the researcher face and how could she or he avoid them?

4. What ethical issues should the researcher in item 3 take into account in designing the survey?

Additional Readings

Hamnett, Michael P., Douglas J. Porter, Amarjit Singh, and Krishna Kumar. 1984. *Ethics, Politics, and International Social Science Research.* Honolulu: University of Hawaii Press. Discussions of research ethics typically focus on the interests of the individual participants in research projects, but this book raises the level of the discussion to include the rights of whole societies.

Homan, Roger. 1991. *The Ethics of Social Research.* London: Longman. A thoughtful analysis of the ethical issues of social science research, by a practicing British social researcher.

Lee, Raymond. 1993. *Doing Research on Sensitive Topics.* Newbury Park, CA: Sage. This book examines the conflicts between scientific research needs and the rights of the people involved—with guidelines for dealing with such conflicts.

Sweet, Stephen. 1999. "Using a Mock Institutional Review Board to Teach Ethics in Sociological Research." *Teaching Sociology* 27 (January): 55–59. Though written for professors, this article provides some research examples that challenge your ethical instincts.

Multimedia Resources

The Wadsworth Sociology Resource Center: Virtual Society
http://sociology.wadsworth.com/
Visit the companion Web site for the second edition of *The Basics of Social Research* to access a wide range of student resources. Begin by clicking on the Student Resources section of the book's Web site to access the following study tools:

- eBabbie Resource Center
- Planning a Research Project
- Doing Data Analysis
- Statistics Review
- Flash Cards
- Internet Links and Exercises
- InfoTrac College Edition: Exercises
- Quizzes

Visit the **eBabbie Resource Center** for an overview of each chapter and helpful online tutorials. Find information on budgeting and step-by-step examples of model research projects at **Planning a Research Project.** Learn how to use quantitative and qualitative data analysis programs at **Doing Data Analysis,** and brush up on your statistics at **Statistics Review.** You can also further your study by accessing **Internet Links and Exercises** related to chapter materials, **Flash Cards, Quizzes,** and many other learning tools.

InfoTrac College Edition
http://www.infotrac-college.com/
wadsworth/access.html
Access the latest news and journal articles with InfoTrac College Edition, an easy-to-use online database of reliable, full-length articles from hundreds of top academic journals. Conduct an electronic search using the following search terms:

Anonymity
Code of ethics
Confidentiality
Informed consent
Institutional review board
Research ethics

Kathy Sloane/Photo Researchers, Inc.

Part Two

THE STRUCTURING OF INQUIRY

Posing problems properly is often more difficult than answering them. Indeed, a properly phrased question often seems to answer itself. You may have discovered the answer to a question just in the process of making the question clear to someone else.

Part 2 deals with what should be observed; that is, Part 2 considers the posing of proper scientific questions, the structuring of inquiry. Part 3 will describe some of the specific methods of social scientific observation.

Chapter 4 addresses the beginnings of research. It examines some of the purposes of inquiry, units of analysis, and the reasons scientists get involved in research projects.

Chapter 5 deals with the specification of what it is you want to measure—the processes of conceptualization and operationalization. It looks at some of the terms that you and I use quite casually in everyday life—prejudice, liberalism, happiness, and so forth—and shows how essential it is to clarify what we really mean by such terms when we do research. This process of clarification is called *conceptualization.*

Once we clarify what we mean by certain terms, we can then measure the referents of those terms. The process of devising steps or operations for measuring what we want to study is called *operationalization.* Chapter 5 deals with the topic of operationalization in general, paying special attention to the framing of questions for interviews and questionnaires.

To complete the introduction to measurement, Chapter 6 breaks with the chronological discussion of how research is conducted. In this chapter, we'll examine techniques for measuring variables in quantitative research through the combination of several indicators: indexes, scales, and typologies. As an example, we might ask survey respondents five different questions about their attitudes toward gender equality and then combine the answers to all five questions into a composite measure of gender-based egalitarianism. Although such composite measures are constructed during the analysis of data (see Part 4), the raw materials for them must be provided for in the design and execution of data collection.

Finally, we'll look at how social researchers select people or things for observation. Chapter 7, on sampling, addresses the fundamental scientific issue of generalizability. As you'll see, we can select a few people or things for observation and then apply what we observe to a much larger group. For example, by surveying 2,000 U.S. citizens about whom they favor for president of the United States, we can accurately predict how tens of millions will vote. In this chapter, we'll examine techniques that increase the generalizability of what we observe.

What you learn in Part 2 will bring you to the verge of making controlled social scientific observations. Part 3 will then show you how to take that next step.

4 RESEARCH DESIGN

Photographers Library LTD/eStock Photography/Picture Quest

What You'll Learn in This Chapter

Here you'll see the wide variety of research designs available to social researchers as well as how to design a study—that is, specifying exactly who or what is to be studied, when, how, and for what purpose.

In this chapter . . .

INTRODUCTION

Science is an enterprise dedicated to "finding out." No matter what you want to find out, though, there will likely be a great many ways of doing it. That's true in life generally. Suppose, for example, that you want to find out whether a particular automobile—say, the new Burpo-Blasto—would be a good car for you. You could, of course, buy one and find out that way. Or you could talk to a lot of B-B owners or to people who considered buying one and didn't. You might check the classified ads to see if there are a lot of B-Bs being sold cheap. You could read a consumer magazine evaluation of Burpo-Blastos. A similar situation occurs in scientific inquiry.

Ultimately, scientific inquiry comes down to making observations and interpreting what you've observed, the subjects of Parts 3 and 4 of this book. Before you can observe and analyze, however, you need a plan. You need to determine what you're going to observe and analyze: why and how. That's what research design is all about.

Although the details vary according to what you wish to study, you face two major tasks in any research design. First, you must specify as clearly as possible what it is you want to find out. Second,

---

I n the following letter published in a college newspaper, the Provost objects to data that had been previously reported.

Provost says percentage was wrong

I am writing to clarify a misstatement in an editorial in the April 19 *The Panther*. As recently as last fall, the concept behind this statement was presented to your staff.

This current use of erroneous numbers demands correction.

The figure used in the statement, "With about 52 percent of the faculty being part time . . ." is absolutely incorrect.

Since the thrust of the editorial is Chapman's ability to live up to its desire to "nurture and help develop students," a proper measure of the difference between full-time faculty presence and that of part-time faculty is how many credits or courses are taught.

For the past four years, full-time faculty have taught about 70 percent of the credits in which students enroll each semester.

Thus, a large majority of our faculty are here full-time; teaching classes, advising students, attending meetings, interacting with students in the hallways and dining rooms.

Once again, I welcome the opportunity to present the truth.

Might I suggest that a future edition of The Panther be devoted to the contributions of part-time faculty.

Harry L. Hamilton
Provost

Which side is correct: the original newspaper report or the Provost's account? Or are both sides correct? If so, why?

---

you must determine the best way to do it. Interestingly, if you can handle the first consideration fully, you'll probably handle the second in the same process. As mathematicians say, a properly framed question contains the answer.

Let's say you're interested in studying corruption in government. That's certainly a worthy and appropriate topic for social research. But what *specifically* are you interested in? What do you mean by "corruption"? Specifically, what kinds of behavior do you have in mind? And what do you mean by "government"? Whom do you want to study: all public employees? elected officials? civil servants? Finally, what is your purpose? Do you want to find out *how much* corruption there is, or do you want to learn *why* it exists? These are the kinds of questions that need to be answered in the course of research design.

This chapter provides a general introduction to research design, while the other chapters in Part 2 elaborate on specific aspects. In practice, all aspects of research design are interrelated. As you read through Part 2, the interrelationships among parts will become clearer.

We'll start by briefly examining the main purposes of social research. Then, we'll consider units of analysis—the what or whom you want to study. Next we'll consider alternative ways of handling time in social research or how to study a moving target that changes over time.

With these ideas in hand, we'll turn to how to design a research project. This overview of the research process serves two purposes: Besides describing how you might go about designing a study, it provides a map of the remainder of this book.

Finally, we'll look at the elements of research proposals. Often the actual conduct of research needs to be preceded by a detailing of your intentions—to obtain funding for a major project or perhaps to get your instructor's approval for a class project. You'll see that the research proposal provides an excellent opportunity for you to consider all aspects of your research in advance.

THREE PURPOSES OF RESEARCH

Social research can serve many purposes. Three of the most common and useful purposes are exploration, description, and explanation. Although a given study can have more than one of these purposes—and most do—examining them separately is useful because each has different implications for other aspects of research design.

Exploration

Much of social research is conducted to explore a topic, that is, to start to familiarize a researcher with that topic. This approach typically occurs when a researcher examines a new interest or when the subject of study itself is relatively new.

As an example, let's suppose that widespread taxpayer dissatisfaction with the government erupts into a taxpayers' revolt. People begin refusing to pay their taxes, and they organize themselves around that issue. You might like to learn more about the movement: How widespread is it? What levels and degrees of support are there within the community? How is the movement organized? What kinds of people are active in it? An exploratory study could help you find at least approximate answers to some of these questions. You might check figures with tax-collecting officials, collect and study the literature of the movement, attend meetings, and interview leaders.

Exploratory studies are also appropriate for more persistent phenomena. Suppose you're unhappy with your college's graduation requirements and want to help change them. You might study the history of such requirements at the college and meet with college officials to learn the reasons for the current standards. You could talk to several students to get a rough idea of their sentiments on the subject. Though this last activity would not necessarily yield an accurate picture of student opinion, it could suggest what the results of a more extensive study might be.

Sometimes exploratory research is pursued through the use of focus groups, or guided small-group discussions. This technique is frequently used in market research; we'll examine it further in Chapter 10.

Exploratory studies are most typically done for three purposes: (1) to satisfy the researcher's curiosity and desire for better understanding, (2) to test the feasibility of undertaking a more extensive study, and (3) to develop the methods to be employed in any subsequent study.

A while back, for example, I became aware of the growing popularity of something called "channeling," in which a person known as a channel or medium enters a trance state and begins speaking with a voice that claims it originates outside the channel. Some of the voices say they come from a spirit world of the dead; some say they are from other planets; and still others say they exist on dimensions of reality difficult to explain in ordinary human terms.

The channeled voices, often referred to as entities, sometimes use the metaphor of radio or television for the phenomenon they represent. "When you watch the news," one told me in the course of an interview, "you don't believe Dan Rather is really inside the television set. The same is true of me. I use this medium's body the way Dan Rather uses your television set."

The idea of channeling interested me from several perspectives, not the least of which was the methodological question of how to study scientifically something that violates so much of what we take for granted, including scientific staples such as space, time, causation, and individuality.

Lacking any rigorous theory or precise expectations, I merely set out to learn more. Using some of the techniques of qualitative field research discussed in Chapter 10, I began amassing information and forming categories for making sense of what I observed. I read books and articles about the phenomenon and talked to people who had attended channeling sessions. I then attended channeling sessions myself, observing those who attended as well as the channel and entity. Next, I conducted personal interviews with numerous channels and entities.

In most interviews, I began by asking the human channels questions about how they first began channeling, what it was like, and why they continued, as well as standard biographical questions. The channel would then go into a trance, whereby the interview continued with the entity speaking. "Who are you?" I might ask. "Where do you come from?" "Why are you doing this?" "How can I tell if you are real or a fake?" Although I went into these interview sessions with several questions prepared in advance, each of the interviews followed whatever course seemed appropriate in the light of answers given.

This example of exploration illustrates where social research often begins. Whereas researchers working from deductive theories have the key variables laid out in advance, one of my first tasks was to identify some of the possibly relevant variables. For example, I noted a channel's gender, age, education, religious background, regional origins, and previous participation in things metaphysical. I chose most of these variables because they commonly affect behavior.

I also noted differences in the circumstances of channeling sessions. Some channels said they must go into deep trances, some use light trances, and others remain conscious. Most sit down while channeling, but others stand and walk about. Some channels operate under pretty ordinary conditions; others seem to require metaphysical props such as dim lights, incense, and chanting. Many of these differences became apparent to me only in the course of my initial observations.

Regarding the entities, I have been interested in classifying where they say they come from. Over the course of my interviews, I've developed a set of questions about specific aspects of "reality," attempting to classify the answers they give. Similarly, I ask each to speak about future events.

Over the course of this research, my examination of specific topics has become increasingly focused as I've identified variables that seem worth pursuing: gender, education, and religion, for example. Note, however, that I began with a reasonably blank slate.

Exploratory studies are quite valuable in social scientific research. They're essential whenever a researcher is breaking new ground, and they almost always yield new insights into a topic for research. Exploratory studies are also a source of grounded theory, as discussed in Chapter 2.

The chief shortcoming of exploratory studies is that they seldom provide satisfactory answers to research questions, though they can hint at the answers and can suggest which research methods could provide definitive answers. The reason exploratory studies are seldom definitive in themselves has to do with representativeness; that is, the people you study in your exploratory research may not be typical of the larger population that interests you. Once you understand representativeness, you'll be able to know whether a given exploratory study actually answered its research problem or only pointed the way toward an answer. (Representativeness is discussed at length in Chapter 7.)

Description

A major purpose of many social scientific studies is to describe situations and events. The researcher observes and then describes what was observed. Because scientific observation is careful and deliberate, however, scientific descriptions are typically more accurate and precise than are casual ones.

The U.S. Census is an excellent example of descriptive social research. The goal of the census is to describe accurately and precisely a wide variety of characteristics of the U.S. population, as well as the populations of smaller areas such as states and counties. Other examples of descriptive studies are the computation of age-gender profiles of populations done by demographers, the computation of crime rates for different cities, and a product-marketing survey that describes the people who use, or would use, a particular product. A researcher who carefully chronicles the events that take place on a labor union picket line has, or at least serves, a descriptive purpose. A researcher who computes and reports the number of times individual legislators voted for or against organized labor also fulfills a descriptive purpose.

Many qualitative studies aim primarily at description. An anthropological ethnography, for example, may try to detail the particular culture of some preliterate society. At the same time, such studies are seldom limited to a merely descriptive purpose. Researchers usually go on to examine *why* the observed patterns exist and what they imply.

Explanation

The third general purpose of social scientific research is to explain things. Descriptive studies answer questions of what, where, when, and how; explanatory questions, of why. So when William Sanders (1994) set about describing the varieties of gang violence, he also wanted to reconstruct the process that brought about violent episodes among the gangs of different ethnic groups.

Reporting the voting intentions of an electorate is descriptive, but reporting why some people plan to vote for Candidate A and others for Candidate B is explanatory. Reporting why some cities have higher crime rates than others involves explanation; identifying variables that explain why some cities have higher crime rates than others involves explanation. A researcher who sets out to know why an antiabortion demonstration ended in a violent confrontation with police, as opposed to simply describing what happened, has an explanatory purpose.

SPSS MC	Programs like SPSS or MicroCase and data such as those provided by the General Social Survey (GSS) provide powerful vehicles for description and explanation in social research. The examples presented here should give you some idea of these aspects of research. We'll also look at other examples from time to time throughout the book.*

---

*Each time the SPSS and MicroCase icons appear, they indicate that the topic under discussion could be pursued through the use of these software programs; whenever you see the GSS icon, you'll be given some tips on using this valuable resource.

Let's look at a specific case. What factors do you suppose might shape people's attitudes toward the legalization of marijuana? To answer this, you might first consider whether men and women differ in their opinions. An explanatory analysis of the 1998 GSS data indicates that 34 percent of men and 22 percent of women said marijuana should be legalized.

What about political orientation? The GSS data show that 43 percent of liberals said marijuana should be legalized, compared with 26 percent of moderates and 18 percent of conservatives. Further, 31 percent of Democrats, compared with 21 percent of both Independents and Republicans, supported legalization.

Given these statistics, you might begin to develop an explanation for attitudes toward marijuana legalization. Further study of gender and political orientation might then lead to a deeper explanation of these attitudes.

THE LOGIC OF NOMOTHETIC EXPLANATION

The preceding examination of what factors might cause attitudes about legalizing marijuana illustrates nomothetic explanation, as discussed in Chapter 1. Recall that in this model, we try to find a few factors (independent variables) that can account for many of the variations in a given phenomenon. This explanatory model stands in contrast to the idiographic model, in which we seek a complete, in-depth understanding of a single case.

In our example, an idiographic approach would suggest all the reasons that one person was opposed to legalization: involving what her parents, teachers, clergy told her about it, any bad experiences experimenting with it, and so forth. When we understand something idiographically, we feel we *really* understand it. When we know all the reasons why someone opposed legalizing marijuana, we couldn't imagine that person having any other attitude.

In contrast, a nomothetic approach might suggest that overall political orientations account for much of the difference of opinion about legalizing marijuana. Because this model is inherently probabilistic, it is more open than the idiographic model to misunderstanding and misinterpretation. Let's examine what social researchers mean when they say one variable (nomothetically) causes another. Then, we'll look at what they *don't* mean.

Criteria for Nomothetic Causality

There are three main criteria for nomothetic causal relationships in social research: (1) the variables must be correlated, (2) the cause takes place before the effect, and (3) the variables are nonspurious.

Correlation Unless some actual relationship—or **correlation**—is found between two variables, we can't say that a causal relationship exists. Our analysis of GSS data suggested that political orientation was a cause of attitudes about legalizing marijuana. Had the same percentage of liberals and conservatives supported legalization, we could hardly say that political orientations caused the attitude. Though this criterion is obvious, it emphasizes the need to base social research assertions on actual observations rather than assumptions.

Time Order Next, we can't say a causal relationship exists unless the cause precedes the effect in time. Notice that it makes more sense to say that most children's religious affiliations are caused by those of their parents than to say that parents' affiliations are caused by those of their children—even though it would be possible for you to change your religion and for your parents to follow suit. Remember, nomothetic explanation deals with "most cases" but not all.

In our marijuana example, it would make sense to say that gender causes, to some extent, attitudes toward legalization, whereas it would make no sense to say that opinions about marijuana determine a person's gender. Notice, however, that the time order connecting political orientations and attitudes about legalization is less clear, though we sometimes reason that general orientations cause specific opinions. As we'll see in the next chapter, this can be a complex matter.

Nonspurious The third requirement for a causal relationship is that the effect cannot be explained in terms of some third variable. For example, there is a correlation between ice cream sales and deaths due to drowning: the more ice cream sold, the more drownings, and vice versa. There is, however, no direct link between ice cream and drowning. The third variable at work here is season or temperature. Most drowning deaths occur during summer—the peak period for ice-cream sales.

Here are a couple of other examples of **spurious relationships,** or ones that aren't genuine. There is a negative relationship between the number of mules and the number of Ph.D.'s in towns and cities: the more mules, the fewer Ph.D.'s and vice versa. Perhaps you can think of another variable that would explain this apparent relationship. The answer is rural versus urban settings. There are more mules (and fewer Ph.D.'s) in rural areas, whereas the opposite is true in cities.

Or, consider the positive correlation between shoe size and math ability among schoolchildren. Here, the third variable that explains the puzzling relationship is age. Older children have bigger feet and more highly developed math skills, on average, than do younger children. See Figure 4-1 for an illustration of this spurious relationship. Observed associations are indicated with thin arrows, while causal relationships are indicated with thick ones. Notice, too, that observed associations go in both directions. That is, as one variable occurs or changes, so does the other.

The list goes on. Areas with many storks have high birth rates. Those with few storks have low birth rates. Do storks really deliver babies? Birth rates are higher in the country than in the city; more storks live in the country than the city. The third variable here is urban/rural areas.

Finally, the more fire trucks that put out a fire, the more damage to the structure. Can you guess what the third variable is? In this case, it is the size of the fire.

Thus, when social researchers say there is a

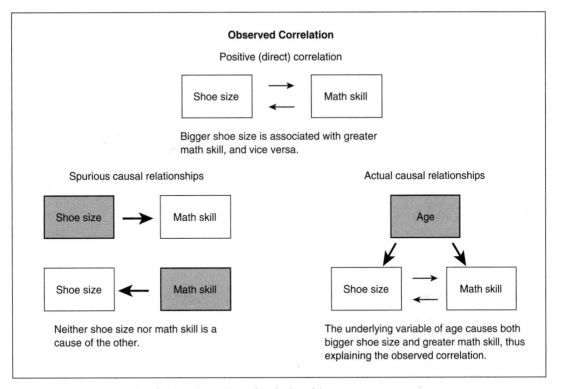

FIGURE 4-1   An Example of a Spurious Causal Relationship

causal relationship between, say, education and racial tolerance, they mean (1) there is a statistical correlation between the two variables, (2) a person's educational level occurred before their current level of tolerance or prejudice, and (3) there is no third variable that can explain away the observed correlation as spurious.

False Criteria for Nomothetic Causality

Because notions of cause and effect are well entrenched in everyday language and logic, it's important to specify some of the things social researchers do *not* mean when they speak of causal relationships. When they say that one variable causes another, they do not necessarily mean to suggest complete causation, to account for exceptional cases, or to claim that the causation exists in a majority of cases.

Complete Causation   Whereas an idiographic explanation of causation is relatively complete, a nomothetic explanation is probabilistic and usually incomplete. As we've seen, social researchers may say that political orientations cause attitudes toward legalizing marijuana even though not all liberals approve nor all conservatives disapprove. Thus, we say that political orientation is one of the causes of the attitude, but not the only one.

Exceptional Cases   In nomothetic explanations, exceptions do not disprove a causal relationship. For example, it is consistently found that women are more religious than men in the United States. Thus, gender may be a cause of religiosity, even if your uncle is a religious zealot or you know a woman who is an avowed atheist. Those exceptional cases do not disprove the overall, causal pattern.

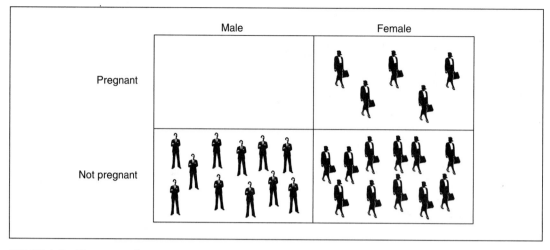

	Male	Female
Pregnant		
Not pregnant		

FIGURE 4-2 **Necessary Cause**

Majority of Cases Causal relationships can be true even if they don't apply in a majority of cases. For example, we say that children who are not supervised after school are more likely to become delinquent than are those who are supervised; hence, lack of supervision is a cause of delinquency. This causal relationship holds true even if only a small percentage of those not supervised become delinquent. As long as they are *more likely* than those who are supervised to be delinquent, we say there is a causal relationship.

The social scientific view of causation may vary from what you are accustomed to, since people commonly use the term *cause* to mean something that completely causes another thing. The somewhat different standard used by social researchers can be seen more clearly in a terms of necessary and sufficient causes.

NECESSARY AND SUFFICIENT CAUSES

A *necessary cause* represents a condition that *must* be present for the effect to follow. For example, it is necessary for you to take college courses in order to get a degree. Take away the courses, and the degree never happens. However, simply taking the courses is not a sufficient cause of getting a degree. You need to take the right ones and pass them.

Similarly, being female is a necessary condition of becoming pregnant, but it is not a sufficient cause. Otherwise, all women would get pregnant.

Figure 4-2 illustrates this relationship between the variables of gender and pregnancy as a matrix showing the possible outcomes of combining these variables.

A *sufficient cause,* on the other hand, represents a condition that, if it is present, guarantees the effect in question. This is not to say that a sufficient cause is the *only* possible cause of a particular effect. For example, skipping an exam in this course would be a sufficient cause for failing it, though students could fail it other ways as well. Thus, a cause can be sufficient, but not necessary. Figure 4-3 illustrates the relationship between taking or not taking the exam and either passing or failing it.

The discovery of a cause that is both necessary *and* sufficient is, of course, the most satisfying outcome in research. If juvenile delinquency were the effect under examination, it would be nice to discover a single condition that (1) must be present for delinquency to develop and (2) always results in delinquency. In such a case, you would surely feel that you knew precisely what caused juvenile delinquency.

Unfortunately, we never discover single causes that are absolutely necessary and absolutely sufficient when analyzing the nomothetic relationships

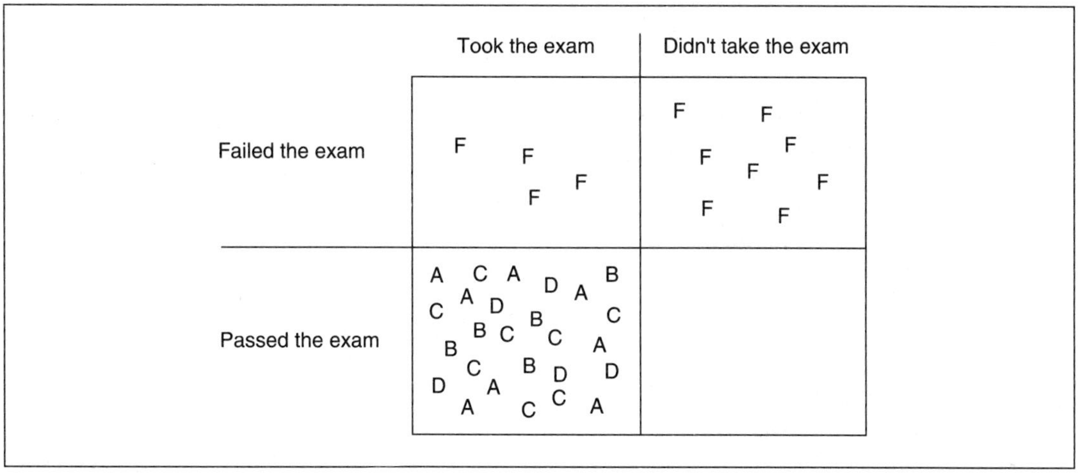

	Took the exam	Didn't take the exam
Failed the exam	F F F F	F F F F F F F F F
Passed the exam	A C A D B C A D B A C B B C B C A D C A B D D A C C A	

FIGURE 4-3 **Sufficient Cause**

among variables. It is not uncommon, however, to find causal factors that are either 100-percent necessary (you must be female to become pregnant) or 100-percent sufficient (skipping every exam will inevitably cause you to fail it).

In the idiographic analysis of single cases, you may reach a depth of explanation from which it is reasonable to assume that things could not have turned out differently, suggesting you have determined the *sufficient* causes for a particular result. (Anyone with all the same details of your genetic inheritance, upbringing, and subsequent experiences would have ended up going to college.) At the same time, there could always be other causal paths to the same result. Thus, the idiographic causes are sufficient but not necessary.

UNITS OF ANALYSIS

In social research, there is virtually no limit to what or whom can be studied, or the **units of analysis.** This topic is relevant to all forms of social research, although its implications are clearest in the case of nomothetic, quantitative studies.

The idea for units of analysis may seem slippery at first, because research—especially nomothetic research—often studies large collections of people or things, or aggregates. It's important to distinguish between the unit of analysis and the aggregates that we generalize about. For instance, a researcher may study a class of people, such as Democrats, college undergraduates, African-American women under 30, or some other collection. But if the researcher is interested in exploring, describing, or explaining how different groups of individuals behave *as individuals,* the unit of analysis is the individual, not the group. This is so even though the researcher then proceeds to generalize about aggregates of individuals, as in saying that more Democrats than Republicans favor legalizing marijuana. Think of it this way: Having an attitude about marijuana is something that can only be an attribute of an individual, not a group; that is, there is no one group "mind" that can have an attitude. So even when we generalize about Democrats, we're generalizing about an attribute they possess as individuals.

In contrast, we may sometimes want to study groups, considered as individual "actors" or entities that have attributes as groups. For instance, we might want to compare the characteristics of different types of street gangs. In that case our unit of analysis would be gangs (not members of gangs), and we might proceed to make generalizations about different types of gangs.

Social researchers perhaps most typically choose individual people as their units of analysis.

You may note the characteristics of individual people—gender, age, region of birth, attitudes, and so forth. You can then combine these descriptions to provide a composite picture of the group the individuals represent, whether a street-corner gang or a whole society.

For example, you may note the age and gender of each student enrolled in Political Science 110 and then characterize the group of students as being 53 percent men and 47 percent women and as having a mean age of 18.6 years. Although the final description would be of the class as a whole, the description is based on characteristics that members of the class have as individuals.

The same distinction between units of analysis and aggregations occurs in explanatory studies. Suppose you wished to discover whether students with good study habits received better grades in Political Science 110 than did students with poor study habits. You would operationalize the variable *study habits* and measure this variable, perhaps in terms of hours of study per week. You might then aggregate students with good study habits and those with poor study habits and see which group received the best grades in the course. The purpose of the study would be to explain why some groups of students do better in the course than do others, but the unit of analysis is still individual students.

Units of analysis in a study are usually also the units of observation. Thus, to study success in a political science course, we would observe individual students. Sometimes, however, we "observe" our units of analysis indirectly. For example, suppose we want to find out whether disagreements about the death penalty tend to cause divorce. In this case, we might "observe" individual husbands and wives by asking them about their attitudes about capital punishment, in order to distinguish couples who agree and disagree on this issue. In this case, our units of observation are individual wives and husbands, but our units of analysis (the things we want to study) are couples.

Units of analysis, then, are those things we examine in order to create summary descriptions of all such units and to explain differences among them. In most research projects, the unit of analysis will probably be clear to you. When the unit of analysis is not clear, however, it's essential to determine what it is; otherwise, you cannot determine what observations are to be made about whom or what.

Some studies try to describe or explain more than one unit of analysis. In these cases, the researcher must anticipate what conclusions she or he wishes to draw with regard to which units of analysis. For example, we may want to discover what kinds of college students (individuals) are most successful in their careers; we may also want to learn what kinds of colleges (organizations) produce the most successful graduates.

To make this discussion more concrete, let's consider several common units of analysis in social research.

Individuals

As mentioned, individual human beings are perhaps the most typical units of analysis for social research. We tend to describe and explain social groups and interactions by aggregating and manipulating the descriptions of individuals.

Any type of individual may be the unit of analysis for social research. This point is more important than it may seem at first. The norm of generalized understanding in social research should suggest that scientific findings are most valuable when they apply to all kinds of people. In practice, however, social researchers seldom study all kinds of people. At the very least, their studies are typically limited to the people living in a single country, though some comparative studies stretch across national boundaries. Often, though, studies are quite circumscribed.

Examples of classes of individuals that might be chosen for study include students, gays and lesbians, auto workers, voters, single parents, and faculty members. Note that each of these terms implies some population of individuals. Descriptive studies with individuals as their units of analysis typically aim to describe the population that comprises those individuals, whereas explanatory studies aim to discover the social dynamics operating within that population.

As the units of analysis, individuals may be characterized in terms of their membership in social groupings. Thus, an individual may be described as belonging to a rich family or to a poor one, or a person may be described as having a college-educated mother or not. We might examine in a research project whether people with college-educated mothers are more likely to attend college than are those with non-college-educated mothers or whether high school graduates in rich families are more likely than those in poor families to attend college. In each case, the unit of analysis —the "thing" whose characteristics we are seeking to describe or explain—is the individual. We then aggregate these individuals and make generalizations about the population they belong to.

Groups

Social groups can also be units of analysis in social research. That is, we may be interested in characteristics that belong to one group, considered as a single entity. If you were to study the members of a criminal gang to learn about criminals, the individual (criminal) would be the unit of analysis; but if you studied all the gangs in a city to learn the differences, say, between big gangs and small ones, between "uptown" and "downtown" gangs, and so forth, you would be interested in gangs rather than their individual members. In this case, the unit of analysis would be the gang, a social group.

Here's another example. Suppose you were interested in the question of access to computers in different segments of society. You might describe families in terms of total annual income and according to whether or not they had computers. You could then aggregate families and describe the mean income of families and the percentage with computers. You would then be in a position to determine whether families with higher incomes were more likely to have computers than were those with lower incomes. In this case, the unit of analysis would be families.

As with other units of analysis, we can derive the characteristics of social groups from those of their individual members. Thus, we might describe a family in terms of the age, race, or education of its head. In a descriptive study, we might find the percentage of all families that have a college-educated head of family. In an explanatory study, we might determine whether such families have, on average, more or fewer children than do families headed by people who have not graduated from college. In each of these examples, the family is the unit of analysis. In contrast, had we asked whether college-educated individuals have more or fewer children than do their less-educated counterparts, then the individual would have been the unit of analysis.

Other units of analysis at the group level could be friendship cliques, married couples, census blocks, cities, or geographical regions. As with individuals, each of these terms implies some population. *Street gangs* implies some population that includes all street gangs, perhaps in a given city. You might then describe this population by generalizing from your findings about individual gangs. For instance, you might describe the geographical distribution of gangs throughout a city. In an explanatory study of street gangs, you might discover whether large gangs are more likely than small ones to engage in intergang warfare. Thus, you would arrive at conclusions about the population of gangs by using individual groups as your unit of analysis.

Organizations

Formal social organizations may also be the units of analysis in social research. For example, a researcher might study corporations, by which he or she implies a population of all corporations. Individual corporations might be characterized in terms of their number of employees, net annual profits, gross assets, number of defense contracts, percentage of employees from racial or ethnic minority groups, and so forth. We might determine whether large corporations hire a larger or smaller percentage of minority group employees than do small corporations. Other examples of formal social organizations suitable as units of analysis include church congregations, colleges, army divisions, academic departments, and supermarkets.

Figure 4-4 provides a graphic illustration of some different units of analysis and the statements that might be made about them.

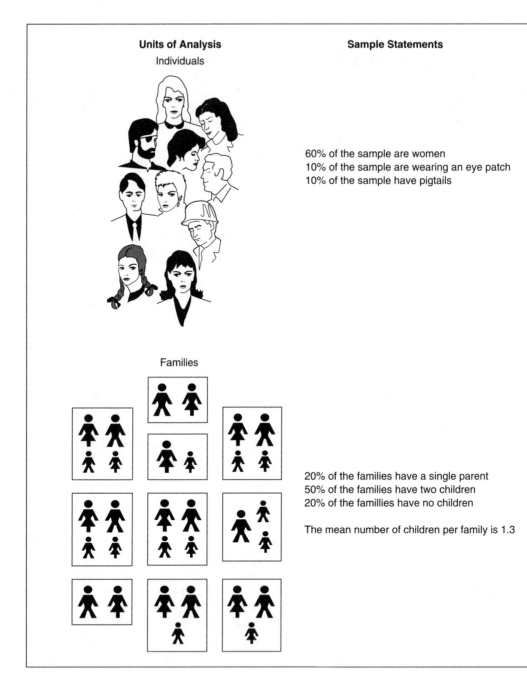

FIGURE 4-4 **Illustrations of Units of Analysis**

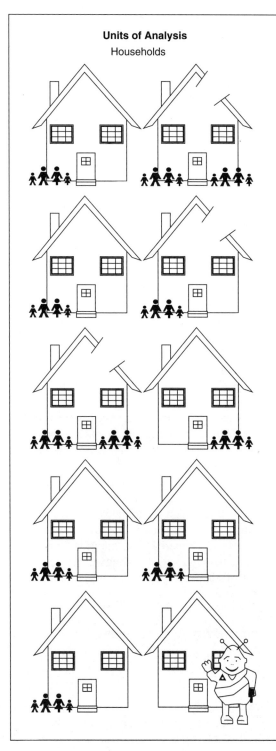

Units of Analysis
Households

Sample Statements

20% of the households are occupied by more than one family

30% of the households have holes in their roofs

10% of the households are occupied by aliens

Notice also that 33% of the families live in multiple-family households with family as the unit of analysis

FIGURE 4-4 **Illustrations of Units of Analysis (continued)**

Social Artifacts

Another unit of analysis is the **social artifact,** or any product of social beings or their behavior. One class of artifacts includes concrete objects such as books, poems, paintings, automobiles, buildings, songs, pottery, jokes, student excuses for missing exams, and scientific discoveries.

For example, Lenore Weitzman and her associates (1972) were interested in learning how gender roles are taught. They chose children's picture books as their unit of analysis. Specifically they examined books that had received the Caldecott Medal. Their results were as follows:

> We found that females were underrepresented in the titles, central roles, pictures, and stories of every sample of books we examined. Most children's books are about boys, men, male animals, and deal exclusively with male adventures. Most pictures show men singly or in groups. Even when women can be found in the books, they often play insignificant roles, remaining both inconspicuous and nameless.
> — (WEITZMAN ET AL. 1972:1128).

In a more recent study, Roger Clark, Rachel Lennon, and Leana Morris (1993) concluded that male and female characters are now portrayed less stereotypically than before, observing a clear progress toward portraying men and women in nontraditional roles. However, they did not find total equality between the sexes.

As this example suggests, just as people or social groups imply populations, each social object implies a set of all objects of the same class: all books, all novels, all biographies, all introductory sociology textbooks, all cookbooks, all press conferences. In a study using books as the units of analysis, an individual book might be characterized by its size, weight, length, price, content, number of pictures, number sold, or description of its author. Then the population of all books or of a particular kind of book could be analyzed for the purpose of description or explanation: what kinds of books sell best and why, for example.

Similarly, a social researcher could analyze whether paintings by Russian, Chinese, or U.S. artists showed the greatest degree of working-class consciousness, taking paintings as the units of analysis and describing each, in part, by the nationality of its creator. Or you might examine a newspaper's editorials regarding a local university for the purpose of describing, or perhaps explaining, changes in the newspaper's editorial position on the university over time. In this example, individual editorials would be the units of analysis.

Social interactions form another class of social artifacts suitable for social research. For example, we might characterize weddings as racially or religiously mixed or not, as religious or secular in ceremony, as resulting in divorce or not, or by descriptions of one or both of the marriage partners (such as, "previously married," "Oakland Raider fan," "wanted by the FBI"). When a researcher reports that weddings between partners of different religions are more likely to be performed by secular authorities than those between partners of the same religion, the weddings are the units of analysis, not the individuals involved.

Other social interactions that might be units of analysis are friendship choices, court cases, traffic accidents, divorces, fistfights, ship launchings, airline hijackings, race riots, final exams, student demonstrations, and congressional hearings. Congressional hearings, for instance, could be characterized by whether or not they occurred during an election campaign, whether the committee chairs were running for a higher office, whether they had received campaign contributions from interested parties, and so on. Notice that even if we characterized and compared the hearings in terms of the committee chairs, the hearings themselves—not the individual chairpersons—would be our units of analysis.

Units of Analysis in Review

The examples in this section should suggest the nearly infinite variety of possible units of analysis in social research. Although individual human beings are typical objects of study, many research questions can be answered more appropriately through the examination of other units of analysis. Indeed, social researchers can study just about anything that bears on social life.

Moreover, the types of units of analysis named in this section don't begin to exhaust the possibilities. Morris Rosenberg (1968:234–48), for example, speaks of individual, group, organizational, institutional, spatial, cultural, and societal units of analysis. John and Lyn Lofland (1995:103–13) speak of practices, episodes, encounters, roles, relationships, groups, organizations, settlements, social worlds, lifestyles, and subcultures as suitable units of study. The important thing here is to grasp the logic of units of analysis. Once you do, the possibilities for fruitful research are limited only by your imagination.

Categorizing possible units of analysis may make the concept seem more complicated than it needs to be. What you call a given unit of analysis—a group, a formal organization, or a social artifact—is irrelevant. The key is to be clear about what your unit of analysis is. When you embark on a research project, you must decide whether you're studying marriages or marriage partners, crimes or criminals, corporations or corporate executives. Otherwise, you run the risk of drawing invalid conclusions because your assertions about one unit of analysis are actually based on the examination of another. We'll see an example of this issue as we look at the ecological fallacy in the next section.

Faulty Reasoning about Units of Analysis: The Ecological Fallacy and Reductionism

At this point, it's appropriate to introduce two types of faulty reasoning that you should be aware of: the ecological fallacy and reductionism. Each represents a potential pitfall regarding units of analysis, and either of which can occur in doing research and drawing conclusions from the results.

The Ecological Fallacy   In this context, "ecological" refers to groups or sets or systems: something larger than individuals. The ecological fallacy is the assumption that something learned about an ecological unit says something about the individuals making up that unit. Let's consider a hypothetical illustration of this fallacy.

Suppose we're interested in learning something about the nature of electoral support received by a female political candidate in a recent citywide election. Let's assume we have the vote tally for each precinct so we can tell which precincts gave her the greatest support and which the least. Assume also that we have census data describing some characteristics of these precincts. Our analysis of such data might show that precincts with relatively young voters gave the female candidate a greater proportion of their votes than did precincts with older voters. We might be tempted to conclude from these findings that young voters are more likely to vote for female candidates than are older voters—in other words, that age affects support for the woman. In reaching such a conclusion, we run the risk of committing the ecological fallacy because it may have been the older voters in those "young" precincts who voted for the woman. Our problem is that we have examined *precincts* as our units of analysis but wish to draw conclusions about *voters*.

The same problem would arise if we discovered that crime rates were higher in cities having large African-American populations than in those with few African Americans. We would not know if the crimes were actually committed by African Americans. Or if we found suicide rates higher in Protestant countries than in Catholic ones, we still could not know for sure that more Protestants than Catholics committed suicide.

In spite of these hazards, social researchers very often have little choice but to address a particular research question through an ecological analysis. Perhaps the most appropriate data are simply not available. For example, the precinct vote tallies and the precinct characteristics mentioned in our initial example might be easy to obtain, but we may not have the resources to conduct a postelection survey of individual voters. In such cases, we may reach a tentative conclusion, recognizing and noting the risk of an ecological fallacy.

While you should be careful not to commit the ecological fallacy, don't let these warnings lead you into committing what we might call the individualistic fallacy. Some people who approach social research for the first time have trouble reconciling general patterns of attitudes and actions with individual exceptions. But generalizations and probabilistic statements are not invalidated by individual exceptions. Your knowing a rich Democrat, for example, doesn't deny the fact that most rich people vote Republican—as a general pattern. Similarly, if you know someone who has gotten rich without any formal education, that doesn't deny the general pattern of higher education relating to higher income.

The ecological fallacy deals with something else altogether—confusing units of analysis in such a way that we draw conclusions about individuals based solely on the observation of groups. Although the patterns observed between variables at the level of groups may be genuine, the danger lies in reasoning from the observed attributes of groups to the attributes of the individuals who made up those groups when we have not actually observed individuals.

Reductionism   A second type of potentially faulty reasoning related to units of analysis is reductionism. Reductionism means seeing and explaining complex phenomena in terms of a single, narrow concept or set of concepts. Thus, we "reduce" to a simple explanation what in reality is complex.

For instance, scientists from different disciplines tend to look at different types of answers and ignore the others. Sociologists tend to consider only sociological variables (such as values, norms, and roles), economists only economic variables (such as supply and demand, marginal value), and psychologists only psychological variables (such as personality types and traumas). Explaining all or most human behavior in terms of economic factors is called economic reductionism; explaining all or most human behavior in terms of psychological factors is called psychological reductionism; and so forth. Notice how this issue relates to the discussion of theoretical paradigms in Chapter 2.

In another example, suppose we ask what caused the American Revolution. Was it a shared commitment to the value of individual liberty? The economic plight of the colonies in relation to Britain? The megalomania of the founders? As soon as we inquire about *the* single cause, we run the risk of reductionism.

Reductionism of any type tends to suggest that particular units of analysis or variables are more relevant than others. If we were to regard shared values as the cause of the American Revolution, our unit of analysis would be the individual colonist. An economist, though, might choose the 13 colonies as units of analysis and examine the economic organizations and conditions of each. A psychologist might choose individual leaders as the units of analysis for purposes of examining their personalities.

Like the ecological fallacy, reductionism can occur when we use inappropriate units of analysis. The appropriate unit of analysis for a given research question, however, is not always clear. Social researchers, especially across disciplinary boundaries, often debate this issue.

THE TIME DIMENSION

So far in this chapter, we have regarded research design as a process for deciding what aspects we shall observe, of whom, and for what purpose. Now we must consider a set of time-related options that cuts across each of these earlier considerations. We can choose to make observations more or less at one time or over a long period.

Time plays many roles in the design and execution of research, quite aside from the time it takes to do research. Earlier we noted that the time sequence of events and situations is critical to determining causation (a point we'll return to in Part 4). Time also affects the generalizability of research findings. Do the descriptions and explanations resulting from a particular study accurately represent the situation of ten years ago, ten years from now, or only the present? Researchers have two principal options available to deal with the issue of time in the design of their research: cross-sectional studies and longitudinal studies.

Cross-Sectional Studies

A **cross-sectional study** involves observations of a sample, or cross section, of a population or phenomenon that are made at one point in time. Exploratory and descriptive studies are often cross-sectional. A single U.S. Census, for instance, is a study aimed at describing the U.S. population at a given time.

Many explanatory studies are also cross-sectional. A researcher conducting a large-scale national survey to examine the sources of racial and religious prejudice would, in all likelihood, be dealing with a single time frame—taking a snapshot, so to speak, of the sources of prejudice at a particular point in history.

Explanatory cross-sectional studies have an inherent problem. Although their conclusions are based on observations made at only one time, typically they aim at understanding causal processes that occur over time. This problem is somewhat akin to that of determining the speed of a moving object on the basis of a high-speed, still photograph that freezes the movement of the object.

Yanjie Bian, for example, conducted a survey of workers in Tianjin, China, for the purpose of studying stratification in contemporary, urban Chinese society. In undertaking the survey in 1988, however, he was conscious of the important changes brought about by a series of national campaigns, such as the Great Proletarian Cultural Revolution, dating from the Chinese Revolution in 1949 (which brought the Chinese Communists into power) and continuing into the present.

> These campaigns altered political atmospheres and affected people's work and nonwork activities. Because of these campaigns, it is difficult to draw conclusions from a cross-sectional social survey, such as the one presented in this book, about general patterns of Chinese workplaces and their effects on workers. Such conclusions may be limited to one period of time and are subject to further tests based on data collected at other times. — (1994:19)

The problem of generalizations about social life from a "snapshot" is one this book repeatedly addresses. One solution is suggested by Bian's final comment—about data collected "at other times": Social research often involves revisiting phenomena and building on the results of earlier research.

Longitudinal Studies

In contrast to cross-sectional studies, a **longitudinal study** is designed to permit observations of the same phenomenon over an extended period.

For example, a researcher can participate in and observe the activities of a UFO cult from its inception to its demise. Other longitudinal studies use records or artifacts to study changes over time. In analyses of newspaper editorials or Supreme Court decisions over time, for example, the studies are longitudinal whether the researcher's actual observations and analyses were made at one time or over the course of the actual events under study.

Many field research projects, involving direct observation and perhaps in-depth interviews, are naturally longitudinal. Thus, for example, when Ramona Asher and Gary Fine (1991) studied the life experiences of the wives of alcoholic men, they were in a position to examine the evolution of their troubled marital relationships over time, sometimes even including the reactions of the subjects to the research itself.

In the classic study *When Prophecy Fails* (1956), Leon Festinger, Henry Reicker, and Stanley Schachter were specifically interested in learning what happened to a flying saucer cult when their predictions of an alien encounter failed to come true. Would the cult members close down the group or would they become all the more committed to their beliefs? A longitudinal study was required to provide an answer. (They redoubled their efforts to get new members.)

Longitudinal studies can be more difficult for quantitative studies such as large-scale surveys. Nonetheless, they are often the best way to study changes over time. There are three special types of longitudinal studies that you should know about: trend studies, cohort studies, and panel studies.

Trend Studies A **trend study** is a type of longitudinal study that examines changes within a population over time. A simple example is a comparison of U.S. Censuses over a period of decades, showing shifts in the makeup of the national population. A similar use of archival data was made by Michael Carpini and Scott Keeter (1991), who wanted to know whether contemporary U.S. citizens were better or more poorly informed about politics than were citizens of an earlier generation. To find out, they compared the results of several Gallup Polls conducted during the 1940s and 1950s with a 1989 survey that asked several of the same questions tapping political knowledge.

Overall, the analysis suggested that contemporary citizens were slightly better informed than were earlier generations. In 1989, 74 percent of the sample could name the vice president of the United States, compared with 67 percent in 1952. Substantially higher percentages could explain presidential vetoes and congressional overrides of vetoes than could people in 1947. On the other hand, more of the 1947 sample could identify their U.S. representative (38 percent) than could the 1989 sample (29 percent).

An in-depth analysis, however, indicates that the slight increase in political knowledge resulted from the fact that the people in the 1989 sample were more highly educated than were those from earlier samples. When educational levels were taken into account, the researchers concluded that political knowledge has actually declined within specific educational groups.

Cohort Studies In a **cohort study,** a researcher examines specific subpopulations, or cohorts, as they change over time. Typically, a cohort is an age group, such as those people born during the 1950s, but it can also be some other time grouping, such as people born during the Vietnam War, people who got married in 1994, and so forth. An example of a cohort study would be a series of national surveys, conducted perhaps every 20 years, to study the attitudes of the cohort born during World War II toward U.S. involvement in global affairs. A sample of people 15–20 years of age might be surveyed in 1960, another sample of those 35–40 years of age in 1980, and another sample of those 55–60 years of age in 2000. Although the specific set of people studied in each survey would differ, each sample would represent the cohort born between 1940 and 1945.

James Davis (1992) turned to a cohort analysis in an attempt to understand shifting political orientations during the 1970s and 1980s in the United States. Overall, he found a liberal trend on issues such as race, gender, religion, politics, crime, and free speech. But did this trend represent people in general getting a bit more liberal, or did it merely

TABLE 4-1 **Age and Political Liberalism**

Survey Dates	1972 to 1974	1977 to 1980	1982 to 1984	1987 to 1989
Age of Cohort	20–24	25–29	30–34	35–39
Percent Who Would Let the Communist Speak	72%	68%	73%	73%

reflect more liberal younger generations replacing the conservative older ones?

To answer this question, Davis examined national surveys conducted in four time periods, five years apart. In each survey, he grouped the respondents into age groups, also five years apart. This strategy allowed him to compare different age groups at any given point in time as well as follow the political development of each age group over time.

One of the questions he examined was whether a person who admitted to being a Communist should be allowed to speak in the respondents' communities. Consistently, the younger respondents in each period of time were more willing to let the Communist speak than were the older ones. Among those aged 20–40 in the first set of the survey, for example, 72 percent took this liberal position, contrasted with 27 percent among respondents 80 and older. What Davis found when he examined the youngest cohort over time is shown in Table 4-1.

This pattern of a slight, conservative shift in the 1970s, followed by a liberal rebound in the 1980s, typifies the several cohorts Davis analyzed (J. Davis 1992:269).

Panel Studies Though similar to trend and cohort studies, a **panel study** examines the same set of people each time. For example, we could interview the same sample of voters every month during an election campaign, asking for whom they intended to vote. Though such a study would allow us to analyze overall trends in voter preferences for different candidates, it would also show the precise patterns of persistence and change in intentions.

For example, a trend study that showed that Candidates A and B each had exactly half of the voters on September 1 and on October 1 as well could indicate that none of the electorate had changed voting plans, that all of the voters had changed their intentions, or something in between. A panel study would eliminate this confusion by showing what kinds of voters switched from A to B and what kinds switched from B to A, as well as other facts.

Joseph Veroff, Shirley Hatchett, and Elizabeth Douvan (1992) wanted to learn about marital adjustment among newlyweds, looking for differences between white and African-American couples. To get subjects for study, they selected a sample of couples who applied for marriage licenses in Wayne County, Michigan, April through June 1986.

Concerned about the possible impact their research might have on the couples' marital adjustment, the researchers divided their sample in half at random: an *experimental* group and a *control* group (concepts we'll explore further in Chapter 8). Couples in the former group were intensively interviewed over a four-year period, whereas the latter group was contacted only briefly each year.

By studying the same couples over time, the researchers could follow the specific problems that arose and the way the couples dealt with them. As a by-product of their research, they found that those studied the most intensely seemed to achieve a somewhat better marital adjustment. The researchers felt that the interviews may have forced couples to discuss matters they may have otherwise buried.

Comparing the Three Types of Longitudinal Studies To reinforce the distinctions among trend, cohort, and panel studies, let's contrast the three study designs in terms of the same variable: political party affiliation. A trend study might look at shifts in U.S. religious affiliations over time, as the Gallup Poll does on a regular basis. A cohort study might follow shifts in religious affiliations among "the Depression generation," specifically, say, people who were between 20 and 30 in 1932. We could study a sample of people 30–40 years old

in 1942, a new sample of people aged 40–50 in 1952, and so forth. A panel study could start with a sample of the whole population or of some special subset and study those specific individuals over time. Notice that only the panel study would give a full picture of the shifts among the various categories of affiliations, including "none." Cohort and trend studies would uncover only net changes.

Longitudinal studies have an obvious advantage over cross-sectional ones in providing information describing processes over time. But this advantage often comes at a heavy cost in both time and money, especially in a large-scale survey. Observations may have to be made at the time events are occurring, and the method of observation may require many research workers.

Panel studies, which offer the most comprehensive data on changes over time, face a special problem: panel attrition. Some of the respondents studied in the first wave of the survey may not participate in later waves. (This is comparable to the problem of experimental mortality discussed in Chapter 8.) The danger is that those who drop out of the study may not be typical, thereby distorting the results of the study. Thus, when Carol S. Aneshensel and colleagues conducted a panel study of adolescent girls (comparing Latinas and non-Latinas), they looked for and found differences in characteristics of survey dropouts among Latinas born in the United States and those born in Mexico. These differences needed to be taken into account to avoid misleading conclusions about differences between Latinas and non-Latinas (Aneshensel et al. 1989).

Approximating Longitudinal Studies

Longitudinal studies do not always provide a feasible or practical means of studying processes that take place over time. Fortunately, researchers often can draw approximate conclusions about such processes even when only cross-sectional data are available. Here are some ways to do that.

Sometimes cross-sectional data imply processes over time on the basis of simple logic. For example, in the study of student drug use conducted at the University of Hawaii that I mentioned in Chapter 2, students were asked to report whether they had ever tried each of several illegal drugs. The study found that some students had tried both marijuana and LSD, some had tried only one, and others had tried neither. Because these data were collected at one time, and because some students presumably would experiment with drugs later on, it would appear that such a study could not tell whether students were more likely to try marijuana or LSD first.

A closer examination of the data showed, however, that although some students reported having tried marijuana but not LSD, there were no students in the study who had tried only LSD. From this finding it was inferred—as common sense suggested—that marijuana use preceded LSD use. If the process of drug experimentation occurred in the opposite time order, then a study at a given time should have found some students who had tried LSD but not marijuana, and it should have found no students who had tried only marijuana.

Researchers can also make logical inferences whenever the time order of variables is clear. If we discovered in a cross-sectional study of college students that those educated in private high schools received better college grades than did those educated in public high schools, we would conclude that the type of high school attended affected college grades, not the other way around. Thus, even though our observations were made at only one time, we would feel justified in drawing conclusions about processes taking place across time.

Very often, age differences discovered in a cross-sectional study form the basis for inferring processes across time. Suppose you're interested in the pattern of worsening health over the course of the typical life cycle. You might study the results of annual checkups in a large hospital. You could group health records according to the ages of those examined and rate each age group in terms of several health conditions—sight, hearing, blood pressure, and so forth. By reading across the age-group ratings for each health condition, you would have something approximating the health history of individuals. Thus, you might conclude that the average person develops vision problems before hearing problems. You would need to be cautious in this

assumption, however, since the differences might reflect societywide trends. Perhaps improved hearing examinations instituted in the schools had affected only the young people in your study.

Asking people to recall their pasts is another common way of approximating observations over time. Researchers use that method when they ask people where they were born or when they graduated from high school or whom they voted for in 1988. Qualitative researchers often conduct in-depth "life history" interviews. For example, C. Lynn Carr (1998) used this technique in a study of "tomboyism." Her respondents, aged 25 to 40, were asked to reconstruct aspects of their lives from childhood on, including experiences of identifying themselves as tomboys.

The danger in this technique is evident. Sometimes people have faulty memories; sometimes they lie. When people are asked in postelection polls whom they voted for, the results inevitably show more people voting for the winner than actually did so on election day. As part of a series of in-depth interviews, such a report can be validated in the context of other reported details; however, results based on a single question in a survey must be regarded with caution.

This discussion of the ways that time figures into social research suggest several questions you should confront in your own research projects. In designing any study, be sure to look at both the explicit and the implicit assumptions you're making about time. Are you interested in describing some process that occurs over time, or are you simply going to describe what exists now? If you want to describe a process occurring over time, will you be able to make observations at different points in the process, or will you have to approximate such observations by drawing logical inferences from what you can observe now? If you opt for a longitudinal design, which method best serves your research purposes?

Examples of Research Strategies

As the preceding discussions have implied, social research follows many paths. The following short excerpts further illustrate this point. As you read each excerpt, take note of both the content of each study and the method used to study the chosen topic. Does the study seem to be exploring, describing, or explaining (or some combination of these)? What are the sources of data in each study? Can you identify the unit of analysis? Is the dimension of time relevant? If so, how will it be handled?

- This case study of unobtrusive mobilizing by Southern California Rape Crisis Center uses archival, observational, and interview data to explore how a feminist organization worked to change police, schools, prosecutors, and some state and national organizations from 1974 to 1994. (Schmitt and Martin 1999:364)
- Using life history narratives, the present study investigates processes of agency and consciousness among 14 women who identified themselves as tomboys. (Carr 1998:528)
- By drawing on interviews with activists in the former Estonian Soviet Socialist Republic, we specify the conditions by which accommodative and oppositional subcultures exist and are successfully transformed into social movements. (Johnston and Snow 1998:473)
- This paper presents the results of an ethnographic study of an AIDS service organization located in a small city. It is based on a combination of participant observation, interviews with participants, and review of organizational records. (Kilburn 1998:89)
- Using interviews obtained during fieldwork in Palestine in 1992, 1993, and 1994, and employing historical and archival records, I argue that Palestinian feminist discourses were shaped and influenced by the sociopolitical context in which Palestinian women acted and with which they interacted. (Abdulhadi 1998:649)
- This article reports on women's experiences of breastfeeding in public as revealed through in-depth interviews with 51 women. (Stearns 1999:308)
- Using interview and observational field data, I demonstrate how a system of temporary employment in a participative workplace both exploited and shaped entry-level

workers' aspirations and occupational goals. (V. Smith 1998:411)
- I collected data [on White Separatist Rhetoric] from several media of public discourse, including periodicals, books, pamphlets, transcripts from radio and television talk shows, and newspaper and magazine accounts. (Berbrier 1998:435)
- In the analysis that follows, racial and gender inequality in employment and retirement will be analyzed, using a national sample of persons who began receiving Social Security Old Age benefits in 1980–81. (Hogan and Perrucci 1998:528)
- Drawing from interviews with female crack dealers, this paper explores the techniques they use to avoid arrest. (Jacobs and Miller 1998:550)

HOW TO DESIGN A RESEARCH PROJECT

You've now seen some of the options available to social researchers in designing projects. I know there are a lot of pieces, and the relationships among them may not be totally clear, so here's a way of pulling the parts together. Let's assume you were to undertake research. Where would you start? Then, where would you go?

Although research design occurs at the beginning of a research project, it involves all the steps of the subsequent project. This discussion, then, provides both guidance on how to start a research project and an overview of the topics that follow in later chapters of this book.

Figure 4-5 presents a schematic view of the social research process. I present this view reluctantly, because it may suggest more of a step-by-step order to research than actual practice bears out. Nonetheless, this idealized overview of the process provides a context for the specific details of particular components of social research. Essentially, it is another and more detailed picture of the scientific process presented in Chapter 2.

At the top of the diagram are interests, ideas, and theories, the possible beginning points for a

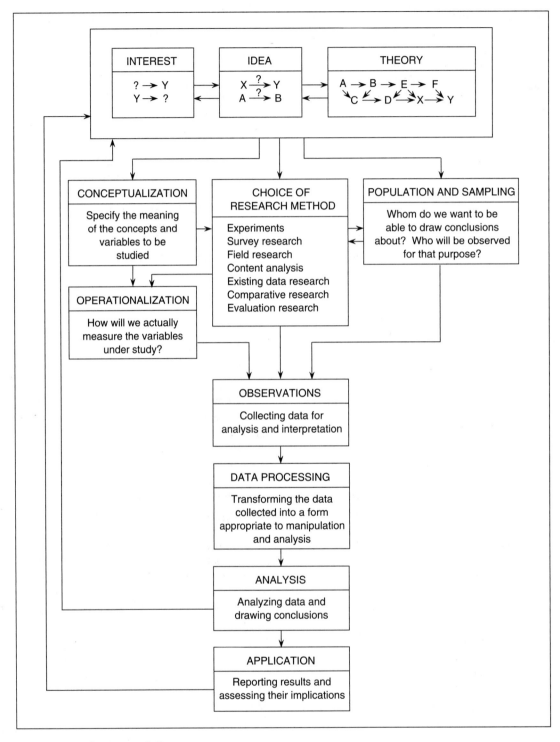

FIGURE 4-5 **The Research Process**

line of research. The letters (A, B, X, Y, and so forth) represent variables or concepts such as prejudice or alienation. Thus, you might have a general interest in finding out what causes some people to be more prejudiced than others, or you might want to know some of the consequences of alienation. Alternatively, your inquiry might begin with a specific idea about the way things are. For example, you might have the idea that working on an assembly line causes alienation. The question marks in the diagram indicate that you aren't sure things are the way you suspect they are—that's why you're doing the research. Notice that a theory is represented as a set of complex relationships among several variables.

The double arrows between "interest," "idea," and "theory" are meant to suggest that there is often a movement back and forth across these several possible beginnings. An initial interest may lead to the formulation of an idea, which may be fit into a larger theory, and the theory may produce new ideas and create new interests.

Any or all of these three may suggest the need for empirical research. The purpose of such research can be to explore an interest, test a specific idea, or validate a complex theory. Whatever the purpose, the researcher needs to make a variety of decisions, as indicated in the remainder of the diagram.

To make this discussion more concrete, let's take a specific research example. Suppose you're concerned with the issue of abortion and have a special interest in learning why some college students support abortion rights and others oppose them. Going a step further, let's say you've formed the impression that students in the humanities and social sciences seem generally more inclined to support the idea of abortion rights than do those in the natural sciences. (That kind of thinking often leads people to design and conduct social research.)

In terms of the options we've discussed in this chapter, you probably have both descriptive and explanatory interests: What percentage of the student body supports a woman's right to an abortion (descriptive), and what causes some to support it and others to oppose it (explanation)? The units of analysis are individuals: college students. You might decide that a cross-sectional study would suit your purposes. Let's assume you'd be satisfied to learn something about the way things are now. You might then decide that a cross-sectional study would suit your purposes. Although this would provide you with no direct evidence of processes taking place over time, you might be able to approximate some longitudinal analyses if you pursue changes in students' attitudes over time.

Getting Started

At the outset of your project, your interests would probably be exploratory. At this point, you might choose among several possible activities in pursuing your interest in student attitudes about abortion rights. To begin with, you might want to read something about the issue. If you have a hunch that attitudes are somehow related to college major, you might find out what other researchers may have written about that. Appendix A of this book will help you make use of your college library. In addition, you would probably talk to some people who support abortion rights and some who don't. You might attend meetings of abortion-related groups. All these activities could help prepare you to handle the various decisions of research design we are about to examine.

Before designing your study, you must define the purpose of your project. What kind of study will you undertake—exploratory, descriptive, explanatory? Do you plan to write a research paper to satisfy a course or thesis requirement? Is your purpose to gain information that will support you in arguing for or against abortion rights? Do you want to write an article for the campus newspaper or an academic journal? In reviewing the previous research literature regarding abortion rights, you should note the design decisions other researchers have made, always asking whether the same decisions would satisfy your purpose.

Usually, your purpose for undertaking research can be expressed as a report. A good first step in designing your project is to outline such a report (see Appendix B for help). Although your final report may not look much like your initial image of it, this exercise will help you figure out which research designs are most appropriate. During this step, clearly describe the kinds of statements you want to make when the research is complete. Here are some examples of such statements: "Students frequently mentioned abortion rights in the context of discussing social issues that concerned them personally." "X percent of State U. students favor a woman's right to choose an abortion." "Engineers are (more/less) likely than sociologists to favor abortion rights."

Conceptualization

Once you have a well-defined purpose and a clear description of the kinds of outcomes you want to achieve, you can proceed to the next step in the design of your study—conceptualization. We often talk pretty casually about social science concepts such as prejudice, alienation, religiosity, and liberalism, but it's necessary to clarify what we mean by these concepts in order to draw meaningful conclusions about them. Chapter 5 examines this process of conceptualization in depth. For now, let's see what it might involve in the case of our hypothetical example.

If you're going to study how college students feel about abortion and why, the first thing you'll have to specify is what you mean by "the right to an abortion." Since support for abortion probably varies according to the circumstances, you'll want to pay attention to the different conditions under which people might approve or disapprove of abortion: for example, when the woman's life is in danger, in the case of rape or incest, or simply as a matter of personal choice.

Similarly, you'll need to specify exact meanings for all the other concepts you plan to study. If you want to study the relationship of opinion about abortion to college major, you'll have to decide whether you want to consider only officially declared majors or to include students' intentions as well. What will you do with those who have no major?

In surveys and experiments, it's necessary to specify such concepts in advance. In less tightly structured research, such as open-ended interviews, an important part of the research may involve the discovery of different dimensions, aspects, or nuances of concepts. In such cases, the research itself may uncover and report aspects of social life that were not evident at the outset of the project.

Choice of Research Method

As we'll discuss in Part 3, each research method has its strengths and weaknesses, and certain concepts are more appropriately studied by some methods than by others. In our study of attitudes toward abortion rights, a survey might be the most appropriate method: either interviewing students or asking them to fill out a questionnaire. Surveys are particularly well suited to the study of public opinion. This is not to say that you couldn't make good use of the other methods presented in Part 3. For example, you might use the method of content analysis to examine letters to the editor and analyze the different images letter writers have of abortion. Field research would provide an avenue to understanding how people interact with one another regarding the issue of abortion, how they discuss it, and how they change their minds. Other research methods introduced in Part 3 could also be used in studying this topic. Usually, the best study design uses more than one research method, taking advantage of their different strengths. If you look back at the brief examples of actual studies at the end of the preceding section, you'll see several instances where the researchers used many methods in a single study.

Operationalization

Once you've specified the concepts to be studied and chosen a research method, the next step is operationalization, or deciding on your measurement techniques (discussed further in Chapters 5 and 6). The meaning of variables in a study is determined in part by how they are measured. Part of the task here is deciding how the desired data will be collected: direct observation, review of official documents, a questionnaire, or some other technique.

If you decided to use a survey to study attitudes toward abortion rights, part of operationalization is determining the wording of questionnaire items. For example, you might operationalize your main variable by asking respondents whether they would approve a woman's right to have an abortion under each of the conditions you've conceptualized: in the case of rape or incest, if her life were threatened by the pregnancy, and so forth. You'd design the questionnaire so that it asked respondents to express approval or disapproval for each situation. Similarly, you would specify exactly how respondents would indicate their college major and what choices to provide those who have not declared a major.

Population and Sampling

In addition to refining concepts and measurements, you must decide whom or what to study. The *population* for a study is that group (usually of people) about whom we want to draw conclusions. We're almost never able to study all the members of the population that interests us, however, and we can never make every possible observation of them. In every case, then, we will select a *sample* from among the data that might be collected and studied. The sampling of information, of course, occurs in everyday life and often produces biased observations. (Recall the discussion of "selective observation" in Chapter 1). Social researchers are more deliberate in their sampling of what will be observed.

Chapter 7 describes methods for selecting samples that adequately reflect the whole population that interests us. Notice in Figure 4-5 that decisions about population and sampling are related to decisions about the research method to be used. Whereas probability sampling techniques would be relevant to a large-scale survey or a content analysis, a field researcher might need to select only those informants who will yield a balanced picture of the situation under study, and an experimenter might assign subjects to experimental and control groups in a manner that creates comparability.

In your hypothetical study of abortion attitudes, the relevant population would be the student population of your college. As you'll discover in Chapter 7, however, selecting a sample will require you to get more specific than that. Will you include part-time as well as full-time students? Only degree candidates or everyone? International students as well as U.S. citizens? Undergraduates, graduate students, or both? There are many such questions—each of which must be answered in terms of your research purpose. If your purpose is to predict how students would vote in a local referendum on abortion, you might want to limit your population to those eligible and likely to vote.

Observations

Having decided what to study among whom by what method, you're now ready to make observations—to collect empirical data. The chapters of Part 3, which describe the various research methods, give the different observation techniques appropriate to each.

To conduct a survey on abortion, you might want to print questionnaires and mail them to a sample selected from the student body. Alternatively, you could arrange to have a team of interviewers conduct the survey over the telephone. The relative advantages and disadvantages of these and other possibilities are discussed in Chapter 9.

Data Processing

Depending on the research method chosen, you'll have amassed a volume of observations in a form that probably isn't immediately interpretable. If you've spent a month observing a street-corner gang firsthand, you'll now have enough field notes to fill a book. In a historical study of ethnic diversity at your school, you may have amassed volumes of official documents, interviews with administrators and others, and so forth. Chapter 14 describes some of the ways social scientific data are processed or transformed for quantitative or qualitative analysis.

In the case of a survey, the "raw" observations are typically in the form of questionnaires with boxes checked, answers written in spaces, and the like. The data-processing phase for a survey typically involves the classification (coding) of written-in answers and the transfer of all information to a computer.

Analysis

Once the collected data are in a suitable form, you're ready to interpret them for the purpose of drawing conclusions that reflect the interests, ideas, and theories that initiated the inquiry. Chapters 13 and 14 describe a few of the many options available to you in analyzing data. In Figure 4-5, notice that the results of your analyses feed back into your initial interests, ideas, and theories. Often this feedback represents the beginning of another cycle of inquiry.

In the survey of student attitudes about abortion rights, the analysis phase would pursue both descriptive and explanatory aims. You might begin by calculating the percentages of students who favored or opposed each of the several different versions of abortion rights. Taken together, these several percentages would provide a good picture of student opinion on the issue.

Moving beyond simple description, you might describe the opinions of subsets of the student body, such as different college majors. Provided that your design called for trapping other information about respondents, you could also look at men versus women; frosh, sophomores, juniors, seniors, and graduate students; or other categories that you've included. The description of subgroups could then lead you into an explanatory analysis.

Application

The final stage of the research process involves the uses made of the research you've conducted and the conclusions you've reached. To start, you'll probably want to communicate your findings so that others will know what you've learned. It may be appropriate to prepare—and even publish—a written report. Perhaps you'll make oral presentations, such as papers delivered to professional and scientific meetings. Other students would also be interested in hearing what you've learned about them.

You may want to go beyond simply reporting what you've learned to discussing the implications of your findings. Do they say anything about actions that might be taken in support of policy goals? Both the proponents and the opponents of abortion rights would be interested.

Finally, be sure to consider what your research suggests in regard to further research on your subject. What mistakes should be corrected in future studies? What avenues—opened up slightly in your study—should be pursued further?

Research Design in Review

As this overview shows, research design involves a set of decisions regarding what topic is to be studied among what population with what research methods for what purpose. Although you'll want to consider many ways of studying a subject—and use your imagination as well as your knowledge of a variety of methods—research design is the process of focusing your perspective for the purposes of a particular study.

If you're doing a research project for one of your courses, many aspects of research design may be specified for you in advance, including the method (such as an experiment) or the topic (as in a course on a particular subject, such as prejudice). The following summary assumes that you're free to choose both your topic and your research strategy.

In designing a research project, you'll find it useful to begin by assessing three things: your interests, your abilities, and the available resources. Each of these considerations will suggest a large number of possible studies.

Simulate the beginning of a somewhat conventional research project: Ask yourself what you're interested in understanding. Surely you have several questions about social behavior and attitudes. Why are some people politically liberal and others politically conservative? Why are some people more religious than others? Why do people join militia groups? Do colleges and universities still discriminate against minority faculty members? Why would a woman stay in an abusive relationship? Spend some time thinking about the kinds of questions that interest and concern you.

Once you have a few questions you'd be interested in answering for yourself, think about the kind of information needed to answer them. What research units of analysis would provide the most relevant information: college students, voters, cities, or corporations? This question will probably be inseparable in your thoughts from the question of research topics. Then ask which aspects of the units of analysis would provide the information you need to answer your research question.

Once you have some ideas about the kind of information relevant to your purpose, ask yourself how you might go about getting that information. Are the relevant data likely to be already available somewhere (say, in a government publication), or would you have to collect them yourself? If you think you would have to collect them, how would you go about doing it? Would you need to survey a large number of people or interview a few people in depth? Could you learn what you need to know by attending meetings of certain groups? Could you glean the data you need from books in the library?

As you answer these questions, you'll find yourself well into the process of research design. Keep in mind your own research abilities and the resources available to you. There's little point in designing a perfect study that you can't actually carry out. You may want to try a research method you have not used before so you can learn from it, but be careful not to put yourself at too great a disadvantage.

Once you have a general idea of what you want to study and how, carefully review previous research in journals and books to see how other researchers have addressed the topic and what they have learned about it. Your review of the literature may lead you to revise your research design: Perhaps you'll decide to use a previous researcher's method or even replicate an earlier study. A standard procedure in the physical sciences, the independent replication of research projects is just as important in the social sciences, although social researchers tend to overlook that. Or, you might want to go beyond replication and study some aspect of the topic that you feel previous researchers have overlooked.

Here's another approach you might take. Suppose a topic has been studied previously using field research methods. Can you design an experiment that would test the findings those earlier research-

ers produced? Or, can you think of existing statistics that could be used to test their conclusions? Did a mass survey yield results that you'd like to explore in greater detail through on-the-spot observations and in-depth interviews? The use of several different research methods to test the same finding is sometimes called *triangulation,* and you should always keep it in mind as a valuable research strategy. Because each research method has particular strengths and weaknesses, there is always a danger that research findings will reflect, at least in part, the method of inquiry. In the best of all worlds, your own research design should bring more than one research method to bear on the topic.

THE RESEARCH PROPOSAL

Quite often, in the design of a research project, you'll have to lay out the details of your plan for someone else's review and/or approval. In the case of a course project, for example, your instructor might very well want to see a "proposal" before you set off to work. Later in your career, if you wanted to undertake a major project, you might need to obtain funding from a foundation or government agency, who would most definitely want a detailed proposal that describes how you would spend their money. You may respond to a Request for Proposals (RFP), which both public and private agencies often circulate in search of someone to do research for them.

This chapter concludes with a brief discussion of how you might prepare a research proposal. This will give you one more overview of the whole research process that the rest of this book details.

Elements of a Research Proposal

Although some funding agencies (or your instructor, for that matter) may have specific requirements for the elements or structure of a research proposal, here are some basic elements you should include.

Problem or Objective What exactly do you want to study? Why is it worth studying? Does the proposed study have practical significance? Does it contribute to the construction of social theories?

Literature Review What have others said about this topic? What theories address it and what do they say? What previous research exists? Are there consistent findings, or do past studies disagree? Are there flaws in the body of existing research that you think you can remedy?

Subjects for Study Whom or what will you study in order to collect data? Identify the subjects in general, theoretical terms; in specific, more concrete terms, identify who is available for study and how you'll reach them. Will it be appropriate to select a sample? If so, how will you do that? If there is any possibility that your research will affect those you study, how will you insure that the research does not harm them?

Measurement What are the key variables in your study? How will you define and measure them? Do your definitions and measurement methods duplicate or differ from those of previous research on this topic? If you have already developed your measurement device (a questionnaire, for example) or will be using something previously developed by others, it might be appropriate to include a copy in an appendix to your proposal.

Data Collection Methods How will you actually collect the data for your study? Will you conduct an experiment or a survey? Will you undertake field research or will you focus on the reanalysis of statistics already created by others? Perhaps you will use more than one method.

Analysis Indicate the kind of analysis you plan to conduct. Spell out the purpose and logic of your analysis. Are you interested in precise description? Do you intend to explain why things are the way they are? Do you plan to account for variations in some quality: for example, why some students are more liberal than others? What possible explanatory variables will your analysis consider,

---

A QUANDARY **REVISITED**

When the Provost and the student newspaper seemed to disagree over the extent of part-time faculty teaching, they used different units of analysis. The newspaper said 52 percent of the *faculty* were part-time; the provost said about 70 percent of the *credits* were taught by full-time faculty. The table below demonstrates how they could both be right, given that the typical full-time faculty member teaches three courses, or 9 credits, whereas the typical part-time faculty teaches one course for 3 credits. For simplicity, I've assumed that there are 100 faculty members.

Faculty Status	Number	Credits Taught by Each	Total Credits Taught
Full-time	48	9	432
Part-time	52	3	156
			Total = 588

In this hypothetical illustration, full-time faculty taught 432 of the 588 credits, or 73 percent. As you can see, it's important to be clear about what the unit of analysis is.

---

and how will you know if you've explained variations adequately?

Schedule It's often appropriate to provide a schedule for the various stages of research. Even if you don't do this for the proposal, do it for yourself. Unless you have a timeline for accomplishing the several stages of research and keeping track of how you're doing, you may end up in trouble.

Budget When you ask someone to cover the costs of your research, you need to provide a budget that specifies where the money will go. Large, expensive projects include budgetary categories such as personnel, equipment, supplies, telephones, and postage. Even for a project you'll pay for yourself, it's a good idea to spend some time anticipating expenses: office supplies, photocopying, computer disks, telephone calls, transportation, and so on.

As you can see, if you're interested in conducting a social research project, it's a good idea to prepare a research proposal for your own purposes, even if you aren't required to do so by your instructor or a funding agency. If you're going to invest your time and energy in such a project, you should do what you can to insure a return on that investment.

Now that you've had a broad overview of social research, let's move on to the remaining chapters in this book and learn exactly how to design and execute each specific step. If you've found a research topic that really interests you, you'll want to keep it in mind as you see how you might go about studying it.

Main Points

- The principal purposes of social research include exploration, description, and explanation. Research studies often combine more than one purpose.

- Exploration is the attempt to develop an initial, rough understanding of some phenomenon.

- Description is the precise measurement and reporting of the characteristics of some population or phenomenon under study.

- Explanation is the discovery and reporting of relationships among different aspects of the phenomenon under study. Whereas descriptive studies answer the question "What's so?" explanatory ones tend to answer the question "Why?"

- Both idiographic and nomothetic models of explanation rest on the idea of causation. The idiographic model aims at a complete understanding of a particular phenomenon, using all relevant causal factors. The nomothetic model aims at a general understanding—not necessarily complete—of a class of phenomena, using a small number of relevant causal factors.

- There are three basic criteria for establishing causation in nomothetic analyses: (1) The variables must be empirically associated, or correlated, (2) the causal variable must occur earlier in time than the variable it is said to affect, and (3) the observed effect cannot be explained as the effect of a different variable.

- Mere association, or correlation, does not in itself establish causation. A spurious causal relationship is an association that in reality is caused by one or more other variables.

- Units of analysis are the people or things whose characteristics social researchers observe, describe, and explain. Typically, the unit of analysis in social research is the individual person, but it may also be a social group, a formal organization, a social artifact, or some other phenomenon such as lifestyles or social interactions.

- The ecological fallacy involves conclusions drawn from the analysis of groups (e.g., corporations) that are then assumed to apply to individuals (e.g., the employees of corporations).

- Reductionism is the attempt to understand a complex phenomenon in terms of a narrow set of concepts, such as attempting to explain the American Revolution solely in terms of economics (or political idealism or psychology).

- Research into processes that occur over time presents social challenges that can be addressed through cross-sectional studies or longitudinal studies.

- Cross-sectional studies are based on observations made at one time. Although such studies are limited by this characteristic, researchers can sometimes make inferences about processes that occur over time.

❑ In longitudinal studies, observations are made at many times. Such observations may be made of samples drawn from general populations (trend studies), samples drawn from more specific subpopulations (cohort studies), or the same sample of people each time (panel studies).

❑ Research design starts with an initial interest, idea, or theoretical expectation and proceeds through a series of interrelated steps to narrow the focus of the study so that concepts, methods, and procedures are well defined. A good research plan accounts for all these steps in advance.

❑ At the outset, a researcher specifies the meaning of the concepts or variables to be studied (conceptualization), chooses a research method or methods (e.g., experiments versus surveys), and specifies the population to be studied and, if applicable, how it will be sampled.

❑ The researcher operationalizes the concepts to be studied by stating precisely how variables in the study will be measured. Research then proceeds through observation, processing the data, analysis, and application, such as reporting the results and assessing their implications.

❑ A research proposal provides a preview of why a study will be undertaken and how it will be conducted. A research project is often required to get permission or necessary resources. Even when not required, a proposal is a useful device for planning.

Key Terms

correlation
spurious relationship
units of analysis
social artifact
ecological fallacy
reductionism

cross-sectional study
longitudinal study
trend study
cohort study
panel study

Review Questions

1. One example in this chapter suggested that political orientations cause attitudes toward legalizing marijuana. Can you make an argument that the time-order is just the opposite of what was assumed?

2. Here are some examples of real research topics. For each one, can you name the unit of analysis? (The answers are at the end of this chapter.)

   a. Women watch TV more than men because they are likely to work fewer hours outside the home than men. . . . Black people watch an average of approximately three-quarters of an hour more television per day than white people. (Hughes 1980:290)

   b. Of the 130 incorporated U.S. cities with more than 100,000 inhabitants in 1960, 126 had at least two short-term nonproprietary general hospitals accredited by the American Hospital Association. (Turk 1980:317)

   c. The early TM [transcendental meditation] organizations were small and informal. The Los Angeles group, begun in June 1959, met at a member's house where, incidentally, Maharishi was living. (Johnston 1980:337)

   d. However, it appears that the nursing staffs exercise strong influence over . . . a decision to change the nursing care system. . . . Conversely, among those decisions dominated by the administration and the medical staffs . . . (Comstock 1980:77)

   e. Though 667,000 out of 2 million farmers in the United States are women, women historically have not been viewed as farmers, but rather, as the farmer's wife. (Votaw 1979:8)

   f. The analysis of community opposition to group homes for the mentally handicapped . . . indicates that deteriorating neighborhoods are most likely to organize in opposition, but that upper-middle class neighborhoods are most likely to enjoy private access to local officials. (Graham and Hogan 1990:513)

g. Some analysts during the 1960s predicted that the rise of economic ambition and political militancy among blacks would foster discontent with the "otherworldly" black mainline churches. (Ellison and Sherkat 1990:551)

h. This analysis explores whether propositions and empirical findings of contemporary theories of organizations directly apply to both private product-producing organizations (PPOs) and public human service organizations (PSOs). (Schiflett and Zey 1990:569)

i. This paper examines variations in job title structures across work roles. Analyzing 3,173 job titles in the California civil service system in 1985, we investigate how and why lines of work vary in the proliferation of job categories that differentiate ranks, functions, or particular organizational locations. (Strang and Baron 1990:479)

3. Review the logic of spuriousness. Can you think up an example where an observed relationship between two variables could actually be explained away by a third variable?

4. Make up a research example—different from those discussed in the text—that illustrates a researcher falling into the trap of the ecological fallacy. How would you modify the example to avoid this trap?

Additional Readings

Bart, Pauline, and Linda Frankel. 1986. *The Student Sociologist's Handbook.* Morristown, NJ: General Learning Press. A handy little reference book to help you get started on a research project. Written from the standpoint of a student term paper, this volume offers a particularly good guide to the periodical literature of the social sciences that's available in a good library.

Casley, D. J., and D. A. Lury. 1987. *Data Collection in Developing Countries.* Oxford: Clarendon Press. This book discusses the special problems of research in the developing world.

Cooper, Harris M. 1989. *Integrating Research: A Guide for Literature Reviews.* Newbury Park, CA: Sage. The author leads you through each step in the literature review process.

Hunt, Morton. 1985. *Profiles of Social Research: The Scientific Study of Human Interactions.* New York: Basic Books. An engaging and informative series of project biographies: James Coleman's study of segregated schools is presented, as well as several other major projects that illustrate the elements of social research in actual practice.

Iversen, Gudmund R. 1991. *Contextual Analysis.* Newbury Park, CA: Sage. Contextual analysis examines the impact of socioenvironmental factors on individual behavior. Durkheim's study of suicide offers a good example of this, identifying social contexts that affect the likelihood of self-destruction.

Maxwell, Joseph A. 1996. *Qualitative Research Design: An Interactive Approach.* Newbury Park, CA: Sage. Maxwell covers many of the same topics that this chapter does but with attention devoted specifically to qualitative research projects.

Menard, Scott. 1991. *Longitudinal Research.* Newbury Park, CA: Sage. Beginning by explaining why researchers conduct longitudinal research, the author goes on to detail a variety of study designs as well as suggestions for the analysis of longitudinal data.

Miller, Delbert. 1991. *Handbook of Research Design and Social Measurement.* Newbury Park, CA: Sage. A useful reference book for introducing or reviewing numerous issues involved in design and measurement. In addition, the book contains a wealth of practical information relating to foundations, journals, and professional associations.

Multimedia Resources

The Wadsworth Sociology Resource Center: Virtual Society
http://sociology.wadsworth.com/

Visit the companion Web site for the second edition of *The Basics of Social Research* to access a wide range of student resources. Begin by clicking on the Student Resources section of the book's Web site to access the following study tools:

- eBabbie Resource Center
- Planning a Research Project
- Doing Data Analysis
- Statistics Review

- Flash Cards
- Internet Links and Exercises
- InfoTrac College Edition: Exercises
- Quizzes

Visit the **eBabbie Resource Center** for an overview of each chapter and helpful online tutorials. Find information on budgeting and step-by-step examples of model research projects at **Planning a Research Project.** Learn how to use quantitative and qualitative data analysis programs at **Doing Data Analysis,** and brush up on your statistics at **Statistics Review.** You can also further your study by accessing **Internet Links and Exercises** related to chapter materials, **Flash Cards, Quizzes,** and many other learning tools.

InfoTrac College Edition
http://www.infotrac-college.com/
wadsworth/access.html
Access the latest news and journal articles with InfoTrac College Edition, an easy-to-use online database of reliable, full-length articles from hundreds of top academic journals. Conduct an electronic search using the following search terms:

Cross-sectional study
Ecological fallacy
Longitudinal study
Panel study
Reductionism
Trend study
Unit of analysis

Answers to Units of Analysis Quiz, Review Question #2

a. Men and women, black and white people (individuals)

b. Incorporated U.S. cities (groups)

c. Transcendental meditation organizations (groups)

d. Nursing staffs (groups)

e. Farmers (individuals)

f. Neighborhoods (groups)

g. Blacks (individuals)

h. Service and production organizations (formal organizations)

i. Job titles (artifacts)

5 CONCEPTUALIZATION, OPERATIONALIZATION, AND MEASUREMENT

David M. Grossman/Photo Researchers, Inc.

What You'll Learn in This Chapter

We'll see how the interrelated steps of conceptualization, operationalization, and measurement allow researchers to turn a general idea for a research topic into useful and valid measurements in the real world. We'll also see that an essential part of this process involves transforming the relatively vague terms of ordinary language into precise objects of study will well-defined and measurable meanings.

In this chapter . . .

AN OPENING **QUANDARY**

People sometimes doubt the social researcher's ability to measure and study things that matter. For many people, for example, religious faith is a deep and important part of life. Yet both the religious and the nonreligious might question the social researcher's ability to measure how religious a given person or group is and—even more difficult—why some people are religious and others not.

This chapter will show that social researchers can't say definitively who is religious and who is not, nor what percentage of people in a particular population are religious, but they can to an extent determine the causes of religiosity. How can that be?

INTRODUCTION

This chapter and the next deal with how researchers move from a general idea about what they want to study to effective and well-defined measurements in the real world. This chapter discusses the interrelated processes of conceptualization, operationalization, and measurement. Chapter 6 builds on this foundation to discuss types of measurements that are more complex.

We begin this chapter by confronting the hidden concern people sometimes have about whether it's truly possible to measure the stuff of life: love, hate, prejudice, religiosity, radicalism, alienation. Over the course of the next few pages, we'll see that researchers can measure *anything that exists*. Once that point has been established, we'll turn to the steps involved in actual measurement.

MEASURING ANYTHING THAT EXISTS

Earlier in this book, I said that one of the two pillars of science is observation. Because this word can suggest a casual, passive activity, scientists often use the term *measurement* instead, meaning careful, deliberate observations of the real world for the purpose of describing objects and events in terms of the attributes composing a variable.

Like the people in the Opening Quandary, you may have some reservations about the ability of science to measure the really important aspects of human social existence. If you've read research reports dealing with something like liberalism or religion or prejudice, you may have been dissatisfied with the way the researchers measured whatever they were studying. You may have felt that they were too superficial, missing the aspects that really matter most. Maybe they measured religiosity as the number of times a person went to church, or liberalism by how people voted in a single election. Your dissatisfaction would surely have increased if you had found yourself being misclassified by the measurement system.

Your dissatisfaction reflects an important fact about social research: Most of the variables we want to study don't actually exist in the way that rocks exist. Indeed, they are made up. Moreover, they seldom have a single, unambiguous meaning.

To see what I mean, suppose we want to study *political party affiliation*. To measure this variable, we might consult the list of registered voters to note whether the people we were studying were registered as Democrats or Republicans and take that as a measure of their party affiliation. But we could also simply ask someone what party they identify with and take their response as our measure. Notice that these two different measurement possibilities reflect somewhat different definitions of "political party affiliation." They might even produce different results: Someone may have registered as a Democrat years ago but gravitated more and more toward a Republican philosophy over time. Or someone who is registered with neither political party may, when asked, say she is affiliated with the one she feels the most kinship with.

Similar points apply to *religious affiliation*. Sometimes this variable refers to official membership in a particular church; other times it simply means whatever religion, if any, you identify yourself with. Perhaps to you it means something else, such as church attendance.

The truth is that neither *party affiliation* nor *religious affiliation* has any real meaning, if by "real" we mean corresponding to some objective aspect of reality. These variables do not exist in nature. They are merely terms we have made up and assigned specific meanings to for some purpose, such as doing social research.

But, you might object, "political affiliation" and "religious affiliation"—and a host of other things social researchers are interested in, such as prejudice or compassion—have some reality. After all, we make statements about them, such as "In Happytown, 55 percent of the adults affiliate with the Republican Party, and 45 percent of them are Episcopalians. Overall, people in Happytown are low in prejudice and high in compassion." Even ordinary people, not just social researchers, have been known to make statements like that. If these things don't exist in reality, what is it that we're measuring and talking about?

What indeed? Let's take a closer look by considering a variable of interest to many social researchers (and many other people as well)—*prejudice*.

Conceptions, Concepts, and Reality

As you and I wandered down the road of life, we observed a lot of things and knew they were real through our observations, and we heard reports from other people that seemed real. For example:

- We personally heard people say nasty things about minority groups.
- We heard people say women were inferior to men.
- We read about African Americans being lynched.
- We read that women and minorities earned less for the same work.
- We learned about "ethnic cleansing" and wars in which one ethnic group tried to eradicate another.

With additional experience, we noticed something more. People who participated in lynching were also quite likely to call African Americans ugly names. A lot of them, moreover, seemed to want women to "stay in their place." Eventually it dawned on us that these several tendencies often appeared together in the same people and also had something in common. At some point, someone had a bright idea: "Let's use the word *prejudiced* for people like that. We can use the term even if they don't do all those things—as long as they're pretty much like that."

Being basically agreeable and interested in efficiency, we agreed to go along with the system. That's where "prejudice" came from. We never observed it. We just agreed to use it as a shortcut, a name that represents a collection of apparently related phenomena that we've each observed in the course of life. In short, we made it up.

Here's another clue that prejudice isn't something that exists apart from our rough agreement to use the term in a certain way. Each of us develops our own mental image of what the set of real phenomena we've observed represents in general and what these phenomena have in common. When I say the word *prejudice,* it evokes a mental image in your mind, just as it evokes one in mine. It's as though file drawers in our minds contained thousands of sheets of paper, with each sheet of paper labeled in the upper right-hand corner. A sheet of paper in each of our minds has the term *prejudice* on it. On your sheet are all the things you've been told about prejudice and everything you've observed that seems to be an example of it. My sheet has what I've been told about it plus all the things I've observed that seem examples of it—and mine isn't the same as yours.

The technical term for those mental images, those sheets of paper in our mental file drawers, is *conception.* That is, I have a conception of prejudice, and so do you. We can't communicate these mental images directly, so we use the terms written in the upper right-hand corner of our own mental sheets of paper as a way of communicating about our conceptions and the things we observe that are related to those conceptions. These terms make it possible for us to communicate and eventually agree on what we will specifically mean by those

terms. In social research, the process of coming to an agreement about what terms mean is *conceptualization,* and the result is called a *concept*.

Let's take another example of a conception. Suppose that I'm going to meet someone named Pat, whom you already know. I ask you what Pat is like. Now suppose that you've seen Pat help lost children find their parents and put a tiny bird back in its nest. Pat got you to take turkeys to poor families on Thanksgiving and to visit a children's hospital on Christmas. You've seen Pat weep through a movie about a mother overcoming adversities to save and protect her child. As you search through your mental files, you may find all or most of those phenomena recorded on a single sheet labeled "compassionate." You look over the other entries on the page, and you find they seem to provide an accurate description of Pat. So you say, "Pat is compassionate."

Now I leaf through my own mental file drawer until I find a sheet marked "compassionate." I then look over the things written on my sheet, and I say, "Oh, that's nice." I now feel I know what Pat is like, but my expectations reflect the entries on my file sheet, not yours. Later, when I meet Pat, I happen to find that my own experiences correspond to the entries I have on my "compassionate" file sheet, and I say that you sure were right. But suppose my observations of Pat contradict the things I have on my file sheet. I tell you that I don't think Pat is very compassionate, and we begin to compare notes.

You say, "I once saw Pat weep through a movie about a mother overcoming adversity to save and protect her child." I look at my "compassionate sheet" and can't find anything like that. Looking elsewhere in my file, I locate that sort of phenomenon on a sheet labeled "sentimental." I retort, "That's not compassion. That's just sentimentality."

To further strengthen my case, I tell you that I saw Pat refuse to give money to an organization dedicated to saving whales from extinction. "That represents a lack of compassion," I argue. You search through your files and find saving the whales on two sheets—"environmental activism" and "cross-species dating"—and you say so. Eventually, we set about comparing the entries we have on our respective sheets labeled "compassionate."

We then discover that we have many differing mental images corresponding to that term.

In the big picture, language and communication work only to the extent that the kinds of entries you and I have on our corresponding mental file sheets overlap considerably. The similarities we have on those sheets represent the agreements existing in our society. As we grow up, we're told approximately the same thing when we're first introduced to a particular term. Dictionaries formalize the agreements our society has about such terms. Each person, then, shapes his or her mental images to correspond with such agreements. But because all of us have different experiences and observations, no two people end up with exactly the same set of entries on any sheet in their file systems. If we want to measure "prejudice" or "compassion," we must first stipulate what, exactly, counts as prejudice or compassion for our purposes.

Returning to the assertion made at the outset of this chapter, we can measure anything that's real. We can measure, for example, whether Pat actually puts the little bird back in its nest, visits the hospital on Christmas, weeps at the movie, or refuses to contribute to saving the whales. All of those behaviors exist, so we can measure them. But is Pat really compassionate? We can't answer that question; we can't measure compassion in any objective sense, because compassion doesn't exist the way those things I just described exist. Compassion exists only in the form of the agreements we have about how to use the term in communicating about things that are real.

Concepts as Constructs

Abraham Kaplan (1964) distinguishes three classes of things that scientists measure. The first class is *direct observables:* those things we can observe rather simply and directly, like the color of an apple or the check mark made in a questionnaire. The second class, *indirect observables,* require "relatively more subtle, complex, or indirect observations" (1964:55). When we note a person's check mark beside "female" in a questionnaire, we've indirectly observed that person's gender. History books or minutes of corporate board meetings provide indirect observations of past social actions. Finally, the third class of observables consists of *constructs*—theoretical creations that are based on observations but that cannot be observed directly or indirectly. A good example is intelligence quotient, or IQ. It is constructed mathematically from observations of the answers given to a large number of questions on an IQ test. No one can directly or indirectly observe IQ. It is no more a "real" characteristic of people than is compassion or prejudice.

Kaplan (1964:49) defines *concept* as a "family of conceptions." A concept is, as Kaplan notes, a construct, something we create. Concepts like compassion and prejudice are constructs created from your conception of them, my conception of them, and the conceptions of all those who have ever used these terms. They cannot be observed directly or indirectly, because they don't exist. We made them up.

To summarize, concepts are constructs derived by mutual agreement from mental images (conceptions). Our conceptions summarize collections of seemingly related observations and experiences. Although the observations and experiences are real, at least subjectively, conceptions, and the concepts derived from them, are only mental creations. The terms associated with concepts are merely devices created for the purposes of filing and communication. A term like *prejudice* is, objectively speaking, only a collection of letters. It has no intrinsic reality beyond that. It has only the meaning we agree to give it.

Usually, however, we fall into the trap of believing that terms for constructs do have intrinsic meaning, that they name real entities in the world. That danger seems to grow stronger when we begin to take terms seriously and attempt to use them precisely. Further, the danger is all the greater in the presence of experts who appear to know more than we do about what the terms really mean: It's easy to yield to authority in such a situation.

Once we assume that terms like *prejudice* and *compassion* have real meanings, we begin the tortured task of discovering what those real meanings are and what constitutes a genuine measurement of them. Regarding constructs as real is called reification. The reification of concepts in day-to-day life is quite common. In science, we want to be quite clear about what it is we are actually measuring.

Does this discussion imply that compassion, prejudice, and similar constructs can't be measured? Interestingly, the answer is no. (And a good thing, too, or a lot of us social researcher types would be out of work.) I've said that we can measure anything that's real. Constructs aren't real in the way that trees are real, but they do have another important virtue: They are useful. That is, they help us organize, communicate about, and understand things that are real. They help us make predictions about real things. Some of those predictions even turn out to be true. Constructs can work this way because, while not real or observable in themselves, they have a definite relationship to things that are real and observable. The bridge from direct and indirect observables to useful constructs is the process called conceptualization.

CONCEPTUALIZATION

As we've seen, day-to-day communication usually occurs through a system of vague and general agreements about the use of terms. Although you and I do not agree completely about the use of the term compassionate, I'm probably safe in assuming that Pat won't pull the wings off flies. A wide range of misunderstandings and conflict—from the interpersonal to the international—is the price we pay for our imprecision, but somehow we muddle through. Science, however, aims at more than muddling; it cannot operate in a context of such imprecision.

The process through which we specify what we mean when we use particular terms in research is called **conceptualization.** Suppose we want to find out, for example, whether women are more compassionate than men. I suspect many people assume this is the case, but it might be interesting to find out if it's really so. We can't meaningfully study the question, let alone agree on the answer, without some working agreements about the meaning of *compassion.* They are "working" agreements in the sense that they allow us to work on

the question. We don't need to agree or even pretend to agree that a particular specification is ultimately the best one.

Conceptualization, then, produces a specific, agreed-upon meaning for a concept for the purposes of research. This process of specifying exact meaning involves describing the indicators we'll be using to measure our concept and the different aspects of the concept, called dimensions.

Indicators and Dimensions

Conceptualization gives definite meaning to a concept by specifying one or more indicators of what we have in mind. An **indicator** is a sign of the presence or absence of the concept we're studying. Here's an example.

We might agree that visiting children's hospitals during Christmas and Hanukkah is an indicator of compassion. Putting little birds back in their nests might be agreed on as another indicator, and so forth. If the unit of analysis for our study is the individual person, we can then observe the presence or absence of each indicator for each person under study. Going beyond that, we can add up the number of indicators of compassion observed for each individual. We might agree on ten specific indicators, for example, and find six present in our study of Pat, three for John, nine for Mary, and so forth.

Returning to our question about whether men or women are more compassionate, we might calculate that the women we studied displayed an average of 6.5 indicators of compassion, the men an average of 3.2. On the basis of our quantitative analysis of group difference, we might therefore conclude that women are, on the whole, more compassionate than men.

Usually, though, it's not that simple. Imagine you're interested in understanding a small fundamentalist religious cult, particularly their harsh views on various groups: gays, nonbelievers, feminists, and others. In fact, they suggest that anyone who refuses to join their group and abide by its teachings will "burn in hell." In the context of your interest in compassion, they don't seem to have much. And yet, the group's literature often speaks of their compassion for others. You want to explore this seeming paradox.

To pursue this research interest, you might arrange to interact with cult members, getting to know them and learning more about their views. You could tell them you were a social researcher interested in learning about their group, or perhaps you would just express an interest in learning more without saying why.

In the course of your conversations with group members and perhaps attendance of religious services, you would put yourself in situations where you could come to understand what the cult members mean by *compassion*. You might learn, for example, that members of the group were so deeply concerned about sinners burning in hell that they were willing to be aggressive, even violent, to make people change their sinful ways. Within their own paradigm, then, cult members would see beating up gays, prostitutes, and abortion doctors as acts of compassion.

Social researchers focus their attention on the meanings given to words and actions by the people under study. Doing so can often clarify the behaviors observed: At least now you understand how the cult can see violent acts as compassionate. On the other hand, paying attention to what words and actions mean to the people under study almost always complicated the concepts researchers are interested in. (We'll return to this issue when we discuss the validity of measures, toward the end of this chapter.)

Whenever we take our concepts seriously and set about specifying what we mean by them, we discover disagreements and inconsistencies. Not only do you and I disagree, but each of us is likely to find a good deal of muddiness within our own mental images. If you take a moment to look at what you mean by compassion, you'll probably find that your image contains several kinds of compassion. That is, the entries on your file sheet can be combined into groups and subgroups, say, compassion toward friends, co-religionists, humans, and birds. You may also find several different strategies for making combinations. For example, you might group the entries into feelings and actions.

The technical term for such groupings is **dimension:** a specifiable aspect of a concept. For instance, we might speak of the feeling dimension of compassion and the action dimension of compassion. In a different grouping scheme, we might distinguish compassion for humans from compassion for animals. Or we might see compassion as helping people have what we want for them versus what they want for themselves. Still differently, we might distinguish compassion as forgiveness from compassion as pity.

Thus, we could subdivide compassion into several clearly defined dimensions. A complete conceptualization involves both specifying dimensions and identifying the various indicators for each.

Specifying the different dimensions of a concept often paves the way for a more sophisticated understanding of what we're studying. We might observe, for example, that women are more compassionate in terms of feelings, and men more so in terms of actions—or vice versa. Whichever turned out to be the case, we would not be able to say whether men or women are really more compassionate. Our research would have shown that there is no single answer to the question. That alone represents an advance in our understanding of reality.

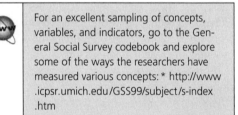

For an excellent sampling of concepts, variables, and indicators, go to the General Social Survey codebook and explore some of the ways the researchers have measured various concepts: * http://www.icpsr.umich.edu/GSS99/subject/s-index.htm

The Interchangeability of Indicators

There is another way that the notion of indicators can help us in our attempts to understand reality by means of "unreal" constructs. Suppose, for the moment, that you and I have compiled a list of 100 indicators of compassion and its various dimensions. Suppose further that we disagree widely on which indicators give the clearest evidence of compassion or its absence. If we pretty much agree on

---

*Each time the Internet icon appears, you'll be given helpful leads for searching the World Wide Web; whenever you see the GSS icon, you'll be given some tips on using this valuable resource.

some indicators, we could focus our attention on those, and we would probably agree on the answer they provided. We would then be able to say that some people are more compassionate than others in some dimension. But suppose we don't really agree on any of the possible indicators. Surprisingly, we can still reach an agreement on whether men or women are the more compassionate. How we do that has to do with the interchangeability of indicators.

The logic works like this. If we disagree totally on the value of the indicators, one solution would be to study all of them. Suppose that women turn out to be more compassionate than men on all 100 indicators—on all the indicators you favor and on all of mine. Then we would be able to agree that women are more compassionate than men even though we still disagree on exactly what compassion means in general.

The interchangeability of indicators means that if several different indicators all represent, to some degree, the same concept, then all of them will behave the same way that the concept would behave if it were real and could be observed. Thus, given a basic agreement about what "compassion" is, if women are generally more compassionate than men, we should be able to observe that difference by using any reasonable measure of compassion. If, on the other hand, women are more compassionate than men on some indicators but not on others, we should see if the two sets of indicators represent different dimensions of compassion.

You have now seen the fundamental logic of conceptualization and measurement. The discussions that follow are mainly refinements and extensions of what you've just read. Before turning to a technical elaboration of measurement, however, we need to fill out the picture of conceptualization by looking at some of the ways social researchers provide the meanings of terms with standards, consistency, and commonality.

Real, Nominal, and Operational Definitions

As we have seen, the design and execution of social research requires us to clear away the confusion over concepts and reality. To this end, logi-

cians and scientists have distinguished three kinds of definitions: real, nominal, and operational.

The first of these reflects the reification of terms. As Carl Hempel has cautioned,

> A "real" definition, according to traditional logic, is not a stipulation determining the meaning of some expression but a statement of the "essential nature" or the "essential attributes" of some entity. The notion of essential nature, however, is so vague as to render this characterization useless for the purposes of rigorous inquiry.
> — (1952:6)

In other words, trying to specify the "real" meaning of concepts only leads to a quagmire: It mistakes a construct for a real entity.

The specification of concepts in scientific inquiry depends instead on nominal and operational definitions. A *nominal definition* is one that is simply assigned to a term without any claim that the definition represents a "real" entity. Nominal definitions are arbitrary—I could define compassion as "plucking feathers off helpless birds" if I wanted to—but they can be more or less useful. For most purposes, especially communication, that last definition of compassion would be pretty useless. Most nominal definitions represent some consensus, or convention, about how a particular term is to be used.

An *operational definition,* as you may recall from Chapter 2, specifies precisely how a concept will be measured—that is, the operations we will perform. An operational definition is nominal rather than real, but it has the advantage of achieving maximum clarity about what a concept means in the context of a given study. In the midst of disagreement and confusion over what a term "really" means, we can specify a working definition for the purposes of an inquiry. Wishing to examine socioeconomic status (SES) in a study, for example, we may simply specify that we are going to treat SES as a combination of income and educational attainment. In this decision, we rule out other possible aspects of SES: occupational status, money in the bank, property, lineage, lifestyle, and so forth. Our findings will then be interesting to the extent that our definition of SES is useful for our purpose.

Creating Conceptual Order

The clarification of concepts is a continuing process in social research. Catherine Marshall and Gretchen Rossman (1995:18) speak of a "conceptual funnel" through which a researcher's interest becomes increasingly focused. Thus, a general interest in social activism could narrow to "individuals who are committed to empowerment and social change" and further focus on discovering "what experiences shaped the development of fully committed social activists." This focusing process is inescapably linked to the language we use.

In some forms of qualitative research, the clarification of concepts is a key element in the collection of data. Suppose you were conducting interviews and observations in a radical political group devoted to combating oppression in U.S. society. Imagine how the meaning of oppression would shift as you delved more and more deeply into the members' experiences and worldviews. For example, you might start out thinking of oppression in physical and perhaps economic terms. The more you learned about the group, however, the more you might appreciate the possibility of psychological oppression.

The same point applies even to contexts where meanings might seem more fixed. In the analysis of textual materials, for example, social researchers sometimes speak of the "hermeneutic circle," a cyclical process of ever deeper understanding.

> The understanding of a text takes place through a process in which the meaning of the separate parts is determined by the global meaning of the text as it is anticipated. The closer determination of the meaning of the separate parts may eventually change the originally anticipated meaning of the totality, which again influences the meaning of the separate parts, and so on. — (KVALE 1996:47)

Consider the concept "prejudice." Suppose you needed to write a definition of the term. You might start out thinking about racial/ethnic prejudice. At some point you would realize you should probably allow for gender prejudice, religious prejudice, antigay prejudice, and the like in your definition. Examining each of these specific types of prejudice would affect your overall understanding of the general concept. As your general understanding changed, however, you would likely see each of the individual forms somewhat differently.

The continual refinement of concepts occurs in all social research methods. Often you will find yourself refining the meaning of important concepts even as you write up your final report.

Although conceptualization is a continuing process, it is vital to address it specifically at the beginning of any study design, especially rigorously structured research designs such as surveys and experiments. In a survey, for example, operationalization results in a commitment to a specific set of questionnaire items that will represent the concepts under study. Without that commitment, the study could not proceed further.

Even in less-structured research methods, however, it's important to begin with an initial set of anticipated meanings that can be refined during data collection and interpretation. No one seriously believes we can observe life with no preconceptions; for this reason, scientific observers must be conscious of and explicit about these conceptual starting points.

Let's explore initial conceptualization the way it applies to structured inquiries such as surveys and experiments. Though specifying nominal definitions focuses our observational strategy, it does not allow us to observe. As a next step we must specify exactly what we are going to observe, how we will observe it, and how we will interpret various possible observations. All these further specifications make up the operational definition of the concept.

In the example of socioeconomic status, we might decide to measure SES in terms of income and educational attainment. We might then ask survey respondents two questions:

1. What was your total family income during the past 12 months?
2. What is the highest level of school you completed?

To organize our data, we would probably want to specify a system for categorizing the answers people give us. For income, we might use categories such as "under $5,000," "$5,000 to $10,000," and so on. Educational attainment might be similarly grouped in categories: less than high school, high school, college, graduate degree. Finally, we would specify the way a person's responses to these two questions would be combined to create a measure of SES.

In this way we would create a working and workable definition of SES. Although others might disagree with our conceptualization and operationalization, the definition would have one essential scientific virtue: It would be absolutely specific and unambiguous. Even if someone disagreed with our definition, that person would have a good idea how to interpret our research results, because what we meant by SES—reflected in our analyses and conclusions—would be precise and clear.

Here is a diagram showing the progression of measurement steps from our vague sense of what a term means to specific measurements in a fully structured scientific study:

Conceptualization
↓
Nominal Definition
↓
Operational Definition
↓
Measurements in the Real World

An Example of Conceptualization: The Concept of Anomie

To bring this discussion of conceptualization in research together, let's look briefly at the history of a specific social scientific concept. Researchers studying urban riots are often interested in the part played by feelings of powerlessness. Social researchers sometimes use the word *anomie* in this context. This term was first introduced into social science by Emile Durkheim, the great French sociologist, in his classic 1897 study, *Suicide.*

Using only government publications on suicide rates in different regions and countries, Durkheim produced a work of analytical genius. To determine the effects of religion on suicide, he compared the suicide rates of predominantly Protestant countries with those of predominantly Catholic ones, Protestant regions of Catholic countries with Catholic regions of Protestant countries, and so forth. To determine the possible effects of the weather, he compared suicide rates in northern and southern countries and regions, and he examined the different suicide rates across the months and seasons of the year. Thus, he could draw conclusions about a supremely individualistic and personal act without having any data about the individuals engaging in it.

At a more general level, Durkheim suggested that suicide also reflects the extent to which a society's agreements are clear and stable. Noting that times of social upheaval and change often present individuals with grave uncertainties about what is expected of them, Durkheim suggested that such uncertainties cause confusion, anxiety, and even self-destruction. To describe this societal condition of normlessness, Durkheim chose the term *anomie.* Durkheim did not make this word up. Used in both German and French, it literally meant "without law." The English term *anomy* had been used for at least three centuries before Durkheim to mean disregard for divine law. However, Durkheim created the social scientific concept of anomie.

In the years that have followed the publication of *Suicide,* social scientists have found anomie a useful concept, and many have expanded on Durkheim's use. Robert Merton, in a classic article entitled "Social Structure and Anomie" (1938), concluded that anomie results from a disparity between the goals and means prescribed by a society. Monetary success, for example, is a widely shared goal in our society, yet not all individuals have the resources to achieve it through acceptable means. An emphasis on the goal itself, Merton suggested, produces normlessness, because those denied the traditional avenues to wealth go about getting it through illegitimate means. Merton's discussion, then, could be considered a further conceptualization of the concept of anomie.

Although Durkheim originally used the concept of anomie as a characteristic of societies, as did Merton after him, other social scientists have used it to describe individuals. To clarify this distinction, some scholars have chosen to use *anomie* in reference to its original, societal meaning and to use the term *anomia* in reference to the individual characteristic. In a given society, then, some individuals experience anomia, and others do not. Elwin Powell, writing 20 years after Merton, provided the following conceptualization of anomia (though using the term *anomie*) as a characteristic of individuals:

> When the ends of action become contradictory, inaccessible or insignificant, a condition of anomie arises. Characterized by a general loss of orientation and accompanied by feelings of "emptiness" and apathy, anomie can be simply conceived as meaninglessness. — (1957:132)

Powell went on to suggest there were two distinct kinds of anomia and to examine how the two rose out of different occupational experiences to result at times in suicide. In his study, however, Powell did not measure anomia per se; he studied the relationship between suicide and occupation, making inferences about the two kinds of anomia. Thus, the study did not provide an operational definition of anomia, only a further conceptualization.

Although many researchers have offered operational definitions of anomia, one name stands out over all. Two years before Powell's article appeared, Leo Srole (1956) published a set of questionnaire items that he said provided a good measure of anomia as experienced by individuals. It consists of five statements that subjects were asked to agree or disagree with:

1. In spite of what some people say, the lot of the average man is getting worse.
2. It's hardly fair to bring children into the world with the way things look for the future.
3. Nowadays a person has to live pretty much for today and let tomorrow take care of itself.
4. These days a person doesn't really know who he can count on.
5. There's little use writing to public officials because they aren't really interested in the problems of the average man. — (1956:713)

In the decades following its publication, the Srole scale has become a research staple for social scientists. You'll likely find this particular operationalization of anomia used in many of the research projects reported in academic journals. Srole touches on this in the accompanying box, "The Origins of Anomia," which he prepared for this book before his death.

This abbreviated history of *anomie* and *anomia* as social scientific concepts illustrates several points. First, it is a good example of the process through which general concepts become operationalized measurements. This is not to say that the issue of how to operationalize anomie/anomia has been resolved once and for all. Scholars will surely continue to reconceptualize and reoperationalize these concepts for years to come, continually seeking more-useful measures.

The Srole scale illustrates another important point. Letting conceptualization and operationalization be open-ended does not necessarily produce anarchy and chaos, as you might expect. Order often emerges. For one thing, although we could define *anomia* any way we chose—in terms of, say, shoe size—we're likely to define it in ways not too different from other people's mental images. If you were to use a really offbeat definition, people would probably ignore you.

A second source of order is that, as researchers discover the utility of a particular conceptualization and operationalization of a concept, they're likely to adopt it, which leads to standardized definitions of concepts. Besides the Srole scale, examples include IQ tests and a host of demographic and economic measures developed by the U.S. Census Bureau. Using such established measures has two advantages: They have been extensively pretested and debugged, and studies using the same scales can be compared. If you and I do separate studies of two different groups and use the Srole scale, we can compare our two groups on the basis of anomia.

Social scientists, then, can measure anything that's real; through conceptualization and operationalization, they can even do a pretty good job of measuring things that aren't. Granting that such concepts as socioeconomic status, prejudice, compassion, and anomia aren't ultimately real, social scientists can create order in handling them. This

THE ORIGINS OF ANOMIA

by Leo Srole

My career-long fixation on anomie began with reading Durkheim's *Le Suicide* as a Harvard undergraduate. Later, as a graduate student at Chicago, I studied under two Durkheimian anthropologists: William Lloyd Warner and Alfred Radcliffe-Brown. Radcliffe-Brown had carried on a lively correspondence with Durkheim, making me a collateral "descendant" of the great French sociologist.

For me, the early impact of Durkheim's work on suicide was mixed but permanent. On the one hand, I had serious reservations about his strenuous, ingenious, and often awkward efforts to force the crude, bureaucratic records on suicide rates to fit with his unidirectional sociological determinism. On the other hand, I was moved by Durkheim's unswerving preoccupation with the moral force of the interpersonal ties that bind us to our time, place, and past, and also his insights about the lethal consequences that can follow from shrinkage and decay in those ties.

My interest in anomie received an eyewitness jolt at the finale of World War II, when I served with the United Nations Relief and Rehabilitation Administration, helping to rebuild a war-torn Europe. At the Nazi concentration camp of Dachau, I saw first-hand the depths of dehumanization that macro-social forces, such as those that engaged Durkheim, could produce in individuals like Hitler, Eichmann, and the others serving their dictates at all levels in the Nazi death factories.

Returning from my UNRRA post, I felt most urgently that the time was long overdue to come to an understanding of the dynamics underlying disintegrated social bonds. We needed to work expeditiously, deemphasizing proliferation of

order is based on utility, however, not on ultimate truth.

DEFINITIONS IN DESCRIPTIVE AND EXPLANATORY STUDIES

As described in Chapter 4, two general purposes of research are description and explanation. The distinction between them has important implications for definition and measurement. If it seems that description is simpler than explanation, you may be surprised to learn that definitions are more problematic for descriptive research than for explanatory research. Before we turn to other aspects of measurement, you'll need a basic understanding of why this is so (we'll discuss this point more fully in Part 4).

It's easy to see the importance of clear and precise definitions for descriptive research. If we want to describe and report the unemployment rate in a city, our definition of "being unemployed" is obviously critical. That definition will depend on our definition of another term: the labor force. If it seems patently absurd to regard a three-year-old child as being unemployed, it is because such a child is not considered a member of the labor force. Thus, we might follow the U.S. Census Bureau's convention and exclude all people under 14 years of age from the labor force.

This convention alone, however, would not give us a satisfactory definition, because it would count as unemployed such people as high school students, the retired, the disabled, and homemakers. We might follow the Census convention further by defining the labor force as "all persons 14 years of age and over who are employed, looking for work, or waiting to be called back to a job from which they have been laid off or furloughed." If a student, homemaker, or retired person is not looking for macro-level theory in favor of a direct explor-atory encounter with individuals, using newly developed state-of-the-art survey research methodology. Such research, I also felt, should focus on a broader spectrum of behavioral pathologies than suicide.

My initial investigations were a diverse effort. In 1950, for example, I was able to interview a sample of 401 bus riders in Springfield, Mass. Four years later, the Midtown Manhattan Men-tal Health Study provided a much larger popula-tion reach. These and other field projects gave me scope to expand and refine my measure-ments of that quality in individuals which re-flected the macro-social quality Durkheim had called *anomie.*

While I began by using Durkheim's term in my own work, I soon decided that it was neces-sary to limit the use of that concept to its macro-social meaning and to sharply segregate it from its individual manifestations. For the latter pur-pose, the cognate but hitherto obsolete Greek term, *anomia,* readily suggested itself.

I first published the anomia construct in a 1956 article in the *American Sociological Review,** describing ways of operationalizing it, and pre-senting the results of its initial field applica-tion research. By 1982, the Science Citation In-dex and Social Science Citation Index had listed some 400 publications in political science, psy-chology, social work, and sociology journals here and abroad that had cited use of that ar-ticle's instruments or findings, warranting the American Institute for Scientific Information to designate it a "citation classic."

*Leo Srole, "Social Integration and Certain Corollar-ies: An Exploratory Study," *American Sociological Re-view* 21 (1956): 709–16.

work, such a person would not be included in the labor force. Unemployed people, then, would be those members of the labor force, as defined, who are not employed.

But what does "looking for work" mean? Must a person register with the state employment service or go from door to door asking for employment? Or would it be sufficient to want a job or be open to an offer of employment? Conventionally, "looking for work" is defined operationally as saying yes in re-sponse to an interviewer's asking "Have you been looking for a job during the past seven days?" (Seven days is the period most often specified, but for some research purposes it might make more sense to shorten or lengthen it.)

As you can see, the conclusion of a descriptive study about the unemployment rate depends di-rectly on how each issue of definition is resolved. Increasing the period during which people are counted as looking for work would add more un-employed people to the labor force as defined, thereby increasing the reported unemployment rate. If we follow another convention and speak of the civilian labor force and the civilian unemploy-ment rate, we are excluding military personnel; that, too, increases the reported unemployment rate, because military personnel would be em-ployed—by definition. Thus the descriptive state-ment that the unemployment rate in a city is 3 per-cent, or 9 percent, or whatever it might be, depends directly on the operational definitions used.

This example is relatively clear because there are several accepted conventions relating to the la-bor force and unemployment. Now, consider how difficult it would be to get agreement about the definitions you would need in order to say, "Forty-five percent of the students at this institution are politically conservative." Like the unemployment rate, this percentage would depend directly on the definition of what is being measured—in this case,

THE IMPORTANCE OF VARIABLE NAMES

by Patricia Fisher
Graduate School of Planning, University of Tennessee

Operationalization is one of those things that's easier said than done. It is quite simple to explain to someone the purpose and importance of operational definitions for variables, and even to describe how operationalization typically takes place. However, until you've tried to operationalize a rather complex variable, you may not appreciate some of the subtle difficulties involved. Of considerable importance to the operationalization effort is the particular name that you have chosen for a variable. Let's consider an example from the field of Urban Planning.

A variable of interest to planners is *citizen participation*. Planners are convinced that partic-ipation in the planning process by citizens is important to the success of plan implementation. Citizen participation is an aid to planners' understanding of the real and perceived needs of a community, and such involvement by citizens tends to enhance their cooperation with and support for planning efforts. Although many different conceptual definitions might be offered by different planners, there would be little misunderstanding over what is meant by *citizen participation*. The name of the variable seems adequate.

However, if we asked different planners to provide very simple operational measures for citizen participation, we are likely to find a variety among their responses that does generate confusion. One planner might keep a tally of attendance by private citizens at city commission

political conservatism. A different definition might result in the conclusion "Five percent of the student body are politically conservative."

Ironically, definitions are less problematic in the case of explanatory research. Let's suppose we're interested in explaining political conservatism. Why are some people conservative and others not? More specifically, let's suppose we're interested in whether conservatism increases with age. What if you and I have 25 different operational definitions of *conservative,* and we can't agree on which definition is best? As we saw in the discussion of indicators, this is not necessarily an insurmountable obstacle to our research. Suppose we found old people to be more conservative than young people in terms of all 25 definitions. Clearly, the exact definition would be of small consequence. Suppose we found old people to be more conservative than young people by every reasonable definition of conservatism we could think of. It wouldn't matter what our definition was. We would conclude that old people are generally more conservative than young people—even though we couldn't agree about exactly what *conservative* means.

In practice, explanatory research seldom results in findings quite as unambiguous as this example suggests; nonetheless, the general pattern is quite common in actual research. There are consistent patterns of relationships in human social life that result in consistent research findings. However, such consistency does not appear in a descriptive situation. Changing definitions almost inevitably result in different descriptive conclusions. The box "The Importance of Variable Names" explores this issue in connection with the variable *citizen participation.*

OPERATIONALIZATION CHOICES

In discussing conceptualization, I frequently have referred to operationalization, for the two are intimately linked. To recap: Conceptualization is the refinement and specification of abstract concepts, and operationalization is the development of specific research procedures (operations) that will result in empirical observations representing those concepts in the real world.

and other local government meetings; another might maintain a record of the different topics addressed by private citizens at similar meetings; while a third might record the number of local government meeting attendees, letters, and phone calls received by the mayor and other public officials, and meetings held by special interest groups during a particular time period. As skilled researchers, we can readily see that each planner would be measuring (in a very simplistic fashion) a different dimension of citizen participation: extent of citizen participation, issues prompting citizen participation, and form of citizen participation. Therefore, the original *naming* of our variable, *citizen participation,* which was quite satisfactory from a conceptual point of view, proved inadequate for purposes of operationalization.

The precise and exact naming of variables is important in research. It is both essential to and a result of good operationalization. Variable names quite often evolve from an iterative process of forming a conceptual definition, then an operational definition, then renaming the concept to better match what can or will be measured. This looping process continues (our example above illustrates only one iteration), resulting in a gradual refinement of the variable name and its measurement until a reasonable fit is obtained. Sometimes the concept of the variable that you end up with is a bit different from the original one that you started with, but at least you are measuring what you are talking about, if only because you are talking about what you are measuring!

As with the methods of data collection, social researchers have a variety of choices when operationalizing a concept. Although the several choices are intimately interconnected, I've separated them for the sake of discussion. Realize, though, that operationalization does not proceed through a systematic checklist.

Range of Variation

In operationalizing any concept, researchers must be clear about the range of variation that interests them. The question is, to what extent are we willing to combine attributes in fairly gross categories?

Let's suppose you want to measure people's incomes in a study by collecting the information from either records or interviews. The highest annual incomes people receive run into the millions of dollars, but not many people get that much. Unless you're studying the very rich, keeping track of extremely high categories probably won't add much to your study. Depending on whom you study, you'll probably want to establish a highest income category with a much lower floor—maybe

$100,000 or more. Although this decision will lead you to throw together people who earn a trillion dollars a year with paupers earning a mere $100,000, they'll survive it, and that mixing probably won't hurt your research any, either. The same decision faces you at the other end of the income spectrum. In studies of the general U.S. population, a bottom category of $5,000 or less usually works fine.

In studies of attitudes and orientations, the question of range of variation has another dimension. Unless you're careful, you may end up measuring only half an attitude without really meaning to. Here's an example of what I mean.

Suppose you're interested in people's attitudes toward expanding the use of nuclear power generators. If you reasonably guess or have experienced that some people consider nuclear power the greatest thing since the wheel, whereas other people have absolutely no interest in it, it makes sense to ask people how much they favor expanding the use of nuclear energy and to give them answer categories ranging from "Favor it very much" to "Don't favor it at all."

This operationalization, however, conceals half the attitudinal spectrum regarding nuclear energy. Many people have feelings that go beyond simply not favoring it: They are, with greater or lesser degrees of intensity, actively opposed to it. In this instance, there is considerable variation on the left side of zero. Some oppose it a little, some quite a bit, and others a great deal. To measure the full range of variation, then, you'd want to operationalize attitudes toward nuclear energy with a range from favoring it very much, through no feelings one way or the other, to opposing it very much.

This consideration applies to many of the variables social researchers study. Virtually any public issue involves both support and opposition, each in varying degrees. Political orientations range from very liberal to very conservative, and depending on the people you're studying, you may want to allow for radicals on one or both ends. Similarly, people are not just more or less religious; some are positively antireligious.

The point is not that you must measure the full range of variation in every case. You should, however, consider whether you need to, given your particular research purpose. If the difference between not religious and antireligious isn't relevant to your research, forget it. Someone has defined pragmatism as "any difference that makes no difference is no difference." Be pragmatic.

Finally, decisions about the range of variation should be governed by the expected distribution of attributes among the subjects of the study. In a study of college professors' attitudes toward the value of higher education, you could probably stop at no value and not worry about those who might consider higher education dangerous to students' health. (If you were studying students, however . . . .)

Variations between the Extremes

Degree of precision is a second consideration in operationalizing variables. What it boils down to is how fine you will make the distinctions among the various possible attributes composing a given variable. Does it matter for your purposes whether a person is 17 or 18 years old, or could you conduct your inquiry by throwing them together in a group labeled 10 to 19 years old? Don't answer too quickly. If you wanted to study rates of voter registration and participation, you'd definitely want to know whether the people you studied were old enough to vote. In general, if you're going to measure age, you must look at the purpose and procedures of your study and decide whether fine or gross differences in age are important to you. In a survey, you'll need to make these decisions in order to design an appropriate questionnaire. In the case of in-depth interviews, these decisions will condition the extent to which you probe for details.

The same thing applies to other variables. If you measure political affiliation, will it matter to your inquiry whether a person is a conservative Democrat rather than a liberal Democrat, or will it be sufficient to know the party? In measuring religious affiliation, is it enough to know that a person is a Protestant, or do you need to know the denomination? Do you simply need to know whether or not a person is married, or will it make a difference to know if he or she has never married or is separated, widowed, or divorced?

There is, of course, no general answer to such questions. The answers come out of the purpose of a given study, or why we are making a particular measurement. I can give you a useful guideline, though. Whenever you're not sure how much detail to pursue in a measurement, go after too much rather than too little. When a subject in an in-depth interview volunteers that she is 37 years old, record "37" in your notes, not "in her thirties." When you're analyzing the data, you can always combine precise attributes into more general categories, but you can never separate any variations you lumped together during observation and measurement.

A Note on Dimensions

We've already discussed dimensions as a characteristic of concepts. When researchers get down to the business of creating operational measures of variables, they often discover—or worse, never notice—that they haven't been exactly clear about which dimensions of a variable they're really interested in. Here's an example.

Let's suppose you're studying people's attitudes toward government, and you want to include an examination of how people feel about corruption. Here are just a few of the dimensions you might examine:

- Do people think there is corruption in government?
- How much corruption do they think there is?
- How certain are they in their judgment of how much corruption there is?
- How do they feel about corruption in government as a problem in society?
- What do they think causes it?
- Do they think it's inevitable?
- What do they feel should be done about it?
- What are they willing to do personally to eliminate corruption in government?
- How certain are they that they would be willing to do what they say they would do?

The list could go on and on—how people feel about corruption in government has many dimensions. It's essential to be clear about which ones are important in our inquiry; otherwise, you may measure how people feel about corruption when you really wanted to know how much they think there is, or vice versa.

Once you've determined how you're going to collect your data (for example, survey, field research) and decided on the relevant range of variation, the degree of precision needed between the extremes of variation, and the specific dimensions of the variables that interest you, you may have another choice: a mathematical-logical one. That is, you may need to decide what level of measurement to use. To discuss this point, we need to take another look at attributes and their relationship to variables.

Defining Variables and Attributes

An attribute, you'll recall, is a characteristic or quality of something. "Female" is an example. So is "old" or "student." Variables, on the other hand, are logical sets of attributes. Thus, *gender* is a variable composed of the attributes female and male.

The conceptualization and operationalization processes can be seen as the specification of variables and the attributes composing them. Thus, in the context of a study of unemployment, *employment status* is a variable having the attributes employed and unemployed; the list of attributes could also be expanded to include the other possibilities discussed earlier, such as homemaker.

Every variable must have two important qualities. First, the attributes composing it should be exhaustive. For the variable to have any utility in research, we must be able to classify every observation in terms of one of the attributes composing the variable. We'll run into trouble if we conceptualize the variable *political party affiliation* in terms of the attributes Republican and Democrat, because some of the people we set out to study will identify with the Green Party, the Reform Party, or some other organization, and some (often a large percentage) will tell us they have no party affiliation. We could make the list of attributes exhaustive by adding "other" and "no affiliation." Whatever we do, we must be able to classify every observation.

At the same time, attributes composing a variable must be mutually exclusive. That is, we must be able to classify every observation in terms of one and only one attribute. For example, we need to define "employed" and "unemployed" in such a way that nobody can be both at the same time. That means being able to classify the person who is working at a job but is also looking for work. (We might run across a fully employed mud wrestler who is looking for the glamour and excitement of being a social researcher.) In this case, we might define the attributes so that employed takes precedence over unemployed, and anyone working at a job is employed regardless of whether he or she is looking for something better.

Levels of Measurement

Attributes operationalized as mutually exclusive and exhaustive may be related in other ways as well. For example, the attributes composing variables may represent different levels of measurement. In this section, we'll examine four levels of measurement: nominal, ordinal, interval, and ratio.

Nominal Measures Variables whose attributes have only the characteristics of exhaustiveness and mutual exclusiveness are nominal measures. Examples include *gender, religious affiliation, political party affiliation, birthplace, college major,* and *hair color.* Although the attributes composing each of these variables—as male and female compose the variable *gender*—are distinct from one another (and exhaust the possibilities of gender among people), they have no additional structures. Nominal measures merely offer names or labels for characteristics.

Imagine a group of people characterized in terms of one such variable and physically grouped by the applicable attributes. For example, say we've asked a large gathering of people to stand together in groups according to the states in which they were born: all those born in Vermont in one group, those born in California in another, and so forth. The variable is place of birth; the attributes are *born in California, born in Vermont,* and so on. All the people standing in a given group have at least one thing in common and differ from the people in all other groups in that same regard. Where the individual groups form, how close they are to one another, or how the groups are arranged in the room is irrelevant. All that matters is that all the members of a given group share the same state of birth and that each group has a different shared state of birth. All we can say about two people in terms of a nominal variable is that they are either the same or different.

Ordinal Measures Variables with attributes we can logically rank-order are **ordinal measures.** The different attributes of ordinal variables represent relatively more or less of the variable. Variables of this type are *social class, conservatism, alienation, prejudice, intellectual sophistication,* and the like. In addition to saying whether two people are the same or different in terms of an ordinal variable, you can also say one is "more" than the other—that is, more conservative, more religious, older, and so forth.

In the physical sciences, hardness is the most frequently cited example of an ordinal measure. We may say that one material (for example, dia-

mond) is harder than another (say, glass) if the former can scratch the latter and not vice versa. By attempting to scratch various materials with other materials, we might eventually be able to arrange several materials in a row, ranging from the softest to the hardest. We could never say how hard a given material was in absolute terms; we could only say how hard in relative terms—which materials it is harder than and which softer than.

Let's pursue the earlier example of grouping the people at a social gathering. This time imagine that we ask all the people who have graduated from college to stand in one group, all those with only a high school diploma to stand in another group, and all those who have not graduated from high school to stand in a third group. This manner of grouping people satisfies the requirements for exhaustiveness and mutual exclusiveness discussed earlier. In addition, however, we might logically arrange the three groups in terms of the relative amount of formal education (the shared attribute) each had. We might arrange the three groups in a row, ranging from most to least formal education. This arrangement would provide a physical representation of an ordinal measure. If we knew which groups two individuals were in, we could determine that one had more, less, or the same formal education as the other.

Notice in this example that it doesn't matter how close or far apart the educational groups are from one another. The college and high school groups might be 5 feet apart, and the less-than-high-school group 500 feet farther down the line. These actual distances don't have any meaning. The high school group, however, should be between the less-than-high-school group and the college group, or else the rank order will be incorrect.

Interval Measures For the attributes composing some variables, the actual distance separating those attributes does have meaning. Such variables are interval measures. For these, the logical distance between attributes can be expressed in meaningful standard intervals.

For example, in the Fahrenheit temperature scale, the difference, or distance, between 80 degrees and 90 degrees is the same as that between 40 degrees and 50 degrees. However, 80 degrees Fahrenheit is not twice as hot as 40 degrees, because in both the Fahrenheit and Celsius scales, "zero" is arbitrary; that is, zero degrees does not really mean total lack of heat. Similarly, minus 30 degrees on either scale doesn't represent 30 degrees less than no heat. (In contrast, the Kelvin scale is based on an absolute zero, which does mean a complete lack of heat.)

About the only interval measures commonly used in social scientific research are constructed measures such as standardized intelligence tests that have been more or less accepted. The interval separating IQ scores of 100 and 110 may be regarded as the same as the interval separating scores of 110 and 120 by virtue of the distribution of observed scores obtained by many thousands of people who have taken the tests over the years. But it would be incorrect to infer that someone with an IQ of 150 is 50 percent more intelligent than someone with an IQ of 100. (A person who received a score of 0 on a standard IQ test could not be regarded, strictly speaking, as having no intelligence, although we might feel he or she was unsuited to be a college professor or even a college student.)

When comparing two people in terms of an interval variable, we can say they are different from one another (nominal), and that one is more than another (ordinal). In addition, we can say "how much" more.

Ratio Measures Most of the social scientific variables meeting the minimum requirements for interval measures also meet the requirements for ratio measures. In ratio measures, the attributes composing a variable, besides having all the structural characteristics mentioned previously, are based on a true zero point. The Kelvin temperature scale is one such measure. Examples from social research include age, length of residence in a given place, number of organizations belonged to, number of times attending church during a particular period of time, number of times married, and number of Arab friends.

Returning to the illustration of methodological party games, we might ask a gathering of people to group themselves by age. All the one-year-olds would stand (or sit or lie) together, the two-year-olds together, the three-year-olds, and so forth. The fact that members of a single group share the same age and that each different group has a different shared age satisfies the minimum requirements for a nominal measure. Arranging the several groups in a line from youngest to oldest meets the additional requirements of an ordinal measure and lets us determine if one person is older than, younger than, or the same age as another. If we space the groups equally far apart, we satisfy the additional requirements of an interval measure and will be able to say how much older one person is than another. Finally, because one of the attributes included in age represents a true zero (babies carried by women about to give birth), the phalanx of hapless partygoers also meets the requirements of a ratio measure, permitting us to say that one person is twice as old as another. (Remember this in case you're asked about it in a workbook assignment.) Another example of a ratio measure is income, which extends from an absolute zero to approximately infinity, if you happen to be the founder of Microsoft.

Comparing two people in terms of a ratio variable, then, allows us to determine (1) that they are different (or the same), (2) that one is more than the other, (3) how much they differ, and (4) the ratio of one to another. Figure 5-1 summarizes this discussion by presenting a graphic illustration of the four levels of measurement.

Implications of Levels of Measurement Because it's unlikely that you'll undertake the physical grouping of people just described (try it once, and you won't be invited to many parties), I should draw your attention to some of the practical implications of the differences that have been distinguished. These implications appear primarily in the analysis of data (discussed in Part 4), but you need to anticipate such implications when you're structuring any research project.

Certain quantitative analysis techniques require variables that meet certain minimum levels of measurement. To the extent that the variables to be examined in a research project are limited to

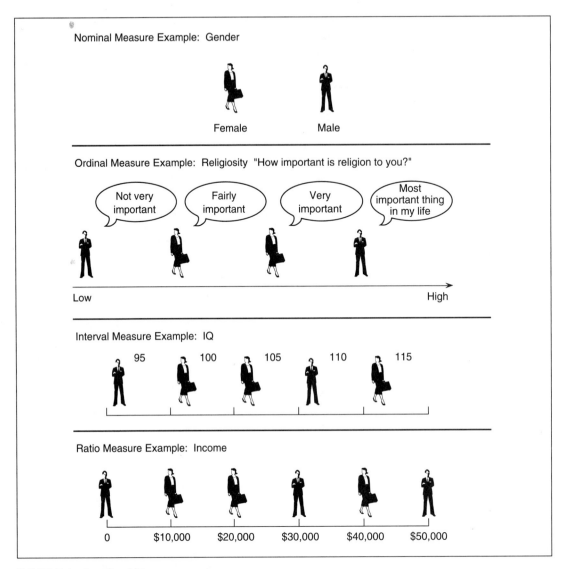

FIGURE 5-1   **Levels of Measurement**

a particular level of measurement—say, ordinal—you should plan your analytical techniques accordingly. More precisely, you should anticipate drawing research conclusions appropriate to the levels of measurement used in your variables. For example, you might reasonably plan to determine and report the mean age of a population under study (add up all the individual ages and divide by the number of people), but you should not plan to report the mean religious affiliation, because that is a nominal variable, and the mean requires ratio-level data. (You could report the modal—the most common—religious affiliation.)

At the same time, you can treat some variables as representing different levels of measurement. Ratio measures are the highest level, descending through interval and ordinal to nominal, the lowest level of measurement. A variable representing a higher level of measurement—say, ratio—may also be treated as representing a lower level of measurement—say, ordinal. Recall, for example, that age is a ratio measure. If you wished to examine only the relationship between age and some ordinal-level variable—say, self-perceived religiosity: high, medium, and low—you might choose to treat age as an ordinal-level variable as well. You might characterize the subjects of your study as being young, middle-aged, and old, specifying what age range composed each of these groupings. Finally, age might be used as a nominal-level variable for certain research purposes. People might be grouped as being born during the depression of the 1930s or not. Another nominal measurement, based on birth date rather than just age, would be the grouping of people by astrological signs.

The level of measurement you'll seek, then, is determined by the analytical uses you've planned for a given variable, keeping in mind that some variables are inherently limited to a certain level. If a variable is to be used in a variety of ways, requiring different levels of measurement, the study should be designed to achieve the highest level required. For example, if the subjects in a study are asked their exact ages, they can later be organized into ordinal or nominal groupings.

You don't necessarily need to measure variables at their highest level of measurement, however. If you're sure to have no need for ages of people at higher than the ordinal level of measurement, you may simply ask people to indicate their age range, such as 20 to 29, 30 to 39, and so forth. In a study of the wealth of corporations, rather than seek more precise information, you may use Dun & Bradstreet ratings to rank corporations. Whenever your research purposes are not altogether clear, however, seek the highest level of measurement possible. Again, although ratio measures can later be reduced to ordinal ones, you cannot convert an ordinal measure to a ratio one. More generally, you cannot convert a lower-level measure to a higher-level one. That is a one-way street worth remembering.

Typically a research project will tap variables at different levels of measurement. For example, William and Denise Bielby (1999) set out to examine the world of film and television, using a nomothetic, longitudinal approach (take a moment to remind yourself what that means). In what they referred to as the "culture industry," the authors found that *reputation* (an ordinal variable) is the

APPLYING THE RESULTS

Say you want to be a Hollywood screenwriter. How might you use the results of the Bielby and Bielby (1999) study to enhance your career? Say you didn't do so well and instead started a school for screenwriters. How could the results of the study be used to plan courses? Finally, how might the results be useful to you if you were a social activist committed to fighting discrimination in the "culture industry"?

best predictor of screenwriters' future productivity. More interestingly, they found that screenwriters who were represented by "core" (or elite) agencies were far more likely not only to find jobs (a nominal variable) but also to find jobs that paid more (a ratio variable). In other words, the researchers found that an agency's reputation (ordinal) was a key independent variable for predicting a screenwriter's success. The researchers also found that being older (ratio), being female (nominal), belonging to an ethnic minority (nominal), and having more years of experience (ratio) were disadvantageous for a screenwriter. On the other hand, higher earnings from previous years (measured in ordinal categories) led to more success in the future. In the researchers' terms, "success breeds success" (Bielby and Bielby 1999:80).

Single or Multiple Indicators

With so many alternatives for operationalizing social research variables, you may find yourself worrying about making the right choices. To counter this feeling, let me add a momentary dash of certainty and stability.

Many social research variables have fairly obvious, straightforward measures. No matter how you cut it, *gender* usually turns out to be a matter of male or female: a nominal-level variable that can be measured by a single observation—either looking (well, not always) or asking a question (usually). In a study involving the size of families, you'll want to think about adopted and foster children as well as blended families, but it's usually pretty easy to find out how many children a family has. For most research purposes, the resident population of a country is the resident population of that country—you can look it up in an almanac and know the answer. A great many variables, then, have obvious single indicators. If you can get one piece of information, you have what you need.

Sometimes, however, there is no single indicator that will give you the measure of a chosen variable. As discussed earlier in this chapter, many concepts are subject to varying interpretations—each with several possible indicators. In these cases, you'll want to make several observations for a given variable. You can then combine the several pieces of information you've collected to create a composite measurement of the variable in question. Chapter 6 is devoted to ways of doing that, so here let's look at just one simple illustration.

Consider the concept "college performance." All of us have noticed that some students perform well in college courses and others don't. In studying these differences, we might ask what characteristics and experiences are related to high levels of performance (many researchers have done just that). How should we measure overall performance? Each grade in any single course is a potential indicator of college performance, but it also may not typify the student's general performance. The solution to this problem is so firmly established that it is, of course, obvious: the grade point average (GPA). We assign numerical scores to each letter grade, total the points earned by a given student, and divide by the number of courses taken to obtain a composite measure. (If the courses vary in number of credits, we adjust the point values accordingly.) It's often appropriate to create such composite measures in social research.

Some Illustrations of Operationalization Choices

To bring together all the operationalization choices available to the social researcher and to show the potential in those possibilities, let's look at some of the distinct ways you might address various research problems. The alternative ways of operationalizing the variables in each case should demonstrate the opportunities that social research can present to our ingenuity and imaginations. To simplify matters, I have not attempted to describe all the research conditions that would make one alternative superior to the others, though in a given situation they would not all be equally appropriate.

1. Are women more compassionate than men?
    a. Select a group of subjects for study, with equal numbers of men and women. Present them with hypothetical situations that involve someone's being in trouble. Ask them what they would do if they were confronted with that situation. What would they do, for example, if they came across a small child who was lost and crying for his or her parents? Consider any answer that involves helping or comforting the child as an indicator of compassion. See whether men or women are more likely to indicate they would be compassionate.
    b. Set up an experiment in which you pay a small child to pretend that he or she is lost. Put the child to work on a busy sidewalk and observe whether men or women are more likely to offer assistance. Also be sure to count the total number of men and women who walk by, because there may be more of one than the other. If that's the case, simply calculate the percentage of men and the percentage of women who help.
    c. Select a sample of people and do a survey in which you ask them what organizations they belong to. Calculate whether women or men are more likely to belong to those that seem to reflect compassionate feelings. To take account of men who belong to more organizations than do women in general—or vice versa—do this: For each person you study, calculate the percentage of his or her organiza-

tional memberships that reflect compassion. See if men or women have a higher average percentage.

2. Are sociology students or accounting students better informed about world affairs?

   a. Prepare a short quiz on world affairs and arrange to administer it to the students in a sociology class and in an accounting class at a comparable level. If you want to compare sociology and accounting majors, be sure to ask students what they are majoring in.

   b. Get the instructor of a course in world affairs to give you the average grades of sociology and accounting students in the course.

   c. Take a petition to sociology and accounting classes that urges that "the United Nations headquarters be moved to New York City." Keep a count of how many in each class sign the petition and how many inform you that the UN headquarters is already located in New York City.

3. Do people consider New York or California the better place to live?

   a. Consulting the *Statistical Abstract of the United States* or a similar publication, check the migration rates into and out of each state. See if you can find the numbers moving directly from New York to California and vice versa.

   b. The national polling companies—Gallup, Harris, Roper, and so forth—often ask people what they consider the best state to live in. Look up some recent results in the library or through your local newspaper.

   c. Compare suicide rates in the two states.

4. Who are the most popular instructors on your campus—those in the social sciences, the natural sciences, or the humanities?

   a. If your school has formal student evaluations of instructors, review some recent results and compute the average ratings of each group.

   b. Begin visiting the introductory courses given in each group of disciplines and

measure the attendance rate of each class.

   c. In December, select a group of faculty in each of the three divisions and ask them to keep a record of the numbers of holiday greeting cards and presents they receive from admiring students. See who wins.

The point of these examples is not necessarily to suggest respectable research projects but to illustrate the many ways variables can be operationalized.

Operationalization Goes On and On

Although I've discussed conceptualization and operationalization as activities that precede data collection and analysis—for example, you must design questionnaire items before you send out a questionnaire—these two processes continue throughout any research project, even if the data have been collected in a structured mass survey. As we've seen, in less-structured methods such as field research, the identification and specification of relevant concepts is inseparable from the ongoing process of observation.

As a researcher, always be open to reexamining your concepts and definitions. The ultimate purpose of social research is to clarify the nature of social life. The validity and utility of what you learn in this regard doesn't depend on when you first figured out how to look at things any more than it matters whether you got the idea from a learned textbook, a dream, or your brother-in-law.

CRITERIA OF MEASUREMENT QUALITY

This chapter has come some distance. It began with the bald assertion that social scientists can measure anything that exists. Then we discovered that most of the things we might want to measure and study don't really exist. Next we learned that it's possible to measure them anyway. Now we conclude the chapter with a discussion of some of the yardsticks against which we judge our relative success or failure in measuring things—even things that don't exist.

Precision and Accuracy

To begin, measurements can be made with varying degrees of precision. As we saw in the discussion of operationalization, precision concerns the fineness of distinctions made between the attributes that compose a variable. The description of a woman as "43 years old" is more precise than "in her forties." Saying a street-corner gang was formed in the summer of 1996 is more precise than saying "during the 1990s."

As a general rule, precise measurements are superior to imprecise ones, as common sense suggests. There are no conditions under which imprecise measurements are intrinsically superior to precise ones. Even so, exact precision is not always necessary or desirable. If knowing that a woman is in her forties satisfies your research requirements, then any additional effort invested in learning her precise age is wasted. The operationalization of concepts, then, must be guided partly by an understanding of the degree of precision required. If your needs are not clear, be more precise rather than less.

Don't confuse precision with accuracy, however. Describing someone as "born in New England" is less precise than "born in Stowe, Vermont"—but suppose the person in question was actually born in Boston. The less-precise description, in this instance, is more accurate, a better reflection of the real world.

Precision and accuracy are obviously important qualities in research measurement, and they probably need no further explanation. When social researchers construct and evaluate measurements, however, they pay special attention to two technical considerations: reliability and validity.

Reliability

In the abstract, **reliability** is a matter of whether a particular technique, applied repeatedly to the same object, yields the same result each time. Let's say you want to know how much I weigh. (No, I don't know why.) As one technique, say you ask two different people to estimate my weight. If the first person estimates 150 pounds and the other estimates 300, we have to conclude the technique of having people estimate my weight isn't very reliable.

Suppose, as an alternative, that you use a bathroom scale as your measurement technique. I step on the scale twice, and you note the result each time. The scale has presumably reported the same weight both times, indicating that the scale provides a more reliable technique for measuring a person's weight than does asking people to estimate it.

Reliability, however, does not ensure accuracy any more than precision does. Suppose I've set my bathroom scale to shave five pounds off my weight just to make me feel better. Although you would (reliably) report the same weight for me each time, you would always be wrong. This new element, called *bias,* is discussed in Chapter 7. For now, just be warned that reliability does not ensure accuracy.

Let's suppose we're interested in studying morale among factory workers in two different kinds of factories. In one set of factories, workers have specialized jobs, reflecting an extreme division of labor. Each worker contributes a tiny part to the overall process performed on a long assembly line. In the other set of factories, each worker performs many tasks, and small teams of workers complete the whole process.

How should we measure morale? Following one strategy, we could observe the workers in each factory, noticing such things as whether they joke with one another, whether they smile and laugh a lot, and so forth. We could ask them how they like their work and even ask them whether they think they would prefer their current arrangement or the other one being studied. By comparing what we observed in the different factories, we might reach a conclusion about which assembly process produces the higher morale.

Now let's look at some reliability problems inherent in this method. First, how you and I are feeling when we do the observing will likely color what we see. We may misinterpret what we see. We may see workers kidding each other but think they're having an argument. We may catch them on an off day. If we were to observe the same group of workers several days in a row, we might arrive at different evaluations on each day. If several observers evaluated the same behavior, on the other hand, they similarly might arrive at different conclusions about the workers' morale.

Here's another strategy for assessing morale. Suppose we check the company records to see how many grievances have been filed with the union during some fixed period. Presumably this would be an indicator of morale: the more grievances, the lower the morale. This measurement strategy would appear to be more reliable: Counting up the grievances over and over, we should keep arriving at the same number.

If you find yourself thinking that the number of grievances doesn't necessarily measure morale, you're worrying about validity, not reliability. We'll discuss validity in a moment. The point for now is that the last method is more like my bathroom scale—it gives consistent results.

In social research, reliability problems crop up in many forms. Reliability is a concern every time a single observer is the source of data, because we have no certain guard against the impact of that observer's subjectivity. We can't tell for sure how much of what's reported originated in the situation observed and how much came from the observer.

Subjectivity is not only a problem with single observers, however. Survey researchers have known for a long time that different interviewers, because of their own attitudes and demeanors, get different answers from respondents. Or, if we were to conduct a study of newspapers' editorial positions on some public issue, we might create a team of coders to take on the job of reading hundreds of editorials and classifying them in terms of their position on the issue. Unfortunately, different coders will code the same editorial differently. Or we might want to classify a few hundred specific occupations in terms of some standard coding scheme, say a set of categories created by the Department of Labor or by the Census Bureau. You and I would not place all those occupations in the same categories.

Each of these examples illustrates problems of reliability. Similar problems arise whenever we ask people to give us information about themselves. Sometimes we ask questions that people don't know the answers to: How many times have you been to church this year? Sometimes we ask people about things they consider totally irrelevant: Are you satisfied with China's current relationship with Albania? In such cases, people will answer differently at different times because they're making up answers as they go. Sometimes we explore issues so complicated that a person who had a clear opinion in the matter might arrive at a different interpretation of the question when asked a second time.

So how do you create reliable measures? If your research design calls for asking people for information, you can be careful to ask only about things the respondents are likely to know the answer to. Ask about things relevant to them, and be clear in what you're asking. Of course, these techniques don't solve every possible reliability problem. Fortunately, social researchers have developed several techniques for cross-checking the reliability of the measures they devise.

Test-Retest Method Sometimes it's appropriate to make the same measurement more than once, a technique called the *test-retest method.* If you don't expect the information being sought to change, then you should expect the same response both times. If answers vary, the measurement method may, to the extent of that variation, be unreliable. Here's an illustration.

In their research on Health Hazard Appraisal (HHA), a part of preventive medicine, Jeffrey Sacks, W. Mark Krushat, and Jeffrey Newman (1980) wanted to determine the risks associated with various background and lifestyle factors, making it possible for physicians to counsel their patients appropriately. By knowing patients' life situations, physicians could advise them on their potential for survival and on how to improve it. This purpose, of course, depended heavily on the accuracy of the information gathered about each subject in the study.

To test the reliability of their information, Sacks and his colleagues had all 207 subjects complete a baseline questionnaire that asked about their characteristics and behavior. Three months later, a follow-up questionnaire asked the same subjects for the same information, and the results of the two surveys were compared. Overall, only 15 percent of the subjects reported the same information in both studies.

Sacks and his colleagues reported the following:

> Almost 10 percent of subjects reported a different height at follow-up examination. Parental age was changed by over one in three subjects. One parent reportedly aged 20 chronologic years in three months. One in five ex-smokers and ex-drinkers have apparent difficulty in reliably recalling their previous consumption pattern. — (1980:730)

Some subjects had erased all trace of previously reported heart murmur, diabetes, emphysema, arrest record, and thoughts of suicide. One subject's mother, deceased in the first questionnaire, was apparently alive and well in time for the second. One subject had one ovary missing in the first study but present in the second. In another case, an ovary present in the first study was missing in the second study—and had been for ten years! One subject was reportedly 55 years old in the first study and 50 years old three months later. (You have to wonder whether the physician-counselors could ever have nearly the impact on their patients that their patients' memories had.) Thus, test-retest revealed that this data-collection method was not especially reliable.

Split-Half Method As a general rule, it's always good to make more than one measurement of any subtle or complex social concept, such as prejudice, alienation, or social class. This procedure lays the groundwork for another check on reliability. Let's say you've created a questionnaire that contains ten items you believe measure prejudice against women. Using the split-half technique, you would randomly assign those ten items to two sets of five. As we saw in the discussion of interchangeability of indicators, each set should provide a good measure of prejudice against women, and the two sets should classify respondents the same way. If the two sets of items classify people differently, you

most likely have a problem of reliability in your measure of the variable.

Using Established Measures Another way to help ensure reliability in getting information from people is to use measures that have proven their reliability in previous research. If you want to measure anomia, for example, you might want to follow Srole's lead.

The heavy use of measures, though, does not guarantee their reliability. For example, the Scholastic Aptitude Tests and the Minnesota Multiphasic Personality Inventory (MMPI) have been accepted as established standards in their respective domains for decades. In recent years, though, they've needed fundamental overhauling to reflect changes in society, eliminating outdated topics and gender bias in wording.

Reliability of Research Workers As we've seen, it's also possible for measurement unreliability to be generated by research workers: interviewers and coders, for example. There are several ways to check on reliability in such cases. To guard against interviewer unreliability, it is common practice in surveys to have a supervisor call a subsample of the respondents on the telephone and verify selected pieces of information.

Replication works in other situations also. If you're worried that newspaper editorials or occupations may not be classified reliably, you could have each independently coded by several coders. Those cases that are classified inconsistently can then be evaluated more carefully and resolved.

Finally, clarity, specificity, training, and practice can prevent a great deal of unreliability and grief. If you and I spent some time reaching a clear agreement on how to evaluate editorial positions on an issue—discussing various positions and reading through several together—we could probably do a good job of classifying them in the same way independently.

The reliability of measurements is a fundamental issue in social research, and we'll return to it more than once in the chapters ahead. For now, however, let's recall that even total reliability doesn't ensure that our measures measure what

APPLYING THE RESULTS

Replication in measurement is exactly the issue that was raised in Florida during the 2000 Presidential election. Specifically, given the thousands of ballots rejected by the vote-counting machines, many people questioned the reliability of this method of measuring votes. Had the election been a survey of voting intentions, the researchers would have reviewed the ballots rejected by the machine and sought to make judgments regarding those intentions. Notice that decisions about hanging and pregnant "chads" would have concerned measurement procedures. Much of the debate hinged on how much each type of chad reflected voter intent: What does a bump in a chad "really" mean? Without agreement on what really constituted a vote, researchers would have simply scored ballots in terms of which candidate was apparently selected and why—that is, the basis on which the researcher's decision was made (e.g., "pregnant chad"). Of course, there would first have to be clear operational definitions of what "swinging," "pregnant," and other sorts of chads were. It would certainly have been possible to determine the results in terms of the standards used—that is, how many of each type of vote for each candidate.

we think they measure. Now let's plunge into the question of validity.

Validity

In conventional usage, **validity** refers to the extent to which an empirical measure adequately reflects the real meaning of the concept under consideration. Whoops! I've already committed us to the view that concepts don't have real meanings. How can we ever say whether a particular measure adequately reflects the concept's meaning, then? Ultimately, of course, we can't. At the same time, as

we've already seen, all of social life, including social research, operates on agreements about the terms we use and the concepts they represent. There are several criteria of success in making measurements that are appropriate to these agreed-upon meanings of concepts.

First, there's something called **face validity.** Particular empirical measures may or may not jibe with our common agreements and our individual mental images concerning a particular concept. For example, you and I might quarrel about the adequacy of measuring worker morale by counting the number of grievances filed with the union. Still, we'd surely agree that the number of grievances has something to do with morale. That is, the measure is valid "on its face," whether or not it's adequate. If I were to suggest that we measure morale by finding out how many books the workers took out of the library during their off-duty hours, you'd undoubtedly raise a more serious objection: That measure wouldn't have much face validity.

Second, I've already pointed to many of the more formally established agreements that define some concepts. The Census Bureau, for example, has created operational definitions of such concepts as family, household, and employment status that seem to have a workable validity in most studies using these concepts.

Three additional types of validity also specify particular ways of testing the validity of measures. The first, **criterion-related validity,** sometimes called predictive validity, is based on some external criterion. For example, the validity of College Board exams is shown in its ability to predict students' success in college. The validity of a written driver's test is determined, in this sense, by the relationship between the scores people get on the test and their subsequent driving records. In these examples, college success and driving ability are the criteria.

To test your understanding of criterion-related validity, see whether you can think of behaviors that might be used to validate each of the following attitudes:

Is very religious
Supports equality of men and women
Supports far-right militia groups
Is concerned about the environment

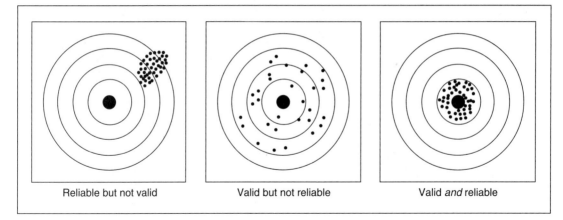

| Reliable but not valid | Valid but not reliable | Valid *and* reliable |

FIGURE 5-2 **An Analogy to Validity and Reliability**

Some possible validators would be, respectively, attends church, votes for women candidates, belongs to the NRA, and belongs to the Sierra Club. Sometimes it's difficult to find behavioral criteria that can be taken to validate measures as directly as in such examples. In those instances, however, we can often approximate such criteria by applying a different test. We can consider how the variable in question ought, theoretically, to relate to other variables. **Construct validity** is based on the logical relationships among variables.

Suppose, for example, that you want to study the sources and consequences of marital satisfaction. As part of your research, you develop a measure of marital satisfaction, and you want to assess its validity.

In addition to developing your measure, you'll have developed certain theoretical expectations about the way the variable *marital satisfaction* relates to other variables. For example, you might reasonably conclude that satisfied husbands and wives will be less likely than dissatisfied ones to cheat on their spouses. If your measure relates to marital fidelity in the expected fashion, that constitutes evidence of your measure's construct validity. If satisfied marriage partners are as likely to cheat on their spouses as are the dissatisfied ones, however, that would challenge the validity of your measure.

Tests of construct validity, then, can offer a weight of evidence that your measure either does or doesn't tap the quality you want it to measure, but this evidence is not definitive proof. Although I have suggested that tests of construct validity are less compelling than tests of criterion validity, there is room for disagreement about which kind of test a particular comparison variable (driving record, marital fidelity) represents in a given situation. It's less important to distinguish the two types of validity tests than to understand the logic of validation that they have in common: If we have been successful in measuring some variable, then our measures should relate in some logical way to other measures.

Finally, **content validity** refers to how much a measure covers the range of meanings included within a concept. For example, a test of mathematical ability cannot be limited to addition alone but also needs to cover subtraction, multiplication, division, and so forth. Or, if we're measuring prejudice, do our measurements reflect all types of prejudice, including prejudice against racial and ethnic groups, religious minorities, women, the elderly, and so on?

Figure 5-2 presents a graphic portrayal of the difference between validity and reliability. If you think of measurement as analogous to repeatedly shooting at the bull's-eye on a target, you'll see that reliability looks like a "tight pattern," regardless of where the shots hit, because reliability is a function of consistency. Validity, on the other hand, is a function of shots being arranged around the bull's-eye. The failure of reliability in the figure is randomly distributed around the target; the failure of validity is systematically off the mark. Notice that neither an unreliable nor an invalid measure is likely to be very useful.

Who Decides What's Valid?

Our discussion of validity began with a reminder that we depend on agreements to determine what's real, and we've just seen some of the ways social scientists can agree among themselves that they have made valid measurements. There is yet another way of looking at validity.

Social researchers sometimes criticize themselves and one another for implicitly assuming they are somewhat superior to those they study. For example, researchers often seek to uncover motivations that the social actors themselves are unaware of. You think you bought that new Burpo-Blasto because of its high performance and good looks, but we know you're really trying to achieve a higher social status.

This implicit sense of superiority would fit comfortably with a totally positivistic approach (the biologist feels superior to the frog on the lab table), but it clashes with the more humanistic and typically qualitative approach taken by many social scientists. We'll explore this issue more deeply in Chapter 10.

In seeking to understand the way ordinary people make sense of their worlds, ethnomethodologists have urged all social scientists to pay more respect to these natural social processes of conceptualization and shared meaning. At the very least, behavior that may seem irrational from the scientist's paradigm may make logical sense when viewed from the actor's point of view.

Ultimately, social researchers should look to both colleagues and subjects as sources of agreement on the most useful meanings and measurements of the concepts they study. Sometimes one

---

A QUANDARY **REVISITED**

Can social scientists measure religiosity and determine its causes? As you've seen, making descriptive statements about religiosity is harder than making explanatory ones. Any assertion about who is or is not religious will depend directly on the definitions used. By one definition, nearly all of the population would be deemed religious; by another definition, few would be so designated.

As the discussion of the interchangeability of indicators suggested, however, we can be more confident and definitive in speaking of the causes of religiosity. For example, it is often reported that U.S. women are more religious than U.S. men. This assertion is based on the observation that women are more religious than men on virtually all indicators of religiosity: church attendance, prayer, beliefs, and so forth. So, even though we may likely disagree on which of these indicators is the best or truest measure of what we mean by the term *religiosity*, women would be more religious than men regardless of the indicator chosen.

---

source will be more useful, sometimes the other. But neither should be dismissed.

Tension between Reliability and Validity

Clearly, we want our measures to be both reliable and valid. Often, however, a tension arises between the criteria of reliability and validity, forcing a trade-off between the two.

Recall the example of measuring morale in different factories. The strategy of immersing yourself in the day-to-day routine of the assembly line, observing what goes on, and talking to the workers would seem to provide a more valid measure of morale than would counting grievances. It just seems obvious that we'd get a clearer sense of whether the morale was high or low using this first method.

As I pointed out earlier, however, the counting strategy would be more reliable. This situation reflects a more general strain in research measurement. Most of the really interesting concepts we want to study have many subtle nuances, and it's hard to specify precisely what we mean by them. Researchers sometimes speak of such concepts as having a "richness of meaning." Although scores of books and articles have been written on the topic of anomie/anomia, for example, they still haven't exhausted its meaning.

Very often, then, specifying reliable operational definitions and measurements seems to rob concepts of their richness of meaning. Positive morale is much more than a lack of grievances filed with the union; anomie is much more than what is measured by the five items created by Leo Srole. Yet, the more variation and richness we allow for a concept, the more opportunity there is for disagreement on how it applies to a particular situation, thus reducing reliability.

To some extent, this dilemma explains the persistence of two quite different approaches to social research: quantitative, nomothetic, structured techniques such as surveys and experiments on the one hand and qualitative, idiographic methods such as field research and historical studies on the other. In the simplest generalization, the former methods tend to be more reliable, the latter more valid.

By being forewarned, you'll be effectively forearmed against this persistent and inevitable dilemma. If there is no clear agreement on how to measure a concept, measure it several different ways. If the concept has several dimensions, measure them all. Above all, know that the concept does not have any meaning other than what you and I give it. The only justification for giving any concept a particular meaning is utility. Measure concepts in ways that help us understand the world around us.

Main Points

- Conceptions are mental images we use as summary devices for bringing together observations and experiences that seem to have something in common. We use terms or labels to reference these conceptions.

- Concepts are constructs; they represent the agreed-upon meanings we assign to terms. Our concepts don't exist in the real world, so they can't be measured directly, but it's possible to measure the things that our concepts summarize.

- Conceptualization is the process of specifying observations and measurements that give concepts definite meaning for the purposes of a research study.

- Conceptualization includes specifying the indicators of a concept and describing its dimensions. Operational definitions specify how variables relevant to a concept will be measured.

- Precise definitions are even more important in descriptive than in explanatory studies. The degree of precision needed varies with the type and purpose of a study.

- Operationalization is an extension of conceptualization that specifies the exact procedures that will be used to measure the attributes of variables.

- Operationalization involves a series of interrelated choices: specifying the range of variation that is appropriate for the purposes of a study, determining how precisely to measure variables, accounting for relevant dimensions of variables, clearly defining the attributes of variables and their relationships, and deciding on an appropriate level of measurement.

- Researchers must choose from four types of measures that capture increasing amounts of information: nominal, ordinal, interval, and ratio. The most appropriate level depends on the purpose of the measurement.

- A given variable can sometimes be measured at different levels. When in doubt, researchers should use the highest level of measurement appropriate to that variable so they can capture the greatest amount of information.

- Operationalization begins in the design phase of a study and continues through all phases of the research project, including the analysis of data.

- Criteria of the quality of measures include precision, accuracy, reliability, and validity.

- Whereas reliability means getting consistent results from the same measure, validity refers to getting results that accurately reflect the concept being measured.

- Researchers can test or improve the reliability of measures through the test-retest method, the split-half method, the use of established measures, and the examination of work performed by research workers.

- The yardsticks for assessing a measure's validity include face validity, criterion-related validity, construct validity, and content validity.

- Creating specific, reliable measures often seems to diminish the richness of meaning our general concepts have. This problem is inevitable. The best solution is to use several different measures, tapping the different aspects of a concept.

Key Terms

conceptualization	reliability
indicator	validity
dimension	face validity
nominal measures	criterion-related validity
ordinal measures	construct validity
interval measures	content validity
ratio measures	

Review Questions

1. Pick a social science concept such as liberalism or alienation, then specify that concept so that it could be studied in a research project. How would you specify the indicators you would use as well as the dimensions you wish to include in and exclude from your conceptualization?

2. This chapter presented some of the questionnaire items commonly used to measure anomia (an individual characteristic) and discussed anomie (a characteristic of society). If you wanted to compare two societies on anomie, or normlessness, what indicators might you look at? For example, what statistical indicators might you examine? What nonstatistical, qualitative indicators?

3. What level of measurement—nominal, ordinal, interval, or ratio—describes each of the following variables?

   a. Race (white, African American, Asian, and so on)

   b. Order of finish in a race (first, second, third, and so on)

   c. Number of children in families

   d. Populations of nations

   e. Attitudes toward nuclear energy (strongly approve, approve, disapprove, strongly disapprove)

   f. Region of birth (Northeast, Midwest, and so on)

   g. Political orientation (very liberal, somewhat liberal, somewhat conservative, very conservative)

4. Suppose you've been contracted by Wadsworth Publishing Company to interview a group of students and evaluate their level of satisfaction with this textbook. How would you measure satisfaction in this case?

Additional Readings

Bohrnstedt, George W. 1983. "Measurement." Pp. 70-121 in *Handbook of Survey Research,* edited by Peter H. Rossi, James D. Wright, and Andy B. Anderson. New York: Academic Press. This essay offers the logical and statistical grounding of reliability and validity in measurement.

Carmines, Edward G., and Richard A. Zeller. 1979. *Reliability and Validity Assessment.* Beverly Hills, CA: Sage.

In this chapter, we've examined the basic logic of validity and reliability in social science measurement. Carmines and Zeller explore these issues in more detail and examine some of the ways to calculate reliability mathematically.

Gould, Julius, and William Kolb. 1964. *A Dictionary of the Social Sciences.* New York: Free Press. A primary reference to the social scientific agreements on various concepts. Although the terms used by social scientists do not have ultimately "true" meanings, this reference book lays out the meanings social scientists have in mind when they use those terms.

Grimes, Michael D. 1991. *Class in Twentieth-Century American Sociology: An Analysis of Theories and Measurement Strategies.* New York: Praeger. This book provides an excellent, long-term view of conceptualization as the author examines a variety of theoretical views of social class and the measurement techniques appropriate to those theories.

Lazarsfeld, Paul F., and Morris Rosenberg, eds. 1955. *The Language of Social Research,* Section I. New York: Free Press of Glencoe. An excellent and diverse classic collection of descriptions of specific measurements in past social research. These 14 articles present useful and readable accounts of actual measurement operations performed by social researchers as well as more conceptual discussions of measurement in general.

Silverman, David. 1993. *Interpreting Qualitative Data: Methods for Analyzing Talk, Text, and Interaction,* Chapter 7. Newbury Park, CA: Sage. This deals with the issues of validity and reliability specifically in regard to qualitative research.

U.S. Department of Health and Human Services. 1992. *Survey Measurement of Drug Use.* Washington, DC: U.S. Government Printing Office. An extensive review of techniques devised and used for measuring various kinds of drug use.

Wallace, Walter. 1971. *The Logic of Science in Sociology,* Chapter 3. Chicago: Aldine-Atherton. A brief and lucid presentation of concept formation within the context of other research steps. This discussion relates conceptualization to observation on the one hand and to generalization on the other.

Multimedia Resources

The Wadsworth Sociology Resource Center: Virtual Society
http://sociology.wadsworth.com/

Visit the companion Web site for the second edition of *The Basics of Social Research* to access a wide range of student resources. Begin by clicking on the Student Resources section of the book's Web site to access the following study tools:

- eBabbie Resource Center
- Planning a Research Project
- Doing Data Analysis
- Statistics Review
- Flash Cards
- Internet Links and Exercises
- InfoTrac College Edition: Exercises
- Quizzes

Visit the **eBabbie Resource Center** for an overview of each chapter and helpful online tutorials. Find information on budgeting and step-by-step examples of model research projects at **Planning a Research Project.** Learn how to use quantitative and qualitative data analysis programs at **Doing Data Analysis,** and brush up on your statistics at **Statistics Review.** You can also further your study by accessing **Internet Links and Exercises** related to chapter materials, **Flash Cards, Quizzes,** and many other learning tools.

InfoTrac College Edition
http://www.infotrac-college.com/ wadsworth/access.html

Access the latest news and journal articles with InfoTrac College Edition, an easy-to-use online database of reliable, full-length articles from hundreds of top academic journals. Conduct an electronic search using the following search terms:

Anomie
Attributes AND variables
Cohort study
Conceptualization
Construct validity
Operationalization
Panel study
Trend study
Validity AND reliability

6 INDEXES, SCALES, AND TYPOLOGIES

Stock Solution./Index Stock Imagery/PictureQuest

What You'll Learn in This Chapter

Now we conclude the discussion of measurement begun in
Chapter 5. Researchers often need to employ multiple indicators
to measure a variable adequately and validly. Indexes, scales, and
typologies are useful composite measures made up of several
indicators of variables. In this chapter you'll learn the logic and
skills of constructing such measures.

In this chapter . . .

AN OPENING **QUANDARY**

Often, data analysis aims at reducing a mass of observations to a more manageable form. Our use of concepts to stand for many similar observations is one example.

Sometimes this sort of reduction can be accomplished in the analysis of quantitative data. You could, for example, ask people to answer five different questions, reduce each person's answers to a single number, and then use that number to reproduce that person's answers. So, if you told me that you had assigned someone a score of 3, I would be able to tell you how they answered each of the original five questions.

How in the world could you do this?

INTRODUCTION

As we saw in Chapter 5, many social scientific concepts have complex and varied meanings. Making measurements that capture such concepts can be a challenge. Recall our discussion of content validity, which concerns whether we have captured all the different dimensions of a concept.

To achieve broad coverage of the various dimensions of a concept, we usually need to make multiple observations pertaining to a given concept. Thus, for example, Bruce Berg (1989:21) advises in-depth interviewers to prepare "essential questions," which are "geared toward eliciting specific, desired information." In addition, the researcher should prepare extra questions: "questions roughly equivalent to certain essential ones, but worded slightly differently."

Multiple indicators are used with quantitative data as well. Though you can sometimes construct a single questionnaire item that captures the variable of interest—"Gender: ☐ Male ☐ Female" is a simple example—other variables are less straightforward and may require you to use several questionnaire items to measure them adequately.

Quantitative data analysts have developed specific techniques for combining indicators into a single measure. This chapter discusses the construction of two types of composite measures of variables—indexes and scales. While scales and indexes can be used in any form of social research, they are most common in survey research and other quantitative methods. A short section at the end of the chapter considers typologies, which are relevant to both qualitative and quantitative research.

Composite measures are frequently used in quantitative research, for several reasons. First, social scientists often wish to study variables that have no clear and unambiguous single indicators. Single indicators do suffice for some variables, such as age. We can determine a survey respondent's age by simply asking: "How old are you?" Similarly, we can determine a newspaper's circulation by merely looking at the figure the newspaper reports. In the case of complex concepts, however, researchers can seldom develop single indicators before they actually do the research. This is especially true with regard to attitudes and orientations. Rarely can a survey researcher, for example, devise single questionnaire items that adequately tap respondents' degrees of prejudice, religiosity, political orientations, alienation, and the like. More likely, the researcher will devise several items, each of which provides some indication of the variables. Taken individually, each of these items is likely to prove invalid or unreliable for many respondents. A composite measure, however, can overcome this problem.

Second, researchers may wish to employ a rather refined ordinal measure of a particular variable—alienation, say—arranging cases in several ordinal categories from very low to very high, for example. A single data item might not have enough categories to provide the desired range of variation. However, an index or scale formed from several items can provide the needed range.

Finally, indexes and scales are efficient devices for data analysis. If considering a single data item gives us only a rough indication of a given variable, considering several data items can give us a more comprehensive and more accurate indication. For example, a single newspaper editorial may give us some indication of the political orientations of that newspaper. Examining several editorials would probably give us a better assessment, but the manipulation of several data items simultaneously could be very complicated. Indexes and scales (especially scales) are efficient data-reduction devices: They allow us to summarize several indicators in a single numerical score, while sometimes nearly maintaining the specific details of all the individual indicators.

Indexes versus Scales

The terms *index* and *scale* are typically used imprecisely and interchangeably in social research literature. The two types of measures do have some characteristics in common, but in this book we'll distinguish between them.

First, let's consider what they have in common. Both scales and indexes are ordinal measures of variables. Both rank-order the units of analysis in terms of specific variables such as *religiosity, alienation, socioeconomic status, prejudice,* or *intellectual sophistication.* A person's score on either a scale or an index of religiosity, for example, gives an indication of his or her relative religiosity vis-à-vis other people.

Further, both scales and indexes are *composite measures of variables:* measurements based on more than one data item. Thus, a survey respondent's score on an index or scale of religiosity is determined by the responses given to several questionnaire items, each of which provides some indication of religiosity. Similarly, a person's IQ score is based on answers to a large number of test questions. The political orientation of a newspaper might be represented by an index or scale score reflecting the newspaper's editorial policy on various political issues.

Despite these shared characteristics, it's useful to distinguish between indexes and scales. In this book we'll distinguish indexes and scales through the manner in which scores are assigned. We construct an **index** simply by accumulating scores assigned to individual attributes. We might measure prejudice, for example, by adding up the number of prejudiced statements each respondent agreed with. We construct a **scale,** however, by assigning scores to patterns of responses, recognizing that

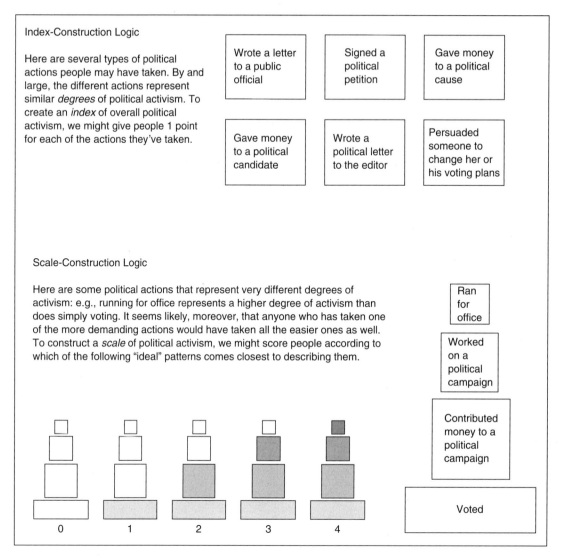

FIGURE 6-1 **Indexes versus Scales**

some items reflect a relatively weak degree of the variable while others reflect something stronger. For example, agreeing that "Women are different from men" is, at best, weak evidence of sexism compared with agreeing that "Women should not be allowed to vote." A scale takes advantage of differences in intensity among the attributes of the same variable to identify distinct patterns of response.

Figure 6-1 provides a graphic illustration of the difference between indexes and scales. Let's as-

sume we want to develop a measure of political activism, distinguishing those people who are very active in political affairs, those who don't participate much at all, and those who are somewhere in between.

The first part of Figure 6-1 illustrates the logic of indexes. The figure shows six different political actions. Although you and I might disagree on some specifics, I think we could agree that the six actions represent roughly the same degree of political activism.

Using these six items, we could construct an index of political activism by giving each person 1 point for each of the actions he or she has taken. If you wrote to a public official and signed a petition, you'd get a total of 2 points. If I gave money to a candidate and persuaded someone to change his or her vote, I'd get the same score as you. Using this approach, we'd conclude that you and I had the same degree of political activism, even though we had taken different actions.

The second part of Figure 6-1 describes the logic of scale construction. In this case, the actions clearly represent different degrees of political activism—ranging from simply voting to running for office. Moreover, it seems safe to assume a pattern of actions in this case. For example, all those who contributed money probably also voted. Those who worked on a campaign probably also gave some money and voted. This suggests that most people will fall into only one of five idealized action patterns, represented by the small illustrations at the bottom of the figure. The discussion of scales, later in this chapter, describes ways of identifying people with the type they most closely represent.

As you might surmise, scales are generally superior to indexes, because scales take into consideration the intensity with which different items reflect the variable being measured. Also, as the example in Figure 6-1 shows, scale scores convey more information than do index scores. Be aware, though, that the term *scale* is commonly misused to refer to measures that are only indexes. Merely calling a given measure a scale instead of an index doesn't make it better.

There are two other misconceptions about scaling that you should know. First, whether the combination of several data items results in a scale almost always depends on the particular sample of observations under study. Certain items may form a scale within one sample but not within another. For this reason, do not assume that a given set of items is a scale simply because it has turned out that way in an earlier study.

Second, the use of specific scaling techniques—such as Guttman scaling, to be discussed—does not ensure the creation of a scale. Rather, such techniques let us determine whether or not a set of items constitutes a scale.

An examination of actual social science research reports will show that researchers use indexes much more than scales. Ironically, however, the methodological literature contains little if any discussion of index construction, while discussions of scale construction abound. There appear to be two reasons for this disparity. First, indexes are more frequently used because scales are often difficult or impossible to construct from the data at hand. Second, methods of index construction seem so obvious and straightforward that they aren't discussed much.

Constructing indexes is not a simple undertaking, however. The general failure to develop index construction techniques has resulted in many bad indexes in social research. With this in mind, I've devoted over half of this chapter to the methods of index construction. With a solid understanding of the logic of this activity, you'll be better equipped to try constructing scales. Indeed, a carefully constructed index may turn out to be a scale.

INDEX CONSTRUCTION

Let's look now at four main steps in the construction of an index: selecting possible items, examining their empirical relationships, scoring the index, and validating it. We'll conclude this discussion by examining the construction of an index that provided interesting findings about the status of women in different countries.

Item Selection

The first step in creating an index is selecting items for a composite index, which is created to measure some variable.

Face Validity The first criterion for selecting items to be included in an index is face validity (or logical validity). If you want to measure political conservatism, for example, each of your items should appear on its face to indicate conservatism (or its opposite, liberalism). Political party affiliation would be one such item. Another would be an item asking people to approve or disapprove of the views of a well-known conservative public figure.

In constructing an index of religiosity, you might consider items such as church attendance, acceptance of certain religious beliefs, and frequency of prayer; each of these appears to offer some indication of religiosity.

Unidimensionality   The methodological literature on conceptualization and measurement stresses the need for unidimensionality in scale and index construction; that is, a composite measure should represent only one dimension of a concept. Thus, items reflecting religiosity should not be included in a measure of political conservatism, even though the two variables might be empirically related to each other.

General or Specific   Although measures should tap the same dimension, the general dimension you're attempting to measure may have many nuances. In the example of religiosity, the indicators mentioned previously—ritual participation, belief, and so on—represent different types of religiosity. If you wished to focus on ritual participation in religion, you should choose items specifically indicating this type of religiosity: church attendance, communion, confession, and the like. If you wished to measure religiosity in a more general way, you would include a balanced set of items, representing each of the different types of religiosity. Ultimately, the nature of the items included will determine how specifically or generally the variable is measured.

Variance   In selecting items for an index, you must also be concerned with the amount of variance they provide. If an item is intended to indicate political conservatism, for example, you should note what proportion of respondents would be identified as conservatives by the item. If a given item identified no one as a conservative or everyone as a conservative—for example, if nobody indicated approval of a radical-right political figure—that item would not be very useful in the construction of an index.

To guarantee variance, you have two options. First, you may select several items the responses to which divide people about equally in terms of the variable; for example, about half conservative and half liberal. Although no single response would justify characterizing a person as very conservative, a person who responded as a conservative on all items might be so characterized.

The second option is to select items differing in variance. One item might identify about half the subjects as conservative, while another might identify few of the respondents as conservatives. Note that this second option is necessary for scaling, and it is reasonable for index construction as well.

Examination of Empirical Relationships

The second step in index construction is to examine the empirical relationships among the items being considered for inclusion. In this, we are anticipating a discussion that will be pursued more fully in Chapter 14. An empirical relationship is established when respondents' answers to one question—in a questionnaire, for example—help us predict how they will answer other questions. If two items are empirically related to each other, we can reasonably argue that each reflects the same variable, and we may include them both in the same index. There are two types of possible relationships among items: bivariate and multivariate.

Bivariate Relationships among Items   A *bivariate relationship* is, simply put, a relationship between two variables. Suppose we want to measure respondents' support for U.S. participation in the United Nations. One indicator of different levels of support might be the question "Do you feel the U.S. financial support of the UN is ☐ Too high ☐ About right ☐ Too low?"

A second indicator of support for the United Nations might be the question "Should the United States contribute military personnel to UN peace-keeping actions? ☐ Strongly approve ☐ Mostly approve ☐ Mostly disapprove ☐ Strongly disapprove."

Both of these questions, on their face, seem to reflect different degrees of support for the United Nations. Nonetheless, some people might feel the United States should give more money but not provide troops. Others might favor sending troops but cutting back on financial support.

If the two items both reflect degrees of the same thing, however, we should expect responses to the two items to generally correspond with one another. Specifically, those who approve of military support should be more likely to favor financial support than would those who disapprove of military support. Conversely, those who favor financial support should be more likely to favor military support than would those disapproving of financial support. If these expectations are met, we say there is a *bivariate relationship* between the two items.

Here's another example. Suppose we want to determine the degree to which respondents feel women have the right to an abortion. We might ask (1) "Do you feel a woman should have the right to an abortion when her pregnancy was the result of rape?" and (2) "Do you feel a woman should have the right to an abortion if continuing her pregnancy would seriously threaten her life?"

Granted, some respondents might agree with item (1) and disagree with item (2); others will do just the reverse. If both items tap into some general opinion people have about the issue of abortion, then the responses to these two items should be related to each other. Those who support the right to an abortion in the case of rape should be more likely to support it if the woman's life is threatened than would those who disapproved of abortion in the case of rape. This would be another example of a bivariate relationship between the two items.

To determine the relative strengths of relationships among the several pairs of items, you should examine all the possible bivariate relationships among the several items being considered for inclusion in an index. Percentage tables or more advanced statistical techniques may be used for this purpose. How we evaluate the strength of the relationships, however, can be rather subtle. The box entitled "'Cause' and 'Effect' Indicators" examines some of these subtleties.

Be wary of items that are not related to one another empirically: It's unlikely they measure the same variable. You should probably drop any item that is not related to several other items.

At the same time, a very strong relationship between two items presents a different problem. If two items are perfectly related to one another, then only one needs to be included in the index; since it completely conveys the indications provided by the other, nothing more would be added by including the other item. (This problem will become even clearer in the next section.)

Here's an example to illustrate the testing of bivariate relationships in index construction. I once conducted a survey of medical school faculty members to find out about the consequences of a "scientific perspective" on the quality of patient care provided by physicians. The primary intent was to determine whether scientifically inclined doctors treated patients more impersonally than did other doctors.

The survey questionnaire offered several possible indicators of respondents' scientific perspectives. Of those, three items appeared to provide especially clear indications of whether the doctors were scientifically oriented:

1. As a medical school faculty member, in what capacity do you feel you can make your greatest *teaching* contribution: as a practicing physician or as a medical researcher?
2. As you continue to advance your own medical knowledge, would you say your ultimate medical interests lie primarily in the direction of total patient management or the understanding of basic mechanisms? [The purpose of this item was to distinguish those who were mostly interested in overall patient care from those mostly interested in biological processes.]
3. In the field of therapeutic research, are you generally more interested in articles reporting evaluations of the effectiveness of various treatments or articles exploring the basic rationale underlying the treatments? [Similarly, I wanted to distinguish those more interested in articles dealing with patient care from those more interested in biological processes.] — (BABBIE 1970:27–31)

For each of these items, we might conclude that those respondents who chose the second answer

"CAUSE" AND "EFFECT" INDICATORS

by Kenneth Bollen
Department of Sociology, University of North Carolina, Chapel Hill

While it often makes sense to expect indicators of the same variable to be positively related to one another, as discussed in the text, this is not always the case.

Indicators should be related to one another if they are essentially "effects" of a variable. For example, to measure self-esteem, we might ask a person to indicate whether he or she agrees or disagrees with the statements (1) "I am a good person" and (2) "I am happy with myself." A person with high self-esteem should agree with both statements while one with low self-esteem would probably disagree with both. Since each indicator depends on or "reflects" self-esteem, we expect them to be positively correlated. More generally, indicators that depend on the same variable should be associated with one another if they are valid measures.

But, this is not the case when the indicators are the "cause" rather than the "effect" of a variable. In this situation the indicators may correlate positively, negatively, or not at all. For example, we could use gender and race as indicators of the variable *exposure to discrimination.*

Being nonwhite or female increases the likelihood of experiencing discrimination, so both are good indicators of the variable. But we would not expect the race and gender of individuals to be strongly associated.

Or, we may measure *social interaction* with three indicators: time spent with friends, time spent with family, and time spent with coworkers. Though each indicator is valid, they need not be positively correlated. Time spent with friends, for instance, may be inversely related to time spent with family. Here, the three indicators "cause" the degree of social interaction.

As a final example, *exposure to stress* may be measured by whether a person recently experienced divorce, death of a spouse, or loss of a job. Though any of these events may indicate stress, they need not correlate with one another.

In short, we expect an association between indicators that depend on or "reflect" a variable, that is, if they are the "effects" of the variable. But if the variable depends on the indicators—if the indicators are the "causes"—those indicators may be either positively or negatively correlated, or even unrelated. Therefore, we should decide whether indicators are causes or effects of a variable before using their intercorrelations to assess their validity.

---

are more scientifically oriented than respondents who chose the first answer. Though this comparative conclusion is reasonable, we should not be misled into thinking that respondents who chose the second answer to a given item are scientists in any absolute sense. They are simply more scientifically oriented than those who chose the first answer to the item.

To see this point more clearly, let's examine the distribution of responses to each item. From the first item—greatest teaching contribution—only about one-third of the respondents appeared scientifically oriented. That is, approximately one-

third said they could make their greatest teaching contribution as medical researchers. In response to the second item—ultimate medical interests—approximately two-thirds chose the scientific answer, saying they were more interested in learning about basic mechanisms than learning about total patient management. In response to the third item—reading preferences—about 80 percent chose the scientific answer.

These three questionnaire items can't tell us how many "scientists" there are in the sample, for none of them is related to a set of criteria for what constitutes being a scientist in any absolute sense.

A.

		Greatest Teaching Contribution	
		Physician	Researcher
Ultimate Medical Interest	Total patient management	49%	13%
	Basic mechanisms	51%	87%
		100% (268)	100% (159)

B.

		Reading Preferences	
		Effectiveness	Rationale
Ultimate Medical Interest	Total patient management	68%	30%
	Basic mechanisms	32%	70%
		100% (78)	100% (349)

C.

		Reading Preferences	
		Effectiveness	Rationale
Greatest Teaching Contribution	Physician	85%	64%
	Researcher	15%	36%
		100% (78)	100% (349)

FIGURE 6-2 **Bivariate Relationships among Scientific Orientation Items**

Using the items for this purpose would present us with the problem of three quite different estimates of how many scientists there were in the sample.

However, these items do provide us with three independent indicators of respondents' relative inclinations toward science in medicine. Each item separates respondents into the more scientific and the less scientific. But each grouping of more or less scientific respondents will have a somewhat different membership from the others. Respondents who seem scientific in terms of one item will not seem scientific in terms of another. Nevertheless, to the extent that each item measures the same general dimension, we should find some correspondence among the several groupings. Re-

spondents who appear scientific in terms of one item should be more likely to appear scientific in their response to another item than would those who appeared nonscientific in their response to the first. In other words, we should find an association or correlation between the responses given to two items.

Figure 6-2 shows the associations among the responses to the three items. Three bivariate tables are presented, showing the distribution of responses for each pair of items. An examination of the three bivariate relationships presented in the figure supports the suggestion that the three items all measure the same variable: *scientific orientation*. To see why this is so, let's begin by looking

Percentage Interested in Basic Mechanisms		Greatest Teaching Contribution	
		Physician	Researcher
Reading Preferences	Effectiveness	27% (66)	58% (12)
	Rationale	58% (219)	89% (130)

FIGURE 6-3 **Trivariate Relationships among Scientific Orientation Items**

at the first bivariate relationship in the table. The table shows that faculty who responded that "researcher" was the role in which they could make their greatest teaching contribution were more likely to identify their ultimate medical interests as "basic mechanisms" (87 percent) than were those who answered "physician" (51 percent). The fact that the "physicians" are about evenly split in their ultimate medical interests is irrelevant for our purposes. It is only relevant that they are less scientific in their medical interests than are the "researchers." The strength of this relationship may be summarized as a 36 percentage point difference.

The same general conclusion applies to the other bivariate relationships. The strength of the relationship between reading preferences and ultimate medical interests may be summarized as a 38 percentage point difference, and the strength of the relationship between reading preferences and the two teaching contributions as a 21 percentage point difference. In summary, then, each single item produces a different grouping of "scientific" and "nonscientific" respondents. However, the responses given to each of the items correspond, to a greater or lesser degree, to the responses given to each of the other items.

Initially, the three items were selected on the basis of face validity—each appeared to give some indication of faculty members' orientations to science. By examining the bivariate relationship between the pairs of items, we have found support for the expectation that they all measure basically the same thing. However, that support does not sufficiently justify including the items in a composite index. Before combining them in a single index, we need to examine the multivariate relationships among the several variables.

Multivariate Relationships among Items
Whereas a bivariate relationship deals with two variables at a time, a multivariate one uses more than two variables. To present the trivariate relationships among the three variables in our example, we would first categorize the sample medical school respondents into four groups according to (1) their greatest teaching contribution and (2) their reading preferences. Figure 6-3 does just that. The numbers in parentheses indicate the number of respondents in each group. Thus, 66 of the faculty members who said they could best teach as physicians also said they preferred articles dealing with the effectiveness of treatments. Then, for each of the four groups, we would determine the percentage of those who say they are ultimately more interested in basic mechanisms. So, for example, of the 66 faculty mentioned, 27 percent are primarily interested in basic mechanisms, as the figure shows.

The arrangement of the four groups is based on a previously drawn conclusion regarding scientific orientations. The group in the upper left corner of the table is presumably the least scientifically oriented, based on greatest teaching contribution and reading preference. The group in the lower right corner is presumably the most scientifically oriented in terms of those items.

Recall that expressing a primary interest in basic mechanisms was also taken as an indication of

Percentage Interested in Basic Mechanisms		Greatest Teaching Contribution	
		Physician	Researcher
Reading Preferences	Effectiveness	51% (66)	87% (12)
	Rationale	51% (219)	87% (130)

FIGURE 6-4 **Hypothetical Trivariate Relationship among Scientific Orientation Items**

scientific orientation. As we should expect, then, those in the lower-right corner are the most likely to give this response (89 percent), and those in the upper-left corner are the least likely (27 percent). The respondents who gave mixed responses in terms of teaching contributions and reading preferences have an intermediate rank in their concern for basic mechanisms (58 percent in both cases).

This figure tells us many things. First, we may note that the original relationships between pairs of items are not significantly affected by the presence of a third item. Recall, for example, that the relationship between teaching contribution and ultimate medical interest was summarized as a 36 percentage point difference. Looking at Figure 6-3, we see that among only those respondents who are most interested in articles dealing with the effectiveness of treatments, the relationship between teaching contribution and ultimate medical interest is 31 percentage points (58 percent minus 27 percent: first row). The same is true among those most interested in articles dealing with the rationale for treatments (89 percent minus 58 percent: second row). The original relationship between teaching contribution and ultimate medical interest is essentially the same as in Figure 6-2, even among those respondents judged as scientific or nonscientific in terms of reading preferences.

We can draw the same conclusion from the columns in Figure 6-3. Recall that the original relationship between reading preferences and ultimate medical interests was summarized as a 38 percentage point difference. Looking only at the "physicians" in Figure 6-3, we see that the relationship

between the other two items is now 31 percentage points. The same relationship is found among the "researchers" in the second column.

The importance of these observations becomes clearer when we consider what might have happened. In Figure 6-4, hypothetical data tell a much different story than do the actual data in Figure 6-3. As you can see, Figure 6-4 shows that the original relationship between teaching contribution and ultimate medical interest persists, even when reading preferences are introduced into the picture. In each row of the table, the "researchers" are more likely to express an interest in basic mechanisms than are the "physicians." Looking down the columns, however, we note that there is no relationship between reading preferences and ultimate medical interest. If we know whether a respondent feels he or she can best teach as a physician or as a researcher, knowing the respondent's reading preference adds nothing to our evaluation of his or her scientific orientation. If something like Figure 6-4 resulted from the actual data, we would conclude that reading preference should not be included in the same index as teaching contribution, since it contributed nothing to the composite index.

This example used only three questionnaire items. If more were being considered, then more-complex multivariate tables would be in order, constructed of four, five, or more variables. The purpose of this step in index construction, again, is to discover the simultaneous interaction of the items in order to determine which should be included in the same index.

SPSS **MC**	Multivariate data analyses are easily accomplished using programs such as SPSS and MicroCase. They are usually referred to as *cross-tabulations.* *

Index Scoring

When you have chosen the best items for the index, you next assign scores for particular responses, thereby creating a single composite index out of the several items. There are two basic decisions to be made in this step.

First, you must decide the desirable range of the index scores. Certainly a primary advantage of an index over a single item is the range of gradations it offers in the measurement of a variable. As noted earlier, political conservatism might be measured from "very conservative" to "not at all conservative" (or "very liberal"). How far to the extremes, then, should the index extend?

In this decision, the question of variance enters once more. Almost always, as the possible extremes of an index are extended, fewer cases are to be found at each end. The researcher who wishes to measure political conservatism to its greatest extreme may find there is almost no one in that category.

The first decision, then, concerns the conflicting desire for (1) a range of measurement in the index and (2) an adequate number of cases at each point in the index. You'll be forced to reach some kind of compromise between these conflicting desires.

The second decision concerns the actual assignment of scores for each particular response. Basically you must decide whether to give each item in the index equal weight or different weights. Although there are no firm rules, I suggest—and practice tends to support this method—that items be weighted equally unless there are compelling reasons for differential weighting. That is, the bur-

den of proof should be on differential weighting; equal weighting should be the norm.

Of course, this decision must be related to the earlier issue regarding the balance of items chosen. If the index is to represent the composite of slightly different aspects of a given variable, then you should give each aspect the same weight. In some instances, however, you may feel that, say, two items reflect essentially the same aspect, and the third reflects a different aspect. If you wished to have both aspects equally represented by the index, you might decide to give the different item a weight equal to the combination of the two similar ones. In such a situation, you might want to assign a maximum score of 2 to the different item and a maximum score of 1 to each of the similar ones.

Although the rationale for scoring responses should take such concerns into account, you'll typically experiment with different scoring methods, examining the relative weights given to different aspects but at the same time worrying about the range and distribution of cases provided. Ultimately, the scoring method chosen will represent a compromise among these several demands. Of course, as in most research activities, such a decision is open to revision on the basis of later examinations. Validation of the index, to be discussed shortly, may lead you to recycle your efforts toward constructing a completely different index.

In the example taken from the medical school faculty survey, I decided to weight the items equally, since I'd chosen them, in part, because they represented slightly different aspects of the overall variable *scientific orientation.* On each of the items, the respondents were given a score of 1 for choosing the "scientific" response to the item and a score of 0 for choosing the "nonscientific" response. Each respondent, then, could receive a score of 0, 1, 2, or 3. This scoring method provided what was considered a useful range of variation— four index categories—and also provided enough cases for analysis in each category.

Here's a similar example of index scoring, from a study of work satisfaction. One of the key variables was *job-related depression,* measured by an index composed of the following four items, which

---

*Each time the SPSS and MicroCase icons appear, they indicate that the topic under discussion could be pursued through the use of these software programs.

asked workers how they felt when thinking about themselves and their jobs:

- "I feel downhearted and blue."
- "I get tired for no reason."
- "I find myself restless and can't keep still."
- "I am more irritable than usual."

The researchers, Amy Wharton and James Baron, reported, "Each of these items was coded: 4 = often, 3 = sometimes, 2 = rarely, 1 = never" (Wharton and Baron 1987:578). They go on to explain how they measured other variables examined in the study:

> Job-related self-esteem was based on four items asking respondents how they saw themselves in their work: happy/sad; successful/not successful; important/not important; doing their best/not doing their best. Each item ranged from 1 to 7, where 1 indicates a self-perception of not being happy, successful, important, or doing one's best. — (P. 578)

As you look through the social research literature, you'll find numerous similar examples of cumulative indexes being used to measure variables. Sometimes the indexing procedures are controversial, as evidenced in the box "What Is the Best College in the United States?"

Handling Missing Data

Regardless of your data-collection method, you'll frequently face the problem of missing data. In a content analysis of the political orientations of newspapers, for example, you may discover that a particular newspaper has never taken an editorial position on one of the issues being studied. In an experimental design involving several retests of subjects over time, some subjects may be unable to participate in some of the sessions. In virtually every survey, some respondents fail to answer some questions (or choose a "don't know" response). Although missing data present problems at all stages of analysis, they're especially troublesome in index construction. There are, however, several methods of dealing with these problems.

First, if there are relatively few cases with missing data, you may decide to exclude them from the construction of the index and the analysis. (I did this in the medical school faculty example.) The primary concerns in this instance are whether the numbers available for analysis will remain sufficient and whether the exclusion will result in a biased sample whenever the index is used in the analysis. The latter possibility can be examined through a comparison—on other relevant variables—of those who would be included and excluded from the index.

Second, you may sometimes have grounds for treating missing data as one of the available responses. For example, if a questionnaire has asked respondents to indicate their participation in a number of activities by checking "yes" or "no" for each, many respondents may have checked some of the activities "yes" and left the remainder blank. In such a case, you might decide that a failure to answer meant "no," and score missing data in this case as though the respondents had checked the "no" space.

Third, a careful analysis of missing data may yield an interpretation of their meaning. In constructing a measure of political conservatism, for example, you may discover that respondents who failed to answer a given question were generally as conservative on other items as were those who gave the conservative answer. As another example, a recent study measuring religious beliefs found that people who answered "don't know" about a given belief were almost identical to the "disbelievers" in their answers about other beliefs. (*Note:* You should not take these examples as empirical guides in your own studies, but only as suggesting general ways to analyze your own data.) Whenever the analysis of missing data yields such interpretations, then, you may decide to score such cases accordingly.

There are many other ways of handling the problem of missing data. If an item has several possible values, you might assign the middle value to cases with missing data; for example, you could assign a 2 if the values are 0, 1, 2, 3, and 4. For a continuous variable such as age, you could

WHAT IS THE BEST COLLEGE IN THE UNITED STATES?

Each year the newsmagazine *U.S. News and World Report* issues a special report ranking the nation's colleges and universities. Their rankings reflect an index, created from several items: educational expenditures per student, graduation rates, selectivity (percentage accepted of those applying), average SAT scores of first-year students, and similar indicators of quality.

Typically, Harvard is ranked the number one school in the nation, followed by Yale and Princeton. However, the 1999 "America's Best Colleges" issue shocked educators, prospective college students, and their parents. The California Institute of Technology had leaped from ninth place in 1998 to first place a year later. While Harvard, Yale, and Princeton still did well, they had been supplanted. What had happened at Caltech to produce such a remarkable surge in quality?

The answer was to be found at *U.S. News and World Report,* not at Caltech. The news-

magazine changed the structure of the ranking index in 1999, which made a big difference in how schools fared.

Bruce Gottlieb (1999) gives this example of how the altered scoring made a difference.

> So, how did Caltech come out on top? Well, one variable in a school's ranking has long been educational expenditures per student, and Caltech has traditionally been tops in this category. But until this year, *U.S. News* considered only a school's ranking in this category—first, second, etc.—rather than how much it spent relative to other schools. It didn't matter whether Caltech beat Harvard by $1 or by $100,000. Two other schools that rose in their rankings this year were MIT (from fourth to third) and Johns Hopkins (from 14th to seventh). All three have high per-student expenditures and all three are especially strong in the hard sciences. Universities are allowed to

similarly assign the mean to cases with missing data (more on this in Chapter 14). Or, missing data can be supplied by assigning values at random. All of these are conservative solutions because they weaken the "purity" of your index and reduce the likelihood that it will relate to other variables in ways you may have hypothesized.

If you're creating an index out of several items, you can sometimes handle missing data by using proportions based on what is observed. Suppose your index is composed of six indicators, and you only have four observations for a particular subject. If the subject has earned 4 points out of a possible 4, you might assign an index score of 6; if the subject has 2 points (half the possible score on four items), you could assign a score of 3 (half the possible score on six observations).

The choice of a particular method to be used depends so much on the research situation that I

can't reasonably suggest a single "best" method or rank the several I have described. Excluding all cases with missing data can bias the representativeness of the findings, but including such cases by assigning scores to missing data can influence the nature of the findings. The safest and best method is to construct the index using alternative methods and see whether the same findings follow from each. Understanding your data is the final goal of analysis anyway.

SPSS The Wadsworth Web site explains how you can use SPSS to create indexes like the ones described in this chapter. Using the Transform → Compute command will allow you to create a new variable and add points to its value based on specific attributes on specified variables.

count their research budgets in their per-student expenditures, though students get no direct benefit from costly research their professors are doing outside of class.

In its "best colleges" issue two years ago, *U.S. News* made precisely this point, saying it considered only the rank ordering of per-student expenditures, rather than the actual amounts, on the grounds that expenditures at institutions with large research programs and medical schools are substantially higher than those at the rest of the schools in the category. In other words, just two years ago, the magazine felt it unfair to give Caltech, MIT, and Johns Hopkins credit for having lots of fancy laboratories that don't actually improve undergraduate education.

Gottlieb reviewed each of the changes in the index and then asked how 1998's ninth-ranked Caltech would have done had the revised index-

ing formula been in place a year earlier. His conclusion: Caltech would have been first in 1998 as well. In other words, the apparent improvement was solely a function of how the index was scored.

Composite measures such as scales and indexes are valuable tools for understanding society. However, it's important that we know how those measures are constructed and what that construction implies.

So, what's *really* the best college in the United States? It depends on how you define "best." There is no "really best," only the various social constructions we can create.

Sources: *U. S. News and World Report,* "America's Best Colleges," August 30, 1999; Bruce Gottlieb, "Cooking the School Books: How *U.S. News* Cheats in Picking Its 'Best American Colleges,'" article posted by Slate [Online], August 31, 1999, available: http://www.slate.com/crapshoot/99-08-31/crapshoot.asp.

Index Validation

Up to this point, we've discussed all the steps in the selection and scoring of items that result in a composite index purporting to measure some variable. If each of the preceding steps is carried out carefully, the likelihood of the index actually measuring the variable is enhanced. To demonstrate success, however, there must be validation of the index. Following the basic logic of validation, we assume that the index provides a measure of some variable; that is, the scores on the index arrange cases in a rank order in terms of that variable. An index of political conservatism rank-orders people in terms of their relative conservatism. If the index does that successfully, then people scored as relatively conservative on the index should appear relatively conservative in all other indications of political orientation, such as their responses to other

questionnaire items. There are several methods of validating an index.

Item Analysis   The first step in index validation is an internal validation called **item analysis.** In item analysis, you examine the extent to which the composite index is related to (or predicts responses to) the individual items it comprises. Here's an illustration of this step.

In the index of scientific orientations among medical school faculty, for example, index scores ranged from 0 (most interested in patient care) to 3 (most interested in research). Now let's consider one of the items in the index: whether respondents wanted to advance their own knowledge more with regard to total patient management or more in the area of basic mechanisms. The latter were treated as being more scientifically oriented than the former. The following empty table shows how we would examine the relationship between the index and the individual item.

Index of Scientific Orientations

	0	1	2	3
Percentage who said they were more interested in basic mechanisms	??	??	??	??

If you take a minute to reflect on the table, you may see that we already know the numbers that go in two of the cells. To get a score of 3 on the index, respondents had to say "basic mechanisms" in response to this question and give the "scientific" answers to the other two items as well. Thus, 100 percent of the 3's on the index said "basic mechanisms." By the same token, all the 0's had to answer this item with "total patient management." Thus, 0 percent of those respondents said "basic mechanisms." Here's how the table looks with the information we already know.

Index of Scientific Orientations

	0	1	2	3
Percentage who said they were more interested in basic mechanisms	0	??	??	100

If the individual item is a good reflection of the overall index, we should expect the 1's and 2's to fill in a progression between 0 percent and 100 percent. More of the 2's should choose "basic mechanisms" than 1's. This is not guaranteed by the way the index was constructed, however; it is an empirical question—one we answer in an item analysis. Here's how this particular item analysis turned out.

Index of Scientific Orientations

	0	1	2	3
Percentage who said they were more interested in basic mechanisms	0	16	91	100

As you can see, in accord with our assumption that the 2's are more scientifically oriented than the 1's, we find that a higher percentage of the 2's (91 percent) than the 1's (16 percent) say "basic mechanisms."

An item analysis of the other two components of the index yields similar results, as shown below.

Index of Scientific Orientations

	0	1	2	3
Percentage who said they could teach best as medical researchers	0	4	14	100
Percentage who said they preferred reading about rationales	0	80	97	100

Each of the items, then, seems an appropriate component in the index. Each seems to reflect the same quality that the index as a whole measures.

In a complex index containing many items, this step provides a convenient test of the independent contribution of each item to the index. If a given item is found to be poorly related to the index, it may be assumed that other items in the index cancel out the contribution of that item. If the item in question contributes nothing to the index's power, it should be excluded.

Although item analysis is an important first test of the index's validity, it is scarcely a sufficient test. If the index adequately measures a given variable, it should successfully predict other indications of that variable. To test this, we must turn to items not included in the index.

External Validation People scored as politically conservative on an index should appear conservative by other measures as well, such as their responses to other items in a questionnaire. Of course, we're talking about relative conservatism, because we can't make an absolute definition of what constitutes conservatism. However, those respondents scored as the most conservative on the index should be the most conservative in answering other questions. Those scored as the least con-

TABLE 6-1 **Validation of Scientific Orientation Index**

	Index of Scientific Orientation			
	Low			High
	0	1	2	3
Percentage interested in attending scientific lectures at the medical school	34	42	46	65
Percentage who say faculty members should have experience as medical researchers	43	60	65	89
Percentage who would prefer faculty duties involving research activities only..........................	0	8	32	66
Percentage who engaged in research during the preceding academic year.............................	61	76	94	99

servative on the index should be the least conservative on other items. Indeed, the ranking of groups of respondents on the index should predict the ranking of those groups in answering other questions dealing with political orientations.

In our example of the scientific orientation index, several questions in the questionnaire offered the possibility of such **external validation.** Table 6-1 presents some of these items, which provide several lessons regarding index validation. First, we note that the index strongly predicts the responses to the validating items in the sense that the rank order of scientific responses among the four groups is the same as the rank order provided by the index itself. That is, the percentages reflect greater scientific orientation as you read across the rows of the table. At the same time, each item gives a different description of scientific orientations overall. For example, the last validating item indicates that the great majority of all faculty were engaged in research during the preceding year. If this were the only indicator of scientific orientation, we would conclude that nearly all faculty were scientific. Nevertheless, those scored as more scientific on the index are more likely to have engaged in research than are those who were scored as relatively less scientific. The third validating item provides a different descriptive picture: Only a minority of the faculty overall say they would prefer duties limited exclusively to research. Nevertheless, the percentages giving this answer correspond to the scores assigned on the index.

Bad Index versus Bad Validators Nearly every index constructor at some time must face the apparent failure of external items to validate the in-

dex. If the internal item analysis shows inconsistent relationships between the items included in the index and the index itself, something is wrong with the index. But if the index fails to predict strongly the external validation items, the conclusion to be drawn is more ambiguous. You must choose between two possibilities: (1) the index does not adequately measure the variable in question, or (2) the validation items do not adequately measure the variable and thereby do not provide a sufficient test of the index.

Having worked long and conscientiously on the construction of an index, you'll likely find the second conclusion compelling. Typically, you'll feel you have included the best indicators of the variable in the index; the validating items are, therefore, second-rate indicators. Nevertheless, you should recognize that the index is purportedly a very powerful measure of the variable; thus, it should be somewhat related to any item that taps the variable even poorly.

When external validation fails, you should reexamine the index before deciding that the validating items are insufficient. One way is to examine the relationships between the validating items and the individual items included in the index. If you discover that some of the index items relate to the validators and others do not, you'll have improved your understanding of the index as it was initially constituted.

There is no cookbook solution to this dilemma; it is an agony serious researchers must learn to survive. Ultimately, the wisdom of your decision to accept an index will be determined by the usefulness of that index in your later analyses. Perhaps you'll initially decide that the index is a good one and that the validators are defective, but you'll later find that the variable in question (as measured by the index) is not related to other variables in the ways you expected. You may then have to compose a new index.

The Status of Women: An Illustration of Index Construction

For the most part, I've talked about index construction in the context of survey research, but other types of research also lend themselves to this kind of composite measure. For example, when the United Nations (1995) set about examining the status of women in the world, they chose to create two indexes, reflecting two different dimensions.

The Gender-related Development Index (GDI) compared women to men in terms of three indicators: life expectancy, education, and income. These indicators are commonly used in monitoring the status of women in the world. The Scandinavian countries of Norway, Sweden, Finland, and Denmark ranked highest on this measure.

The second index, the Gender Empowerment Measure (GEM), aimed more at power issues and comprised three different indicators:

- The proportion of parliamentary seats held by women
- The proportion of administrative, managerial, professional, and technical positions held by women
- A measure of access to jobs and wages

Once again, the Scandinavian countries ranked high but were joined by Canada, New Zealand, the Netherlands, the United States, and Austria. Having two different measures of gender equality allowed the researchers to make more-sophisticated distinctions. For example, in several countries, most notably Greece, France, and Japan, women fared relatively well on the GDI but quite poorly on the GEM; thus while they were doing fairly well in terms of income, education, and life expectancy, they were still denied access to power. And while the GDI scores were higher in the wealthier nations than in the poorer ones, GEM scores showed that women's empowerment did not seem to depend on national wealth, with many poor, developing

APPLYING THE RESULTS

In our discussion of the Gender Empowerment Measure (GEM), we analyze the status of women in countries around the world. How might you use the logic of this analysis to examine and assess the status of women in a particular organization, such as the college you attend or a corporation you are familiar with?

countries outpacing some rich, industrial ones in regard to such empowerment.

By examining several different dimensions of the variables involved in their study, the UN researchers also uncovered an aspect of women's earnings that generally goes unnoticed. Population Communications International (1996:1) has summarized the finding nicely:

Every year, women make an invisible contribution of eleven trillion U.S. dollars to the global economy, the UNDP [United Nations Development Programme] report says, counting both unpaid work and the underpayment of women's work at prevailing market prices. This "underevaluation" of women's work not only undermines their purchasing power, says the 1995 HDR [Human Development Report], but also reduces their already low social status and affects their ability to own property and use credit. Mahbub ul Haq, the principal author of the report, says that, "if women's work were accurately reflected in national statistics, it would shatter the myth that men are the main breadwinners of the world." The UNDP report finds that women work longer hours than men in almost every country, including both paid and unpaid duties. In developing countries, women do approximately 53% of all work and spend two-thirds of their work time on unremunerated activities. In industrialized countries, women do an average of 51% of the total work, and—like their counterparts in the developing world—perform about two-thirds of their total labor without pay. Men in industrialized countries are compensated for two-thirds of their work.

As you can see, indexes can be constructed from many different kinds of data for a variety of purposes. Now we'll turn our attention from the construction of indexes to an examination of scaling techniques.

SCALE CONSTRUCTION

Good indexes provide an ordinal ranking of cases on a given variable. All indexes are based on this kind of assumption: A senator who voted for seven conservative bills is considered to be more conservative than one who voted for only four of them. What an index may fail to take into account, however, is that not all indicators of a variable are equally important or equally strong. The first senator might have voted in favor of seven mildly conservative bills, whereas the second senator might have voted in favor of four extremely conservative bills. (The second senator might have considered the other seven bills too liberal and voted against them.)

Scales offer more assurance of ordinality by tapping the intensity structures among the indicators. The several items going into a composite measure may have different intensities in terms of the variable. Many methods of scaling are available. To illustrate the variety of techniques available, we'll look at four scaling procedures, along with a technique called the semantic differential. Although these examples focus on questionnaires, the logic of scaling, like that of indexing, applies to other research methods as well.

Bogardus Social Distance Scale

Let's suppose you're interested in the extent to which U.S. citizens are willing to associate with, say, sex offenders. You might ask the following questions:

1. Are you willing to let sex offenders live in your country?
2. Are you willing to let sex offenders live in your community?
3. Are you willing to let sex offenders live in your neighborhood?
4. Would you be willing to let a sex offender live next door to you?
5. Would you let your child marry a sex offender?

These questions increase in terms of how closely the respondents want to associate with sex offenders. Beginning with the original concern to measure willingness to associate with sex offenders, you have thus developed several questions indicating differing degrees of intensity on this variable. The kinds of items presented constitute a **Bogardus social distance scale** (created by Emory Bogardus). This scale is a measurement technique for determining the willingness of people to participate in social relations—of varying degrees of closeness—with other kinds of people.

The clear differences of intensity suggest a structure among the items. Presumably, if a person is willing to accept a given kind of association, he or she would be willing to accept all those preceding it in the list—those with lesser intensities. For example, the person who is willing to permit sex offenders to live in the neighborhood will surely accept them in the community and the nation but may or may not be willing to accept them as next-door neighbors or relatives. This, then, is the logical structure of intensity inherent among the items.

Empirically, one would expect to find the largest number of people accepting co-citizenship and the fewest accepting intermarriage. In this sense, we speak of "easy items" (for example, residence in the United States) and "hard items" (for example, intermarriage). More people agree to the easy items than to the hard ones. With some inevitable exceptions, logic demands that once a person has refused a relationship presented in the scale, he or she will also refuse all the harder ones that follow it.

The Bogardus social distance scale illustrates the important economy of scaling as a data-reduction device. By knowing *how many* relationships with sex offenders a given respondent will accept, we know *which* relationships were accepted. Thus, a single number can accurately summarize five or six data items without a loss of information.

Thurstone Scales

Often the inherent structure of the Bogardus social distance scale is not appropriate to the variable being measured. Indeed, such a logical structure among several indicators is seldom apparent. A **Thurstone scale** (a format created by Louis Thurstone) is an attempt to develop a format for generating groups of indicators of a variable that have at least an empirical structure among them.

One of the basic formats is that of "equal-appearing intervals." A group of judges is given perhaps a hundred items felt to be indicators of a given variable. Each judge is then asked to estimate how strong an indicator of a variable each item is by assigning scores of perhaps 1 to 13. If the variable were *prejudice,* for example, the judges would be asked to assign the score of 1 to the very weakest indicators of prejudice, the score of 13 to the strongest indicators, and intermediate scores to those felt to be somewhere in between.

Once the judges have completed this task, the researcher examines the scores assigned to each item by all the judges to determine which items produced the greatest agreement among the judges. Those items on which the judges disagreed broadly would be rejected as ambiguous. Among those items producing general agreement in scoring, one or more would be selected to represent each scale score from 1 to 13.

The items selected in this manner might then be included in a survey questionnaire. Respondents who appeared prejudiced on those items representing a strength of 5 would then be expected to appear prejudiced on those having lesser strengths, and if some of those respondents did not appear prejudiced on the items with a strength of 6, it would be expected that they would also not appear prejudiced on those with greater strengths.

If the Thurstone scale items were adequately developed and scored, the economy and effectiveness of data reduction inherent in the Bogardus social distance scale would appear. A single score might be assigned to each respondent (the strength of the hardest item accepted), and that score would adequately represent the responses to several questionnaire items. And as is true of the Bogardus scale, a respondent scored 6 might be regarded as more prejudiced than one scored 5 or less.

Thurstone scaling is not often used in research today, primarily because of the tremendous expenditure of energy and time required to have 10 to 15 judges score the items. Because the quality of their judgments would depend on their experience with the variable under consideration, professional researchers might be needed. Moreover, the meanings conveyed by the several items indicating a given variable tend to change over time. Thus an item might have a given weight at one time and quite a different weight later on. For a Thurstone scale to be effective, it would have to be updated periodically.

Likert Scaling

You may sometimes hear people refer to a questionnaire item containing response categories such as "strongly agree," "agree," "disagree," and "strongly disagree" as a Likert scale. This is technically a misnomer, although Rensis Likert (pronounced 'LICK-ert') did create this commonly used question format.

The particular value of this format is the unambiguous ordinality of response categories. If respondents were permitted to volunteer or select such answers as "sort of agree," "pretty much agree," "really agree," and so forth, the researcher would find it impossible to judge the relative strength of agreement intended by the various respondents. The Likert format resolves this problem.

Likert had something more in mind, however. He created a method by which this question format could be used to determine the relative intensity of different items. As a simple example, suppose we wish to measure prejudice against women. To do this, we create a set of 20 statements, each of which reflects that prejudice. One of the items might be "Women can't drive as well as men." Another might be "Women shouldn't be allowed to vote." Likert's scaling technique would demon-

	Very Much	Some-what	Neither	Some-what	Very Much	
Enjoyable	☐	☐	☐	☐	☐	Unenjoyable
Simple	☐	☐	☐	☐	☐	Complex
Discordant	☐	☐	☐	☐	☐	Harmonic
Traditional	☐	☐	☐	☐	☐	Modern
			etc.			

FIGURE 6-5 **Semantic Differential: Feelings about Musical Selections**

strate the difference in intensity between these items as well as pegging the intensity of the other 18 statements.

Let's suppose we ask a sample of people to agree or disagree with each of the 20 statements. Simply giving one point for each of the indicators of prejudice against women would yield the possibility of index scores ranging from 0 to 20. A **Likert scale** goes one step beyond that and calculates the average index score for those agreeing with each of the individual statements. Let's say that all those who agreed that women are poorer drivers than were men had an average index score of 1.5 (out of a possible 20). Those who agreed that women should be denied the right to vote might have an average index score of, say, 19.5—indicating the greater degree of prejudice reflected in that response.

As a result of this item analysis, respondents could be rescored to form a scale: 1.5 points for agreeing that women are poorer drivers, 19.5 points for saying women shouldn't vote, and points for other responses reflecting how those items related to the initial, simple index. If those who disagreed with the statement "I might vote for a woman for president" had an average index score of 15, then the scale would give 15 points to people disagreeing with that statement.

In practice, Likert scaling is seldom used today. I don't know why; maybe it seems too complex. The item format devised by Likert, however, is one of the most commonly used in contemporary questionnaire design. Typically, it's now used in the creation of simple indexes. With, say, five response categories, scores of 0 to 4 or 1 to 5 might be assigned, taking the direction of the items into account (for example, assign a score of 5 to "strongly agree" for positive items and to "strongly disagree" for negative items). Each respondent would then be assigned an overall score representing the summation of the scores he or she received for responses to the individual items.

Semantic Differential

Like the Likert format, the **semantic differential** asks respondents to choose between two opposite positions. Here's how it works.

Suppose you're evaluating the effectiveness of a new music appreciation lecture on subjects' appreciation of music. As a part of your study, you want to play some musical selections and have the subjects report their feelings about them. A good way to tap those feelings would be to use a semantic differential format.

To begin, you must determine the dimensions along which subjects should judge each selection. Then you need to find two opposite terms, representing the polar extremes along each dimension. Let's suppose one dimension that interests you is simply whether subjects enjoyed the piece or not. Two opposite terms in this case could be "enjoyable" and "unenjoyable." Similarly, you might want to know whether they regarded the individual selections as "complex" or "simple," "harmonic" or "discordant," and so forth.

Once you have determined the relevant dimensions and have found terms to represent the extremes of each, you might prepare a rating sheet each subject would complete for each piece of music. Figure 6-5 shows what it might look like.

On each line of the rating sheet, the subject would indicate how he or she felt about the piece of music: whether it was enjoyable or unenjoyable, for example, and whether it was "somewhat" that way or "very much" so. To avoid creating a biased pattern of responses to such items, it's a good idea to vary the placement of terms that are likely to be related to each other. Notice, for example, that "discordant" and "traditional" are on the left side of the sheet, with "harmonic" and "modern" on the right. Most likely, those selections scored as "discordant" would also be scored as "modern" as opposed to "traditional."

Both the Likert and semantic differential formats have a greater rigor and structure than do other question formats. As I've indicated earlier, these formats produce data suitable to both indexing and scaling.

Guttman Scaling

Researchers today often use the scale developed by Louis Guttman. Like Bogardus, Thurstone, and Likert scaling, Guttman scaling is based on the fact that some items under consideration may prove to be more extreme indicators of the variable than others. One example should suffice to illustrate this pattern.

In the earlier example of measuring scientific orientations among medical school faculty members, a simple index was constructed. As it happens, however, the three items included in the index essentially form a Guttman scale.

The construction of a **Guttman scale** would begin with some of the same steps that initiate index construction. You would begin by examining the face validity of items available for analysis. Then, you would examine the bivariate and perhaps multivariate relations among those items. In scale construction, however, you would also look for relatively "hard" and "easy" indicators of the variable being examined.

Earlier, when we talked about attitudes regarding a woman's right to have an abortion, we discussed several conditions that can affect people's opinions: whether the woman is married, whether

her health is endangered, and so forth. These differing conditions provide an excellent illustration of Guttman scaling.

Here are the percentages of the people in the 1996 GSS sample who supported a woman's right to an abortion, under three different conditions:

Woman's health is seriously endangered	92%
Pregnant as a result of rape	86%
Woman is not married	48%

The different percentages supporting abortion under the three conditions suggest something about the different *levels* of support that each item indicates. For example, if someone would support abortion when the mother's life is seriously endangered, that's not a very strong indicator of general support for abortion, because almost everyone agreed with that. Supporting abortion for unmarried women seems a much stronger indicator of support for abortion in general—fewer than half the sample took that position.

Guttman scaling is based on the notion that anyone who gives a strong indicator of some variable will also give the weaker indicators. In this case, we would assume that anyone who supported abortion for unmarried women would also support it in the case of rape or of the woman's health being threatened. Table 6-2 tests this assumption by presenting the number of respondents who gave each of the possible response patterns.

The first four response patterns in the table compose what we would call the *scale types:* those patterns that form a scalar structure. Following those respondents who supported abortion under all three conditions (line 1), we see (line 2) that those with only two pro-choice responses have chosen the two easier ones; those with only one such response (line 3) chose the easiest of the three (the woman's health being endangered). And finally, there are some respondents who opposed abortion in all three circumstances (line 4).

The second part of the table presents those response patterns that violate the scalar structure of the items. The most radical departures from the

TABLE 6-2   **Scaling Support for Choice of Abortion**

	Women's Health	Result of Rape	Woman Unmarried	Number of Cases
	+	+	+	612
	+	+	−	448
Scale Types	+	−	−	92
	−	−	−	79
			Total = 1231	
	−	+	−	15
Mixed Types	+	−	+	5
	−	−	+	2
	−	+	+	5
			Total = 27	

+ = favors woman's right to choose; − = opposes woman's right to choose

scalar structure are the last two response patterns: those who accepted only the hardest item and those who rejected only the easiest one.

The final column in the table indicates the number of survey respondents who gave each of the response patterns. The great majority (1,231, or 98 percent) fit into one of the scale types. The presence of mixed types, however, indicates that the items do not form a perfect Guttman scale.

Recall at this point that one of the chief functions of scaling is efficient data reduction. Scales provide a technique for presenting data in a summary form while maintaining as much of the original information as possible. When the scientific orientation items were formed into an index in our earlier discussion, respondents were given one point for each scientific response they gave. If these same three items were scored as a Guttman scale, some respondents would be assigned scale scores that would permit the most accurate reproduction of their original responses to all three items.

In the present example of attitudes regarding abortion, respondents fitting into the scale types would receive the same scores as were assigned in the index construction. Persons selecting all three pro-choice responses would still be scored 3, those who selected pro-choice responses to the two easier items and were opposed on the hardest item would be scored 2, and so on. For each of the four scale types we could predict accurately all the ac-

tual responses given by all the respondents based on their scores.

The mixed types in the table present a problem, however. The first mixed type (− + −) was scored 1 on the index to indicate only one pro-choice response. But, if 1 were assigned as a scale score, we would predict that the 15 respondents in this group had chosen only the easiest item (approving abortion when the woman's life was endangered), and we would be making two errors for each such respondent. Scale scores are assigned, therefore, with the aim of minimizing the errors that would be made in reconstructing the original responses.

Table 6-3 illustrates the index and scale scores that would be assigned to each of the response patterns in our example. Note that one error is made for each respondent in the mixed types. This is the minimum we can hope for in a mixed type pattern. In the first mixed type, for example, we would erroneously predict a pro-choice response to the easiest item for each of the 15 respondents in this group, making a total of 15 errors.

The extent to which a set of empirical responses form a Guttman scale is determined by the accuracy with which the original responses can be reconstructed from the scale scores. For each of the 1,258 respondents in this example, we will predict three questionnaire responses, for a total of 3,774 predictions. Table 6-3 indicates that we will make 27 errors using the scale scores assigned.

TABLE 6-3 **Index and Scale Scores**

	Response Pattern	Number of Cases	Index Scores	Scale Scores*	Total Scale Errors
Scale Types	+ + +	612	3	3	0
	+ + −	448	2	2	0
	+ − −	92	1	1	0
	− − −	79	0	0	0
Mixed Types	− + −	15	1	2	15
	+ − +	5	2	3	5
	− − +	2	1	0	2
	− + +	5	2	3	5

Total Scale Errors = 27

$$\text{Coefficient of reproducibility} = 1 - \frac{\text{number of errors}}{\text{number of guesses}}$$

$$= 1 - \frac{27}{1258 \times 3} = \frac{27}{3774}$$

$$= .993 = 99.3\%$$

*This table presents one common method for scoring mixed types, but you should be advised that other methods are also used.

The percentage of correct predictions is called the *coefficient of reproducibility:* the percentage of original responses that could be reproduced by knowing the scale scores used to summarize them. In the present example, the coefficient of reproducibility is 3,747/3,774 or 99.3 percent.

Except for the case of perfect (100 percent) reproducibility, there is no way of saying that a set of items does or does not form a Guttman scale in any absolute sense. Virtually all sets of such items *approximate* a scale. As a general guideline, however, coefficients of 90 or 95 percent are the commonly used standards in this regard. If the observed reproducibility exceeds the level you've set, you'll probably decide to score and use the items as a scale.

The decision concerning criteria in this regard is, of course, arbitrary. Moreover, a high degree of reproducibility does not insure that the scale constructed in fact measures the concept under consideration, although it increases confidence that all the component items measure the same thing. Also, you should realize that a high coefficient of reproducibility is most likely when few items are involved.

One concluding remark should be made in regard to Guttman scaling: It is based on the structure observed among the actual data under examination. This is an important point that is often misunderstood. It does not make sense to say that a set of questionnaire items (perhaps developed and used by a previous researcher) constitutes a Guttman scale. Rather, we can say only that they form a scale within a given body of data being analyzed. Scalability, then, is a sample-dependent, empirical matter. Although a set of items may form a Guttman scale among one sample of survey respondents, for example, there is no guarantee that this set will form such a scale among another sample. In this sense, then, a set of questionnaire items in and of themselves never forms a scale, but a set of empirical observations may.

This concludes our discussion of indexing and scaling. Like indexes, scales are composite measures of a variable, typically broadening the meaning of the variable beyond what might be captured by a single indicator. Both scales and indexes seek to measure variables at the ordinal level of measurement. Unlike indexes, however, scales take advantage of any intensity structure that may be present among the individual indicators. To the extent that such an intensity structure is found and the data from the people or other units of analysis comply with the logic of that intensity structure, we can have confidence that we have created an ordinal measure.

TYPOLOGIES

We conclude this chapter with a short discussion of typology construction and analysis. Recall that indexes and scales are constructed to provide ordinal measures of given variables. We attempt to assign index or scale scores to cases in such a way as to indicate a rising degree of prejudice, religiosity, conservatism, and so forth. In such cases, we're dealing with single dimensions.

Often, however, the researcher wishes to summarize the intersection of two or more variables, thereby creating a set of categories or types, which we call a **typology.** You may, for example, wish to examine the political orientations of newspapers separately in terms of domestic issues and foreign policy. The fourfold presentation in Table 6-4 describes such a typology.

Newspapers in cell A of the table are conservative on both foreign policy and domestic policy; those in cell D are liberal on both. Those in cells B and C are conservative on one and liberal on the other.

Frequently, you arrive at a typology in the course of an attempt to construct an index or scale.

TABLE 6-4   **A Political Typology of Newspapers**

		Foreign Policy	
		Conservative	Liberal
Domestic Policy	Conservative	A	B
	Liberal	C	D

A QUANDARY **REVISITED**

If I were to tell you that we had given each respondent one point for every relationship they were willing to have with sex offenders, and I told you further that a particular respondent had been given a score of 3, you would be able to reproduce each of these five answers.

1. Are you willing to let sex offenders live in your country?   YES
2. Are you willing to let sex offenders live in your community?   YES
3. Are you willing to let sex offenders live in your neighborhood?   YES
4. Would you be willing to let a sex offender live next door to you?   NO
5. Would you let your child marry a sex offender?   NO

While this logic is very clear in the case of the Bogardus social distance scale, we've also seen how social researchers approximate that structure in creating other types of scales, such as Thurstone and Guttman scales, which also take account of differing intensities among the indicators of a variable.

The items that you felt represented a single variable appear to represent two. You might have been attempting to construct a single index of political orientations for newspapers but discovered—empirically—that foreign and domestic politics had to be kept separate.

In any event, you should be warned against a difficulty inherent in typological analysis. Whenever the typology is used as the independent variable, there will probably be no problem. In the preceding example, you might compute the percentages of newspapers in each cell that normally endorse Democratic candidates; you could then easily examine the effects of both foreign and domestic policies on political endorsements.

It's extremely difficult, however, to analyze a typology as a dependent variable. If you want to discover *why* newspapers fall into the different cells of typology, you're in trouble. That becomes apparent when we consider the ways you might construct and read your tables. Assume, for example, that you want to examine the effects of community size on political policies. With a single dimension, you could easily determine the percentages of rural and urban newspapers that were scored conservative and liberal on your index or scale.

With a typology, however, you would have to present the distribution of the urban newspapers in your sample among types A, B, C, and D. Then you would repeat the procedure for the rural ones in the sample and compare the two distributions. Let's suppose that 80 percent of the rural newspapers are scored as type A (conservative on both dimensions) as compared with 30 percent of the urban ones. Moreover, suppose that only 5 percent of the rural newspapers are scored as type B (conservative only on domestic issues) as compared with 40 percent of the urban ones. It would be incorrect to conclude from an examination of type B that urban newspapers are more conservative on domestic issues than are rural ones, since 85 percent of the rural newspapers, compared with 70 percent of the urban ones, have this characteristic. The relative sparsity of rural newspapers in type B is due to their concentration in type A. It should be apparent that an interpretation of such data would be very difficult for anything other than description.

In reality, you'd probably examine two such dimensions separately, especially if the dependent variable has more categories of responses than does the example given.

Don't think that typologies should always be avoided in social research; often they provide the most appropriate device for understanding the data. To examine the "pro-life" orientation in depth, you might create a typology involving both abortion and capital punishment. Libertarianism could be seen in terms of both economic and social permissiveness. You have been warned, however, against the special difficulties involved in using typologies as dependent variables.

Main Points

- Single indicators of variables seldom capture all the dimensions of a concept, have sufficiently clear validity to warrant their use, or permit the desired range of variation to allow ordinal rankings. Composite measures, such as scales and indexes, solve these problems by including several indicators of a variable in one summary measure.

- Although both indexes and scales are intended as ordinal measures of variables, scales typically satisfy this intention better than indexes.

- Whereas indexes are based on the simple cumulation of indicators of a variable, scales take advantage of any logical or empirical intensity structures that exist among a variable's indicators.

- The principal steps in constructing an index include selecting possible items, examining their empirical relationships, scoring the index, and validating it.

- Criteria of item selection include face validity, unidimensionality, the degree of specificity with which a dimension is to be measured, and the amount of variance provided by the items.

- If different items are indeed indicators of the same variable, then they should be related empirically to one another. In constructing an index, the researcher needs to examine bivariate and multivariate relationships among the items.

- Index scoring involves deciding the desirable range of scores and determining whether items will have equal or different weights.

- There are various techniques that allow items to be used in an index in spite of missing data.

- Item analysis is a type of internal validation, based on the relationship between individual items in the composite measure and the measure itself. External validation refers to the relationships between the composite measure and other indicators of the variable—indicators not included in the measure.

❑ Four types of scaling techniques are represented by the Bogardus social distance scale, a device for measuring the varying degrees to which a person would be willing to associate with a given class of people; Thurstone scaling, a technique that uses judges to determine the intensities of different indicators; Likert scaling, a measurement technique based on the use of standardized response categories; and Guttman scaling, a method of discovering and using the empirical intensity structure among several indicators of a given variable. Guttman scaling is probably the most popular scaling technique in social research today.

❑ The semantic differential is a question format that asks respondents to make ratings that lie between two extremes, such as "very positive" and "very negative."

❑ A typology is a nominal composite measure often used in social research. Typologies may be used effectively as independent variables, but interpretation is difficult when they are used as dependent variables.

Key Terms

index	Thurstone scale
scale	Likert scale
item analysis	semantic differential
external validation	Guttman scale
Bogardus social distance scale	typology

Review Questions

1. In your own words, what is the difference between an index and a scale?

2. Suppose you wanted to create an index for rating the quality of colleges and universities.

What are three data items that might be included in such an index?

3. Why do you suppose Thurstone scales have not been used more widely in the social sciences?

4. What would be some questionnaire items that could measure attitudes toward nuclear power and that would probably form a Guttman scale?

Additional Readings

Anderson, Andy B., Alexander Basilevsky, and Derek P. J. Hum. 1983. "Measurement: Theory and Techniques." Pp. 231–87 in *Handbook of Survey Research,* edited by Peter H. Rossi, James D. Wright, and Andy B. Anderson. New York: Academic Press. The logic of measurement is analyzed in the context of composite measures.

Bobo, Lawrence, and Frederick C. Licari. 1989. "Education and Political Tolerance: Testing the Effects of Cognitive Sophistication and Target Group Effect." *Public Opinion Quarterly* 53:285–308. The authors use a variety of techniques for determining how best to measure tolerance toward different groups in society.

Indrayan, A., M. J. Wysocki, A. Chawla, R. Kumar, and N. Singh. 1999. "Three-Decade Trend in Human Development Index in India and Its Major States." *Social Indicators Research* 46 (1): 91–120. The authors use several human development indexes to compare the status of different states in India.

Lazarsfeld, Paul, Ann Pasanella, and Morris Rosenberg, eds. 1972. *Continuities in the Language of Social Research.* New York: Free Press. See especially Section 1. An excellent collection of conceptual discussions and concrete illustrations. The construction of composite measures is presented within the more general area of conceptualization and measurement.

McIver, John P., and Edward G. Carmines. 1981. *Unidimensional Scaling.* Newbury Park, CA: Sage. Here's an excellent way to pursue Thurstone, Likert, and Guttman scaling in further depth.

Miller, Delbert. 1991. *Handbook of Research Design and Social Measurement.* Newbury Park, CA: Sage. An excellent compilation of frequently used and semistandardized scales. The many illustrations reported in Part 4 of the Miller book may be directly adaptable to studies or at least suggestive of modified measures. Studying the several different illustrations, moreover, may also give you a better understanding of the logic of composite measures in general.

Multimedia Resources

The Wadsworth Sociology Resource Center: Virtual Society
http://sociology.wadsworth.com/
 Visit the companion Web site for the second edition of *The Basics of Social Research* to access a wide range of student resources. Begin by clicking on the Student Resources section of the book's Web site to access the following study tools:

- eBabbie Resource Center
- Planning a Research Project
- Doing Data Analysis
- Statistics Review
- Flash Cards
- Internet Links and Exercises
- InfoTrac College Edition: Exercises
- Quizzes

Visit the **eBabbie Resource Center** for an overview of each chapter and helpful online tutorials. Find information on budgeting and step-by-step examples of model re-

search projects at **Planning a Research Project.** Learn how to use quantitative and qualitative data analysis programs at **Doing Data Analysis,** and brush up on your statistics at **Statistics Review.** You can also further your study by accessing **Internet Links and Exercises** related to chapter materials, **Flash Cards, Quizzes,** and many other learning tools.

InfoTrac College Edition
http://www.infotrac-college.com/
wadsworth/access.html
 Access the latest news and journal articles with InfoTrac College Edition, an easy-to-use online database of reliable, full-length articles from hundreds of top academic journals. Conduct an electronic search using the following search terms:

 Composite AND measure
 Intelligence tests AND validity
 Likert
 Sociological AND index
 Sociological AND scale
 Thurstone

THE LOGIC OF SAMPLING

Bob Daemmrich/PhotoEdit

What You'll Learn in This Chapter

Now you'll see how social scientists can select a few people for study—and discover things that apply to hundreds of millions of people not studied.

In this chapter . . .

AN OPENING QUANDARY

In 1936, the *Literary Digest* collected the voting intentions of two million voters in order to predict whether Franklin D. Roosevelt or Alf Landon would be elected president of the United States. During more recent election campaigns, with many more voters going to the polls, national polling firms have typically sampled around 2,000 voters across the country.

Which technique do you think is the most effective? Why?

INTRODUCTION

One of the most visible uses of survey sampling lies in the political polling that is subsequently tested by the election results. While some people doubt the accuracy of sample surveys, others complain that political polls take all the suspense out of campaigns by foretelling the result.

Going into the 2000 presidential elections, however, pollsters generally agreed that the election was "too close to call." Robert Worcester has compiled the national polls completed during the two days before the election. Despite some variations, the overall picture they present is amazingly consistent.

TABLE 7-1   Election Eve Polls Reporting Percentage of Population Voting for U.S. Presidential Candidates, 2000

	Gore	Bush	Nader	Buchanan*
11/5: Hotline [Polling Co/GSG]	43%	51%	4%	1%
11/5: Marist College	46	51	2	1
11/5: Fox [Opinion Dynamics]	47	47	3	2
11/5: Newsweek [PRSA]	46	49	6	0
11/5: NBC/Wall St. Journal [Hart/Teeter]	45	48	4	2
11/5: Pew	46	49	3	1
11/5: ICR	44	46	7	2
11/5: Harris	47	47	5	1
11/5: Harris (online)	47	47	4	2
11/5: ABC/Washington Post [TNSI]	46	49	3	1
11/6: IDB/CSM [TIPP]	47	49	4	0
11/6: CBS	48	47	4	1
11/6: Portrait of America [Rasmussen]	43	52	4	1
11/6: CNN/USA Today [Gallup]	46	48	4	1
11/6: Reuters/MSNBC [Zogby]	48	46	5	1
11/6: Voter.com [Lake/Goeas]	45	51	4	0
November 7, 2000 Election Results	**48%**	**48%**	**3%**	**1%**

Source: Adapted from Robert Worcester, *WAPOR Newsletter,* Winter 2001 (in press).

* "Don't knows" have been apportioned so the totals equal 100%. (Rounding error may result in totals of 99% or 101%.)

As we now know, the election was so close that even the election officials were unable to declare the result unambiguously, and the matter had to be settled in the Supreme Court. That cliffhanger proved once and for all that the accuracy of modern polling doesn't have to take all the suspense out of elections.

Now, how many interviews do you suppose it took each of these pollsters to come within a couple of percentage points in estimating the behavior of about a hundred million voters? Often fewer than 2,000! In this chapter, we're going to find out how social researchers can pull off such wizardry.

Political polling, like other forms of social research, rests on observations. But neither pollsters nor other social researchers can observe everything that might be relevant to their interests. A critical part of social research, then, is deciding what to observe and what not. If you want to study voters, for example, which voters should you study?

The process of selecting observations is called *sampling.* Although sampling can mean any procedure for selecting units of observation—for example, interviewing every tenth passerby on a busy street—the key to generalizing from a sample to a larger population is *probability sampling,* which involves the important idea of *random selection.*

Much of this chapter is devoted to the logic and skills of probability sampling. This topic is more rigorous and precise than some of the other topics in this book. Whereas social research as a whole is both art and science, sampling leans toward science. Although this subject is somewhat technical, the basic logic of sampling is not difficult to understand. In fact, the logical neatness of this topic can make it easier to comprehend than, say, conceptualization.

Although probability sampling is central to social research today, we'll take some time to examine a variety of nonprobability methods as well. These methods have their own logic and can provide useful samples for social inquiry.

Before we discuss the two major types of sampling, I'll introduce you to some basic ideas by way of a brief history of sampling. As you'll see, the pollsters who correctly predicted the election cliffhanger of 2000 did so in part because researchers had learned to avoid some pitfalls that earlier pollsters had avoided.

A BRIEF HISTORY OF SAMPLING

Sampling in social research has developed hand in hand with political polling. This is the case, no doubt, because political polling is one of the few opportunities social researchers have to discover the accuracy of their estimates. On election day, they find out how well or how poorly they did.

President Alf Landon

President Alf Landon? Who's he? Did you sleep through an entire presidency in your U.S. history class? No—but Alf Landon would have been president if a famous poll conducted by the *Literary Digest* had proved to be accurate. The *Literary Digest* was a popular newsmagazine published between 1890 and 1938. In 1920, *Digest* editors mailed postcards to people in six states, asking them whom they were planning to vote for in the presidential campaign between Warren Harding and James Cox. Names were selected for the poll from telephone directories and automobile registration lists. Based on the postcards sent back, the *Digest* correctly predicted that Harding would be elected. In the elections that followed, the *Literary Digest* expanded the size of its poll and made correct predictions in 1924, 1928, and 1932.

In 1936, the *Digest* conducted its most ambitious poll: Ten million ballots were sent to people listed in telephone directories and on lists of automobile owners. Over two million people responded, giving the Republican contender, Alf Landon, a stunning 57 to 43 percent landslide over the incumbent, President Franklin Roosevelt. The editors modestly cautioned,

> We make no claim to infallibility. We did not coin the phrase "uncanny accuracy" which has been so freely applied to our Polls. We know only too well the limitations of every straw vote, however enormous the sample gathered, however scientific the method. It would be a miracle if every State of the forty-eight behaved on Election Day exactly as forecast by the Poll. — (*LITERARY DIGEST* 1936A:6)

Two weeks later, the *Digest* editors knew the limitations of straw polls even better: The voters gave Roosevelt a second term in office by the largest landslide in history, with 61 percent of the vote. Landon won only 8 electoral votes to Roosevelt's 523.

The editors were puzzled by their unfortunate turn of luck. A part of the problem surely lay in the 22 percent return rate garnered by the poll. The editors asked,

> Why did only one in five voters in Chicago to whom the *Digest* sent ballots take the trouble to reply? And why was there a preponderance of Republicans in the one-fifth that did reply? . . . We were getting better cooperation in what we have always regarded as a public service from Republicans than we were getting from Democrats. Do Republicans live nearer to mailboxes? Do Democrats generally disapprove of straw polls? — (*LITERARY DIGEST* 1936B:7)

Actually, there was a better explanation—what is technically called the *sampling frame* used by the *Digest*. In this case the sampling frame consisted of telephone subscribers and automobile owners. In the context of 1936, this design selected a disproportionately wealthy sample of the voting population, especially coming on the tail end of the worst economic depression in the nation's history. The sample effectively excluded poor people, and the poor voted predominantly for Roosevelt's New Deal recovery program. The *Digest*'s poll may or may not have correctly represented the voting intentions of telephone subscribers and automobile owners. Unfortunately for the editors, it decidedly did not represent the voting intentions of the population as a whole.

You may be able to find the *Literary Digest* in your library. You can find traces of it by searching the Web. As an alternative, go to http://www.eBay.com and see how many old issues are available for sale.*

President Thomas E. Dewey

The 1936 election also saw the emergence of a young pollster whose name would become synonymous with public opinion. In contrast to the *Literary Digest,* George Gallup correctly predicted that Roosevelt would beat Landon. Gallup's success in 1936 hinged on his use of something called quota sampling, which we'll look at more closely later in the chapter. For now, it's enough to know that quota sampling is based on a knowledge of the characteristics of the population being sampled: what proportion are men, what proportion are women, what proportions are of various incomes, ages, and so on. Quota sampling selects people to match a set of these characteristics: the right number of poor, white, rural men; the right number of rich, African-American, urban women; and so on. The quotas are based on those variables most relevant to the study. In the case of Gallup's poll, the sample selection was based on levels of income; the selection procedure ensured the right proportion of respondents at each income level.

Gallup and his American Institute of Public Opinion used quota sampling to good effect in 1936, 1940, and 1944—correctly picking the presidential winner each of those years. Then, in 1948, Gallup and most political pollsters suffered the embarrassment of picking Governor Thomas Dewey of New York over the incumbent, President Harry Truman. The pollsters' embarrassing miscue continued right up to election night. A famous photograph shows a jubilant Truman—whose followers' battle cry was "Give 'em hell, Harry!"—holding aloft a newspaper with the banner headline, "Dewey Defeats Truman."

Several factors accounted for the pollsters' failure in 1948. First, most pollsters stopped polling in early October despite a steady trend toward Truman during the campaign. In addition, many voters were undecided throughout the campaign, and they went disproportionately for Truman when they stepped in the voting booth.

More important, Gallup's failure rested on the unrepresentativeness of his samples. Quota sampling—which had been effective in earlier years—was Gallup's undoing in 1948. This technique requires that the researcher know something about the total population (of voters in this instance). For national political polls, such information came primarily from census data. By 1948, however, World War II had produced a massive movement from the country to cities, radically changing the character of the U.S. population from what the 1940 census showed, and Gallup relied on 1940 census data. City dwellers, moreover, tended to vote Democratic; hence, the overrepresentation of rural voters in his poll had the effect of underestimating the number of Democratic votes.

Two Types of Sampling Methods

By 1948, some academic researchers had already been experimenting with a form of sampling based on probability theory. This technique involves the selection of a "random sample" from a list containing the names of everyone in the population being sampled. By and large, the probability sampling methods used in 1948 were far more accurate than quota sampling techniques.

Today, probability sampling remains the primary method of selecting large, representative samples for social research, including national political polls. At the same time, probability sampling can be impossible or inappropriate in many research situations. Accordingly, before turning to the logic and techniques of probability sampling,

---

*Each time the Internet icon appears, you'll be given helpful leads for searching the World Wide Web.

we'll first take a look at techniques for nonprobability sampling and how they're used in social research.

NONPROBABILITY SAMPLING

Social research is often conducted in situations that do not permit the kinds of probability samples used in large-scale social surveys. Suppose you wanted to study homelessness: There is no list of all homeless individuals, nor are you likely to create such a list. Moreover, as you'll see, there are times when probability sampling wouldn't be appropriate even if it were possible. Many such situations call for **nonprobability sampling.**

In this section, we'll examine four types of nonprobability sampling: reliance on available subjects, purposive or judgmental sampling, snowball sampling, and quota sampling. We'll conclude with a brief discussion of techniques for obtaining information about social groups through the use of informants.

Reliance on Available Subjects

Relying on available subjects, such as stopping people at a street corner or some other location, is an extremely risky sampling method; even so, it's used all too frequently. Clearly, this method does not permit any control over the representativeness of a sample. It's justified only if the researcher wants to study the characteristics of people passing the sampling point at specified times or if less risky sampling methods are not feasible. Even when this method is justified on grounds of feasibility, researchers must exercise great caution in generalizing from their data. Also, they should alert readers to the risks associated with this method.

University researchers frequently conduct surveys among the students enrolled in large lecture classes. The ease and frugality of such a method explains its popularity, but it seldom produces data of any general value. It may be useful for pretesting a questionnaire, but such a sampling method should not be used for a study purportedly describing students as a whole.

Consider this report on the sampling design in an examination of knowledge and opinions about nutrition and cancer among medical students and family physicians:

> The fourth-year medical students of the University of Minnesota Medical School in Minneapolis comprised the student population in this study. The physician population consisted of all physicians attending a "Family Practice Review and Update" course sponsored by the University of Minnesota Department of Continuing Medical Education. — (COOPER-STEPHENSON AND THEOLOGIDES 1981:472)

After all is said and done, what will the results of this study represent? They do not provide a meaningful comparison of medical students and family physicians in the United States or even in Minnesota. Who were the physicians who attended the course? We can guess that they were probably more concerned about their continuing education than were other physicians, but we can't say for sure. While such studies can be the source of useful insights, we must take care not to overgeneralize from them.

Purposive or Judgmental Sampling

Sometimes it's appropriate to select a sample on the basis of knowledge of a population, its elements, and the purpose of the study. This type of sampling is called **purposive** or **judgmental sampling.** In the initial design of a questionnaire, for example, you might wish to select the widest variety of respondents to test the broad applicability of questions. Although the study findings would not represent any meaningful population, the test run might effectively uncover any peculiar defects in your questionnaire. This situation would be considered a pretest, however, rather than a final study.

In some instances, you may wish to study a small subset of a larger population in which many members of the subset are easily identified, but the enumeration of them all would be nearly impossible. For example, you might want to study the leadership of a student protest movement; many of the leaders are easily visible, but it would not be feasible to define and sample *all* leaders. In studying all or a sample of the most visible leaders, you may collect data sufficient for your purposes.

Or let's say you want to compare left-wing and right-wing students. Because you may not be able to enumerate and sample from all such students, you might decide to sample the memberships of left- and right-leaning groups, such as the Green Party and the Young Americans for Freedom. Although such a sample design would not provide a good description of either left-wing or right-wing students as a whole, it might suffice for general comparative purposes.

Field researchers are often particularly interested in studying deviant cases—cases that don't fit into fairly regular patterns of attitudes and behaviors—in order to improve their understanding of the more regular pattern. For example, you might gain important insights into the nature of school spirit, as exhibited at a pep rally, by interviewing people who did not appear to be caught up in the emotions of the crowd or by interviewing students who did not attend the rally at all. Selecting deviant cases for study is another example of purposive study.

Snowball Sampling

Another nonprobability sampling technique, which some consider to be a form of accidental sampling, is called **snowball sampling.** This procedure is appropriate when the members of a special population are difficult to locate, such as homeless individuals, migrant workers, or undocumented immigrants. In snowball sampling, the researcher collects data on the few members of the target population he or she can locate, then asks those individuals to provide the information needed to locate other members of that population whom they happen to know. "Snowball" refers to the process of accumulation as each located subject suggests other subjects. Because this procedure also results in samples with questionable repre-

sentativeness, it's used primarily for exploratory purposes.

Suppose you wish to learn a community organization's pattern of recruitment over time. You might begin by interviewing fairly recent recruits, asking them who introduced them to the group. You might then interview the people named, asking them who introduced *them* to the group. You might then interview those people named, asking, in part, who introduced *them*. Or, in studying a loosely structured political group, you might ask one of the participants who he or she believes to be the most influential members of the group. You might interview those people and, in the course of the interviews, ask who *they* believe to be the most influential. In each of these examples, your sample would "snowball" as each of your interviewees suggested other people to interview.

Quota Sampling

Quota sampling is the method that helped George Gallup avoid disaster in 1936—and set up the disaster of 1948. Like probability sampling, quota sampling addresses the issue of representativeness, although the two methods approach the issue quite differently.

Quota sampling begins with a matrix, or table, describing the characteristics of the target population. Depending on your research purposes, you may need to know what proportion of the population is male and what proportion female as well as what proportions of each gender fall into various age categories, educational levels, ethnic groups, and so forth. In establishing a national quota sample, you might need to know what proportion of the national population is urban, eastern, male, under 25, white, working class, and the like, and all the possible combinations of these attributes.

Once you've created such a matrix and assigned a relative proportion to each cell in the matrix, you proceed to collect data from people having all the characteristics of a given cell. You then assign to all the people in a given cell a weight appropriate to their portion of the total population. When all the sample elements are so weighted, the overall data should provide a reasonable representation of the total population.

Although quota sampling resembles probability sampling, it has several inherent problems. First, the *quota frame* (the proportions that different cells represent) must be accurate, and it is often difficult to get up-to-date information for this purpose. The Gallup failure to predict Truman as the presidential victor in 1948 was due partly to this problem. Second, the selection of sample elements within a given cell may be biased even though its proportion of the population is accurately estimated. Instructed to interview five people who meet a given, complex set of characteristics, an interviewer may still avoid people living at the top of seven-story walk-ups, having particularly run-down homes, or owning vicious dogs.

In recent years, attempts have been made to combine probability and quota sampling methods, but the effectiveness of this effort remains to be seen. At present, you would be advised to treat quota sampling warily if your purpose is statistical description.

At the same time, the logic of quota sampling can sometimes be applied usefully to a field research project. In the study of a formal group, for example, you might wish to interview both leaders and nonleaders. In studying a student political organization, you might want to interview radical, moderate, and conservative members of that group. You may be able to achieve sufficient representativeness in such cases by using quota sampling to ensure that you interview both men and women, both younger and older people, and so forth.

Selecting Informants

When field research involves the researcher's attempt to understand some social setting—a juvenile gang or local neighborhood, for example—much of that understanding will come from a collaboration with some members of the group being studied. Whereas social researchers speak of *respondents* as people who provide information about themselves, allowing the researcher to construct a composite picture of the group those respondents represent, an **informant** is a member of the group who can talk directly about the group per se.

Especially important to anthropologists, informants are important to other social researchers as well. If you wanted to learn about informal social networks in a local public housing project, for example, you would do well to locate individuals who could understand what you were looking for and help you find it.

When Jeffrey Johnson (1990) set out to study a salmon fishing community in North Carolina, he used several criteria to evaluate potential informants. Did their positions allow them to interact regularly with other members of the camp, for example, or were they isolated? (He found that the carpenter had a wider range of interactions than did the boat captain.) Was their information about the camp pretty much limited to their specific jobs, or did it cover many aspects of the operation? These and other criteria helped determine how useful the potential informants might be.

Usually, you'll want to select informants somewhat typical of the groups you're studying. Otherwise, their observations and opinions may be misleading. Interviewing only physicians will not give you a well-rounded view of how a community medical clinic is working, for example. Along the same lines, an anthropologist who interviews only men in a society where women are sheltered from outsiders will get a biased view. Similarly, while informants fluent in English are convenient for English-speaking researchers from the United States, they do not typify the members of many societies and even many subgroups within English-speaking countries.

Simply because they're the ones willing to work with outside investigators, informants will almost always be somewhat "marginal" or atypical within their group. Sometimes this is obvious. Other times, however, you'll learn about their marginality only in the course of your research.

In Jeffrey Johnson's study, the county agent identified one fisherman who seemed squarely in the mainstream of the community. Moreover, he was cooperative and helpful to Johnson's research. The more Johnson worked with the fisherman, however, the more he found the man to be a marginal member of the fishing community.

> First, he was a Yankee in a southern town. Second, he had a pension from the Navy [so he was not seen as a "serious fisherman" by others in the community]. . . . Third, he was a major Republican activist in a mostly Democratic village. Finally, he kept his boat in an isolated anchorage, far from the community harbor. — (1990:56)

Informants' marginality may not only bias the view you get, but their marginal status may also limit their access (and hence yours) to the different sectors of the community you wish to study.

These comments should give you some sense of the concerns involved in nonprobability sampling, typically used in qualitative research projects. I conclude with the following injunction:

> Your overall goal is to collect the *richest possible data.* Rich data mean, ideally, a wide and diverse range of information collected over a relatively prolonged period of time. Again, ideally, you achieve this through direct, face-to-face contact with, and prolonged immersion in, some social location or circumstance. — (LOFLAND AND LOFLAND 1995:16)

 To see some practical implications of choosing and using informants, visit the Web site of Canada's Community Adaptation and Sustainable Livelihoods (CASL) Program: http://iisd.ca/casl/CASLGuide/KeyInformEx.htm

In other words, nonprobability sampling does have its uses, particularly in qualitative research projects. But researchers must take care to acknowledge the limitations of nonprobability sampling, especially regarding accurate and precise representations of populations. This point will be-

come clearer as we discuss the logic and techniques of probability sampling.

THE THEORY AND LOGIC OF PROBABILITY SAMPLING

While appropriate to some research purposes, nonprobability sampling methods cannot guarantee that the sample we observed is representative of the whole population. When researchers want precise, statistical descriptions of large populations—for example, the percentage of the population who are unemployed, plan to vote for Candidate X, or feel a rape victim should have the right to an abortion—they turn to **probability sampling.** All large-scale surveys use probability sampling methods.

Although the application of probability sampling involves some sophisticated use of statistics, the basic logic of probability sampling is not difficult to understand. If all members of a population were identical in all respects—all demographic characteristics, attitudes, experiences, behaviors, and so on—there would be no need for careful sampling procedures. In this extreme case of perfect homogeneity, in fact, any single case would suffice as a sample to study characteristics of the whole population.

In fact, of course, the human beings who compose any real population are quite heterogeneous, varying in many ways. Figure 7-1 offers a simplified illustration of a heterogeneous population: The 100 members of this small population differ by gender and race. We'll use this hypothetical micropopulation to illustrate various aspects of probability sampling.

The fundamental idea behind probability sampling is this: To provide useful descriptions of the total population, a sample of individuals from a population must contain essentially the same variations that exist in the population. This isn't as simple as it might seem, however. Let's take a minute to look at some of the ways researchers might go astray. Then, we'll see how probability sampling provides an efficient method for selecting a sample

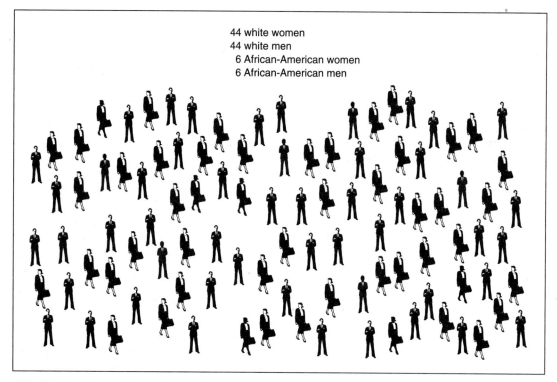

44 white women
44 white men
6 African-American women
6 African-American men

FIGURE 7-1 **A Population of 100 Folks**

that should adequately reflect variations that exist in the population.

Conscious and Unconscious Sampling Bias

At first glance, it may look as though sampling is pretty straightforward. To select a sample of 100 university students, you might simply interview the first 100 students you find walking around campus. This kind of sampling method is often used by untrained researchers, but it runs a very high risk of introducing biases into the samples.

In connection with sampling, *bias* simply means that those selected are not typical or representative of the larger populations they have been chosen from. This kind of bias does not have to be intentional. In fact, it is virtually inevitable when you pick people by the seat of your pants.

Figure 7-2 illustrates what can happen when researchers simply select people who are convenient

for study. Although women are only 50 percent of our micropopulation, those closest to the researcher (in the upper-right corner) happen to be 70 percent women, and although the population is 12 percent black, none was selected into the sample.

Beyond the risks inherent in simply studying people who are convenient, other problems can arise. To begin with, the researcher's personal leanings may affect the sample to the point where it does not truly represent the student population. Suppose you're a little intimidated by students who look particularly "cool," feeling they might ridicule your research effort. You might consciously or unconsciously avoid interviewing such people. Or, you might feel that the attitudes of "super-straight-looking" students would be irrelevant to your research purposes and so avoid interviewing them.

Even if you sought to interview a "balanced" group of students, you wouldn't know the exact

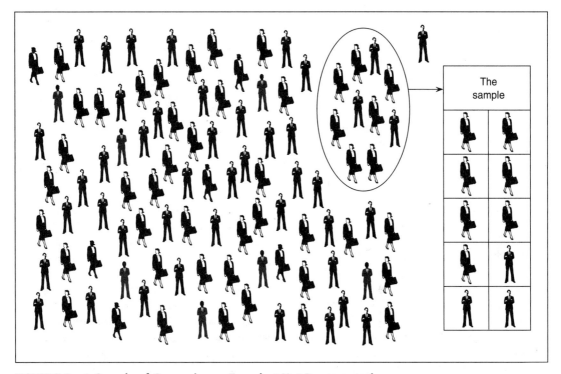

FIGURE 7-2 **A Sample of Convenience: Easy, but Not Representative**

proportions of different types of students making up such a balance, and you wouldn't always be able to identify the different types just by watching them walk by.

Even if you made a conscientious effort to interview, say, every tenth student entering the university library, you could not be sure of a representative sample, because different types of students visit the library with different frequencies. Your sample would overrepresent students who visit the library more often than do others.

Similarly, the "public opinion" call-in polls—in which radio stations or newspapers ask people to call specified telephone numbers to register their opinions—cannot be trusted to represent general populations. At the very least, not everyone in the population will even be aware of the poll. This problem also invalidates polls by magazines and newspapers who publish coupons for readers to complete and mail in. Even among those who are aware of such polls, not all will express an opinion,

especially if doing so will cost them a stamp, an envelope, or a telephone charge. Similar considerations apply to polls taken over the Internet.

Ironically, the failure of such polls to represent all opinions equally was inadvertently acknowledged by Philip Perinelli (1986), a staff manager of AT&T Communications' DIAL-IT 900 Service, which offers a call-in poll facility to organizations. Perinelli attempted to counter criticisms by saying, "The 50-cent charge assures that only interested parties respond and helps assure also that no individual 'stuffs' the ballot box." We cannot determine general public opinion while considering "only interested parties." This excludes those who don't care 50-cents' worth, as well as those who recognize that such polls are not valid. Both types of people may have opinions and may even vote on election day. Perinelli's assertion that the 50-cent charge will prevent ballot stuffing actually means that only those who can afford it will engage in ballot stuffing.

The possibilities for inadvertent sampling bias are endless and not always obvious. Fortunately there are techniques that help us avoid bias.

Representativeness and Probability of Selection

Although the term **representativeness** has no precise, scientific meaning, it carries a common-sense meaning that makes it useful here. For our purpose, a sample is representative of the population from which it is selected if the aggregate characteristics of the sample closely approximate those same aggregate characteristics in the population. If, for example, the population contains 50 percent women, then a sample must contain "close to" 50 percent women to be representative. Later, we'll discuss "how close" in detail.

Note that samples need not be representative in all respects; representativeness is limited to those characteristics that are relevant to the substantive interests of the study. However, you may not know in advance which characteristics are relevant.

A basic principle of probability sampling is that a sample will be representative of the population from which it is selected if all members of the population have an equal chance of being selected in the sample. (We'll see shortly that the size of the sample selected also affects the degree of representativeness.) Samples that have this quality are often labeled **EPSEM** samples (EPSEM stands for "equal probability of selection method"). Later we'll discuss variations of this principle, which forms the basis of probability sampling.

Moving beyond this basic principle, we must realize that samples—even carefully selected EPSEM samples—seldom if ever perfectly represent the populations from which they are drawn. Nevertheless, probability sampling offers two special advantages.

First, probability samples, although never perfectly representative, are typically more representative than other types of samples, because the biases previously discussed are avoided. In practice, a probability sample is more likely than a nonprobability sample to be representative of the population from which it is drawn.

Second, and more important, probability theory permits us to estimate the accuracy or representativeness of the sample. Conceivably, an uninformed researcher might, through wholly haphazard means, select a sample that nearly perfectly represents the larger population. The odds are against doing so, however, and we would be unable to estimate the likelihood that he or she has achieved representativeness. The probability sampler, on the other hand, can provide an accurate estimate of success or failure. We'll shortly see exactly how this estimate can be achieved.

I've said that probability sampling ensures that samples are representative of the population we wish to study. As we'll see in a moment, probability sampling rests on the use of a random selection procedure. To develop this idea, though, we need to give more precise meaning to two important terms: *element* and *population.**

An **element** is that unit about which information is collected and that provides the basis of analysis. Typically, in survey research, elements are people or certain types of people. However, other kinds of units can constitute the elements for social research: Families, social clubs, or corporations might be the elements of a study. In a given study, elements and units of analysis are often the same as units of analysis, though the former are used in sample selection and the latter in data analysis.

Up to now we've used the term *population* to mean the group or collection that we're interested in generalizing about. More formally, a **population** is the theoretically specified aggregation of study elements. Whereas the vague term *Americans* might be the target for a study, the delineation of the population would include the definition of the element *Americans* (for example, citizenship, residence) and the time referent for the study (Americans as of when?). Translating the abstract "adult New Yorkers" into a workable population would require a specification of the age defining *adult* and the boundaries of New York. Specifying

---

*I would like to acknowledge a debt to Leslie Kish and his excellent textbook *Survey Sampling*. Although I've modified some of the conventions used by Kish, his presentation is easily the most important source of this discussion.

the term *college student* would include a consideration of full- and part-time students, degree candidates and nondegree candidates, undergraduate and graduate students, and so forth.

A **study population** is that aggregation of elements from which the sample is actually selected. As a practical matter, researchers are seldom in a position to guarantee that every element meeting the theoretical definitions laid down actually has a chance of being selected in the sample. Even where lists of elements exist for sampling purposes, the lists are usually somewhat incomplete. Some students are always inadvertently omitted from student rosters. Some telephone subscribers request that their names and numbers be unlisted.

Often, researchers decide to limit their study populations more severely than indicated in the preceding examples. National polling firms may limit their national samples to the 48 adjacent states, omitting Alaska and Hawaii for practical reasons. A researcher wishing to sample psychology professors may limit the study population to those in psychology departments, omitting those in other departments. Whenever the population under examination is altered in such fashions, you must make the revisions clear to your readers.

Random Selection

With these definitions in hand, we can define the ultimate purpose of sampling: to select a set of elements from a population in such a way that descriptions of those elements accurately portray the total population from which the elements are selected. Probability sampling enhances the likelihood of accomplishing this aim and also provides methods for estimating the degree of probable success.

Random selection is the key to this process. In **random selection,** each element has an equal chance of selection independent of any other event in the selection process. Flipping a coin is the most frequently cited example: Provided that the coin is perfect (that is, not biased in terms of coming up heads or tails), the "selection" of a head or a tail is independent of previous selections of heads or tails. No matter how many heads turn up in a row,

the chance that the next flip will produce "heads" is exactly 50–50. Rolling a perfect set of dice is another example.

Such images of random selection, while useful, seldom apply directly to sampling methods in social research. More typically, social researchers use tables of random numbers or computer programs that provide a random selection of sampling units. A **sampling unit** is that element or set of elements considered for selection in some stage of sampling. In Chapter 9, on survey research, we'll see how computers are used to select random telephone numbers for interviewing, a technique called *random-digit dialing.*

The reasons for using random selection methods are twofold. First, this procedure serves as a check on conscious or unconscious bias on the part of the researcher. The researcher who selects cases on an intuitive basis might very well select cases that would support his or her research expectations or hypotheses. Random selection erases this danger. More important, random selection offers access to the body of probability theory, which provides the basis for estimating the characteristics of the population as well as estimates of the accuracy of samples. Let's now examine probability theory in greater detail.

Probability Theory, Sampling Distributions, and Estimates of Sample Error

Probability theory is a branch of mathematics that provides the tools researchers need to devise sampling techniques that produce representative samples and to analyze the results of their sampling statistically. More formally, probability theory provides the basis for estimating the parameters of a population. A **parameter** is the summary description of a given variable in a population. The mean income of all families in a city is a parameter; so is the age distribution of the city's population. When researchers generalize from a sample, they're using sample observations to estimate population parameters. Probability theory enables them both to make these estimates and to arrive at a judgment of how likely the estimates will accurately

FIGURE 7-3 **A Population of Ten People with $0–$9**

represent the actual parameters in the population. So, for example, probability theory allows pollsters to infer from a sample of 2,000 voters how a population of 100 million voters is likely to vote—and to specify exactly what the probable margin of error in the estimates is.

Probability theory accomplishes these seemingly magical feats by way of the concept of sampling distributions. A single sample selected from a population will give an estimate of the population parameter. Other samples would give the same or slightly different estimates. Probability theory tells us about the distribution of estimates that would be produced by a large number of such samples. To see how this works, we'll look at two examples of sampling distributions, beginning with a simple example in which our population consists of just ten cases.

The Sampling Distribution of Ten Cases Suppose there are ten people in a group, and each has a certain amount of money in his or her pocket. To simplify, let's assume that one person has no money, another has one dollar, another has two dollars, and so forth up to the person with nine dol-

lars. Figure 7-3 presents the population of ten people.*

Our task is to determine the average amount of money one person has: specifically, the mean number of dollars. If you simply add up the money shown in Figure 7-3, you'll find that the total is $45, so the mean is $4.50. Our purpose in the rest of this exercise is to estimate that mean without actually observing all ten individuals. We'll do that by selecting random samples from the population and using the means of those samples to estimate the mean of the whole population.

To start, suppose we were to select—at random—a sample of only one person from the ten. Our ten possible samples thus consist of the ten cases shown in Figure 7-3.

The ten dots shown on the graph in Figure 7-4 represent these ten samples. Since we're taking samples of only one, they also represent the "means" we would get as estimates of the population. The distribution of the dots on the graph is called the *sampling distribution*. Obviously, it

---

*I want to thank Hanan Selvin for suggesting this method of introducing probability sampling.

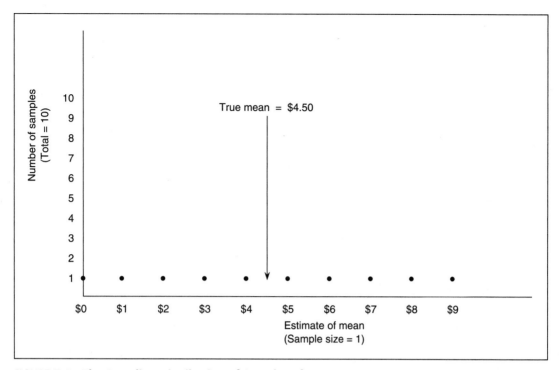

FIGURE 7-4 **The Sampling Distribution of Samples of 1**

wouldn't be a very good idea to select a sample of only one, since we stand a very good chance of missing the true mean of $4.50 by quite a bit.

Now suppose if we take a sample of two. As shown in Figure 7-5, increasing the sample size improves our estimations. There are now 45 possible samples: [$0 $1], [$0 $2], . . . [$7 $8], [$8 $9]. Moreover, some of those samples produce the same means. For example, [$0 $6], [$1 $5], and [$2 $4] all produce means of $3. In Figure 7-5, the three dots shown above the $3 mean represent those three samples.

Moreover, the 45 samples are not evenly distributed, as they were when the sample size was only one. Rather, they are somewhat clustered around the true value of $4.50. Only two possible samples deviate by as much as $4 from the true value ([$0 $1] and [$8 $9]), whereas five of the samples would give the true estimate of $4.50; another eight samples miss the mark by only 50 cents (plus or minus).

Now suppose we select even larger samples. What do you suppose that will do to our estimates of the mean? Figure 7-6 presents the sampling distributions of samples of 3, 4, 5, and 6.

The progression of sampling distributions is clear. Every increase in sample size improves the distribution of estimates of the mean. The limiting case in this procedure, of course, is to select a sample of ten. There would be only one possible sample (everyone) and it would give us the true mean of $4.50.

Sampling Distribution and Estimates of Sampling Error Let's turn now to a more realistic sampling situation involving a much larger population and see how the notion of sampling distribution applies. Assume that we wish to study the student population of State University (SU) to determine the percentage of students who approve or disapprove of a student conduct code proposed by the administration. The study population will be

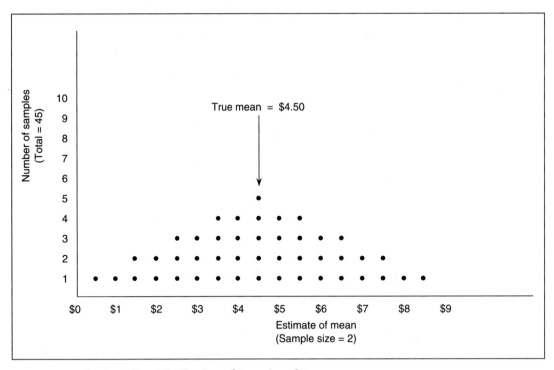

FIGURE 7-5 **The Sampling Distribution of Samples of 2**

the aggregation of, say, 20,000 students contained in a student roster: the sampling frame. The elements will be the individual students at SU. We'll select a random sample of, say, 100 students for the purposes of estimating the entire student body. The variable under consideration will be *attitudes toward the code,* a binomial variable: *approve* and *disapprove.* (The logic of probability sampling applies to the examination of other types of variables, such as mean income, but the computations are somewhat more complicated. Consequently, this introduction focuses on binomials.)

The horizontal axis of Figure 7-7 presents all possible values of this parameter in the population—from 0 percent to 100 percent approval. The midpoint of the axis—50 percent—represents half the students approving of the code and the other half disapproving.

To choose our sample, we give each student on the student roster a number and select 100 random numbers from a table of random numbers. Then we interview the 100 students whose numbers have been selected and ask for their attitudes toward the student code: whether they approve or disapprove. Suppose this operation gives us 48 students who approve of the code and 52 who disapprove. This summary description of a variable in a sample is called a **statistic.** We present this statistic by placing a dot on the *x* axis at the point representing 48 percent.

Now let's suppose we select another sample of 100 students in exactly the same fashion and measure their approval or disapproval of the student code. Perhaps 51 students in the second sample approve of the code. We place another dot in the appropriate place on the *x* axis. Repeating this process once more, we may discover that 52 students in the third sample approve of the code.

Figure 7-8 presents the three different sample statistics representing the percentages of students in each of the three random samples who approved of the student code. The basic rule of random sampling is that such samples drawn from a population give estimates of the parameter that

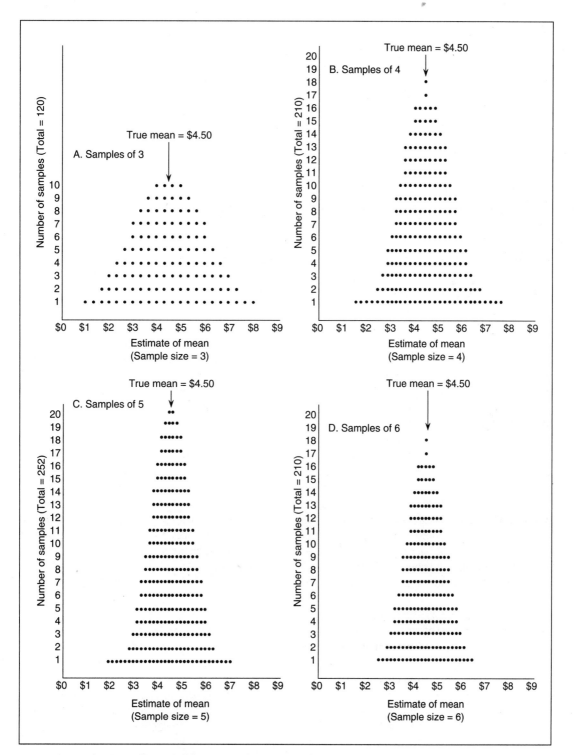

FIGURE 7-6 **The Sampling Distributions of Samples of 3, 4, 5, and 6**

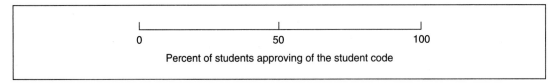

FIGURE 7-7   **Range of Possible Sample Study Results**

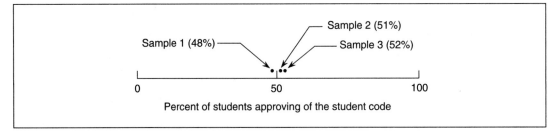

FIGURE 7-8   **Results Produced by Three Hypothetical Studies**

exists in the total population. Each of the random samples, then, gives us an estimate of the percentage of students in the total student body who approve of the student code. Unhappily, however, we have selected three samples and now have three separate estimates.

To retrieve ourselves from this problem, let's draw more and more samples of 100 students each, question each of the samples concerning their approval or disapproval of the code, and plot the new sample statistics on our summary graph. In drawing many such samples, we discover that some of the new samples provide duplicate estimates, as in the illustration of ten cases. Figure 7-9 shows the sampling distribution of, say, hundreds of samples. This is often referred to as a *normal curve.*

Note that by increasing the number of samples selected and interviewed, we have also increased the range of estimates provided by the sampling operation. In one sense we have increased our dilemma in attempting to guess the parameter in the population. Probability theory, however, provides certain important rules regarding the sampling distribution presented in Figure 7-9.

First, if many independent random samples are selected from a population, the sample statistics provided by those samples will be distributed around the population parameter in a known way. Thus, although Figure 7-9 shows a wide range of estimates, more of them are in the vicinity of 50 percent than elsewhere in the graph. Probability theory tells us, then, that the true value is in the vicinity of 50 percent.

Second, probability theory gives us a formula for estimating how closely the sample statistics are clustered around the true value. To put it another way, probability theory enables us to estimate the **sampling error**—the degree of error to be expected for a given sample design. This formula contains three factors: the parameter, the sample size, and the standard error (a measure of sampling error):

$$s = \sqrt{\frac{P \times Q}{n}}$$

The symbols P and Q in the formula equal the population parameters for the binomial: If 60 percent of the student body approve of the code and 40 percent disapprove, P and Q are 60 percent and 40 percent, respectively, or .6 and .4. Note that $Q = 1 - P$ and $P = 1 - Q$. The symbol n equals the number of cases in each sample, and s is the standard error.

Let's assume that the population parameter in the student example is 50 percent approving of the

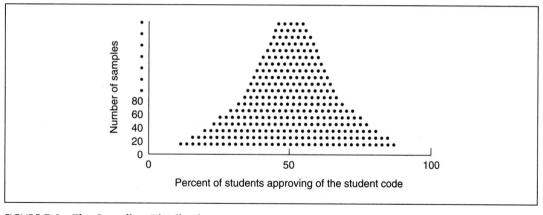

FIGURE 7-9   **The Sampling Distribution**

code and 50 percent disapproving. Recall that we've been selecting samples of 100 cases each. When these numbers are put into the formula, we find that the standard error equals .05, or 5 percent.

In probability theory, the standard error is a valuable piece of information because it indicates the extent to which the sample estimates will be distributed around the population parameter. (If you're familiar with the standard deviation in statistics, you may recognize that the standard error, in this case, is the standard deviation of the sampling distribution.) Specifically, probability theory indicates that certain proportions of the sample estimates will fall within specified increments—each equal to one standard error—from the population parameter. Approximately 34 percent (.3413) of the sample estimates will fall within one standard error increment above the population parameter, and another 34 percent will fall within one standard error below the parameter. In our example, the standard error increment is 5 percent, so we know that 34 percent of our samples will give estimates of student approval between 50 percent (the parameter) and 55 percent (one standard error above); another 34 percent of the samples will give estimates between 50 percent and 45 percent (one standard error below the parameter). Taken together, then, we know that roughly two-thirds (68 percent) of the samples will give estimates within 65 percent of the parameter.

Moreover, probability theory dictates that roughly 95 percent of the samples will fall within

plus or minus two standard errors of the true value, and 99.9 percent of the samples will fall within plus or minus three standard errors. In our present example, then, we know that only one sample out of a thousand would give an estimate lower than 35 percent approval or higher than 65 percent.

The proportion of samples falling within one, two, or three standard errors of the parameter is constant for any random sampling procedure such as the one just described, providing that a large number of samples are selected. The size of the standard error in any given case, however, is a function of the population parameter and the sample size. If we return to the formula for a moment, we note that the standard error will increase as a function of an increase in the quantity P times Q. Note further that this quantity reaches its maximum in the situation of an even split in the population. If $P = .5$, $PQ = .25$; if $P = .6$, $PQ = .24$; if $P = .8$, $PQ = .16$; if $P = .99$, $PQ = .0099$. By extension, if P is either 0.0 or 1.0 (either 0 percent or 100 percent approve of the student code), the standard error will be 0. If everyone in the population has the same attitude (no variation), then every sample will give exactly that estimate.

The standard error is also a function of the sample size—an inverse function. As the sample size increases, the standard error decreases. As the sample size increases, the several samples will be clustered nearer to the true value. Another general guideline is evident in the formula: Because of the square root formula, the standard error is reduced by half if the sample size is quadrupled. In our present example, samples of 100 produce a standard error of 5 percent; to reduce the standard error to 2.5 percent, we must increase the sample size to 400.

All of this information is provided by established probability theory in reference to the selection of large numbers of random samples. (If you've taken a statistics course, you may know this as the "Central Tendency Theorem.") If the population parameter is known and many random samples are selected, we can predict how many of the samples will fall within specified intervals from the parameter.

Recognize that this discussion illustrates only the logic of probability sampling; it does not describe the way research is actually conducted. Usually, we don't know the parameter: The very reason we conduct a sample survey is to estimate that value. Moreover, we don't actually select large numbers of samples: We select only one sample. Nevertheless, the preceding discussion of probability theory provides the basis for inferences about the typical social research situation. Knowing what it would be like to select thousands of samples allows us to make assumptions about the one sample we do select and study.

Confidence Levels and Confidence Intervals

Whereas probability theory specifies that 68 percent of that fictitious large number of samples would produce estimates falling within one standard error of the parameter, we can turn the logic around and infer that any single random sample has a 68-percent chance of falling within that range. This observation leads us to the two key components of sampling error estimates: **confidence level** and **confidence interval.** We express the accuracy of our sample statistics in terms of a *level of confidence* that the statistics fall within a specified *interval* from the parameter. For example, we may say we are 95 percent confident that our sample statistics (for example, 50 percent favor the new student code) are within plus or minus 5 percentage points of the population parameter. As the confidence interval is expanded for a given statistic, our confidence increases. For ex-

ample, we may say that we are 99.9 percent confident that our statistic falls within three standard errors of the true value.

Although we may be confident (at some level) of being within a certain range of the parameter, we've already noted that we seldom know what the parameter is. To resolve this problem, we substitute our sample estimate for the parameter in the formula; that is, lacking the true value, we substitute the best available guess.

The result of these inferences and estimations is that we can estimate a population parameter and also the expected degree of error on the basis of one sample drawn from a population. Beginning with the question, "What percentage of the student body approves of the student code?" you could select a random sample of 100 students and interview them. You might then report that your best estimate is that 50 percent of the student body approves of the code and that you are 95-percent confident that between 40 and 60 percent (plus or minus two standard errors) approve. The range from 40 to 60 percent is the confidence interval. (At the 68-percent confidence level, the confidence interval would be 45–55 percent.)

The logic of confidence levels and confidence intervals also provides the basis for determining the appropriate sample size for a study. Once you've decided on the degree of sampling error you can tolerate, you'll be able to calculate the number of cases needed in your sample. Thus, for example, if you want to be 95 percent confident that your study findings are accurate within 5 percentage points of the population parameters, you should select a sample of at least 400. (Appendix F is a convenient guide in this regard.)

This then is the basic logic of probability sampling. Random selection permits the researcher to link findings from a sample to the body of probability theory so as to estimate the accuracy of those findings. All statements of accuracy in sampling must specify both a confidence level and a confidence interval. The researcher must report that he or she is *x* percent confident that the population parameter is between two specific values.

Two cautions are in order before we conclude this discussion of the basic logic of probability sam-

APPLYING THE RESULTS

Here's how George Gallup (1984:7) described his sampling error in a newspaper report of a Gallup Poll regarding attitudes of children and parents:

> The adult findings are based on in-person interviews with 1520 adults, 18 and older, conducted in more than 300 scientifically selected localities across the nation during the period October 26–29. For results based on samples of this size, one can say with 95 percent confidence that the error attributable to sampling and other random effects could be three percentage points in either direction.

And here is what the *New York Times* had to say about a poll it conducted on religious opinions ("How the Poll Was Conducted," 1995:15):

> In theory, in 19 cases out of 20 the results based on such samples will differ by no more than three percentage points in either direction, from what would have been obtained by seeking out all American adults.

When you read statements like these in the popular media, be warned that such statements are sometimes made when they aren't warranted (though this isn't true in our examples). And be especially wary of survey or poll results that fail to represent confidence levels and confidence intervals—without these specifications, the "findings" are of dubious value.

example, mathematically assumes an infinitely large population, an infinite number of samples, and sampling with replacement—that is, every sampling unit selected is "thrown back into the pot" and could be selected again. Second, our discussion has greatly oversimplified the inferential jump from the distribution of several samples to the probable characteristics of one sample.

I offer these cautions to provide perspective on the uses of probability theory in sampling. Social researchers often appear to overestimate the precision of estimates produced by the use of probability theory. As I'll mention elsewhere in this chapter and throughout the book, variations in sampling techniques and nonsampling factors may further reduce the legitimacy of such estimates. For example, those selected in a sample who fail or refuse to participate further detract from the representativeness of the sample.

Nevertheless, the calculations discussed in this section can be extremely valuable to you in understanding and evaluating your data. Although the calculations do not provide as precise estimates as some researchers might assume, they can be quite valid for practical purposes. They are unquestionably more valid than less rigorously derived estimates based on less rigorous sampling methods. Most important, being familiar with the basic logic underlying the calculations can help you react sensibly both to your own data and to those reported by others.

POPULATIONS AND SAMPLING FRAMES

The preceding section introduced the theoretical model for social research sampling. Although as students, research consumers, and researchers we need to understand that theory, it is no less important to appreciate the less-than-perfect conditions that exist in the field. In this section we'll look at one aspect of field conditions that requires a compromise with idealized theoretical conditions and assumptions: the congruence of or disparity between populations of sampling frames.

pling. First, the survey uses of probability theory as discussed here are technically not wholly justified. The theory of sampling distribution makes assumptions that almost never apply in survey conditions. The exact proportion of samples contained within specified increments of standard errors, for

Simply put, a **sampling frame** is the list or quasi list of elements from which a probability sample is selected. If a sample of students is selected from a student roster, the roster is the sampling frame. If the primary sampling unit for a complex population sample is the census block, the list of census blocks composes the sampling frame—in the form of a printed booklet, a magnetic tape file, or some other computerized record. Here are some reports of sampling frames appearing in research journals:

> The data for this research were obtained from a random sample of *parents of children in the third grade in public and parochial schools in Yakima County, Washington.* — (PETERSEN AND MAYNARD 1981:92)

> The sample at Time 1 consisted of 160 names drawn randomly from the *telephone directory of Lubbock, Texas.* — (TAN 1980:242)

> The data reported in this paper . . . were gathered from a probability sample of *adults aged 18 and over residing in households in the 48 contiguous United States.* Personal interviews with 1,914 respondents were conducted by the Survey Research Center of the University of Michigan during the fall of 1975. — (JACKMAN AND SENTER 1980:345)

In each example I've italicized the actual sampling frames.

Properly drawn samples provide information appropriate for describing the population of elements composing the sampling frame—nothing more. I emphasize this point in view of the all-too-common tendency for researchers to select samples from a given sampling frame and then make assertions about a population similar to, but not identical to, the population defined by the sampling frame.

For example, take a look at this report, which discusses the drugs most frequently prescribed by U.S. physicians:

> Information on prescription drug sales is not easy to obtain. But Rinaldo V. DeNuzzo, a professor of pharmacy at the Albany College of Pharmacy, Union University, Albany, NY, has been tracking prescription drug sales for 25 years by polling nearby drugstores. He publishes the results in an industry trade magazine, *MM&M.*
>
> DeNuzzo's latest survey, covering 1980, is based on reports from 66 pharmacies in 48 communities in New York and New Jersey. Unless there is something peculiar about that part of the country, his findings can be taken as representative of what happens across the country. — (MOSKOWITZ 1981:33)

What is striking in the excerpt is the casual comment about whether there is anything peculiar about New York and New Jersey. There is. The lifestyle in these two states hardly typifies the other 48. We cannot assume that residents in these large, urbanized, Eastern seaboard states necessarily have the same drug-use patterns as do residents of Mississippi, Nebraska, or Vermont.

Does the survey even represent prescription patterns in New York and New Jersey? To determine that, we would have to know something about the way the 48 communities and the 66 pharmacies were selected. We should be wary in this regard, in view of the reference to "polling nearby drugstores." As we'll see, there are several methods for selecting samples that ensure representativeness, and unless they're used, we shouldn't generalize from the study findings.

A sampling frame, then, must be consonant with the population we wish to study. In the simplest sample design, the sampling frame is a list of the elements composing the study population. In practice, though, existing sampling frames often define the study population rather than the other way around. That is, we often begin with a population in mind for our study; then we search for possible sampling frames. Having examined and evaluated the frames available for our use, we decide which frame presents a study population most appropriate to our needs.

Studies of organizations are often the simplest from a sampling standpoint because organizations typically have membership lists. In such cases, the list of members constitutes an excellent sampling frame. If a random sample is selected from a membership list, the data collected from that sample may be taken as representative of all members—if all members are included in the list.

Populations that can be sampled from good organizational lists include elementary school, high school, and university students and faculty; church members; factory workers; fraternity or sorority members; members of social, service, or political clubs; and members of professional associations.

The preceding comments apply primarily to local organizations. Often, statewide or national organizations do not have a single membership list. There is, for example, no single list of Episcopalian church members. However, a slightly more complex sample design could take advantage of local church membership lists by first sampling churches and then subsampling the membership lists of those churches selected. (More about that later.)

Other lists of individuals may be especially relevant to the research needs of a particular study. Government agencies maintain lists of registered voters, for example, that might be used if you wanted to conduct a preelection poll or an in-depth examination of voting behavior—but you must insure that the list is up-to-date. Similar lists contain the names of automobile owners, welfare recipients, taxpayers, business permit holders, licensed professionals, and so forth. Although it may be difficult to gain access to some of these lists, they provide excellent sampling frames for specialized research purposes.

Realizing that the sampling elements in a study need not be individual persons, we may note that the lists of other types of elements also exist: universities, businesses of various types, cities, academic journals, newspapers, unions, political clubs, professional associations, and so forth.

Telephone directories are frequently used for "quick and dirty" public opinion polls. Undeniably they're easy and inexpensive to use—no doubt the reason for their popularity. And, if you want to make assertions about telephone subscribers, the directory is a fairly good sampling frame. (Realize, of course, that a given directory will not include new subscribers or those who have requested unlisted numbers. Sampling is further complicated by the directories' inclusion of nonresidential listings.) Unfortunately, telephone directories are all too often used as a listing of a city's population or of its voters. Of the many defects in this reasoning, the chief one involves a social-class bias. Poor people are less likely to have telephones; rich people may have more than one line. A telephone directory sample, therefore, is likely to have a middle- or upper-class bias.

The class bias inherent in telephone directory samples is often hidden. Preelection polls conducted in this fashion are sometimes quite accurate, perhaps because of the class bias evident in voting itself: Poor people are less likely to vote. Frequently, then, these two biases nearly coincide, so that the results of a telephone poll may come very close to the final election outcome. Unhappily, you never know for sure until after the election. And sometimes, as in the case of the 1936 *Literary Digest* poll, you may discover that the voters have not acted according to the expected class biases. The ultimate disadvantage of this method, then, is the researcher's inability to estimate the degree of error to be expected in the sample findings.

Street directories and tax maps are often used for easy samples of households, but they may also suffer from incompleteness and possible bias. For example, in strictly zoned urban regions, illegal housing units are unlikely to appear on official records. As a result, such units could not be selected, and sample findings could not be representative of those units, which are often poorer and more overcrowded than the average.

Though the preceding comments apply to the United States, the situation is quite different in some other countries. In Japan, for example, the government maintains quite accurate population registration lists. Moreover, citizens are required by law to keep their information up-to-date, such as changes in residence or births and deaths in the household. As a consequence, you can select simple random samples of the Japanese population more easily than in the United States. Such a registration list in the United States would conflict directly with this country's norms regarding individual privacy.

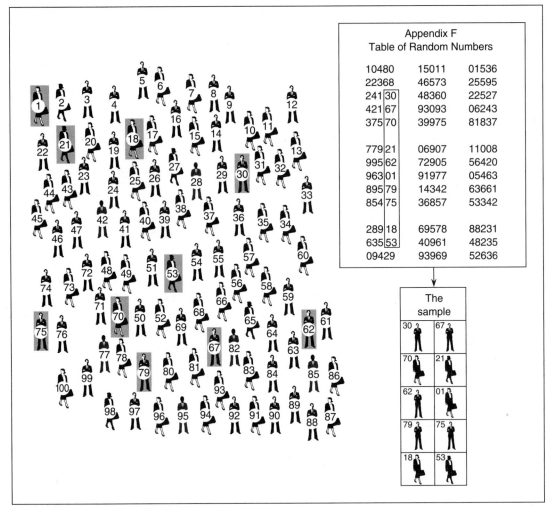

FIGURE 7-10 **A Simple Random Sample**

Review of Populations and Sampling Frames

Because social research literature gives surprisingly little attention to the issues of populations and sampling frames, I've devoted special attention to them here. Here is a summary of the main guidelines to remember:

1. Findings based on a sample can be taken as representing only the aggregation of elements that compose the sampling frame.

2. Often, sampling frames do not truly include all the elements their names might imply. Omissions are almost inevitable. Thus, a first concern of the researcher must be to assess the extent of the omissions and to correct them if possible. (Of course, the researcher may feel that he or she can safely ignore a small number of omissions that cannot easily be corrected.)

3. To be generalized even to the population composing the sampling frame, all elements

must have equal representation in the frame. Typically, each element should appear only once. Elements that appear more than once will have a greater probability of selection, and the sample will, overall, overrepresent those elements.

Other, more practical matters relating to populations and sampling frames will be treated elsewhere in this book. For example, the form of the sampling frame—such as a list in a publication, a 3-by-5 card file, computer disks, or magnetic tapes—can affect how easy it is to use. And ease of use may often take priority over scientific considerations: An "easier" list may be chosen over a "harder" one, even though the latter is more appropriate to the target population. We should not take a dogmatic position in this regard, but every researcher should carefully weigh the relative advantages and disadvantages of such alternatives.

TYPES OF SAMPLING DESIGNS

Up to this point, we've focused on simple random sampling (SRS). Indeed, the body of statistics typically used by social researchers assumes such a sample. As you'll see shortly, however, you have several options in choosing your sampling method, and you'll seldom if ever choose simple random sampling. There are two reasons for this. First, with all but the simplest sampling frame, simple random sampling is not feasible. Second, and probably surprisingly, simple random sampling may not be the most accurate method available. Let's turn now to a discussion of simple random sampling and the other options available.

Simple Random Sampling

As noted, **simple random sampling** is the basic sampling method assumed in the statistical computations of social research. Because the mathematics of random sampling are especially complex, we'll detour around them in favor of describing the ways of employing this method in the field.

Once a sampling frame has been properly established, to use simple random sampling the researcher assigns a single number to each element in the list, not skipping any number in the process. A table of random numbers (Appendix C) is then used to select elements for the sample. The box entitled "Using a Table of Random Numbers" explains its use.

If your sampling frame is in a machine-readable form, such as computer disk or magnetic tape, a simple random sample can be selected automatically by computer. (In effect, the computer program numbers the elements in the sampling frame, generates its own series of random numbers, and prints out the list of elements selected.)

Figure 7-10 offers a graphic illustration of simple random sampling. Note that the members of our hypothetical micropopulation have been numbered from 1 to 100. Moving to Appendix C, we decide to use the last two digits of the first column and to begin with the third number from the top. This yields person number 30 as the first one selected into the sample. Number 67 is next, and so forth. (Person 100 would have been selected if "00" had come up in the list.)

Systematic Sampling

Simple random sampling is seldom used in practice. As you'll see, it's not usually the most efficient method, and it can be laborious if done manually. Typically, simple random sampling requires a list of elements. When such a list is available, researchers usually employ systematic sampling instead.

In **systematic sampling,** every kth element in the total list is chosen (systematically) for inclusion in the sample. If the list contained 10,000 elements and you wanted a sample of 1,000, you would select every tenth element for your sample. To ensure against any possible human bias in using this method, you should select the first element at random. Thus, in the preceding example, you would begin by selecting a random number between one and ten. The element having that number is included in the sample, plus every tenth element following it. This method is technically referred to as

USING A TABLE OF RANDOM NUMBERS

In social research, it's often appropriate to select a set of random numbers from a table such as the one in Appendix C. Here's how to do that.

Suppose you want to select a simple random sample of 100 people (or other units) out of a population totaling 980.

1. To begin, number the members of the population: in this case, from 1 to 980. Now the problem is to select 100 random numbers. Once you've done that, your sample will consist of the people having the numbers you've selected. (*Note:* It's not essential to actually number them, as long as you're sure of the total. If you have them in a list, for example, you can always count through the list after you've selected the numbers.)

2. The next step is to determine the number of digits you'll need in the random numbers you select. In our example, there are 980 members of the population, so you'll need three-digit numbers to give everyone a chance of selection. (If there were 11,825 members of the population, you'd need to select five-digit numbers.) Thus, we want to select 100 random numbers in the range from 001 to 980.

3. Now turn to the first page of Appendix C. Notice there are several rows and columns of five-digit numbers, and there are several pages. The table represents a series of random numbers in the range from 00001 to 99999. To use the table for your hypothetical sample, you have to answer these questions:

   a. How will you create three-digit numbers out of five-digit numbers?
   b. What pattern will you follow in moving through the table to select your numbers?
   c. Where will you start?

   Each of these questions has several satisfactory answers. The key is to create a plan and follow it. Here's an example.

4. To create three-digit numbers from five-digit numbers, let's agree to select five-digit numbers from the table but consider only the left-most three digits in each case. If we picked the first number on the first page—10480—we would only consider the 104. (We could agree to take the digits farthest to the right, 480, or the middle three digits, 048, and any of these plans would work.) They key is to make a plan and stick with it. For convenience, let's use the left-most three digits.

a *systematic sample with a random start.* Two terms are frequently used in connection with systematic sampling. The **sampling interval** is the standard distance between elements selected in the sample: ten in the preceding sample. The **sampling ratio** is the proportion of elements in the population that are selected: $1/10$ in the example.

$$\text{sampling interval} = \frac{\text{population size}}{\text{sample size}}$$

$$\text{sampling ratio} = \frac{\text{sample size}}{\text{population size}}$$

In practice, systematic sampling is virtually identical to simple random sampling. If the list of elements is indeed randomized before sampling, one might argue that a systematic sample drawn from that list is in fact a simple random sample. By now, debates over the relative merits of simple random sampling and systematic sampling have been resolved largely in favor of the latter, simpler method. Empirically, the results are virtually identical. And, as you'll see in a later section, systematic sampling, in some instances, is slightly more accurate than simple random sampling.

There is one danger involved in systematic sampling. The arrangement of elements in the list can make systematic sampling unwise. Such an arrangement is usually called *periodicity.* If the list of elements is arranged in a cyclical pattern that

5. We can also choose to progress through the tables any way we want: down the columns, up them, across to the right or to the left, or diagonally. Again, any of these plans will work just fine so long as we stick to it. For convenience, let's agree to move down the columns. When we get to the bottom of one column, we'll go to the top of the next; when we exhaust a given page, we'll start at the top of the first column of the next page.

6. Now, where do we start? You can close your eyes and stick a pencil into the table and start wherever the pencil point lands. (I know it doesn't sound scientific, but it works.) Or, if you're afraid you'll hurt the book or miss it altogether, close your eyes and make up a column number and a row number. ("I'll pick the number in the fifth row of column 2.") Start with that number.

7. Let's suppose we decide to start with the fifth number in column 2. If you look on the first page of Appendix C, you'll see that the starting number is 39975. We have selected 399 as

our first random number, and we have 99 more to go. Moving down the second column, we select 069, 729, 919, 143, 368, 695, 409, 939, and so forth. At the bottom of column 2, we select number 104 and continue to the top of column 3: 015, 255, and so on.

8. See how easy it is? But trouble lies ahead. When we reach column 5, we are speeding along, selecting 816, 309, 763, 078, 061, 277, 988 . . . Wait a minute! There are only 980 students in the senior class. How can we pick number 988? The solution is simple: Ignore it. Any time you come across a number that lies outside your range, skip it and continue on your way: 188, 174, and so forth. The same solution applies if the same number comes up more than once. If you select 399 again, for example, just ignore it the second time.

9. That's it. You keep up the procedure until you've selected 100 random numbers. Returning to your list, your sample consists of person number 399, person number 69, person number 729, and so forth.

coincides with the sampling interval, a grossly biased sample may be drawn. Here are two examples that illustrate this danger.

In a classic study of soldiers during World War II, the researchers selected a systematic sample from unit rosters. Every tenth soldier on the roster was selected for the study. The rosters, however, were arranged in a table of organizations: sergeants first, then corporals and privates, squad by squad. Each squad had ten members. As a result, every tenth person on the roster was a squad sergeant. The systematic sample selected contained only sergeants. It could, of course, have been the case that no sergeants were selected for the same reason.

As another example, suppose we select a sample of apartments in an apartment building. If the sample is drawn from a list of apartments arranged in numerical order (for example, 101, 102, 103, 104, 201, 202, and so on), there is a danger of the sampling interval coinciding with the number of apartments on a floor or some multiple thereof. Then the samples might include only northwest-corner apartments or only apartments near the elevator. If these types of apartments have some other particular characteristic in common (for example, higher rent), the sample will be biased. The same danger would appear in a systematic sample of houses in a subdivision arranged with the same number of houses on a block.

In considering a systematic sample from a list, then, you should carefully examine the nature of that list. If the elements are arranged in any particular order, you should figure out whether that order will bias the sample to be selected, then you should take steps to counteract any possible bias (for example, take a simple random sample from cyclical portions).

Usually, however, systematic sampling is superior to simple random sampling, in convenience if nothing else. Problems in the ordering of elements in the sampling frame can usually be remedied quite easily.

Stratified Sampling

So far we have discussed two methods of sample selection from a list: random and systematic. **Stratification** is not an alternative to these methods; rather, it represents a possible modification of their use.

Simple random sampling and systematic sampling both ensure a degree of representativeness and permit an estimate of the error present. Stratified sampling is a method for obtaining a greater degree of representativeness by decreasing the probable sampling error. To understand this method, we must return briefly to the basic theory of sampling distribution.

Recall that sampling error is reduced by two factors in the sample design. First, a large sample produces a smaller sampling error than does a small sample. Second, a homogeneous population produces samples with smaller sampling errors than does a heterogeneous population. If 99 percent of the population agrees with a certain statement, it's extremely unlikely that any probability sample will greatly misrepresent the extent of agreement. If the population is split 50–50 on the statement, then the sampling error will be much greater.

Stratified sampling is based on this second factor in sampling theory. Rather than selecting your sample from the total population at large, the researcher ensures that appropriate numbers of elements are drawn from homogeneous subsets of that population. To get a stratified sample of university students, for example, you would first organize your population by college class and then draw appropriate numbers of frosh, sophomores, juniors, and seniors. In a nonstratified sample, rep-

resentation by class would be subjected to the same sampling error as would other variables. In a sample stratified by class, the sampling error on this variable is reduced to zero.

More-complex stratification methods are also possible. In addition to stratifying by class, you might also stratify by gender, by GPA, and so forth. In this fashion you might be able to ensure that your sample would contain the proper numbers of male sophomores with a 3.5 average, of female sophomores with a 4.0 average, and so forth.

The ultimate function of stratification, then, is to organize the population into homogeneous subsets (with heterogeneity between subsets) and to select the appropriate number of elements from each. To the extent that the subsets are homogeneous on the stratification variables, they may be homogeneous on other variables as well. Because *age* is related to *college class,* a sample stratified by class will be more representative in terms of age as well, compared with an unstratified sample. Because occupational aspirations still seem to be related to gender, a sample stratified by gender will be more representative in terms of occupational aspirations.

The choice of stratification variables typically depends on what variables are available. Gender can often be determined in a list of names. University lists are typically arranged by class. Lists of faculty members may indicate their departmental affiliation. Government agency files may be arranged by geographical region. Voter registration lists are arranged according to precinct.

In selecting stratification variables from among those available, however, you should be concerned primarily with those that are presumably related to variables you want to represent accurately. Because gender is related to many variables and is often available for stratification, it is often used. Education is related to many variables, but it is often not available for stratification. Geographical location within a city, state, or nation is related to many things. Within a city, stratification by geographical location usually increases representativeness in social class, ethnic group, and so forth. Within a nation, it increases representativeness in a broad

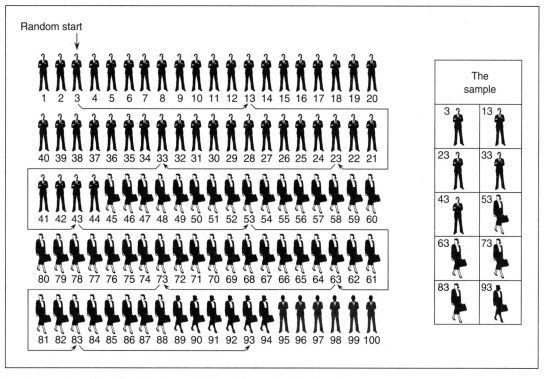

FIGURE 7-11 **A Stratified, Systematic Sample with a Random Start**

range of attitudes as well as in social class and ethnicity.

When you're working with a simple list of all elements in the population, two methods of stratification predominate. In one method, you sort the population elements into discrete groups based on whatever stratification variables are being used. On the basis of the relative proportion of the population represented by a given group, you select—randomly or systematically—several elements from that group constituting the same proportion of your desired sample size. For example, if sophomore men with a 4.0 average compose 1 percent of the student population and you desire a sample of 1,000 students, you would select 10 sophomore men with a 4.0 average.

The other method is to group students as described and then put those groups together in a continuous list, beginning with all frosh men with a 4.0 average and ending with all senior women

with a 1.0 or below. You would then select a systematic sample, with a random start, from the entire list. Given the arrangement of the list, a systematic sample would select proper numbers (within an error range of 1 or 2) from each subgroup. (*Note:* A simple random sample drawn from such a composite list would cancel out the stratification.)

Figure 7-11 offers a graphic illustration of stratified, systematic sampling. As you can see, we lined up our micropopulation according to gender and race. Then, beginning with a random start of "3," we've taken every tenth person thereafter: 3, 13, 23, . . . , 93.

Stratified sampling ensures the proper representation of the stratification variables; this, in turn, enhances the representation of other variables related to them. Taken as a whole, then, a stratified sample is more likely than a simple random sample to be more representative on several variables. Although the simple random sample is still regarded as somewhat sacred, it should now be clear that you can often do better.

Implicit Stratification in Systematic Sampling

I mentioned that systematic sampling can, under certain conditions, be more accurate than simple random sampling. This is the case whenever the arrangement of the list creates an implicit stratification. As already noted, if a list of university students is arranged by class, then a systematic sample provides a stratification by class where a simple random sample would not.

In a study of students at the University of Hawaii, after stratification by school class, the students were arranged by their student identification numbers. These numbers, however, were their social security numbers. The first three digits of the social security number indicate the state in which the number was issued. As a result, within a class, students were arranged by the state in which they were issued a social security number, providing a rough stratification by geographical origin.

An ordered list of elements, therefore, may be more useful to you than an unordered, randomized list. I've stressed this point in view of the unfortunate belief that lists should be randomized before systematic sampling. Only if the arrangement presents the problems discussed earlier should the list be rearranged.

Illustration: Sampling University Students

Let's put these principles into practice by looking at an actual sampling design used to select a sample of university students. The purpose of the study was to survey, with a mail-out questionnaire, a representative cross-section of students attending the main campus of the University of Hawaii. The following sections describes the steps and decisions involved in selecting that sample.

Study Population and Sampling Frame The obvious sampling frame available for use in this sample selection was the computerized file maintained by the university administration. The tape contained students' names, local and permanent addresses, social security numbers, and a variety of other information such as field of study, class, age, and gender.

The computer database, however, contained files on all people who could, by any conceivable definition, be called students, many of whom seemed inappropriate to the purposes of the study. As a result, researchers needed to define the study population in a somewhat more restricted fashion. The final definition included those 15,225 day-program degree candidates registered for the fall semester on the Manoa campus of the university, including all colleges and departments, both undergraduate and graduate students, and both U.S. and foreign students. The computer program used for sampling, therefore, limited consideration to students fitting this definition.

Stratification The sampling program also permitted stratification of students before sample selection. The researchers decided that stratification by college class would be sufficient, although the students might have been further stratified within class, if desired, by gender, college, major, and so forth.

Sample Selection Once the students had been arranged by class, a systematic sample was selected across the entire rearranged list. The sample size for the study was initially set at 1,100. To achieve this sample, the sampling program was set for a $^1/_{14}$ sampling ratio. The program generated a random number between 1 and 14; the student having that number and every fourteenth student thereafter was selected in the sample.

Once the sample had been selected, the computer was instructed to print each student's name and mailing address on self-adhesive mailing labels. These labels were then simply transferred to envelopes for mailing the questionnaires.

Sample Modification This initial design of the sample had to be modified. Before the mailing of questionnaires, the researchers discovered that unexpected expenses in the production of the questionnaires made it impossible to cover the costs of mailing to all 1,100 students. As a result, one-third of the mailing labels were systematically selected (with a random start) for exclusion from the sample. The final sample for the study was thereby reduced to 733 students.

I mention this modification to illustrate the frequent need to alter a study plan in midstream. Because the excluded students were systematically omitted from the initial systematic sample, the remaining 733 students could still be taken as reasonably representing the study population. The reduction in sample size did, of course, increase the range of sampling error.

MULTISTAGE CLUSTER SAMPLING

The preceding sections have dealt with reasonably simple procedures for sampling from lists of elements. Such a situation is ideal. Unfortunately, however, much interesting social research requires the selection of samples from populations that cannot easily be listed for sampling purposes: the population of a city, state, or nation; all university students in the United States; and so forth. In such cases, the sample design must be much more complex. Such a design typically involves the initial sampling of groups of elements—*clusters*—followed by the selection of elements within each of the selected clusters.

Cluster sampling may be used when it's either impossible or impractical to compile an exhaustive list of the elements composing the target population, such as all church members in the United States. Often, however, the population elements are already grouped into subpopulations, and a list of those subpopulations either exists or can be created practically. For example, church members in the United States belong to discrete churches, which are either listed or could be. Following a cluster sample format, then, researchers could sample the list of churches in some manner (for example, a stratified, systematic sample). Next, they would obtain lists of members from each of the selected churches. Each of the lists would then be

sampled, to provide samples of church members for study. (For an example, see Glock, Ringer, and Babbie 1967.)

Another typical situation concerns sampling among population areas such as a city. Although there is no single list of a city's population, citizens reside on discrete city blocks or census blocks. Researchers can, therefore, select a sample of blocks initially, create a list of people living on each of the selected blocks, and take a subsample of the people on each block.

In a more complex design, researchers might sample blocks, list the households on each selected block, sample the households, list the people residing in each household, and, finally, sample the people within each selected household. This multistage sample design leads ultimately to a selection of a sample of individuals but does not require the initial listing of all individuals in the city's population.

Multistage cluster sampling, then, involves the repetition of two basic steps: listing and sampling. The list of primary sampling units (churches, blocks) is compiled and, perhaps, stratified for sampling. Then a sample of those units is selected. The selected primary sampling units are then listed and perhaps stratified. The list of secondary sampling units is then sampled, and so forth.

Multistage cluster sampling makes possible those studies otherwise impossible. Consider, for example, the "Santa Claus" survey described in the box entitled "Sampling Santa's Fans."

Multistage Designs and Sampling Error

Although cluster sampling is highly efficient, the price of that efficiency is a less accurate sample. A simple random sample drawn from a population list is subject to a single sampling error, but a two-stage cluster sample is subject to two sampling errors. First, the initial sample of clusters will represent the population of clusters only within a range of sampling error. Second, the sample of elements selected within a given cluster will represent all the elements in that cluster only within a range of sampling error. Thus, for example, a researcher

SAMPLING SANTA'S FANS

With the approach of Christmas 1985, the *New York Times* thought it would be interesting to survey the nation's children regarding their beliefs in Santa Claus. There being no national registry of those who've been naughty and nice, the *Times* had to use some ingenuity. Here's their description of what they did:

> The latest *New York Times* Poll is based on telephone interviews conducted December 14–18 with 261 children aged 3 through 10 around the United States, excluding Alaska and Hawaii.
>
> The sample of telephone exchanges called was selected by a computer from a complete list of exchanges in the country. The exchanges were chosen to ensure that each region of the country was represented in proportion to its population. For each exchange, the telephone numbers were formed by random digits, thus permitting access to both listed and unlisted residential numbers.
>
> After interviews with 1,358 adults were completed, parents were asked if their children could be interviewed on the subject of Christmas. The results have been weighted to take account of household size and number of residential telephones and to adjust for variations in the sample relation to region, race, sex, age and education.

By the way, 87 percent of the children said they believed in Santa Claus: ranging from 96 percent among those 3–5 down to 69 percent among the 9–10-year-olds.

Source: Sara Rimer, "Poll Sees Landslide for Santa: Of U.S. Children, 87% Believe," *New York Times,* December 24, 1985.

runs a certain risk of selecting a sample of disproportionately wealthy city blocks, plus a sample of disproportionately wealthy households within those blocks. The best solution to this problem lies in the number of clusters selected initially and the number of elements within each cluster.

Typically, researchers are restricted to a total sample size; for example, you may be limited to conducting 2,000 interviews in a city. Given this broad limitation, however, you have several options in designing your cluster sample. At the extremes you could choose one cluster and select 2,000 elements within that cluster, or you could select 2,000 clusters with one element selected within each. Of course, neither approach is advisable, but a broad range of choices lies between them. Fortunately, the logic of sampling distributions provides a general guideline for this task.

Recall that sampling error is reduced by two factors: an increase in the sample size and increased homogeneity of the elements being sampled. These factors operate at each level of a multistage sample design. A sample of clusters will best represent all clusters if a large number are selected and if all clusters are very much alike. A sample of elements will best represent all elements in a given cluster if a large number are selected from the cluster and if all the elements in the cluster are very much alike.

With a given total sample size, however, if the number of clusters is increased, the number of elements within a cluster must be decreased. In this respect, the representativeness of the clusters is increased at the expense of more poorly representing the elements composing each cluster, or vice versa. Fortunately, homogeneity can be used to ease this dilemma.

Typically, the elements composing a given natural cluster within a population are more homogeneous than are all elements composing the total population. The members of a given church are more alike than are all church members; the residents of a given city block are more alike than are the residents of a whole city. As a result, relatively few elements may be needed to represent a given natural cluster adequately, although a larger number of clusters may be needed to represent adequately the diversity found among the clusters. This fact is most clearly seen in the extreme case of very different clusters composed of identical elements within each. In such a situation, a large number of clusters would adequately represent all its members. Although this extreme situation never exists in reality, it's closer to the truth in most cases than its opposite: identical clusters composed of grossly divergent elements.

The general guideline for cluster design, then, is to maximize the number of clusters selected while decreasing the number of elements within each cluster. However, this scientific guideline must be balanced against an administrative constraint. The efficiency of cluster sampling is based on the ability to minimize the listing of population elements. By initially selecting clusters, you need only list the elements composing the selected clusters, not all elements in the entire population. Increasing the number of clusters, however, goes directly against this efficiency factor. A small number of clusters may be listed more quickly and more cheaply than a large number. (Remember that all the elements in a selected cluster must be listed even if only a few are to be chosen in the sample.)

The final sample design will reflect these two constraints. In effect, you'll probably select as many clusters as you can afford. Lest this issue be left too open-ended at this point, here is one general guideline. Population researchers conventionally aim at the selection of 5 households per census block. If a total of 2,000 households are to be interviewed, you would aim at 400 blocks with 5 household interviews on each. Figure 7-12 presents a graphic overview of this process.

Before turning to other, more detailed procedures available to cluster sampling, let me reiterate that this method almost inevitably involves a loss of accuracy. The manner in which this appears, however, is somewhat complex. First, as noted earlier, a multistage sample design is subject to a sampling error at each stage. Because the sample size is necessarily smaller at each stage than the total sample size, the sampling error at each stage will be greater than would be the case for a single-stage random sample of elements. Second, sampling error is estimated on the basis of observed variance among the sample elements. When those elements are drawn from among relatively homogeneous clusters, the estimated sampling error will be too optimistic and must be corrected in the light of the cluster sample design.

Stratification in Multistage Cluster Sampling

Thus far, we've looked at cluster sampling as though a simple random sample were selected at each stage of the design. In fact, stratification techniques can be used to refine and improve the sample being selected.

The basic options here are essentially the same as those in single-stage sampling from a list. In selecting a national sample of churches, for example, you might initially stratify your list of churches by denomination, geographical region, size, rural or urban location, and perhaps by some measure of social class.

Once the primary sampling units (churches, blocks) have been grouped according to the relevant, available stratification variables, either simple random or systematic sampling techniques can be used to select the sample. You might select a specified number of units from each group, or stratum, or you might arrange the stratified clusters in a continuous list and systematically sample that list.

To the extent that clusters are combined into homogeneous strata, the sampling error at this stage will be reduced. The primary goal of stratification, as before, is homogeneity.

There's no reason why stratification couldn't take place at each level of sampling. The elements listed within a selected cluster might be stratified before the next stage of sampling. Typically, however, this is not done. (Recall the assumption of relative homogeneity within clusters.)

Probability Proportionate to Size (PPS) Sampling

This section introduces you to a more sophisticated form of cluster sampling, one that is used in many large-scale survey sampling projects. In the

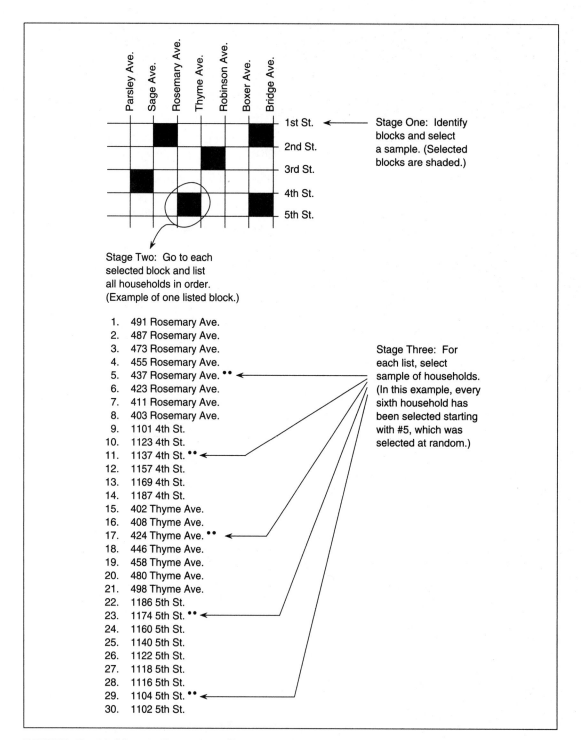

FIGURE 7-12 **Multistage Cluster Sampling**

preceding discussion, I talked about selecting a random or systematic sample of clusters and then a random or systematic sample of elements within each cluster selected. Notice that this produces an overall sampling scheme in which every element in the whole population has the same probability of selection.

Let's say we're selecting households within a city. If there are 1,000 city blocks and we initially select a sample of 100, that means that each block has a $^{100}/_{1,000}$ or .1 chance of being selected. If we next select 1 household in 10 from those residing on the selected blocks, each household has a .1 chance of selection within its block. To calculate the overall probability of a household being selected, we simply multiply the probabilities at the individual steps in sampling. That is, each household has a $^1/_{10}$ chance of its block being selected and a $^1/_{10}$ chance of that specific household being selected if the block is one of those chosen. Each household, in this case, has a $^1/_{10} \times ^1/_{10} = ^1/_{100}$ chance of selection overall. Because each household would have the same chance of selection, the sample so selected should be representative of all households in the city.

There are dangers in this procedure, however. In particular, the variation in the size of blocks (measured in numbers of households) presents a problem. Let's suppose that half the city's population resides in 10 densely packed blocks filled with high-rise apartment buildings, and suppose that the rest of the population lives in single-family dwellings spread out over the remaining 900 blocks. When we first select our sample of $^1/_{10}$ of the blocks, it's quite possible that we'll miss all of the 10 densely packed high-rise blocks. No matter what happens in the second stage of sampling, our final sample of households will be grossly unrepresentative of the city, comprising only single-family dwellings.

Whenever the clusters sampled are of greatly differing sizes, it's appropriate to use a modified sampling design called *probability proportionate to size*—**PPS.** This design guards against the problem I've just described and still produces a final sample in which each element has the same chance of selection.

As the name suggests, each cluster is given a chance of selection proportionate to its size. Thus, a city block with 200 households has twice the chance of selection as one with only 100 households. Within each cluster, however, a fixed number of elements is selected, say, 5 households per block. Notice how this procedure results in each household having the same probability of selection overall.

Let's look at households of two different city blocks. Block A has 100 households, Block B has only 10. In PPS sampling, we would give Block A ten times as good a chance of being selected as Block B. So if, in the overall sample design, Block A has a $^1/_{20}$ chance of being selected, that means Block B would only have a $^1/_{200}$ chance. Notice that this means that all the households on Block A would have a $^1/_{20}$ chance of having their block selected; Block B households have only a $^1/_{200}$ chance.

If Block A is selected and we're taking 5 households from each selected block, then the households on Block A have a $^5/_{100}$ chance of being selected into the block's sample. Since we can multiply probabilities in a case like this, we see that every household on Block A had an overall chance of selection equal to $^1/_{20} \times ^5/_{100} = ^5/_{2000} = ^1/_{400}$.

If Block B happens to be selected, on the other hand, its households stand a much better chance of being among the 5 chosen there: $^5/_{10}$. When this is combined with their relatively poorer chance of having their block selected in the first place, however, they end up with the same chance of selection as those on Block A: $^1/_{200} \times ^5/_{10} = ^5/_{2000} = ^1/_{400}$.

Further refinements to this design make it a very efficient and effective method for selecting large cluster samples. For now, however, it's enough to understand the basic logic involved.

Disproportionate Sampling and Weighting

Ultimately, a probability sample is representative of a population if all elements in the population have an equal chance of selection in that sample. Thus, in each of the preceding discussions, we've noted that the various sampling procedures result in an equal chance of selection—even though the ultimate selection probability is the product of several partial probabilities.

More generally, however, a probability sample is one in which each population element has a known nonzero probability of selection—even though different elements may have different probabilities. If controlled probability sampling procedures have been used, any such sample may be representative of the population from which it is drawn if each sample element is assigned a weight equal to the inverse of its probability of selection. Thus, where all sample elements have had the same chance of selection, each is given the same weight: 1. This is called a *self-weighting* sample.

Sometimes it's appropriate to give some cases more weight than others, a process called **weighting.** Disproportionate sampling and weighting come into play in two basic ways. First, you may sample subpopulations disproportionately to ensure sufficient numbers of cases from each for analysis. For example, a given city may have a suburban area containing one-fourth of its total population. Yet you might be especially interested in a detailed analysis of households in that area and may feel that one-fourth of this total sample size would be too few. As a result, you might decide to select the same number of households from the suburban area as from the remainder of the city. Households in the suburban area, then, are given a disproportionately better chance of selection than those located elsewhere in the city.

As long as you analyze the two area samples separately or comparatively, you need not worry about the differential sampling. If you want to combine the two samples to create a composite picture of the entire city, however, you must take the disproportionate sampling into account. If n is the number of households selected from each area, then the households in the suburban area had a chance of selection equal to n divided by one-fourth of the total city population. Because the total city population and the sample size are the same for both areas, the suburban-area households should be given a weight of $\frac{1}{4}n$, and the remaining households should be given a weight of $\frac{3}{4}n$. This weighting procedure could be simplified

by merely giving a weight of 3 to each of the households selected outside the suburban area. (This procedure gives a proportionate representation to each sample element. The population figure would have to be included in the weighting if population estimates were desired.)

Here's an example of the problems that can be created when disproportionate sampling is not accompanied by a weighting scheme. When the *Harvard Business Review* decided to survey its subscribers on the issue of sexual harassment at work, it seemed appropriate to oversample women because female subscribers were vastly outnumbered by male subscribers. Here's how G. C. Collins and Timothy Blodgett explained the matter:

> We also skewed the sample another way: to ensure a representative response from women, we mailed a questionnaire to virtually every female subscriber, for a male/female ratio of 68% to 32%. This bias resulted in a response of 52% male and 44% female (and 4% who gave no indication of gender)—compared to HBR's U.S. subscriber proportion of 93% male and 7% female.
> — (1981:78)

Notice a couple of things in this quotation. First, it would be nice to know a little more about what "virtually every female" means. Evidently, the authors of the study didn't send questionnaires to all female subscribers, but there's no indication of who was omitted and why. Second, they didn't use the term *representative* in its normal social science usage. What they mean, of course, is that they wanted to get a substantial or "large enough" response from women, and oversampling is a perfectly acceptable way of accomplishing that.

By sampling more women than a straightforward probability sample would have produced, the authors were able to "select" enough women (812) to compare with the men (960). Thus, when they report, for example, that 32 percent of the women and 66 percent of the men agree that "the amount of sexual harassment at work is greatly exaggerated," we know that the female response is based on a substantial number of cases. That's good. There are problems, however.

To begin with, subscriber surveys are always problematic. In this case, the best the researchers can hope to talk about is "what subscribers to *Harvard Business Review* think." In a loose way, it might make sense to think of that population as representing the more sophisticated portion of corporate management. Unfortunately, the overall response rate was 25 percent. Although that's quite good for subscriber surveys, it's a low response rate in terms of generalizing from probability samples.

Beyond that, however, the disproportionate sample design creates a further problem. When the authors state that 73 percent of respondents favor company policies against harassment (Collins and Blodgett, 1981:78), that figure is undoubtedly too high, since the sample contains a disproportionately high percentage of women—who are more likely to favor such policies. And, when the researchers report that top managers are more likely to feel that claims of sexual harassment are exaggerated than are middle- and lower-level managers (1981:81), that finding is also suspect. As the researchers report, women are disproportionately represented in lower management. That alone might account for the apparent differences among levels of management. In short, the failure to take account of the oversampling of women confounds all survey results that don't separate the findings by gender. The solution to this problem would have been to weight the responses by gender, as described earlier in this section.

PROBABILITY SAMPLING IN REVIEW

Much of this chapter has been devoted to the key sampling method used in controlled survey research: probability sampling. In each of the variations examined, we've seen that elements are chosen for study from a population on a basis of random selection with known nonzero probabilities.

Depending on the field situation, probability sampling can be either very simple or extremely difficult, time consuming, and expensive. Whatever

A QUANDARY **REVISITED**

Contrary to common sense, we have seen that the number of people selected in a sample, while important, is less important than *how* people are selected. The *Literary Digest* mailed ballots to ten million people and received two million from voters around the country. However, the people they selected for their enormous sample—auto owners and telephone subscribers—were not representative of the population in 1936, in the aftermath of the Great Depression. Overall, the sample was wealthier than was the voting population at large. Since rich people are more likely than the general public to vote Republican, the *Literary Digest* tallied the voting intentions of a disproportionate number of Republicans.

The probability sampling techniques used today allow researchers to select smaller, more representative samples. Even a couple of thousand respondents, properly selected, can accurately predict the behavior of a hundred million voters.

the situation, however, it remains the most effective method for the selection of study elements. There are two reasons for that.

First, probability sampling avoids researchers' conscious or unconscious biases in element selection. If all elements in the population have an equal (or unequal and subsequently weighted) chance of selection, there is an excellent chance that the sample so selected will closely represent the population of all elements.

Second, probability sampling permits estimates of sampling error. Although no probability sample will be perfectly representative in all respects, controlled selection methods permit the researcher to estimate the degree of expected error.

In this lengthy chapter, we've taken on a basic issue in much social research: selecting observations that will tell us something more general than the specifics we've actually observed. This issue confronts field researchers, who face more action and more actors than they can observe and record fully, as well as political pollsters who want to predict an election but can't interview all voters. As we proceed through the book, we'll see in greater detail how social researchers have found ways to deal with this issue.

Main Points

- Social researchers must select observations that will allow them to generalize to people and events not observed. Often this involves a selection of people to observe.

- Sometimes you can and should select probability samples using precise statistical techniques, but other times nonprobability techniques are more appropriate.

- Nonprobability sampling techniques include relying on available subjects, purposive or judgmental sampling, snowball sampling, and quota sampling. In addition, researchers studying a social group may make use of informants. Each of these techniques has its uses, but none of them ensures that the resulting sample is representative of the population being sampled.

- Probability sampling methods provide an excellent way of selecting representative samples from large, known populations. These methods counter the problems of conscious and unconscious sampling bias by giving each element in the population a known (nonzero) probability of selection.

- The key to probability sampling is random selection.

- The most carefully selected sample will never provide a perfect representation of the population from which it was selected. There will always be some degree of sampling error.

- By predicting the distribution of samples with respect to the target parameter, probability sampling methods make it possible to estimate the amount of sampling error expected in a given sample.

- The expected error in a sample is expressed in terms of confidence levels and confidence intervals.

- A sampling frame is a list or quasi list of the members of a population. It is the resource used in the selection of a sample. A sample's representativeness depends directly on the extent to which a sampling frame contains all the members of the total population that the sample is intended to represent.

- Several sampling designs are available to researchers.

- Simple random sampling is logically the most fundamental technique in probability sampling, but it is seldom used in practice.

- Systematic sampling involves the selection of every kth member from a sampling frame. This method is more practical than simple random sampling, and, with a few exceptions, it is functionally equivalent.

- Stratification, the process of grouping the members of a population into relatively homogeneous strata before sampling, improves the representativeness of a sample by reducing the degree of sampling error.

- Multistage cluster sampling is a relatively complex sampling technique that is frequently used when a list of all the members of a population does not exist. Typically, researchers must balance the number of clusters and the size of each cluster to achieve a given sample size. Stratification can be used to replace the sampling error involved in multistage cluster sampling.

- Probability proportionate to size (PPS) is a special, efficient method for multistage cluster sampling.

❏ If the members of a population have unequal probabilities of selection into the sample, researchers must assign weights to the different observations made in order to provide a representative picture of the total population. Basically, the weight assigned to a particular sample member should be the inverse of its probability of selection.

Key Terms

nonprobability sampling	parameter
purposive (judgmental)	statistic
sampling	sampling error
snowball sampling	confidence level
quota sampling	confidence interval
informant	sampling frame
probability sampling	simple random sampling
representativeness	systematic sampling
EPSEM	sampling interval
element	sampling ratio
population	stratification
study population	cluster sampling
random selection	PPS
sampling unit	weighting

Review Questions

1. Review the discussion of the 1948 Gallup Poll that predicted that Thomas Dewey would defeat Harry Truman for president. What are some ways Gallup could have modified his quota sample design to avoid the error?

2. Using Appendix C of this book, select a simple random sample of 10 numbers in the range from 1 to 9,876. What is each step in the process?

3. What are the steps involved in selecting a multistage cluster sample of students taking first-year English in U.S. colleges and universities?

4. In Chapter 9, we'll discuss surveys conducted on the Internet. Can you anticipate possible problems concerning sampling frames, representativeness, and the like? Do you see any solutions?

Additional Readings

Frankfort-Nachmias, Chava, and Anna Leon-Guerrero. 1997. *Social Statistics for a Diverse Society.* Thousand Oaks, CA: Pine Forge Press. See Chapter 11 especially. This statistics textbook covers many of the topics we've discussed in this chapter but in a more statistical context. It demonstrates the links between probability sampling and statistical analyses.

Kalton, Graham. 1983. *Introduction to Survey Sampling.* Newbury Park, CA: Sage. Kalton goes into more of the mathematical details of sampling than the present chapter does without attempting to be as definitive as Kish, described next.

Kish, Leslie. 1965. *Survey Sampling.* New York: Wiley. Unquestionably the definitive work on sampling in social research. Kish's coverage ranges from the simplest matters to the most complex and mathematical, both highly theoretical and downright practical. Easily readable and difficult passages intermingle as Kish exhausts everything you could want or need to know about each aspect of sampling.

Sudman, Seymour. 1983. "Applied Sampling." Pp. 145–94 in *Handbook of Survey Research,* edited by Peter H. Rossi, James D. Wright, and Andy B. Anderson. New York: Academic Press. An excellent, practical guide to survey sampling.

Multimedia Resources

The Wadsworth Sociology Resource Center: Virtual Society
http://sociology.wadsworth.com/
Visit the companion Web site for the second edition of *The Basics of Social Research* to access a wide range of student resources. Begin by clicking on the Student Resources section of the book's Web site to access the following study tools:

• eBabbie Resource Center
• Planning a Research Project
• Doing Data Analysis
• Statistics Review
• Flash Cards
• Internet Links and Exercises
• InfoTrac College Edition: Exercises
• Quizzes

Visit the **eBabbie Resource Center** for an overview of each chapter and helpful online tutorials. Find information on budgeting and step-by-step examples of model research projects at **Planning a Research Project.** Learn how to use quantitative and qualitative data analysis programs at **Doing Data Analysis,** and brush up on your statistics at **Statistics Review.** You can also further your study by accessing **Internet Links and Exercises** related to chapter materials, **Flash Cards, Quizzes,** and many other learning tools.

InfoTrac College Edition
**http://www.infotrac-college.com/
wadsworth/access.html**

Access the latest news and journal articles with InfoTrac College Edition, an easy-to-use online database of reliable, full-length articles from hundreds of top academic journals. Conduct an electronic search using the following search terms:

Cluster sample
Confidence interval
Confidence level
Multistage sample
Nonprobability sample
Probability sample
Quota sample
Sampling bias
Sampling distribution
Sampling error
Sampling frame
Stratified sample

Paul Conklin/PhotoEdit/PictureQuest

Part Three

MODES OF
OBSERVATION

Having explored the structuring of inquiry in depth, we're now ready to dive into the various observational techniques available to social scientists.

Experiments are usually thought of in connection with the physical sciences. In Chapter 8 we'll see how social scientists use experiments. This is the most rigorously controllable of the methods we'll examine. Understanding experiments is also a useful way to enhance your understanding of the general logic of social scientific research.

Chapter 9 will describe survey research, one of the most popular methods in social science. This type of research involves collecting data by asking people questions—either in self-administered questionnaires or through interviews, which, in turn, can be conducted face-to-face or over the telephone.

Chapter 10, on field research, examines perhaps the most natural form of data collection used by social researchers: the direct observation of social phenomena in natural settings. Some researchers go beyond mere observation to participate in what they're studying, because they want a more intimate view and fuller understanding of it.

Chapter 11 discusses three forms of unobtrusive data collection that take advantage of some of the data available all around us. For example, content analysis is a method of collecting social data through carefully specifying and counting social artifacts such as books, songs, speeches, and paintings. Without making any personal contact with people, you can use this method to examine a wide variety of social phenomena. The analysis of existing statistics offers another way of studying people without having to talk to them. Governments and a variety of private organizations regularly compile great masses of data, which you often can use with little or no modification to answer properly posed questions. Finally, historical documents are a valuable resource for social scientific analysis.

Chapter 12, on evaluation research, looks at a rapidly growing subfield in social science, involving the application of experimental and quasi-experimental models to the testing of social interventions in real life. You might use evaluation research, for example, to test the effectiveness of a drug rehabilitation program or the efficiency of a new school cafeteria. In the same chapter, we'll look briefly at social indicators as a way of assessing broader social processes.

Before we turn to the actual descriptions of the several methods, two points should be made. First, you'll probably discover that you've been using these scientific methods casually in your daily life for as long as you can remember. You use some form of field research every day. You employ a crude form of content analysis every time you judge an author's motivation or orientation from her or his writings. You engage in at least casual experiments frequently. The chapters in Part 3 will show you how to improve your use of these methods so as to avoid the pitfalls of casual, uncontrolled observation.

Second, none of the data-collection methods described in the following chapters is appropriate to all research topics and situations. I've tried to give you some ideas, early in each chapter, regarding when a given method might be appropriate. Still, I could never anticipate all the possible research topics that may one day interest you. In general, you should always try to use a variety of techniques in the study of any topic. Because each method has its weaknesses, the use of several methods can help fill in any gaps; if the different, independent approaches to the topic all yield the same conclusion, you've achieved a form of replication.

8 EXPERIMENTS

Jeff Greenberg/PhotoEdit

What You'll Learn in This Chapter

This chapter examines the experimental method, a mode of
observation that enables researchers to probe causal relationships.
Many experiments in social research are conducted under the
controlled conditions of a laboratory, but experimenters can also
take advantage of natural occurrences to study the effects of events
in the social world.

In this chapter . . .

AN OPENING **QUANDARY**

The impact of the observer raises many serious questions regarding the usefulness of experiments in social research. How can the manipulation of people in a controlled, experimental environment tell us anything about "natural" human behavior? After all is said and done, doesn't an experiment simply tell us how people behave when they participate in an experiment?

INTRODUCTION

This chapter addresses the research method most commonly associated with structured science in general: the experiment. Here we'll focus on the experiment as a mode of scientific observation in social research. At the most basic level, experiments involve (1) taking action and (2) observing the consequences of that action. Social researchers typically select a group of subjects, do something to them, and observe the effect of what was done. In this chapter, we'll examine the logic and some of the techniques of social scientific experiments.

It's worth noting that experiments also are used often in nonscientific human inquiry. In preparing a stew, for example, we add salt, taste, add more salt, and taste again. In defusing a bomb, we clip the red wire, observe whether the bomb explodes, clip another, and . . .

We also experiment copiously in our attempt to develop generalized understanding about the world we live in. All skills are learned through experimentation: eating, walking, riding a bicycle, and so forth. This chapter will discuss some ways social researchers use experiments to develop generalized understandings. We'll see that, like other

217

methods available to the social researcher, experimenting has its special strengths and weaknesses.

TOPICS APPROPRIATE TO EXPERIMENTS

Experiments are more appropriate for some topics and research purposes than others. Experiments are especially well suited to research projects involving relatively limited and well-defined concepts and propositions. In terms of the traditional image of science, discussed earlier in this book, the experimental model is especially appropriate for hypothesis testing. Because experiments focus on determining causation, they're also better suited to explanatory than to descriptive purposes.

Let's assume, for example, we want to discover ways of reducing prejudice against African Americans. We hypothesize that learning about the contribution of African Americans to U.S. history will reduce prejudice, and we decide to test this hypothesis experimentally. To begin, we might test a group of experimental subjects to determine their levels of prejudice against African Americans. Next, we might show them a documentary film depicting the many important ways African Americans have contributed to the scientific, literary, political, and social development of the nation. Finally, we'd measure our subjects' levels of prejudice against African Americans to determine whether the film has actually reduced prejudice.

Experimentation has also been successful in the study of small group interaction. Thus, we might bring together a small group of experimental subjects and assign them a task, such as making recommendations for popularizing car pools. We observe, then, how the group organizes itself and deals with the problem. Over the course of several such experiments, we might systematically vary the nature of the task or the rewards for handling the task successfully. By observing differences in the way groups organize themselves and operate under these varying conditions, we can learn a great deal about the nature of small group interaction and the factors that influence it. For example, attorneys sometimes present evidence in different ways to different mock juries, to see which method is the most effective.

We typically think of experiments as being conducted in laboratories. Indeed, most of the examples in this chapter involve such a setting. This need not be the case, however. Social researchers often study what are called *natural experiments:* "experiments" that occur in the regular course of social events. The latter portion of this chapter deals with such research.

THE CLASSICAL EXPERIMENT

In both the natural and the social sciences, the most conventional type of experiment involves three major pairs of components: (1) independent and dependent variables, (2) pretesting and posttesting, and (3) experimental and control groups. This section looks at each of these components and the way they're put together in the execution of an experiment.

Independent and Dependent Variables

Essentially, an experiment examines the effect of an independent variable on a dependent variable. Typically, the independent variable takes the form of an experimental stimulus, which is either present or absent. That is, the stimulus is a dichotomous variable, having two attributes—present or not present. In this typical model, the experimenter compares what happens when the stimulus is present to what happens when it is not.

In the example concerning prejudice against African Americans, *prejudice* is the dependent variable and *exposure to African-American history* is the independent variable. The researcher's hypothesis suggests that prejudice depends, in part, on a lack of knowledge of African-American history. The purpose of the experiment is to test the validity of this hypothesis by presenting some subjects with an appropriate stimulus, such as a documentary film. In other terms, the independent variable is the cause and the dependent variable is the effect. Thus, we might say that watching the film caused a change in prejudice or that reduced prejudice was an effect of watching the film.

The independent and dependent variables appropriate to experimentation are nearly limitless. Moreover, a given variable might serve as an independent variable in one experiment and as a dependent variable in another. For example, *prejudice* is the dependent variable in the previous example, but it might be the independent variable in an experiment examining the effect of prejudice on voting behavior.

To be used in an experiment, both independent and dependent variables must be operationally defined. Such operational definitions might involve a variety of observation methods. Responses to a questionnaire, for example, might be the basis for defining prejudice. Speaking to or ignoring African-American, or agreeing or disagreeing with them, might be elements in the operational definition of interaction with African Americans in a small group setting.

Conventionally, in the experimental model, dependent and independent variables must be operationally defined before the experiment begins. However, as you'll see in connection with survey research and other methods, it's sometimes appropriate to make a wide variety of observations during data collection and then determine the most useful operational definitions of variables during later analyses. Ultimately, however, experimentation, like other quantitative methods, requires specific and standardized measurements and observations.

Pretesting and Posttesting

In the simplest experimental design, subjects are measured in terms of a dependent variable (**pretesting**), exposed to a stimulus representing an independent variable, and then remeasured in terms of the dependent variable (**posttesting**). Any differences between the first and last measurements on the dependent variable are then attributed to the independent variable.

In the example of prejudice and exposure to African-American history, we'd begin by pretesting the extent of prejudice among our experimental subjects. Using a questionnaire asking about attitudes toward African Americans, for example, we could measure the extent of prejudice exhibited by each individual subject and the average prejudice level of the whole group. After exposing the subjects to the African-American history film, we could administer the same questionnaire again. Responses given in this posttest would permit us to measure the later extent of prejudice for each subject and the average prejudice level of the group as a whole. If we discovered a lower level of prejudice during the second administration of the questionnaire, we might conclude that the film had indeed reduced prejudice.

In the experimental examination of attitudes such as prejudice, we face a special practical problem relating to validity. As you may already have imagined, the subjects might respond differently to the questionnaires the second time even if their attitudes remain unchanged. During the first administration of the questionnaire, the subjects may be unaware of its purpose. By the second measurement, they may have figured out that the researchers are interested in measuring their prejudice. Because no one wishes to seem prejudiced, the subjects may "clean up" their answers the second time around. Thus, the film will seem to have reduced prejudice although, in fact, it has not.

This is an example of a more general problem that plagues many forms of social research: The very act of studying something may change it. The techniques for dealing with this problem in the context of experimentation will be discussed in various places throughout the chapter. The first technique involves the use of control groups.

Experimental and Control Groups

Laboratory experiments seldom if ever involve only the observation of an **experimental group** to which a stimulus has been administered. In addition, the researchers also observe a **control group,** which does not receive the experimental stimulus.

In the example of prejudice and African-American history, we might examine two groups of subjects. To begin, we give each group a questionnaire designed to measure their prejudice against African Americans. Then we show the film only to

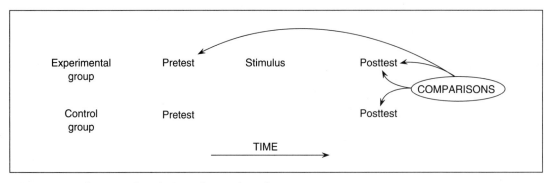

FIGURE 8-1 **Diagram of Basic Experimental Design**

the experimental group. Finally, we administer a posttest of prejudice to both groups. Figure 8-1 illustrates this basic experimental design.

Using a control group allows the researcher to detect any effects of the experiment itself. If the posttest shows that the overall level of prejudice exhibited by the control group has dropped as much as that of the experimental group, then the apparent reduction in prejudice must be a function of the experiment or of some external factor rather than a function of the film. If, on the other hand, prejudice is reduced only in the experimental group, this reduction would seem to be a consequence of exposure to the film, because that's the only difference between the two groups. Alternatively, if prejudice is reduced in both groups but to a greater degree in the experimental group than in the control group, that, too, would be grounds for assuming that the film reduced prejudice.

The need for control groups in social research became clear in connection with a series of studies of employee satisfaction conducted by F. J. Roethlisberger and W. J. Dickson (1939) in the late 1920s and early 1930s. These two researchers were interested in discovering what changes in working conditions would improve employee satisfaction and productivity. To pursue this objective, they studied working conditions in the telephone "bank wiring room" of the Western Electric Works in the Chicago suburb of Hawthorne, Illinois.

To the researchers' great satisfaction, they discovered that making working conditions better increased satisfaction and productivity consistently.

As the workroom was brightened up through better lighting, for example, productivity went up. When lighting was further improved, productivity went up again.

To further substantiate their scientific conclusion, the researchers then dimmed the lights. Whoops—productivity again improved!

At this point it became evident that the wiring-room workers were responding more to the attention given them by the researchers than to improved working conditions. As a result of this phenomenon, often called the *Hawthorne effect,* social researchers have become more sensitive to and cautious about the possible effects of experiments themselves. In the wiring-room study, the use of a proper control group—one that was studied intensively without any other changes in the working conditions—would have pointed to the existence of this effect.

The need for control groups in experimentation has been nowhere more evident than in medical research. Time and again, patients who participate in medical experiments have appeared to improve, but it has been unclear how much of the improvement has come from the experimental treatment and how much from the experiment. In testing the effects of new drugs, then, medical researchers frequently administer a *placebo*—a "drug" with no relevant effect, such as sugar pills—to a control group. Thus, the control-group patients believe they, like the experimental group, are receiving an experimental drug. Often, they improve. If the new drug is effective, however, those receiving the actual drug will improve more than those receiving the placebo.

In social scientific experiments, control groups provide an important guard against not only the effects of the experiments themselves but also the effects of any events outside the laboratory during the experiments. In the example of the study of prejudice, suppose that a very popular African-American leader is assassinated in the middle of, say, a week-long experiment. Such an event might very well horrify the experimental subjects, requiring them to examine their own attitudes toward African Americans, with the result of reduced prejudice. Because such an effect should happen about equally for members of the control and experimental groups, a greater reduction of prejudice among the experimental group would, again, point to the impact of the experimental stimulus: the documentary film.

Sometimes an experimental design will require more than one experimental or control group. In the case of the documentary film, for example, we might also want to examine the impact of reading a book on African-American history. In that case, we might have one group see the film and read the book; another group only see the movie; still another group only read the book; and the control group do neither. With this kind of design, we could determine the impact of each stimulus separately, as well as their combined effect.

The Double-Blind Experiment

Like patients who improve when they merely think they're receiving a new drug, sometimes experimenters tend to prejudge results. In medical research, the experimenters may be more likely to "observe" improvements among patients receiving the experimental drug than among those receiving the placebo. (This would be most likely, perhaps, for the researcher who developed the drug.) A **double-blind experiment** eliminates this possibility, because neither the subjects nor the experimenters know which is the experimental group and which is the control. In the medical case, those researchers responsible for administering the drug and for noting improvements would not be told

which subjects were receiving the drug and which the placebo. Conversely, the researcher who knew which subjects were in which group would not administer the experiment.

In social scientific experiments, as in medical ones, the danger of experimenter bias is further reduced to the extent that the operational definitions of the dependent variables are clear and precise. Thus, medical researchers would be less likely to unconsciously bias their reading of a patient's temperature than they would to bias their assessment of how lethargic the patient was. For the same reason, the small group researcher would be less likely to misperceive which subject spoke, or to whom he or she spoke, than whether the subject's comments sounded cooperative or competitive, a more subjective judgment that's difficult to define in precise behavioral terms.

As I've indicated several times, seldom can we devise operational definitions and measurements that are wholly precise and unambiguous. This is another reason why it can be appropriate to employ a double-blind design in social research experiments.

SELECTING SUBJECTS

In Chapter 7 we discussed the logic of sampling, which involves selecting a sample that is representative of some populations. Similar considerations apply to experiments. Because most social researchers work in colleges and universities, it seems likely that most social research laboratory experiments are conducted with college undergraduates as subjects. Typically, the experimenter asks students enrolled in his or her classes to participate in experiments or advertises for subjects in a college newspaper. Subjects may or may not be paid for participating in such experiments. (See Chapter 3 for more on the ethical issues involved in this situation.)

In relation to the norm of generalizability in science, this tendency clearly represents a potential defect in social research. Simply put, college undergraduates do not typify the public at large.

There is a danger, therefore, that we may learn much about the attitudes and actions of college undergraduates but not about social attitudes and actions in general.

However, this potential defect is less significant in explanatory research than it would be in descriptive research. True, having noted the level of prejudice among a group of college undergraduates in our pretesting, we would have little confidence that the same level existed among the public at large. On the other hand, if we found that a documentary film reduced whatever level of prejudice existed among those undergraduates, we would have more confidence—without being certain—that it would have a similar effect in the community at large. Social processes and patterns of causal relationships appear to be more generalizable and more stable than specific characteristics such as an individual's level of prejudice.

Aside from the question of generalizability, the cardinal rule of subject selection and experimentation concerns the comparability of experimental and control groups. Ideally, the control group represents what the experimental group would have been like if it had *not* been exposed to the experimental stimulus. The logic of experiments requires, therefore, that experimental and control groups be as similar as possible. There are several ways to accomplish this.

Probability Sampling

The discussions of the logic and techniques of probability sampling in Chapter 7 provide one method for selecting two groups of people that are similar to each other. Beginning with a sampling frame composed of all the people in the population under study, the researcher might select two probability samples. If these samples each resemble the total population from which they're selected, they'll also resemble each other.

Recall also, however, that the degree of resemblance (representativeness) achieved by probability sampling is largely a function of the sample size. As a general guideline, probability samples of less than 100 are not likely to be terribly representative, and social scientific experiments seldom involve that many subjects in either experimental or control groups. As a result, then, probability sampling is seldom used in experiments to select subjects from a larger population. Researchers do, however, use the logic of random selection when they assign subjects to groups.

Randomization

Having recruited, by whatever means, a total group of subjects, the experimenter may randomly assign those subjects to either the experimental or the control group. Such **randomization** might be accomplished by numbering all of the subjects serially and selecting numbers by means of a random number table, or the experimenter might assign the odd-numbered subjects to the experimental group and the even-numbered subjects to the control group.

Let's return again to the basic concept of probability sampling. If we recruit 40 subjects all together, in response to a newspaper advertisement, for example, there's no reason to believe that the 40 subjects represent the entire population from which they've been drawn. Nor can we assume that the 20 subjects randomly assigned to the experimental group represent that larger population. We can have greater confidence, however, that the 20 subjects randomly assigned to the experimental group will be reasonably similar to the 20 assigned to the control group.

Following the logic of our earlier discussions of sampling, we can see our 40 subjects as a population from which we select two probability samples—each consisting of half the population. Because each sample reflects the characteristics of the total population, the two samples will mirror each other.

As we saw in Chapter 7, our assumption of similarity in the two groups depends in part on the number of subjects involved. In the extreme case, if we recruited only two subjects and assigned, by the flip of a coin, one as the experimental subject and one as the control, there would be no reason to assume that the two subjects are similar to each other. With larger numbers of subjects, however, randomization makes good sense.

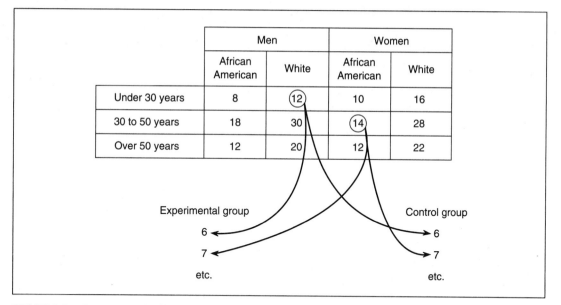

FIGURE 8-2   **Quota Matrix Illustration**

Matching

Another way to achieve comparability between the experimental and control groups is through **matching.** This process is similar to the quota sampling methods discussed in Chapter 7. If 12 of our subjects are young white men, we might assign 6 of those at random to the experimental group and the other 6 to the control group. If 14 are middle-aged African-American women, we might assign 7 to each group.

The overall matching process could be most efficiently achieved through the creation of a quota matrix constructed of all the most relevant characteristics. Figure 8-2 provides a simplified illustration of such a matrix. In this example, the experimenter has decided that the relevant characteristics are race, age, and gender. Ideally, the quota matrix is constructed to result in an even number of subjects in each cell of the matrix. Then, half the subjects in each cell go into the experimental group and half into the control group.

Alternatively, we might recruit more subjects than our experimental design requires. We might then examine many characteristics of the large initial group of subjects. Whenever we discover a pair

of quite similar subjects, we might assign one at random to the experimental group and the other to the control group. Potential subjects who are unlike anyone else in the initial group might be left out of the experiment altogether.

Whatever method we employ, the desired result is the same. The overall average description of the experimental group should be the same as that of the control group. For example, they should have about the same average age, the same gender composition, the same racial composition, and so forth. This test of comparability should be used whether the two groups are created through probability sampling or through randomization.

Thus far, I've referred to the "relevant" variables without saying clearly what those variables are. Of course, I can't give a definite answer to this question, anymore than I could specify in Chapter 7 which variables should be used in stratified sampling. Which variables are relevant ultimately depends on the nature and purpose of an experiment. As a general rule, however, the control and experimental groups should be comparable in terms of those variables most likely to be related to the dependent variable under study. In a study of prejudice, for example, the two groups should be alike in terms of education, ethnicity, and age, among other characteristics. In some cases, moreover, we may delay assigning subjects to experimental and control groups until we have initially measured the dependent variable. Thus, for example, we might administer a questionnaire measuring subjects' prejudice and then match the experimental and control groups to assure ourselves that the two groups exhibit the same overall level of prejudice.

Matching or Randomization?

When assigning subjects to the experimental and control groups, you should be aware of two arguments in favor of randomization over matching. First, you may not be in a position to know in advance which variables will be relevant for the matching process. Second, most of the statistics used to analyze the results of experiments assume randomization. Failure to design your experiment that way, then, makes your later use of those statistics less meaningful.

On the other hand, randomization only makes sense if you have a fairly large pool of subjects, so that the laws of probability sampling apply. With only a few subjects, matching would be a better procedure.

Sometimes researchers can combine matching and randomization. When conducting an experiment in the educational enrichment of young adolescents, for example, Milton Yinger and his colleagues (1977) needed to assign a large number of students, aged 13 and 14, to several different experimental and control groups to ensure the comparability of students composing each of the groups. They achieved this goal by the following method.

Beginning with a pool of subjects, the researchers first created strata of students nearly identical to one another in terms of some 15 variables. From each of the strata, students were randomly assigned to the different experimental and control groups. In this fashion, the researchers actually improved on conventional randomization. Essentially, they used a stratified sampling procedure (recall Chapter 7), except that they employed far more stratification variables than are typically used in, say, survey sampling.

Thus far I've described the classical experiment—the experimental design that best represents the logic of causal analysis in the laboratory. In practice, however, social researchers use a great variety of experimental designs. Let's look at some now.

VARIATIONS ON EXPERIMENTAL DESIGN

Donald Campbell and Julian Stanley (1963), in a classic book on research design, describe some 16 different experimental and quasi-experimental designs. This section describes some of these variations to better show the potential for experimentation in social research.

Preexperimental Research Designs

To begin, Campbell and Stanley discuss three "preexperimental" designs, not to recommend them but because they're frequently used in less-than-professional research. In the first such design—the *one-shot case study*—a single group of subjects is measured on a dependent variable following the administration of some experimental stimulus. Suppose, for example, that we show the African-American history film mentioned earlier to a group of people and then administer a questionnaire that seems to measure prejudice against African Americans. Suppose further that the answers given to the questionnaire seem to represent a low level of prejudice. We might be tempted to conclude that the film reduced prejudice. Lacking a pretest, however, we can't be sure. Perhaps the questionnaire doesn't really represent a very sensitive measure of prejudice, or perhaps the group we're studying was low in prejudice to begin with. In either case, the film might have made no difference, though our experimental results might have misled us into thinking it did.

The second preexperimental design discussed by Campbell and Stanley adds a pretest for the experimental group but lacks a control group. This design—which the authors call the *one-group pretest-posttest design*—suffers from the possibility that some factor other than the independent variable might cause a change between the pretest and posttest results, such as the assassination of a respected African-American leader. Thus, although we can see that prejudice has been reduced, we can't be sure the film caused that reduction.

To round out the possibilities for preexperimental designs, Campbell and Stanley point out that some research is based on experimental and control groups but has no pretests. They call this design the *static-group comparison.* For example, we might show the African-American history film to one group but not to another and then measure prejudice in both groups. If the experimental group had less prejudice at the conclusion of the experiment, we might assume the film was responsible. But unless we had randomized our subjects, we would have no way of knowing that the two groups had the same degree of prejudice initially; perhaps the experimental group started out with less.

Figure 8-3 graphically illustrates these three preexperimental research designs, using a different research question: Does exercise cause weight reduction? To make the several designs clearer, the figure shows individuals rather than groups, but the same logic pertains to group comparisons. Let's review the three preexperimental designs in this new example.

The one-shot study design represents a common form of logical reasoning in everyday life. Asked whether exercise causes weight-reduction, we may bring to mind an example that would seem to support the proposition: someone who exercises and is thin. There are problems with this reasoning, however. Perhaps the person was thin long before beginning to exercise. Or, perhaps he became thin for some other reason, like eating less or getting sick. The observations shown in the diagram do not guard against these other possibilities. Moreover, the observation that the man in the diagram is in trim shape depends on our intuitive idea of what constitutes trim and overweight body shapes. All told, this is very weak evidence for testing the relationship between exercise and weight loss.

The one-group pretest-posttest design offers somewhat better evidence that exercise produces weight loss. Specifically, we've ruled out the possibility that the man was thin before beginning to ex-

ercise. However, we still have no assurance that it was his exercising that caused him to lose weight.

Finally, the static-group comparison eliminates the problem of our questionable definition of what constitutes trim or overweight body shapes. In this case, we can compare the shapes of the man who exercises and the one who does not. This design, however, reopens the possibility that the man who exercises was thin to begin with.

Validity Issues in Experimental Research

At this point I want to present in a more systematic way the factors that affect experimental research—those I've already discussed as well as additional factors. First we'll look at what Campbell and Stanley call the sources of *internal invalidity,* reviewed and expanded in a follow-up book by Thomas Cook and Donald Campbell (1979). Then we'll consider the problem of generalizing experimental results to the "real" world, referred to as *external invalidity.* Having examined these, we'll be in a position to appreciate the advantages of some of the more sophisticated experimental and quasi-experimental designs social science researchers sometimes use.

Sources of Internal Invalidity   The problem of **internal invalidity** refers to the possibility that the conclusions drawn from experimental results may not accurately reflect what has gone on in the experiment itself. The threat of internal invalidity is present whenever anything other than the experimental stimulus can affect the dependent variable.

Campbell and Stanley (1963:5–6) and Cook and Campbell (1979:51–55) point to several sources of internal invalidity. Here are eight, to illustrate this concern:

1. *History.* During the course of the experiment, historical events may occur that confound the experimental results. The assassination of an African-American leader during the course of an experiment on reducing anti–African-American prejudice is one example.

2. *Maturation.* People are continually growing and changing, and such changes affect the results of the experiment. In a long-term experiment, the fact

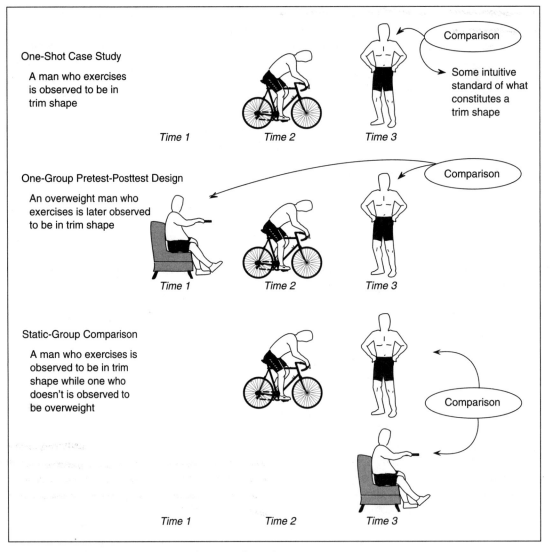

One-Shot Case Study

A man who exercises is observed to be in trim shape

Time 1   *Time 2*   *Time 3*

Comparison

Some intuitive standard of what constitutes a trim shape

One-Group Pretest-Posttest Design

An overweight man who exercises is later observed to be in trim shape

Time 1   *Time 2*   *Time 3*

Comparison

Static-Group Comparison

A man who exercises is observed to be in trim shape while one who doesn't is observed to be overweight

Comparison

Time 1   *Time 2*   *Time 3*

FIGURE 8-3   **Three Preexperimental Research Designs**

that the subjects grow older (and wiser?) may have an effect. In shorter experiments, they may grow tired, sleepy, bored, or hungry or change in other ways that affect their behavior in the experiment.

3. *Testing.* Often the process of testing and retesting influences people's behavior, thereby confounding the experimental results. Suppose we administer a questionnaire to a group as a way of measuring their prejudice. Then we administer an experimental stimulus and remeasure their preju-

dice. By the time we conduct the posttest, the subjects will probably have gotten more sensitive to the issue of prejudice and will be more thoughtful in their answers. In fact, they may have figured out that we're trying to find out how prejudiced they are, and, because few people like to appear prejudiced, they may give answers that they think we want or that will make them look good.

4. *Instrumentation.* The process of measurement in pretesting and posttesting brings in some of the is-

sues of conceptualization and operationalization discussed earlier in the book. If we use different measures of the dependent variable (say, different questionnaires about prejudice), how can we be sure they're comparable to one another? Perhaps prejudice will seem to decrease simply because the pretest measure was more sensitive than the posttest measure. Or if the measurements are being made by the experimenters, their standards or abilities may change over the course of the experiment.

5. *Statistical regression.* Sometimes it's appropriate to conduct experiments on subjects who start out with extreme scores on the dependent variable. If you were testing a new method for teaching math to hard-core failures in math, you'd want to conduct your experiment on people who previously have done extremely poorly in math. But consider for a minute what's likely to happen to the math achievement of such people over time without any experimental interference. They're starting out so low that they can only stay at the bottom or improve: They can't get worse. Even without any experimental stimulus, then, the group as a whole is likely to show some improvement over time. Referring to a *regression to the mean,* statisticians often point out that extremely tall people as a group are likely to have children shorter than themselves, and extremely short people as a group are likely to have children taller than themselves. There is a danger, then, that changes occurring by virtue of subjects starting out in extreme positions will be attributed erroneously to the effects of the experimental stimulus.

6. *Selection biases.* We discussed selection bias earlier when we examined different ways of selecting subjects for experiments and assigning them to experimental and control groups. Comparisons don't have any meaning unless the groups are comparable at the start of an experiment.

7. *Experimental mortality.* Although some social experiments could, I suppose, kill subjects, *experimental mortality* refers to a more general and less extreme problem. Often, experimental subjects will drop out of the experiment before it's completed, and this can affect statistical comparisons and conclusions. In the classical experiment in-

volving an experimental and a control group, each with a pretest and posttest, suppose that the bigots in the experimental group are so offended by the African-American history film that they leave before it's over. Those subjects sticking around for the posttest will have been less prejudiced to start with, so the group results will reflect a substantial "decrease" in prejudice.

8. *Demoralization.* On the other hand, feelings of deprivation within the control group may result in their giving up. In educational experiments, demoralized control-group subjects may stop studying, act up, or get angry.

These, then, are some of the sources of internal invalidity in experiments. cited by Campbell, Stanley, and Cook. Aware of these, experimenters have devised designs aimed at handling them. The classical experiment, if coupled with proper subject selection and assignment, addresses each of these problems. Let's look again at that study design, presented graphically in Figure 8-4, as it applies to our hypothetical study of prejudice.

If we use the experimental design shown in Figure 8-4, we should expect two findings from our African-American history film experiment. For the experimental group, the level of prejudice measured in their posttest should be less than was found in their pretest. In addition, when the two posttests are compared, less prejudice should be found in the experimental group than in the control group.

This design also guards against the problem of history in that anything occurring outside the experiment that might affect the experimental group should also affect the control group. Consequently, there should still be a difference in the two posttest results. The same comparison guards against problems of maturation as long as the subjects have been randomly assigned to the two groups. Testing and instrumentation can't be problems, since both the experimental and control groups are subject to the same tests and experimenter effects. If the subjects have been assigned to the two groups randomly, statistical regression should affect both equally, even if people with extreme scores on prejudice (or whatever the dependent

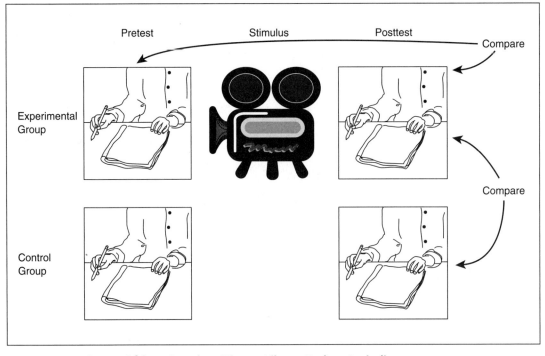

Pretest   Stimulus   Posttest
Compare

Experimental
Group

Control
Group

Compare

FIGURE 8-4 **Using an African-American History Film to Reduce Prejudice**

variable is) are being studied. Selection bias is ruled out by the random assignment of subjects. Experimental mortality is more complicated to handle, but the data provided in this study design offer several ways to deal with it. Slight modifications to the design—administering a placebo (such as a film having nothing to do with African Americans) to the control group, for example—can make the problem even easier to manage.

The remaining five problems of internal invalidity are avoided through the careful administration of a controlled experimental design. The experimental design we've been discussing facilitates the clear specification of independent and dependent variables. Experimental and control subjects can be kept separate, reducing the possibility of diffusion or imitation of treatments. Administrative controls can avoid compensations given to the control group, and compensatory rivalry can be watched for and taken into account in evaluating the results of the experiment, as can the problem of demoralization.

Sources of External Invalidity Internal invalidity accounts for only some of the complications faced by experimenters. In addition, there are problems of what Campbell and Stanley call **external invalidity,** which relates to the generalizability of experimental findings to the "real" world. Even if the results of an experiment are an accurate gauge of what happened during that experiment, do they really tell us anything about life in the wilds of society?

Campbell and Stanley describe four forms of this problem; I'll present one of them to you as an illustration. The generalizability of experimental findings is jeopardized, as the authors point out, if there's an interaction between the testing situation and the experimental stimulus (1963:18). Here's an example of what they mean.

Staying with the study of prejudice and the African-American history film, let's suppose that our experimental group—in the classical experiment—has less prejudice in its posttest than in its pretest and that its posttest shows less prejudice

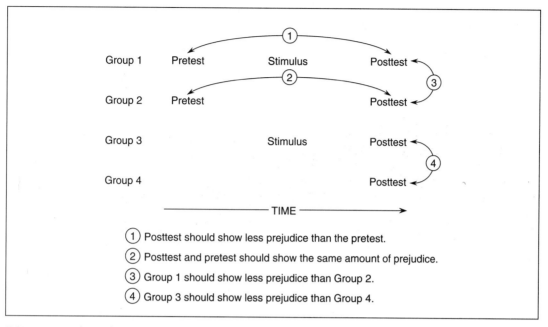

FIGURE 8-5   **The Solomon Four-Group Design**

than that of the control group. We can be confident that the film actually reduced prejudice among our experimental subjects. But would it have the same effect if the film were shown in theaters or on television? We can't be sure, since the film might only be effective when people have been sensitized to the issue of prejudice, as the subjects may have been in taking the pretest. This is an example of interaction between the testing and the stimulus. The classical experimental design cannot control for that possibility. Fortunately, experimenters have devised other designs that can.

The *Solomon four-group design* (D. Campbell and Stanley 1963:24–25) addresses the problem of testing interaction with the stimulus. As the name suggests, it involves four groups of subjects, assigned randomly from a pool. Figure 8-5 presents this design graphically.

Notice that Groups 1 and 2 in Figure 8-5 compose the classical experiment. Group 3 is administered the experimental stimulus without a pretest, and Group 4 is only posttested. This latest experimental design permits four meaningful comparisons. If the African-American history film really re-

duces prejudice—unaccounted for by the problem of internal validity and unaccounted for by an interaction between the testing and the stimulus—we should expect four findings:

1. In Group 1, posttest prejudice should be less than pretest prejudice.
2. There should be less prejudice evident in the Group 1 posttest than in the Group 2 posttest.
3. The Group 3 posttest should show less prejudice than the Group 2 pretest.
4. The Group 3 posttest should show less prejudice than the Group 4 posttest.

Notice that findings (3) and (4) rule out any interaction between the testing and the stimulus. And remember that these comparisons are meaningful only if subjects have been assigned randomly to the different groups, thereby providing groups of equal prejudice initially, even though their preexperimental prejudice is only measured in Groups 1 and 2.

There is a side benefit to this research design, as the authors point out. Not only does the Solomon

APPLYING THE RESULTS

One dramatic application of small group research that also poses the problem of external invalidity is the use of "mock juries." In this application, people are asked to pretend they are juries in a real trial. Lawyers try various strategies on different mock juries to determine which strategy will be the most effective in a real trial. This is a meaningless exercise, however, unless the effects produced in the "laboratory" of mock juries can be replicated in actual court cases.

four-group design rule out interactions between testing and the stimulus, it also provides data for comparisons that will reveal the amount of such interaction that occurs in the classical experimental design. This knowledge would allow a researcher to review and evaluate the value of any prior research that used the simpler design.

The last experimental design I'll mention here is what Campbell and Stanley (1963:25–26) call the *posttest-only control-group design;* it consists of the second half—Groups 3 and 4—of the Solomon design. As the authors argue persuasively, with proper randomization, only Groups 3 and 4 are needed for a true experiment that controls for the problems of internal invalidity as well as for the interaction between testing and stimulus. With randomized assignment to experimental and control groups (which distinguishes this design from the static-group comparison discussed earlier), the subjects will be initially comparable on the dependent variable—comparable enough to satisfy the conventional statistical tests used to evaluate the results—so it's not necessary to measure them. Indeed, Campbell and Stanley suggest the only justification for pretesting in this situation is tradition. Experimenters have simply grown accustomed to pretesting and feel more secure with research designs that include it. Be clear, however, that this point applies only to experiments in which subjects have been assigned to experimental and

control groups randomly, since that's what justifies the assumption that the groups are equivalent—without actually measuring them to find out.

This discussion has introduced the intricacies of experimental design, its problems, and some solutions. There are, of course, a great many other possible experimental designs in use. Some involve more than one stimulus and combinations of stimuli. Others involve several tests of the dependent variable over time and the administration of the stimulus at different times for different groups. If you're interested in pursuing this topic, you might look at the Campbell and Stanley book.

AN ILLUSTRATION OF EXPERIMENTATION

Experiments have been used to study a wide variety of topics in the social sciences. Some experiments have been conducted within laboratory situations; others occur out in the "real world." The following discussion will give you a glimpse of both.

In George Bernard Shaw's well-loved play, *Pygmalion*—the basis for the musical, *My Fair Lady*—Eliza Doolittle speaks of the powers others have in determining our social identity. Here's how she distinguishes the way she's treated by her tutor, Professor Higgins, and by Higgins's friend, Colonel Pickering:

> You see, really and truly, apart from the things anyone can pick up (the dressing and the proper way of speaking, and so on), the difference between a lady and a flower girl is not how she behaves, but how she's treated. I shall always be a flower girl to Professor Higgins, because he always treats me as a flower girl, and always will, but I know I can be a lady to you, because you always treat me as a lady, and always will.
> — (ACT V)

The sentiment Eliza expresses here is basic social science, addressed more formally by sociologists such as Charles Horton Cooley ("looking-glass self") and George Herbert Mead ("the generalized other"). The basic point is that who we think we are—our self-concept—and how we behave is largely a function of how others see and treat us. Related to this, the way others perceive us is largely conditioned by their expectations. If they've been told we're stupid, for example, they're likely to see us that way—and we may come to see ourselves that way and actually act stupidly.

This topic has generally been called the *Pygmalion effect,* and it's nicely suited to controlled experiments. In one of the best-known experiments on this topic, Robert Rosenthal and Lenore Jacobson (1968) administered what they called a "Harvard Test of Inflected Acquisition" to students in a West Coast school. Subsequently, they met with the students' teachers to present the results of the test. In particular, Rosenthal and Jacobson identified certain students as very likely to exhibit a sudden spurt in academic abilities during the coming year, based on the results of the test.

When IQ test scores were compared later, the researchers' predictions proved accurate. The students identified as "spurters" far exceeded their classmates during the following year, suggesting that the predictive test was a powerful one. In fact, the test was a hoax! The researchers had made their predictions randomly among both good and poor students. What they told the teachers did not really reflect students' test scores at all. The progress made by the "spurters" was simply a result of the teachers' expecting the improvement and paying more attention to those students, encouraging them, and rewarding them for achievements. (Notice the similarity between this situation and the Hawthorne effect, discussed earlier in this chapter.)

The Rosenthal-Jacobson study attracted a lot of popular as well as scientific attention. Subsequent experiments have focused on specific aspects of what has become known as the *attribution process,* or the *expectations communication model.* This research, largely conducted by psychologists, parallels research primarily by sociologists, which takes a slightly different focus and is often gathered under the label *expectations-states theory.* The former studies focus on situations in which the expectations of a dominant individual affect the performance of subordinates—as in the case of a teacher and students, or a boss and employees. The socio-

logical research has tended to focus more on the role of expectations among equals in small, task-oriented groups. In a jury, for example, how do jurors initially evaluate each other, and how do those initial assessments affect their later interactions?

 Search the Web for "Pygmalion Effect" to learn more about this phenomenon, including attempts to find practical applications of it.*

Here's an example of an experiment conducted to examine the way our perceptions of our abilities and those of others affect our willingness to accept the other person's ideas. Martha Foschi, G. Keith Warriner, and Stephen Hart (1985) were particularly interested in the role "standards" played in that respect.

> In general terms, by "standards" we mean how well or how poorly a person has to perform in order for an ability to be attributed or denied him/her. In our view, standards are a key variable affecting how evaluations are processed and what expectations result. For example, depending on the standards used, the same level of success may be interpreted as a major accomplishment or dismissed as unimportant.
> — (1985:108–9)

To begin examining the role of standards, the researchers designed an experiment involving four experimental groups and a control. Subjects were told that the experiment involved something called "pattern recognition ability," which was an innate ability some people had and others didn't. The researchers said that subjects would be working in pairs on pattern recognition problems.

In fact, of course, there's no such thing as pattern recognition ability. The object of the experiment was to determine how information about this supposed ability affected subjects' subsequent behavior.

---

*Each time the Internet icon appears, you'll be given helpful leads for searching the World Wide Web.

The first stage of the experiment was to "test" each subject's pattern recognition abilities. If you had been a subject in the experiment, you would have been shown a geometrical pattern for 8 seconds, followed by two more patterns, each of which was similar to but not the same as the first one. Your task would be to choose which of the subsequent set had a pattern closest to the first one you saw. You would be asked to do this 20 times, and a computer would print out your "score." Half the subjects would be told they had gotten 14 correct; the other half would be told they had only gotten 6 correct—regardless of which patterns they matched with which. Depending on the luck of the draw, you would think you had done either quite well or quite badly. Notice, however, that you wouldn't really have any standard for judging your performance—maybe getting 4 correct would be considered a great performance.

At the same time you were given your score, however, you would also be given your "partner's score," although both the "partners" and their "scores" were also computerized fictions. (Subjects were told they would be communicating with their partners via computer terminals but would not be allowed to see each other.) If you were assigned a score of 14, you would be told your partner had a score of 6; if you were assigned 6, you would be told your partner had 14.

This procedure meant that you would enter the teamwork phase of the experiment believing either (1) you had done better than your partner or (2) you had done worse than your partner. This information constituted part of the "standard" you would be operating under in the experiment. In addition, half of each group was told that a score of between 12 and 20 meant that the subject *definitely* had pattern recognition ability; the other subjects were told that a score of 14 wasn't really high enough to prove anything definite. Thus, you would emerge from this with one of the following beliefs:

1. You are *definitely better* at pattern recognition than your partner.
2. You are *possibly better* than your partner.
3. You are *possibly worse* than your partner.
4. You are *definitely worse* than your partner.

The control group for this experiment was told nothing about their own abilities or their partners'. In other words, they had no expectations.

The final step in the experiment was to set the "teams" to work. As before, each subject would be given an initial pattern, followed by a comparison pair to choose from. When you entered your choice in this round, however, you would be told what your partner had answered; you would then be asked to choose again. In your final choice, you could either stick with your original choice or switch. The "partner's" choice was, of course, created by the computer, and, as you can guess, there were often disagreements in the teams: 16 out of 20 times, in fact.

The dependent variable in this experiment was the extent to which subjects would switch their choices to match those of their partners. The researchers hypothesized that the *definitely better* group would switch least often, followed by the *probably better* group, followed by the *control group*, followed by the *probably worse* group, followed by the *definitely worse* group, who would switch most often.

The number of times subjects in the five groups switched their answers follows. Realize that each had 16 opportunities to do so. These data indicate that each of the researchers' expectations was correct—with the exception of the comparison between the *possibly worse* and *definitely worse* groups. The latter group was in fact the more likely to switch, but the difference was too small to be taken as a confirmation of the hypothesis.

Group	Mean Number of Switches
Definitely better	5.05
Possibly better	6.23
Control Group	7.95
Possibly worse	9.23
Definitely worse	9.28

In more detailed analyses, the researchers found that the same basic pattern held for both men and women, though it was somewhat clearer for women than for men. Here are the actual data:

	Mean Number of Switches	
	Women	*Men*
Definitely better	4.50	5.66
Possibly better	6.34	6.10
Control Group	7.68	8.34
Possibly worse	9.36	9.09
Definitely worse	10.00	8.70

Because specific research efforts like this one sometimes seem extremely focused in their scope, you might wonder about their relevance to anything. As part of a larger research effort, however, studies like this one add concrete pieces to our understanding of more general social processes.

It's worth taking a minute or so to consider some of the life situations where "expectation states" might have very real and important consequences. I've mentioned the case of jury deliberations. How about all forms of prejudice and discrimination? Or, consider how expectation states figure into job interviews or meeting your heartthrob's parents. If you think about it, you'll probably see other situations where these laboratory concepts apply in real life.

"NATURAL" EXPERIMENTS

Although we tend to equate the terms *experiment* and *laboratory experiment,* many important social scientific experiments occur outside controlled settings, often in the course of normal social events. Sometimes nature designs and executes experiments that we can observe and analyze; sometimes social and political decision makers serve this natural function.

Imagine, for example, that a hurricane has struck a particular town. Some residents of the town suffer severe financial damages, while others escape relatively lightly. What, we might ask, are the behavioral consequences of suffering a natural disaster? Are those who suffer most more likely to take precautions against future disasters than are those who suffer least? To answer these questions, we might interview residents of the town some

APPLYING THE RESULTS

One example of a natural experiment comes from the annals of social research concerning World War II. After the war ended, social researchers undertook retrospective surveys of wartime morale among civilians in several German cities. Among other things, they wanted to determine the effect of mass bombing on the morale of civilians. They compared the reports of wartime morale of residents in heavily bombed cities and cities that received relatively little bombing. Bombing did not reduce morale.

time after the hurricane. We might question them regarding their precautions before the hurricane and the ones they're currently taking, comparing the people who suffered greatly from the hurricane with those who suffered relatively little. In this fashion, we might take advantage of a natural experiment, which we could not have arranged even if we'd been perversely willing to do so.

Because the researcher must take things pretty much as they occur, natural experiments raise many of the validity problems discussed earlier. Thus when Stanislav Kasl, Rupert Chisolm, and Brenda Eskenazi (1981) chose to study the impact that the Three Mile Island (TMI) nuclear accident in Pennsylvania had on plant workers, they had to be especially careful in the study design:

> Disaster research is necessarily opportunistic, quasi-experimental, and after-the-fact. In the terminology of Campbell and Stanley's classical analysis of research designs, our study falls into the "static-group comparison" category, considered one of the weak research designs. However, the weaknesses are potential and their actual presence depends on the unique circumstances of each study. — (1981:474)

The foundation of this study was a survey of the people who had been working at Three Mile Island on March 28, 1979, when the cooling system failed in the number 2 reactor and began melting the uranium core. The survey was conducted five to six months after the accident. Among other things, the survey questionnaire measured workers' attitudes toward working at nuclear power plants. If they had measured only the TMI workers' attitudes after the accident, the researchers would have had no idea whether attitudes had changed as a consequence of the accident. But they improved their study design by selecting another, nearby—seemingly comparable—nuclear power plant (abbreviated as PB) and surveyed workers there as a control group: hence their reference to a static-group comparison.

Even with an experimental and a control group, the authors were wary of potential problems in their design. In particular, their design was based on the idea that the two sets of workers were equivalent to one another, except for the single fact of the accident. The researchers could have assumed this if they had been able to assign workers to the two plants randomly, but of course they couldn't. Instead, they compared characteristics of the two groups to see whether they were equivalent. Ultimately, the researchers concluded that the two sets of workers were very much alike, and the plant the employees worked at was merely a function of where they lived.

Even granting that the two sets of workers were equivalent, the researchers faced another problem of comparability. They could not contact all the workers who had been employed at TMI at the time of the accident. The researchers discuss the problem as follows:

> One special attrition problem in this study was the possibility that some of the no-contact nonrespondents among the TMI subjects, but not PB subjects, had permanently left the area because of the accident. This biased attrition would, most likely, attenuate the estimated extent of the impact. Using the evidence of disconnected or "not in service" telephone numbers, we estimate this bias to be negligible (1 percent). — (Kasl, Chisolm, and Eskenazi 1981:475)

The TMI example points both to the special problems involved in natural experiments and to the possibility of taking those problems into account. Social research generally requires ingenuity and insight; natural experiments call for a little more than the average.

Earlier in this chapter, we used a hypothetical example of studying whether an African-American history film reduced prejudice. Sandra Ball-Rokeach, Joel Grube, and Milton Rokeach (1981) were able to address that topic in real life through a natural experiment. In 1977, the television dramatization of Alex Haley's *Roots,* a historical saga about African Americans, was presented by ABC on eight consecutive nights. It garnered the largest audiences in television history up to that time. Ball-Rokeach and her colleagues wanted to know if *Roots* changed white Americans' attitudes toward African Americans. Their opportunity arose in 1979, when a sequel—*Roots: The Next Generation*—was televised. Although it would have been nice (from a researcher's point of view) to assign random samples of Americans either to watch or not watch the show, this wasn't possible. Instead, the researchers selected four samples in Washington State and mailed questionnaires (before the broadcast) that measured attitudes toward African Americans. Following the last episode of the show, respondents were called and asked how many, if any, episodes they had watched. Subsequently, questionnaires were sent to respondents, remeasuring their attitudes toward African Americans.

By comparing attitudes before and after for both those who watched the show and those who didn't, the researchers reached several conclusions. For example, they found that people with already egalitarian attitudes were much more likely to watch the show than were those who were more prejudiced toward African Americans: a self-selection phenomenon. Comparing the before and after attitudes of those who watched the show, moreover, suggested that the show itself had little or no effect. Those who watched it were no more egalitarian afterward than they had been before.

This example anticipates the subject of Chapter 12, evaluation research, which can be seen as a special type of natural experiment. As you'll see, evaluation research involves taking the logic of experimentation into the field to observe and evaluate the effects of stimuli in real life. Because this is an increasingly important form of social research, an entire chapter is devoted to it.

STRENGTHS AND WEAKNESSES OF THE EXPERIMENTAL METHOD

Experiments are the primary tool for studying causal relationships. However, like all research methods, experiments have both strengths and weaknesses.

The chief advantage of a controlled experiment lies in the isolation of the experimental variable and its impact over time. This is seen most clearly in terms of the basic experimental model. A group of experimental subjects are found, at the outset of the experiment, to have a certain characteristic; following the administration of an experimental stimulus, they are found to have a different characteristic. To the extent that subjects have experienced no other stimuli, we may conclude that the change of characteristics is caused by the experimental stimulus.

Further, since individual experiments are often rather limited in scope, requiring relatively little time and money and relatively few subjects, we often can replicate a given experiment several times using many different groups of subjects. (This isn't always the case, of course, but it's usually easier to repeat experiments than, say, surveys.) As in all other forms of scientific research, replication of research findings strengthens our confidence in the validity and generalizability of those findings.

The greatest weakness of laboratory experiments lies in their artificiality. Social processes that occur in a laboratory setting might not necessarily

---

A QUANDARY REVISITED

As we've seen, the impact of the experiment itself on subjects' responses is a major concern in social research. Several elements of experimental designs address this concern. First, the use of control groups allows researchers to account for any effects of the experiment that are not related to the stimulus. Second, the Solomon four-group design tests for the possible impact of pretests on the dependent variable. And, finally, some experiments are done in real-life situations, with no experimental constraints, so-called "natural experiments."

Thus, while the impact of observer is a live issue in experiments, researchers have developed methods for addressing it.

---

occur in more natural social settings. For example, an African-American history film might genuinely reduce prejudice among a group of experimental subjects. This would not necessarily mean, however, that the same film shown in neighborhood movie theaters throughout the country would reduce prejudice among the general public. Artificiality is not as much of a problem, of course, for natural experiments as for those conducted in the laboratory.

In discussing several of the sources of internal and external invalidity mentioned by Campbell, Stanley, and Cook, we saw that we can create experimental designs that logically control such problems. This possibility points to one of the great advantages of experiments: They lend themselves to a logical rigor that is often much more difficult to achieve in other modes of observation.

Main Points

❏ Experiments are an excellent vehicle for the controlled testing of causal processes.

❏ The classical experiment tests the effect of an experimental stimulus (the independent variable) on a dependent variable through the pretesting and posttesting of experimental and control groups.

- It is generally less important that a group of experimental subjects be representative of some larger population than that experimental and control groups be similar to each other.

- A double-blind experiment guards against experimenter bias because neither the experimenter nor the subject knows which subjects are in the control and experimental groups.

- Probability sampling, randomization, and matching are all methods of achieving comparability in the experimental and control groups. Randomization is the generally preferred method. In some designs, it can be combined with matching.

- Campbell and Stanley describe three forms of preexperiments: the one-shot case study, the one-group pretest-posttest design, and the static-group comparison.

- Campbell and Stanley list, among others, 8 sources of internal invalidity in experimental design: history, maturation, testing, instrumentation, statistical regression, selection biases, experimental mortality, and demoralization. The classical experiment with random assignment of subjects guards against each of these 8 sources of internal invalidity.

- Experiments also face problems of external invalidity: Experimental findings may not reflect real life.

- The interaction of testing with the stimulus is an example of external invalidity that the classical experiment does not guard against.

- The Solomon four-group design and other variations on the classical experiment can safeguard against external invalidity.

- Campbell and Stanley suggest that, given proper randomization in the assignment of subjects to the experimental and control groups, there is no need for pretesting in experiments.

- Natural experiments often occur in the course of social life in the real world, and social researchers can implement them in somewhat the same way they would design and conduct laboratory experiments.

- Like all research methods, experiments have strengths and weaknesses. Their primary weakness is artificiality: What happens in an experiment may not reflect what happens in the outside world. Their strengths include the isolation of the independent variable, which permits causal inferences; the relative ease of replication; and scientific rigor.

Key Terms

pretesting	randomization
posttesting	matching
experimental group	internal invalidity
control group	external invalidity
double-blind experiment	

Review Questions

1. What are some examples of internal invalidity? Pick 4 of the 8 sources discussed in the book and make up examples (not discussed in the book) to illustrate each.

2. Think of a recent natural disaster you've witnessed or read about. What research question might be studied by treating that disaster as a natural experiment? Take two or three paragraphs to outline how the study might be done.

3. Say you want to evaluate a new operating system or other software. How might you set up an experiment to see what people really think of it? Keep in mind the use of control groups and the placebo effect.

4. Think of a recent, highly publicized trial. How might the attorneys have used mock juries to evaluate different strategies for presenting evidence?

Additional Readings

Campbell, Donald, and Julian Stanley. 1963. *Experimental and Quasi-Experimental Designs for Research*. Chicago: Rand McNally. An excellent analysis of the logic and methods of experimentation in social research. This book is especially useful in its application of the logic of experiments to other social research methods. Though fairly old, this book has attained the status of a classic and is still frequently cited.

Cook, Thomas D., and Donald T. Campbell. 1979. *Quasi-Experimentation: Design and Analysis Issues for Field Settings.* Chicago: Rand McNally. An expanded and updated version of Campbell and Stanley.

Jones, Stephen R. G. 1990. "Worker Independence and Output: The Hawthorne Studies Reevaluated." *American Sociological Review* 55:176–90. This article reviews these classical studies and questions the traditional interpretation (which was presented in this chapter).

Martin, David W. 1996. *Doing Psychology Experiments.* 4th ed. Monterey, CA: Brooks/Cole. Thorough explanations of the logic behind research methods, often in a humorous style. The book emphasizes ideas of particular importance to the beginning researcher, such as getting an idea for an experiment or reviewing the literature.

Ray, William J. 2000. *Methods toward a Science of Behavior and Experience.* 6th ed. Belmont, CA: Wadsworth. A comprehensive examination of social science research methods, with a special emphasis on experimentation. This book is especially strong in the philosophy of science.

Multimedia Resources

The Wadsworth Sociology Resource Center: Virtual Society
http://sociology.wadsworth.com/
Visit the companion Web site for the second edition of *The Basics of Social Research* to access a wide range of student resources. Begin by clicking on the Student Resources section of the book's Web site to access the following study tools:

- eBabbie Resource Center
- Planning a Research Project
- Doing Data Analysis
- Statistics Review
- Flash Cards
- Internet Links and Exercises
- InfoTrac College Edition: Exercises
- Quizzes

Visit the **eBabbie Resource Center** for an overview of each chapter and helpful online tutorials. Find information on budgeting and step-by-step examples of model research projects at **Planning a Research Project.** Learn how to use quantitative and qualitative data analysis programs at **Doing Data Analysis,** and brush up on your statistics at **Statistics Review.** You can also further your study by accessing **Internet Links and Exercises** related to chapter materials, **Flash Cards, Quizzes,** and many other learning tools.

InfoTrac College Edition
http://www.infotrac-college.com/
wadsworth/access.html
Access the latest news and journal articles with InfoTrac College Edition, an easy-to-use online database of reliable, full-length articles from hundreds of top academic journals. Conduct an electronic search using the following search terms:

Double-blind experiment
Experiment AND control group
Experiment AND matching
Experiment AND placebo
Experiment AND randomization
Experiment AND stimulus
Mock jury
Natural experiment
Pretest AND posttest

9 SURVEY RESEARCH

Don Pitcher/Stock Boston

What You'll Learn in This Chapter

Here you'll learn about many of the methods researchers use to collect data through surveys—from mail questionnaires to personal interviews to online surveys conducted over the Internet. You'll also learn how to select an appropriate method and how to implement it effectively.

In this chapter . . .

AN OPENING **QUANDARY**

All of us get telephone calls asking us to participate in a "survey," but some are actually telemarketing calls.

Caller: The first question is "Whose head is on the copper penny?"
You: Karl Marx.
Caller: Close enough. You have just won . . .

This is obviously not legitimate. But subtler techniques can entangle you in a phony sales pitch before you know what's happening. When you get a call announcing that you've been selected for a "survey," how can you tell whether it's genuine?

INTRODUCTION

Surveys are a very old research technique. In the Old Testament, for example, we find the following:

> After the plague the Lord said to Moses and to Eleazar the son of Aaron, the priest, "Take a census of all the congregation of the people of Israel, from twenty old and upward." — (NUMBERS 26:1–2)

Ancient Egyptian rulers conducted censuses to help them administer their domains. Jesus was born away from home because Joseph and Mary were journeying to Joseph's ancestral home for a Roman census.

A little-known survey was attempted among French workers in 1880. A German political sociologist mailed some 25,000 questionnaires to workers to determine the extent of their exploitation by employers. The rather lengthy questionnaire included items such as these:

> Does your employer or his representative resort to trickery in order to defraud you of a part of your earnings?
>
> If you are paid piece rates, is the quality of the article made a pretext for fraudulent deductions from your wages?

The survey researcher in this case was not George Gallup but Karl Marx ([1880] 1956: 208). Though 25,000 questionnaires were mailed out, there is no record of any being returned.

Today, survey research is a frequently used mode of observation in the social sciences. In a typical survey, the researcher selects a sample of respondents and administers a standardized questionnaire to each person in the sample. Chapter 7 discussed sampling techniques in detail. This chapter discusses how to prepare a questionnaire and describes the various options for administering it so that respondents answer your questions adequately.

The chapter concludes with a short discussion of *secondary analysis,* the analysis of survey data collected by someone else. This use of survey results has become an important aspect of survey research in recent years, and it's especially useful for students and others with scarce research funds.

Let's begin by looking at the kinds of topics that researchers can appropriately study by using survey research.

TOPICS APPROPRIATE FOR SURVEY RESEARCH

Surveys may be used for descriptive, explanatory, and exploratory purposes. They are chiefly used in studies that have individual people as the units of analysis. Although this method can be used for other units of analysis, such as groups or interactions, some individual persons must serve as **respondents** or informants. Thus, we could undertake a survey in which divorces were the unit of analysis, but we would need to administer the survey questionnaire to the participants in the divorces (or to some other informants).

Survey research is probably the best method available to the social researcher who is interested in collecting original data for describing a population too large to observe directly. Careful probability sampling provides a group of respondents whose characteristics may be taken to reflect those of the larger population, and carefully constructed standardized questionnaires provide data in the same form from all respondents.

Surveys are also excellent vehicles for measuring attitudes and orientations in a large population. Public opinion polls—for example, Gallup, Harris, Roper, and Yankelovich—are well-known examples of this use. Indeed, polls have become so prevalent that at times the public seems unsure what to think of them. Pollsters are criticized by those who don't think (or want to believe) that polls are accurate (candidates who are "losing" in polls often tell voters not to trust the polls). But polls are also criticized for being *too* accurate—for example, when exit polls on election day are used to predict a winner before the actual voting is complete.

The general attitude toward public opinion research is further complicated by scientifically unsound "surveys" that nonetheless capture people's attention because of the topics they cover and/ or their "findings." A good example is the "Hite Reports" on human sexuality. While enjoying considerable attention in the popular press, Shere Hite was roundly criticized by the research community for her data-collection methods. For example, a 1987 Hite report was based on questionnaires completed by women around the country—but which women? Hite reported that she distributed some 100,000 questionnaires through various organizations, and around 4,500 were returned.

Now 4,500 and 100,000 are large numbers in the context of survey sampling. However, given Hite's research methods, her 4,500 respondents didn't necessarily represent U.S. women any more than the *Literary Digest*'s enormous 1936 sample represented the U.S. electorate when their 2 million sample ballots indicated Alf Landon would bury FDR in a landslide.

Sometimes, people use the pretense of survey research for quite different purposes. For example, you may have received a telephone call indicating you've been selected for a survey, only to find that the first question was "How would you like to make thousands of dollars a week right there in your own home?" Unfortunately, a few unscrupulous telemarketers try to prey on the general cooperation people have given to survey researchers.

By the same token, political parties and charitable organizations have begun conducting phony "surveys." Often under the guise of collecting public opinion about some issue, callers ultimately ask respondents for a monetary contribution.

Recent political campaigns have produced another form of bogus survey, called the "push poll." Here's what the American Association for Public Opinion Polling had to say in condemning this practice:

> A "push poll" is a telemarketing technique in which telephone calls are used to canvass potential voters, feeding them false or misleading "information" about a candidate under the pretense of taking a poll to see how this "information" affects voter preferences. In fact, the intent is not to measure public opinion but to manipulate it—to "push" voters away from one candidate and toward the opposing candidate. Such polls defame selected candidates by spreading false or misleading information about them. The intent is to disseminate campaign propaganda under the guise of conducting a legitimate public opinion poll. — (BEDNARZ 1996)

In short, the labels "survey" and "poll" are sometimes misused. Done properly, however, survey research can be a useful tool of social inquiry. Designing useful (and trustworthy) survey research begins with formulating good questions. Let's turn to that topic now.

GUIDELINES FOR ASKING QUESTIONS

In social research, variables are often operationalized when researchers ask people questions as a way of getting data for analysis and interpretation. Sometimes the questions are asked by an interviewer; sometimes they are written down and given to respondents for completion. In other cases, several general guidelines can help researchers frame and ask questions that serve as excellent operationalizations of variables while avoiding pitfalls that can result in useless or even misleading information.

Surveys include the use of a **questionnaire**—an instrument specifically designed to elicit information that will be useful for analysis. While some of the specific points to follow are more appropriate to structured questionnaires than to the more open-ended questionnaires used in qualitative, in-depth interviewing, the underlying logic is valuable whenever we ask people questions in order to gather data.

Choose Appropriate Question Forms

Let's begin with some of the options available to you in creating questionnaires. These options include using questions or statements and choosing open-ended or closed-ended questions.

Questions and Statements Although the term *questionnaire* suggests a collection of questions, an examination of a typical questionnaire will probably reveal as many statements as questions. This is not without reason. Often, the researcher is interested in determining the extent to which respondents hold a particular attitude or perspective. If you can summarize the attitude in a fairly brief statement, you can present that statement and ask respondents whether they agree or disagree with it. As you may remember, Rensis Likert greatly formalized this procedure through the creation of the Likert scale, a format in which respondents are asked to strongly agree, agree, disagree, or strongly disagree, or perhaps strongly approve, approve, and so forth.

Questions and statements may both be used profitably. Using both in a given questionnaire gives you more flexibility in the design of items and can make the questionnaire more interesting as well.

Open-Ended and Closed-Ended Questions In asking questions, researchers have two options. They may ask **open-ended questions,** in which case the respondent is asked to provide his or her own answer to the question. For example, the respondent may be asked, "What do you feel is the most important issue facing the United States today?" and be provided with a space to write in the answer (or be asked to report it verbally to an interviewer). As we'll see in Chapter 10, in-depth, qualitative interviewing relies almost exclusively on open-ended questions. However, they are also used in survey research.

In the case of **closed-ended questions,** the respondent is asked to select an answer from among a list provided by the researcher. Closed-ended questions are very popular in survey research because they provide a greater uniformity of responses and are more easily processed than open-ended ones.

Open-ended responses must be coded before they can be processed for computer analysis, as will be discussed in Chapter 14. This coding process often requires the researcher to interpret the meaning of responses, opening the possibility of misunderstanding and researcher bias. There is also a danger that some respondents will give answers that are essentially irrelevant to the researcher's intent. Closed-ended responses, on the other hand, can often be transferred directly into a computer format.

The chief shortcoming of closed-ended questions lies in the researcher's structuring of responses. When the relevant answers to a given question are relatively clear, there should be no problem. In other cases, however, the researcher's structuring of responses may overlook some important responses. In asking about "the most important issue facing the United States," for example, his or her checklist of issues might omit certain issues that respondents would have said were important.

The construction of closed-ended questions should be guided by two structural requirements. First, the response categories provided should be exhaustive: They should include all the possible responses that might be expected. Often, researchers ensure this by adding a category such as "Other (Please specify: _____)." Second, the answer categories must be mutually exclusive: The respondent should not feel compelled to select more than one. (In some cases, you may wish to solicit multiple answers, but these may create difficulties in data processing and analysis later on.) To ensure that your categories are mutually exclusive, carefully consider each combination of categories, asking yourself whether a person could reasonably choose more than one answer. In addition, it's useful to add an instruction to the question asking the respondent to select the one best answer, but this technique is not a satisfactory substitute for a carefully constructed set of responses.

Make Items Clear

It should go without saying that questionnaire items should be clear and unambiguous, but the broad proliferation of unclear and ambiguous questions in surveys makes the point worth emphasizing. Often we can become so deeply involved in the topic under examination that opinions and perspectives are clear to us but not to our respondents, many of whom have paid little or no attention to the topic. Or, if we have only a superficial understanding of the topic, we may fail to specify the intent of a question sufficiently. The question "What do you think about the proposed peace plan?" may evoke in the respondent a counterquestion: "Which proposed peace plan?" Questionnaire items should be precise so that the respondent knows exactly what the researcher is asking.

The possibilities for misunderstanding are endless, and no researcher is immune (Polivka and Rothgeb 1993). One of the most established research projects in the United States is the Census Bureau's ongoing "Current Population Survey" or CPS, which measures, among other critical data, the nation's unemployment rate. A part of the measurement of employment patterns focuses on a respondent's activities during "last week," by which the Census Bureau means Sunday through Saturday. Studies undertaken to determine the accuracy of the survey found that more than half the respondents took "last week" to include only Monday through Friday. By the same token, whereas the Census Bureau defines "working full-time" as 35 or more hours a week, the same evaluation studies showed that some respondents used the more traditional definition of 40 hours per week. As a consequence, the wording of these questions in the CPS was modified in 1994 to specify the Census Bureau's definitions.

Similarly, the use of the term *Native American* to mean *American Indian* often produces an overrepresentation of that ethnic group in surveys. Clearly, many respondents understand the term to mean "born in the United States."

Avoid Double-Barreled Questions

Frequently, researchers ask respondents for a single answer to a question that actually has multiple parts. That seems to happen most often when the researcher has personally identified with a complex question. For example, you might ask respondents to agree or disagree with the statement "The United States should abandon its space pro-

gram and spend the money on domestic programs." Although many people would unequivocally agree with the statement and others would unequivocally disagree, still others would be unable to answer. Some would want to abandon the space program and give the money back to the taxpayers. Others would want to continue the space program but also put more money into domestic programs. These latter respondents could neither agree nor disagree without misleading you.

As a general rule, whenever the word *and* appears in a question or questionnaire statement, check whether you're asking a double-barreled question. See the box entitled "Double-Barreled and Beyond" for some imaginative variations on this theme.

Respondents Must Be Competent to Answer

In asking respondents to provide information, you should continually ask yourself whether they can do so reliably. In a study of child rearing, you might ask respondents to report the age at which they first talked back to their parents. Quite aside from the problem of defining *talking back to parents*, it's doubtful that most respondents would remember with any degree of accuracy.

As another example, student government leaders occasionally ask their constituents to indicate how students' fees ought to be spent. Typically, respondents are asked to indicate the percentage of available funds that should be devoted to a long list of activities. Without a fairly good knowledge of the nature of those activities and the costs involved in them, the respondents cannot provide meaningful answers. Administrative costs, for example, will receive little support although they may be essential to the program as a whole.

One group of researchers examining the driving experience of teenagers insisted on asking an open-ended question concerning the number of miles driven since receiving a license. Although consultants argued that few drivers would be able to estimate such information with any accuracy, the question was asked nonetheless. In response,

Even established, professional researchers have sometimes created double-barreled questions and worse. Consider this question, asked of U.S. citizens in April 1986, at a time when the country's relationship with Libya was at an especially low point. Some observers suggested the United States might end up in a shooting war with the small North African nation. The Harris Poll sought to find out what U.S. public opinion was.

> If Libya now increases its terrorist acts against the U.S. and we keep inflicting more damage on Libya, then inevitably it will all end in the U.S. going to war and finally invading that country which would be wrong.

Respondents were given the opportunity of answering "Agree," "Disagree," or "Not sure." Notice the elements contained in the complex statement:

1. Will Libya increase its terrorist acts against the U.S.?
2. Will the U.S. inflict more damage on Libya?
3. Will the U.S. inevitably or otherwise go to war against Libya?
4. Would the U.S. invade Libya?
5. Would that be right or wrong?

These several elements offer the possibility of numerous points of view—far more than the three alternatives offered respondents to the survey. Even if we were to assume hypothetically that Libya would "increase its terrorist attacks" and the United States would "keep inflicting more damage" in return, you might have any one of at least seven distinct expectations about the outcome:

	U.S. will not go to war	War is probable but not inevitable	War is inevitable
U.S. will not invade Libya	1	2	3
U.S. will invade Libya but it would be wrong		4	5
U.S. will invade Libya and it would be right		6	7

The examination of prognoses about the Libyan situation is not the only example of double-barreled questions sneaking into public opinion research. Here are some statements the Harris Poll asked people to agree or disagree with in an attempt to gauge U.S. public opinion about then–Soviet General Secretary Gorbachev:

> He looks like the kind of Russian leader who will recognize that both the Soviets and the Americans can destroy each other with nuclear missiles so it is better to come to verifiable arms control agreements.
>
> He seems to be more modern, enlightened, and attractive, which is a good sign for the peace of the world.
>
> Even though he looks much more modern and attractive, it would be a mistake to think he will be much different from other Russian leaders.

How many elements can you identify in each of the statements? How many possible opinions could people have in each case? What does a simple "agree" or "disagree" really mean in such cases?

Source: Reported in *World Opinion Update,* October 1985 and May 1986, respectively.

some teenagers reported driving hundreds of thousands of miles.

Respondents Must Be Willing to Answer

Often, we would like to learn things from people that they are unwilling to share with us. For example, Yanjie Bian indicates that it has often been difficult to get candid answers from people in China.

> [Here] people are generally careful about what they say on nonprivate occasions in order to survive under authoritarianism. During the Cultural Revolution between 1966 and 1976, for example, because of the radical political agenda and political intensity throughout the country, it was almost impossible to use survey techniques to collect valid and reliable data inside China about the Chinese people's life experiences, characteristics, and attitudes towards the Communist regime. — (1994:19–20)

Sometimes, U.S. respondents may say they're undecided when, in fact, they have an opinion but think they're in a minority. Under that condition, they may be reluctant to tell a stranger (the interviewer) what that opinion is. Given this problem, the Gallup Organization, for example, has used a "secret ballot" format, which simulates actual election conditions, in that the "voter" enjoys complete anonymity. In an analysis of the Gallup Poll election data from 1944 to 1988, Andrew Smith and G. F. Bishop (1992) have found that this technique substantially reduced the percentage of respondents who said they were undecided about how they would vote.

This problem is not limited to survey research, however. Richard Mitchell (1991:100) faced a similar problem in his field research among U.S. survivalists:

> Survivalists, for example, are ambivalent about concealing their identities and inclinations. They realize that secrecy protects them from the ridicule of a disbelieving majority, but enforced sep-

aratism diminishes opportunities for recruitment and information exchange. . . .

> "Secretive" survivalists eschew telephones, launder their mail through letter exchanges, use nicknames and aliases, and carefully conceal their addresses from strangers. Yet once I was invited to group meetings, I found them cooperative respondents.

Questions Should Be Relevant

Similarly, questions asked in a questionnaire should be relevant to most respondents. When attitudes are requested on a topic that few respondents have thought about or really care about, the results are not likely to be useful. Of course, because the respondents may express attitudes even though they have never given any thought to the issue, researchers run the risk of being misled.

This point is illustrated occasionally when researchers ask for responses relating to fictitious people and issues. In one political poll I conducted, I asked respondents whether they were familiar with each of 15 political figures in the community. As a methodological exercise, I made up a name: Tom Sakumoto. In response, 9 percent of the respondents said they were familiar with him. Of those respondents familiar with him, about half reported seeing him on television and reading about him in the newspapers.

When you obtain responses to fictitious issues, you can disregard those responses. But when the issue is real, you may have no way of telling which responses genuinely reflect attitudes and which reflect meaningless answers to an irrelevant question.

Ideally, we would like respondents to simply report that they don't know, have no opinion, or are undecided in those instances where that is the case. Unfortunately, however, they often make up answers.

Short Items Are Best

In the interests of being unambiguous and precise and of pointing to the relevance of an issue, researchers tend to create long and complicated items. That should be avoided. Respondents are often unwilling to study an item in order to understand it. The respondent should be able to read an item quickly, understand its intent, and select or provide an answer without difficulty. In general, assume that respondents will read items quickly and give quick answers. Accordingly, provide clear, short items that will not be misinterpreted under those conditions.

Avoid Negative Items

The appearance of a negation in a questionnaire item paves the way for easy misinterpretation. Asked to agree or disagree with the statement "The United States should not recognize Cuba," a sizable portion of the respondents will read over the word *not* and answer on that basis. Thus, some will agree with the statement when they're in favor of recognition, and others will agree when they oppose it. And you may never know which are which.

Similar considerations apply to other "negative" words. In a study of support for civil liberties, for example, respondents were asked whether they felt "the following kinds of people should be *prohibited* from teaching in public schools" and were presented with a list including such items as a communist, a Ku Klux Klansman, and so forth. The response categories "yes" and "no" were given beside each entry. A comparison of the responses to this item with other items reflecting support for civil liberties strongly suggested that many respondents gave the answer "yes" to indicate willingness for such a person to teach, rather than to indicate that such a person should be prohibited from teaching. (A later study in the series giving as answer categories "permit" and "prohibit" produced much clearer results.)

Avoid Biased Items and Terms

Recall from our discussion of conceptualization and operationalization in Chapter 5 that there are no ultimately true meanings for any of the concepts we typically study in social science. *Prejudice* has no ultimately correct definition; whether a given person is prejudiced depends on our definition of that term. The same general principle applies to the responses we get from people completing a questionnaire.

The meaning of someone's response to a question depends in large part on its wording. This is true of every question and answer. Some questions seem to encourage particular responses more than do other questions. In the context of questionnaires, **bias** refers to any property of questions that encourages respondents to answer in a particular way.

Most researchers recognize the likely effect of a question that begins, "Don't you agree with the President of the United States that . . . " and no reputable researcher would use such an item. Unhappily, the biasing effect of items and terms is far subtler than this example suggests.

The mere identification of an attitude or position with a prestigious person or agency can bias responses. The item "Do you agree or disagree with the recent Supreme Court decision that . . . " would have a similar effect. Such wording may not produce consensus or even a majority in support of the position identified with the prestigious person or agency, but it will likely increase the level of support over what would have been obtained without such identification.

Sometimes the impact of different forms of question wording is relatively subtle. For example, when Kenneth Rasinski (1989) analyzed the results of several General Social Survey studies of attitudes toward government spending, he found that the way programs were identified had an impact on the amount of public support they received. Here are some comparisons:

More Support	Less Support
"Assistance to the poor"	"Welfare"
"Halting rising crime rate"	"Law enforcement"
"Dealing with drug addiction"	"Drug rehabilitation"
"Solving problems of big cities"	"Assistance to big cities"
"Improving conditions of blacks"	"Assistance to blacks"
"Protecting social security"	"Social security"

In 1986, for example, 62.8 percent of the respondents said too little money was being spent on "assistance to the poor," while in a matched survey that year, only 23.1 percent said we were spending too little on "welfare."

In this context, be wary of what researchers call the *social desirability* of questions and answers. Whenever we ask people for information, they answer through a filter of what will make them look good. This is especially true if they're interviewed face-to-face. Thus, for example, a particular man may feel that things would be a lot better if women were kept in the kitchen, excluded from voting, forced to be quiet in public, and so forth. Asked whether he supports equal rights for women, however, he may want to avoid looking like a chauvinist. Recognizing that his views are out of step with current thinking, he may choose to say "yes."

The best way to guard against this problem is to imagine how you would feel giving each of the answers you intend to offer to respondents. If you would feel embarrassed, perverted, inhumane, stupid, irresponsible, or otherwise socially disadvantaged by any particular response, give serious thought to how willing others will be to give those answers.

The biasing effect of particular wording is often difficult to anticipate. In both surveys and experiments, it is sometimes useful to ask respondents to consider hypothetical situations and say how they think they would behave. Because those situations often involve other people, the names used can affect responses. For example, researchers have long known that male names for the hypothetical people may produce different responses than do female names. Research by Joseph Kasof (1993) points to the importance of what the specific names are: whether they generally evoke positive or negative images in terms of attractiveness, age, intelligence, and so forth. Kasof's review of past research suggests there has been a tendency to use more positively valued names for men than for women.

As in all other research, carefully examine the purpose of your inquiry and construct items that will be most useful to it. You should never be misled into thinking there are ultimately "right" and "wrong" ways of asking the questions. When in doubt about the best question to ask, moreover, remember that you should ask more than one.

These, then, are some general guidelines for writing questions to elicit data for analysis and interpretation. Next we look at how to construct questionnaires.

QUESTIONNAIRE CONSTRUCTION

Questionnaires are used in connection with many modes of observation in social research. Although structured questionnaires are essential to and most directly associated with survey research, they are also widely used in experiments, field research, and other data-collection activities. For this reason, questionnaire construction can be an important practical skill for researchers. As we discuss the established techniques for constructing questionnaires, let's begin with some issues of questionnaire format.

General Questionnaire Format

The format of a questionnaire is just as important as the nature and wording of the questions asked. An improperly laid out questionnaire can lead respondents to miss questions, confuse them about the nature of the data desired, and even lead them to throw the questionnaire away.

As a general rule, a questionnaire should be spread out and uncluttered. Inexperienced researchers tend to fear that their questionnaire will look too long; as a result, they squeeze several questions onto a single line, abbreviate questions, and use as few pages as possible. These efforts are ill-advised and even dangerous. Putting more than one question on a line will cause some respondents to miss the second question altogether. Some respondents will misinterpret abbreviated questions. More generally, respondents who find they have spent considerable time on the first page of what seemed a short questionnaire will be more demoralized than respondents who quickly complete the first several pages of what initially seemed a rather long form. Moreover, the latter will have

```
        1. Yes

      (2.) No

        3. Don't know
```

FIGURE 9-1 **Circling the Answer**

made fewer errors and will not have been forced to reread confusing, abbreviated questions. Nor will they have been forced to write a long answer in a tiny space.

The desirability of spreading out questions in the questionnaire cannot be overemphasized. Squeezed-together questionnaires are disastrous, whether completed by the respondents themselves or administered by trained interviewers. And the processing of such questionnaires is another nightmare; I'll have more to say about that in Chapter 14.

Formats for Respondents

In one of the most common types of questionnaire items, the respondent is expected to check one response from a series. For this purpose my experience has been that boxes adequately spaced apart are the best format. Modern word processing makes the use of boxes a practical technique these days; setting boxes in type can also be accomplished easily and neatly. You can approximate boxes by using brackets: [ ], but if you're creating a questionnaire on a computer, you should take the few extra minutes to use genuine boxes that will give your questionnaire a more professional look. Here are some easy examples:

□ ○ ❑

Rather than providing boxes to be checked, you might print a code number beside each response and ask the respondent to circle the appropriate number (see Figure 9-1). This method has the added advantage of specifying the code number to be entered later in the processing stage (see Chapter 14). If numbers are to be circled, however, you should provide clear and prominent instructions to the respondent, because many will be tempted to

cross out the appropriate number, which makes data processing even more difficult. (Note that the technique can be used more safely when interviewers administer the questionnaires, since the interviewers themselves record the responses.)

Contingency Questions

Quite often in questionnaires, certain questions will be relevant to some of the respondents and irrelevant to others. In a study of birth control methods, for instance, you would probably not want to ask men if they take birth control pills.

This sort of situation often arises when researchers wish to ask a series of questions about a certain topic. You may want to ask whether your respondents belong to a particular organization and, if so, how often they attend meetings, whether they have held office in the organization, and so forth. Or, you might want to ask whether respondents have heard anything about a certain political issue and then learn the attitudes of those who have heard of it.

Each subsequent question in series such as these is called a **contingency question:** Whether it is to be asked and answered is contingent on responses to the first question in the series. The proper use of contingency questions can facilitate the respondents' task in completing the questionnaire, because they are not faced with trying to answer questions irrelevant to them.

There are several formats for contingency questions. The one shown in Figure 9-2 is probably the clearest and most effective. Note two key elements in this format. First, the contingency question is isolated from the other questions by being set off to the side and enclosed in a box. Second, an arrow connects the contingency question to the answer on which it is contingent. In the illustration, only those respondents answering yes are expected to answer the contingency question. The rest of the respondents should simply skip it.

Note that the questions shown in Figure 9-2 could have been dealt with in a single question. The question might have read, "How many times, if any, have you smoked marijuana?" The response categories, then, might have read: "Never," "Once,"

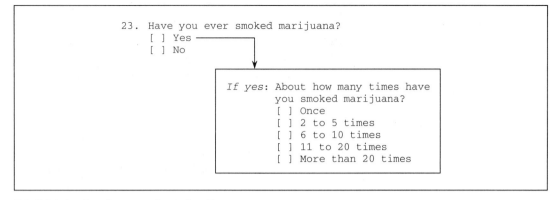

FIGURE 9-2 **Contingency Question Format**

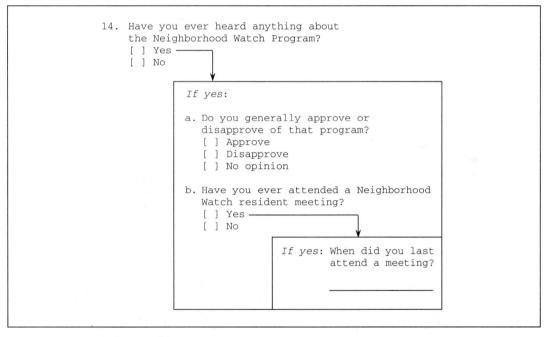

FIGURE 9-3 **Contingency Table**

"2 to 5 times," and so forth. This single question would apply to all respondents, and each would find an appropriate answer category. Such a question, however, might put some pressure on respondents to report having smoked marijuana, because the main question asks how many times they have smoked it, even though it allows for those *exceptional* cases *who have never smoked marijuana even once.* (The emphases used in the

previous sentence give a fair indication of how respondents might read the question.) The contingency question format illustrated in Figure 9-2 should reduce the subtle pressure on respondents to report having smoked marijuana.

Used properly, even rather complex sets of contingency questions can be constructed without confusing the respondent. Figure 9-3 illustrates a more complicated example.

---

13.   Have you ever voted in a national, state, or local election?
[ ] Yes   (Please answer questions 14–25.)
[ ] No   (Please skip questions 14–25. Go directly to question 26 on page 8.)

FIGURE 9-4 **Instructions to Skip**

Sometimes a set of contingency questions is long enough to extend over several pages. Suppose you're studying political activities of college students, and you wish to ask a large number of questions of those students who have voted in a national, state, or local election. You could separate out the relevant respondents with an initial question such as "Have you ever voted in a national, state, or local election?" but it would be confusing to place the contingency questions in a box stretching over several pages. It would make more sense to enter instructions in parentheses after each answer telling respondents to answer or skip the contingency questions. Figure 9-4 provides an illustration of this method.

In addition to these instructions, it is worthwhile to place an instruction at the top of each page containing only the contingency questions. For example, you might say, "This page is only for respondents who have voted in a national, state, or local election." Clear instructions such as these spare respondents the frustration of reading and puzzling over questions that are irrelevant to them and increase the likelihood of responses from those for whom the questions are relevant.

Matrix Questions

Quite often, you'll want to ask several questions that have the same set of answer categories. This is typically the case whenever the Likert response categories are used. In such cases, it is often possible to construct a matrix of items and answers as illustrated in Figure 9-5.

This format offers several advantages over other formats. First, it uses space efficiently. Second, respondents will probably find it faster to complete a set of questions presented in this fash-

ion. In addition, this format may increase the comparability of responses given to different questions for the respondent as well as for the researcher. Because respondents can quickly review their answers to earlier items in the set, they might choose between, say, "strongly agree" and "agree" on a given statement by comparing the strength of their agreement with their earlier responses in the set.

There are some dangers inherent in using this format, however. Its advantages may encourage you to structure an item so that the responses fit into the matrix format when a different, more idiosyncratic set of responses might be more appropriate. Also, the matrix question format can foster a response-set among some respondents: They may develop a pattern of, say, agreeing with all the statements. This would be especially likely if the set of statements began with several that indicated a particular orientation (for example, a liberal political perspective) with only a few later ones representing the opposite orientation. Respondents might assume that all the statements represented the same orientation and, reading quickly, misread some of them, thereby giving the wrong answers. This problem can be reduced somewhat by alternating statements representing different orientations and by making all statements short and clear.

Ordering Items in a Questionnaire

The order in which questionnaire items are presented can also affect responses. First, the appearance of one question can affect the answers given to later ones. For example, if several questions have been asked about the dangers of terrorism to the United States and then a question asks respondents to list (open-ended) things that they believe represent dangers to the United States, terrorism

17. Beside *each* of the statements presented below, please indicate whether you Strongly Agree (SA), Agree (A), Disagree (D), Strongly Disagree (SD), or are Undecided (U).

	SA	A	D	SD	U
a. What this country needs is more law and order.........................	[ ]	[ ]	[ ]	[ ]	[ ]
b. The police should be disarmed in America..........................	[ ]	[ ]	[ ]	[ ]	[ ]
c. During riots, looters should be shot on sight	[ ]	[ ]	[ ]	[ ]	[ ]
etc.					

FIGURE 9-5 **Matrix Question Format**

will receive more citations than would otherwise be the case. In this situation, asking the open-ended question first is best.

Similarly, if respondents are asked to assess their overall religiosity ("How important is your religion to you in general?"), their responses to later questions concerning specific aspects of religiosity will be aimed at consistency with the prior assessment. The converse is true as well. If respondents are first asked specific questions about different aspects of their religiosity, their subsequent overall assessment will reflect the earlier answers.

The impact of item order is not uniform among respondents. When J. Edwin Benton and John Daly (1991) conducted a local government survey, they found that the less-educated respondents were more influenced by the order of questionnaire items than were those with more education.

Some researchers attempt to overcome this effect by randomizing the order of items. This effort is usually futile. In the first place, a randomized set of items will probably strike respondents as chaotic and worthless. The random order also makes it more difficult for respondents to answer, because they must continually switch their attention from one topic to another. Finally, even a randomized ordering of items will have the effect discussed previously—except that researchers will have no control over the effect.

The safest solution is sensitivity to the problem. Although you cannot avoid the effect of item order,

try to estimate what that effect will be so that you can interpret results meaningfully. If the order of items seems especially important in a given study, you might construct more than one version of the questionnaire with different orderings of the items in each. You will then be able to determine the effects by comparing responses to the various versions. At the very least, you should pretest your questionnaire in the different forms. (We'll discuss pretesting in a moment.)

The desired ordering of items differs between interviews and self-administered questionnaires. In the latter, it's usually best to begin the questionnaire with the most interesting set of items. The potential respondents who glance casually over the first few items should want to answer them. Perhaps the items will ask for attitudes they're aching to express. At the same time, however, the initial items should not be threatening. (It might be a bad idea to begin with items about sexual behavior or drug use.) Requests for duller, demographic data (age, gender, and the like) should generally be placed at the end of a self-administered questionnaire. Placing these items at the beginning, as many inexperienced researchers are tempted to do, gives the questionnaire the initial appearance of a routine form, and the person receiving it may not be motivated to complete it.

Just the opposite is generally true for interview surveys. When the potential respondent's door first opens, the interviewer must gain rapport quickly.

After a short introduction to the study, the interviewer can best begin by enumerating the members of the household, getting demographic data about each. Such items are easily answered and generally nonthreatening. Once the initial rapport has been established, the interviewer can then move into more sensitive matters. An interview that began with the question "Do you believe in witchcraft?" would probably end rather quickly.

Questionnaire Instructions

Every questionnaire, whether it is to be completed by respondents or administered by interviewers, should contain clear instructions and introductory comments where appropriate.

It's useful to begin every self-administered questionnaire with basic instructions for completing it. Although many people these days have experience with forms and questionnaires, begin by telling them exactly what you want: that they are to indicate their answers to certain questions by placing a check mark or an X in the box beside the appropriate answer or by writing in their answer when asked to do so. If many open-ended questions are used, respondents should be given some guidelines about whether brief or lengthy answers are expected. If you wish to encourage your respondents to elaborate on their responses to closed-ended questions, that should be noted.

If a questionnaire has subsections—political attitudes, religious attitudes, background data—introduce each with a short statement concerning its content and purpose. For example, "In this section, we would like to know what people consider the most important community problems." Demographic items at the end of a self-administered questionnaire might be introduced thus: "Finally, we would like to know just a little about you so we can see how different types of people feel about the issues we have been examining."

Short introductions such as these help the respondent make sense of the questionnaire. They make the questionnaire seem less chaotic, especially when it taps a variety of data. And they help put the respondent in the proper frame of mind for answering the questions.

APPLYING THE RESULTS

Many of the guidelines for the construction of survey questionnaires apply equally to the construction of forms in general. You've most likely been asked at one time or another to fill out a form that you found particularly frustrating. If you could examine it to see what made it so galling, I suspect you would find that the form violated one or more of these guidelines.

It's useful to see the forms we fill out as a kind of questionnaire, and your ability to construct a professional questionnaire would qualify you as a first-rate form-maker. While you may not regard this as exciting as being an Olympic athlete or rock star (it's not), there are more job openings in making forms.

Some questions may require special instructions to facilitate proper answering. This is especially true if a given question varies from the general instructions pertaining to the whole questionnaire. Some specific examples will illustrate this situation.

Despite attempts to provide mutually exclusive answers in closed-ended questions, often more than one answer will apply for respondents. If you want a single answer, you should make this perfectly clear in the question. An example would be "From the list below, please check the primary reason for your decision to attend college." Often the main question can be followed by a parenthetical note: "Please check the one best answer." If, on the other hand, you want the respondent to check as many answers as apply, you should make this clear.

When a set of answer categories are to be rank-ordered by the respondent, the instructions should indicate this, and a different type of answer format should be used (for example, blanks instead of boxes). These instructions should indicate how many answers are to be ranked (for example: all;

only the first and second; only the first and last; the most important and least important). These instructions should also spell out the order of ranking (for example, "Place a 1 beside the most important item, a 2 beside the next most important, and so forth"). Rank-ordering of responses is often difficult for respondents, however, because they may have to read and reread the list several times, so this technique should only be used in those situations where no other method will produce the desired result.

In multiple-part matrix questions, it's useful to give special instructions unless the same format is used throughout the questionnaire. Sometimes respondents will be expected to check one answer in each column of the matrix; in other questionnaires they'll be expected to check one answer in each row. Whenever the questionnaire contains both formats, it's useful to add an instruction clarifying which is expected in each case.

Pretesting the Questionnaire

No matter how carefully researchers design a data-collection instrument such as a questionnaire, there is always the possibility—indeed the certainty—of error. They will always make some mistake: an ambiguous question, one that people cannot answer, or some other violation of the rules just discussed.

The surest protection against such errors is to pretest the questionnaire in full or in part. Give the questionnaire to the ten people in your bowling league, for example. It's not usually essential that the pretest subjects comprise a representative sample, although you should use people for whom the questionnaire is at least relevant.

By and large, it's better to ask people to complete the questionnaire than to read through it looking for errors. All too often, a question seems to make sense on a first reading but proves to be impossible to answer.

Stanley Presser and Johnny Blair (1994) describe several different pretesting strategies and report on the effectiveness of each. They also provide data on the cost of the various methods.

There are many more tips and guidelines for questionnaire construction, but covering them all would take a book in itself. Now I'll complete this discussion with an illustration of a real questionnaire, showing how some of these comments find substance in practice.

Before turning to the illustration, however, I want to mention a critical aspect of questionnaire design that I discuss in Chapter 14: precoding. Because the information collected by questionnaires is typically transformed into some type of computer format, it's usually appropriate to include data-processing instructions on the questionnaire itself. These instructions indicate where specific pieces of information will be stored in the machine-readable data files. In Chapter 14, I'll discuss the nature of such storage and point out appropriate questionnaire notations. As a preview, however, notice that the following illustration has been precoded with the mysterious numbers that appear near questions and answer categories.

A Composite Illustration

Figure 9-6 is part of a questionnaire used by the University of Chicago's National Opinion Research Center in its General Social Survey. The questionnaire deals with people's attitudes toward the government and is designed to be self-administered.

SELF-ADMINISTERED QUESTIONNAIRES

So far we've discussed how to formulate questions and how to design effective questionnaires. As important as these tasks are, the labor will be wasted unless the questionnaire produces useful data—which means that respondents actually complete the questionnaire. We turn now to the major methods for getting responses to questionnaires.

I've referred several times in this chapter to interviews versus self-administered questionnaires. Actually, there are three main methods of administering survey questionnaires to a sample of respondents: self-administered questionnaires, in which respondents are asked to complete the questionnaire themselves; surveys administered

10. Here are some things the government might do for the economy. Circle one number for each action to show whether you are in favor of it or against it.

> 1. Strongly in favor of
> 2. In favor of
> 3. Neither in favor of nor against
> 4. Against
> 5. Strongly disagree

PLEASE CIRCLE A NUMBER

a. Control of wages by legislation	1	2	3	4	5	28/
b. Control of prices by legislation	1	2	3	4	5	29/
c. Cuts in government spending	1	2	3	4	5	30/
d. Government financing of projects to create new jobs	1	2	3	4	5	31/
e. Less government regulation of business ...	1	2	3	4	5	32/
f. Support for industry to develop new products and technology	1	2	3	4	5	33/
g. Supporting declining industries to protect jobs	1	2	3	4	5	34/
h. Reducing the work week to create more jobs	1	2	3	4	5	35/

11. Listed below are various areas of government spending. Please indicate whether you would like to see more or less government spending in each area. Remember that if you say "much more," it might require a tax increase to pay for it.

> 1. Spend much more
> 2. Spend more
> 3. Spend the same as now
> 4. Spend less
> 5. Spend much less
> 8. Can't choose

PLEASE CIRCLE A NUMBER

a. The environment	1	2	3	4	5	8	36/
b. Health	1	2	3	4	5	8	37/
c. The police and law enforcement	1	2	3	4	5	8	38/
d. Education	1	2	3	4	5	8	39/
e. The military and defense	1	2	3	4	5	8	40/
f. Retirement benefits	1	2	3	4	5	8	41/
g. Unemployment benefits	1	2	3	4	5	8	42/
h. Culture and the arts	1	2	3	4	5	8	43/

12. If the government *had* to choose between keeping down inflation or keeping down unemployment, to which do you think it should give highest priority?
    Keeping down inflation 1   44/
    Keeping down unemployment 2
    Can't choose 8

13. Do you think that labor unions in this country have too much power or too little power?
    Far too much power 1   45/
    Too much power 2
    About the right amount of power 3
    Too little power 4
    Far too little power 5
    Can't choose 8

FIGURE 9-6 **A Sample Questionnaire**

14. How about business and industry, do they have too much power or too little power?

    Far too much power 1   46/
    Too much power 2
    About the right amount of power 3
    Too little power 4
    Far too little power 5
    Can't choose 8

15. And what about the federal government, does it have too much power or too little power?

    Far too much power 1   47/
    Too much power 2
    About the right amount of power 3
    Too little power 4
    Far too little power 5
    Can't choose 8

16. In general, how good would you say labor unions are for the country as a whole?

    Excellent 1   48/
    Very good 2
    Fairly good 3
    Not very good 4
    Not good at all 5
    Can't choose 8

17. What do you think the government's role in each of these industries should be?

    ```
    1. Own it
    2. Control prices and profits
       but not own it
    3. Neither own it nor control its
       prices and profits
    8. Can't choose
    ```

    PLEASE CIRCLE A
    NUMBER

a. Electric power	1	2	3	8	49/
b. The steel industry	1	2	3	8	50/
c. Banking and insurance	1	2	3	8	51/

18. On the whole, do you think it should or should not be the government's responsibility to . . .

    ```
    1. Definitely should be
    2. Probably should be
    3. Probably should not be
    4. Definitely should not be
    8. Can't choose
    ```

    PLEASE CIRCLE A NUMBER

a. Provide a job for everyone who wants one	1	2	3	4	8	52/
b. Keep prices under control	1	2	3	4	8	53/
c. Provide health care for the sick	1	2	3	4	8	54/
d. Provide a decent standard of living for the old	1	2	3	4	8	55/

FIGURE 9-6   **A Sample Questionnaire** (*continued*)

e. Provide industry with the help it needs to grow...............................	1	2	3	4	8	56/
f. Provide a decent standard of living for the unemployed............................	1	2	3	4	8	57/
g. Reduce income differences between the rich and poor..........................	1	2	3	4	8	58/
h. Give financial assistance to college students from low-income families..........	1	2	3	4	8	59/
i. Provide decent housing for those who can't afford it..........................	1	2	3	4	8	60/

19. How interested would you say you personally are in politics?

Very interested 1	61/
Fairly interested 2	
Somewhat interested 3	
Not very interested 4	
Not at all interested 5	
Can't choose 8	

20. Here are some other areas of government spending. Please indicate whether you would like to see more or less government spending in each area. Remember that if you say "much more," it might require a tax increase to pay for it.

> 1. Spend much more
> 2. Spend more
> 3. Spend the same as now
> 4. Spend less
> 5. Spend much less
> 8. Can't choose

<u>PLEASE CIRCLE A NUMBER</u>

a. Prenatal care for pregnant mothers who can't afford it....................	1	2	3	4	5	8	62/
b. Health care for children whose families don't have insurance..........	1	2	3	4	5	8	63/
c. Preschool programs like Head Start for poor children.....................	1	2	3	4	5	8	64/
d. Child care for poor children...........	1	2	3	4	5	8	65/
e. Child care for all children with working parents......................	1	2	3	4	5	8	66/
f. Housing for poor families with children.............................	1	2	3	4	5	8	67/
g. Services for disabled and chronically ill children...........................	1	2	3	4	5	8	68/
h. Drug abuse prevention and treatment for children and youth..................	1	2	3	4	5	8	69/
i. Nutrition programs for poor children and families, such as food stamps and school lunches....................	1	2	3	4	5	8	70/
j. Contraceptive services for teenagers.............................	1	2	3	4	5	8	71/

THANK YOU VERY MUCH FOR COMPLETING THE QUESTIONNAIRE

FIGURE 9-6   **A Sample Questionnaire** (*continued*)

by interviewers in face-to-face encounters; and surveys conducted by telephone. This section and the next two discuss each of these methods in turn.

The most common form of self-administered questionnaire is the mail survey. However, there are several other techniques that are often used as well. At times, it may be appropriate to administer a questionnaire to a group of respondents gathered at the same place at the same time. A survey of students taking introductory psychology might be conducted in this manner during class. High school students might be surveyed during homeroom period.

Some recent experimentation has been conducted with regard to the home delivery of questionnaires. A research worker delivers the questionnaire to the home of sample respondents and explains the study. Then the questionnaire is left for the respondent to complete, and the researcher picks it up later.

Home delivery and the mail can also be used in combination. Questionnaires are mailed to families, and then research workers visit homes to pick up the questionnaires and check them for completeness. Or, research workers can hand deliver questionnaires with a request that the respondents mail the completed questionnaires to the research office.

On the whole, when a research worker either delivers the questionnaire, picks it up, or both, the completion rate seems higher than for straightforward mail surveys. Additional experimentation with this technique is likely to point to other ways to improve completion rates while reducing costs. The remainder of this section, however, is devoted specifically to the mail survey, which is still the typical form of self-administered questionnaire.

Mail Distribution and Return

The basic method for collecting data through the mail has been to send a questionnaire accompanied by a letter of explanation and a self-addressed, stamped envelope for returning the questionnaire. The respondent is expected to complete the questionnaire, put it in the envelope, and return it. If, by any chance, you've received such a questionnaire and failed to return it, it would be valuable to recall the reasons you had for not returning it and keep them in mind any time you plan to send questionnaires to others.

A common reason for not returning questionnaires is that it's too much trouble. To overcome this problem, researchers have developed several ways to make returning them easier. For instance, a self-mailing questionnaire requires no return envelope: When the questionnaire is folded a particular way, the return address appears on the outside. The respondent therefore doesn't have to worry about losing the envelope.

More-elaborate designs are available also. The university student questionnaire described later in this chapter was bound in a booklet with a special, two-panel back cover. Once the questionnaire was completed, the respondent needed only to fold out the extra panel, wrap it around the booklet, and seal the whole thing with the adhesive strip running along the edge of the panel. The foldout panel contained my return address and postage. When I repeated the study a couple of years later, I improved on the design. Both the front and back covers had foldout panels: one for sending the questionnaire out and the other for getting it back —thus avoiding the use of envelopes altogether.

The point here is that anything you can do to make the job of completing and returning the questionnaire easier will improve your study. Imagine receiving a questionnaire that made no provisions for its return to the researcher. Suppose you had to (1) find an envelope, (2) write the address on it, (3) figure out how much postage it required, and (4) put the stamps on it. How likely is it that you would return the questionnaire?

A few brief comments on postal options are in order. You have options for mailing questionnaires out and for getting them returned. On outgoing mail, your choices are essentially between first-class postage and bulk rate. First class is more certain, but bulk rate is far cheaper. (Check your local post office for rates and procedures.) On return mail, your choice is between postage stamps and business-reply permits. Here, the cost differential is more complicated. If you use stamps, you pay for them whether people return their questionnaires or not. With the business-reply permit, you pay for only those that are used, but you pay an additional surcharge of about a nickel. This means that stamps are cheaper if a lot of questionnaires are returned, but business-reply permits are cheaper if fewer are returned (and you won't know in advance how many will be returned).

There are many other considerations involved in choosing among the several postal options. Some researchers, for example, feel that the use of postage stamps communicates more "humanness" and sincerity than do bulk rate and business-reply permits. Others worry that respondents will steam off the stamps and use them for some purpose other than returning the questionnaires. Because both bulk rate and business-reply permits require establishing accounts at the post office, you'll probably find stamps much easier in small surveys.

Monitoring Returns

The mailing of questionnaires sets up a new research question that may prove valuable to a study. Researchers shouldn't sit back idly as questionnaires are returned; instead, they should undertake a careful recording of the varying rates of return among respondents.

An invaluable tool in this activity is a return-rate graph. The day on which questionnaires were mailed is labeled Day 1 on the graph, and every day thereafter the number of returned questionnaires is logged on the graph. It's usually best to compile two graphs. One shows the number returned each day—rising, then dropping. The second reports the cumulative number or percentage. In part, this activity provides the researchers with gratification, as they get to draw a picture of their successful data collection. More important, however, it is their guide to how the data collection is going. If follow-up mailings are planned, the graph provides a clue about when such mailings should be launched. (The dates of subsequent mailings should be noted on the graph.)

As completed questionnaires are returned, each should be opened, scanned, and assigned an identification (ID) number. These numbers should be assigned serially as the questionnaires are returned, even if other ID numbers have already been assigned. Two examples should illustrate the important advantages of this procedure.

Let's assume you're studying attitudes toward a political figure. In the middle of the data collection, the media break the story that the politician is having extramarital affairs. By knowing the date of that public disclosure and the dates when questionnaires were received, you'll be in a position to determine the effects of the disclosure. (Recall the discussion in Chapter 8 of history in connection with experiments.)

In a less sensational way, serialized ID numbers can be valuable in estimating nonresponse biases in the survey. Barring more direct tests of bias, you may wish to assume that those who failed to answer the questionnaire will be more like respondents who delayed answering than like those who answered right away. An analysis of questionnaires received at different points in the data collection might then be used for estimates of sampling bias. For example, if the GPAs reported by student respondents decrease steadily through the data collection, with those replying right away having higher GPAs and those replying later having lower GPAs, you might tentatively conclude that those who failed to answer at all have lower GPAs yet. Although it would not be advisable to make statistical estimates of bias in this fashion, you could take advantage of approximate estimates based on the patterns you've observed.

If respondents have been identified for purposes of follow-up mailing, then preparations for those mailings should be made as the questionnaires are returned. The case study later in this section discusses this process in greater detail.

Follow-up Mailings

Follow-up mailings may be administered in several ways. In the simplest, nonrespondents are simply sent a letter of additional encouragement to participate. A better method, however, is to send a new copy of the survey questionnaire with the follow-up letter. If potential respondents have not returned their questionnaires after two or three weeks, the questionnaires have probably been lost or misplaced. Receiving a follow-up letter might encourage them to look for the original questionnaire, but if they can't find it easily, the letter may go for naught.

The methodological literature strongly suggests that follow-up mailings provide an effective method for increasing return rates in mail surveys. In general, the longer a potential respondent delays replying, the less likely he or she is to do so at all. Properly timed follow-up mailings, then, provide additional stimuli to respond.

The effects of follow-up mailings will be seen in the response-rate curves recorded during data collection. The initial mailings will be followed by a rise and subsequent fall of returns; the follow-up mailings will spur a resurgence of returns; and more follow-ups will do the same. In practice, three mailings (an original and two follow-ups) seem the most efficient.

The timing of follow-up mailings is also important. Here the methodological literature offers less precise guides, but it has been my experience that two or three weeks is a reasonable space between mailings. (This period might be increased by a few days if the mailing time—out and in—is more than two or three days.)

When researchers conduct several surveys of the same population over time, they can develop more-specific guidelines. The Survey Research Office at the University of Hawaii conducts frequent student surveys and has been able to refine the mailing and remailing procedure considerably. Indeed, they have found a consistent pattern of returns that appears to transcend differences of survey content, quality of instrument, and so forth. Within two weeks of the first mailing, approximately 40 percent of the questionnaires are returned; within two weeks of the first follow-up, an additional 20 percent are received; and within two weeks of the final follow-up, an additional 10 percent are received. (These response rates reflect the sending of additional questionnaires, not just letters.) Your results may vary, but this illustration should indicate the value of carefully tabulating return rates for every survey conducted.

If the individuals in the survey sample are not identified on the questionnaires, it may not be possible to remail only to nonrespondents. In such a case, send your follow-up mailing to all members of the sample, thanking those who may have already participated and encouraging those who have not to do so. (The case study reported later describes another method you can use in an anonymous mail survey.)

Acceptable Response Rates

A question that new survey researchers frequently ask concerns the percentage return rate, or the **response rate,** that should be achieved in a mail survey. The body of inferential statistics used in connection with survey analysis assumes that all members of the initial sample complete and return their questionnaires. Because this almost never happens, response bias becomes a concern, with the researcher testing (and hoping) for the possibility that the respondents look essentially like a random sample of the initial sample, and thus a somewhat smaller random sample of the total population. (For more detailed discussions of response bias, you might want to read Donald [1960] and Brownlee [1975].)

Nevertheless, overall response rate is one guide to the representativeness of the sample respondents. If a high response rate is achieved, there is less chance of significant response bias than in a low rate. Conversely, a low response rate is a danger signal, because the nonrespondents are likely to differ from the respondents in ways other than just their willingness to participate in your survey. Richard Bolstein (1991), for example, found that those who did not respond to a preelection political poll were less likely to vote than those who did participate. Estimating the turnout rate from the survey respondents, then, would have overestimated the number who would show up at the polls.

But what is a high or low response rate? A quick review of the survey literature will uncover a wide range of response rates. Each of these may be accompanied by a statement like "This is regarded as a relatively high response rate for a survey of this type." (A U.S. senator made this statement regarding a poll of constituents that achieved a 4-percent return rate.) Even so, it's possible to state some rules of thumb about return rates. I believe that a response rate of 50 percent is adequate for analysis and reporting. A response of 60 percent is good; a response rate of 70 percent is very good. Bear in mind, however, that these are only rough guides; they have no statistical basis, and a demonstrated lack of response bias is far more important than a high response rate. If you want to pursue this matter further, Delbert Miller (1991:145–55) has reviewed several specific surveys to offer a better sense of the variability of response rates.

As you can imagine, one of the more persistent discussions among survey researchers concerns ways of increasing response rates. You'll recall that this was a chief concern in the earlier discussion of options for mailing out and receiving questionnaires. Survey researchers have developed many ingenious techniques addressing this problem. Some have experimented with novel formats. Others have tried paying respondents to participate. The problem with paying, of course, is that it's expensive to make meaningfully high payment to hundreds or thousands of respondents, but some imaginative alternatives have been used. Some researchers have said, "We want to get your two-cents' worth on some issues, and we're willing to pay"—enclosing two pennies. Another enclosed a quarter, suggesting that the respondent make some little child happy. Still others have enclosed paper money.

Don Dillman (1978) provides an excellent review of the various techniques that survey researchers have used to increase return rates on mail surveys, and he evaluates the impact of each. More important, Dillman stresses the necessity of paying attention to all aspects of the study—what he calls the "Total Design Method"—rather than one or two special gimmicks.

More recently, Francis Yammarino, Steven Skinner, and Terry Childers (1991) have undertaken an in-depth analysis of the response rates achieved in many studies using different techniques. Their findings are too complex to summarize easily, but you might find some guidance there for effective survey design.

A Case Study

The steps involved in the administration of a mail survey are many and can best be appreciated in a walk-through of an actual study. Accordingly, this section concludes with a detailed description of how the student survey we discussed in Chapter 7 as an illustration of systematic sampling was administered. This study did not represent the theoretical ideal for such studies, but in that regard it serves present purposes all the better. The study was conducted by the students in my graduate seminar in survey research methods.

As you may recall, 1,100 students were selected from the university registration tape through a stratified, systematic sampling procedure. For each student selected, six self-adhesive mailing labels were printed by the computer.

By the time we were ready to distribute the questionnaires, it became apparent that our meager research funds wouldn't cover several mailings to the entire sample of 1,100 students (questionnaire printing costs were higher than anticipated). As a result, we chose a systematic two-thirds sample of the mailing labels, yielding a subsample of 773 students.

Earlier, we had decided to keep the survey anonymous in the hope of encouraging candid responses to some sensitive questions. (Later surveys of the same issues among the same population indicated that this anonymity was unnecessary.) Thus, the questionnaires would carry no identification of students on them. At the same time, we hoped to reduce the follow-up mailing costs by mailing only to nonrespondents.

To achieve both of these aims, a special post-card method was devised. Each student was mailed a questionnaire that carried no identifying marks, plus a postcard addressed to the research office—with one of the student's mailing labels affixed to the reverse side of the card. The introductory letter asked the student to complete and return the questionnaire—assuring anonymity—and to return the postcard simultaneously. Receiving the postcard would tell us—without indicating which questionnaire it was—that the student had returned his or her questionnaire. This procedure would then facilitate follow-up mailings.

The 32-page questionnaire was printed in booklet form. The three-panel cover described earlier in this chapter permitted the questionnaire to be returned without an additional envelope.

A letter introducing the study and its purposes was printed on the front cover of the booklet. It explained why the study was being conducted (to learn how students feel about a variety of issues), how students had been selected for the study, the importance of each student's responding, and the mechanics of returning the questionnaire.

Students were assured that their responses to the survey were anonymous, and the postcard method was explained. A statement followed about the auspices under which the study was being conducted, and a telephone number was provided for those who might want more information about the study. (Five students called for information.)

By printing the introductory letter on the questionnaire, we avoided the necessity of enclosing a separate letter in the outgoing envelope, thereby simplifying the task of assembling mailing pieces.

The materials for the initial mailing were assembled as follows. (1) One mailing label for each student was stuck on a postcard. (2) Another label was stuck on an outgoing manila envelope. (3) One postcard and one questionnaire were placed in each envelope—with a glance to ensure that the name on the postcard and on the envelope were the same in each case.

The distribution of the survey questionnaires had been set up for a bulk rate mailing. Once the questionnaires had been stuffed into envelopes, they were grouped by zip code, tied in bundles, and delivered to the post office.

Shortly after the initial mailing, questionnaires and postcards began arriving at the research office. Questionnaires were opened, scanned, and assigned identification numbers as described earlier in this chapter. For every postcard received, a search was made for that student's remaining labels, and they were destroyed.

After two or three weeks, the remaining mailing labels were used to organize a follow-up mailing. This time a special, separate letter of appeal was included in the mailing piece. The new letter indicated that many students had returned their questionnaires already, and it was very important for all others to do so as well.

The follow-up mailing stimulated a resurgence of returns, as expected, and the same logging procedures were continued. The returned postcards told us which additional mailing labels to destroy.

Unfortunately, time and financial pressures made it impossible to undertake a third mailing, as had been initially planned, but the two mailings resulted in an overall return rate of 62 percent.

This illustration should give you a fairly good sense of what's involved in the execution of mailed self-administered questionnaires. Let's turn now to the second principal method of conducting surveys, in-person interviews.

INTERVIEW SURVEYS

The **interview** is an alternative method of collecting survey data. Rather than asking respondents to read questionnaires and enter their own answers, researchers send interviewers to ask the questions orally and record respondents' answers. Interviewing is typically done in a face-to-face encounter, but telephone interviewing, discussed in the next section, follows most of the same guidelines.

Most interview surveys require more than one interviewer, although you might undertake a small-scale interview survey yourself. Portions of this section will discuss methods for training and supervising a staff of interviewers assisting you with a survey.

This section deals specifically with survey interviewing. Chapter 10 discusses the less structured, in-depth interviews often conducted in qualitative field research.

The Role of the Survey Interviewer

There are several advantages to having a questionnaire administered by an interviewer rather than a respondent. To begin with, interview surveys typically attain higher response rates than do mail surveys. A properly designed and executed interview survey ought to achieve a completion rate of at least 80 to 85 percent. (Federally funded surveys often require one of these response rates.) Respondents seem more reluctant to turn down an interviewer standing on their doorstep than to throw away a mailed questionnaire.

The presence of an interviewer also generally decreases the number of "don't knows" and "no answers." If minimizing such responses is important to the study, the interviewer can be instructed to probe for answers ("If you had to pick one of the answers, which do you think would come closest to your feelings?").

Interviewers can also serve as a guard against confusing questionnaire items. If the respondent clearly misunderstands the intent of a question, the interviewer can clarify matters, thereby obtaining relevant responses. (As we'll discuss shortly, such clarifications must be strictly controlled through formal specifications.)

Finally, the interviewer can observe respondents as well as ask questions. For example, the interviewer can note the respondent's race if this is considered too delicate a question to ask. Similar observations can be made regarding the quality of the dwelling, the presence of various possessions, the respondent's ability to speak English, the respondent's general reactions to the study, and so forth. In one survey of students, respondents were given a short, self-administered questionnaire to complete—concerning sexual attitudes and behavior—during the course of the interview. While a student completed the questionnaire, the interviewer made detailed notes regarding the dress and grooming of the respondent.

This procedure raises an ethical issue. Some researchers have objected that such practices violate the spirit of the agreement by which the respondent has allowed the interview. Although ethical issues seldom are clear-cut in social research, it's important to be sensitive to them (see Chapter 3).

Survey research is of necessity based on an unrealistic stimulus-response theory of cognition and behavior. Researchers must assume that a questionnaire item will mean the same thing to every respondent, and every given response must mean the same when given by different respondents. Although this is an impossible goal, survey questions are drafted to approximate the ideal as closely as possible.

The interviewer must also fit into this ideal situation. The interviewer's presence should not affect a respondent's perception of a question or the answer given. In other words, the interviewer should be a neutral medium through which questions and answers are transmitted.

As such, different interviewers should obtain exactly the same responses from a given respondent. (Recall our earlier discussions of reliability.) This neutrality has a special importance in area samples. To save time and money, a given interviewer is typically assigned to complete all the interviews in a particular geographical area—a city block or a group of nearby blocks. If the interviewer does anything to affect the responses obtained, the bias thus interjected might be interpreted as a characteristic of that area.

Let's suppose that a survey is being done to determine attitudes toward low-cost housing in order to help in the selection of a site for a new government-sponsored development. An interviewer assigned to a given neighborhood might—through word or gesture—communicate his or her own distaste for low-cost housing developments. Respondents might therefore tend to give responses in general agreement with the interviewer's own position. The results of the survey would indicate that the neighborhood in question strongly resists construction of the development in its area when in fact their apparent resistance simply reflects the interviewer's attitudes.

General Guidelines for Survey Interviewing

The manner in which interviews ought to be conducted will vary somewhat by survey population and will be affected to some degree by the nature of the survey content. Nevertheless, some general guidelines apply to most interviewing situations.

Appearance and Demeanor As a rule, interviewers should dress in a fashion similar to that of the people they'll be interviewing. A richly dressed interviewer will probably have difficulty getting good cooperation and responses from poorer respondents; a poorly dressed interviewer will have similar difficulties with richer respondents. To the extent that the interviewer's dress and grooming differ from those of the respondents, it should be in the direction of cleanliness and neatness in modest apparel. If cleanliness is not next to godliness, it appears to be next to neutrality. Although middle-class neatness and cleanliness may not be accepted by all sectors of U.S. society, they remain the primary norm and are the most likely to be acceptable to the largest number of respondents.

Dress and grooming are typically regarded as signs of a person's attitudes and orientations. At the time this is being written, torn jeans, green hair, and razor blade earrings may communicate—correctly or incorrectly—that the interviewer is politically radical, sexually permissive, favorable to drug use, and so forth. Any of these impressions could bias responses or affect the willingness of people to be interviewed.

In demeanor, interviewers should be pleasant if nothing else. Because they'll be prying into a respondent's personal life and attitudes, they must communicate a genuine interest in getting to know the respondent without appearing to spy. They must be relaxed and friendly without being too casual or clinging. Good interviewers also have the ability to determine very quickly the kind of person the respondent will feel most comfortable with, the kind of person the respondent would most enjoy talking to. Clearly, the interview will be more successful if the interviewer can become the kind of person the respondent is comfortable with. Further, because respondents are asked to volunteer a portion of their time and to divulge personal information, they deserve the most enjoyable experience the researcher and interviewer can provide.

Familiarity with Questionnaire If an interviewer is unfamiliar with the questionnaire, the study suffers and an unfair burden is placed on the respondent. The interview is likely to take more time than necessary and be unpleasant. Moreover, the interviewer cannot acquire familiarity by skimming through the questionnaire two or three times. He or she must study it carefully, question by question, and must practice reading it aloud.

Ultimately, the interviewer must be able to read the questionnaire items to respondents without error, without stumbling over words and phrases. A good model is the actor reading lines in a play or movie. The lines must be read as though they constituted a natural conversation, but that conversation must follow exactly the language set down in the questionnaire.

By the same token, the interviewer must be familiar with the specifications prepared in conjunction with the questionnaire. Inevitably some questions will not exactly fit a given respondent's situation, and the interviewer must determine how the question should be interpreted in that situation. The specifications provided to the interviewer should give adequate guidance in such cases, but the interviewer must know the organization and contents of the specifications well enough to refer to them efficiently. It would be better for the interviewer to leave a given question unanswered than to spend five minutes searching through the specifications for clarification or trying to interpret the relevant instructions.

Following Question Wording Exactly The first part of this chapter discussed the significance of question wording for the responses obtained. A slight change in the wording of a given question may lead a respondent to answer "yes" rather than "no." It follows that interviewers must be instructed to follow the wording of questions exactly. Otherwise all the effort that the developers have put into carefully phrasing the questionnaire items to obtain the information they need and to ensure that respondents interpret items precisely as intended will be wasted.

Recording Responses Exactly Whenever the questionnaire contains open-ended questions, those soliciting the respondent's own answer, the interviewer must record that answer exactly as given. No attempt should be made to summarize, paraphrase, or correct bad grammar.

This exactness is especially important because the interviewer will not know how the responses are to be coded. Indeed, the researchers themselves may not know the coding until they've read a hundred or so responses. For example, the questionnaire might ask respondents how they feel about the traffic situation in their community. One respondent might answer that there are too many cars on the roads and that something should be done to limit their numbers. Another might say that more roads are needed. If the interviewer recorded these two responses with the same summary— "congested traffic"—the researchers would not be able to take advantage of the important differences in the original responses.

Sometimes, verbal responses are too inarticulate or ambiguous to permit interpretation. However, the interviewer may be able to understand the intent of the response through the respondent's gestures or tone. In such a situation, the interviewer should still record the exact verbal response but also add marginal comments giving both the interpretation and the reasons for arriving at it.

More generally, researchers can use any marginal comments explaining aspects of the response not conveyed in the verbal recording, such as the respondent's apparent anger, embarrassment, uncertainty in answering, and so forth. In each case, however, the exact verbal response should also be recorded.

Probing for Responses Sometimes respondents in an interview will give an inappropriate or incomplete answer. In such cases, a **probe,** or request for an elaboration, can be useful. For example, a closed-ended question may present an attitudinal statement and ask the respondent to strongly agree, agree somewhat, disagree somewhat, or strongly disagree. The respondent, however, may reply: "I think that's true." The interviewer should follow this reply with: "Would you say you strongly agree or agree somewhat?" If necessary, interviewers can explain that they must check one or the other of the categories provided. If the respondent adamantly refuses to choose, the interviewer should write in the exact response given by the respondent.

Probes are more frequently required in eliciting responses to open-ended questions. For example, in response to a question about traffic conditions, the respondent might simply reply, "Pretty bad." The interviewer could obtain an elaboration on this response through a variety of probes. Sometimes the best probe is silence; if the interviewer sits quietly with pencil poised, the respondent will probably fill the pause with additional comments. (This technique is used effectively by newspaper reporters.) Appropriate verbal probes might be "How is that?" or "In what ways?" Perhaps the most generally useful probe is "Anything else?"

Often, interviewers need to probe for answers that will be sufficiently informative for analytical purposes. In every case, however, such probes *must* be completely neutral; they must not in any way affect the nature of the subsequent response. Whenever you anticipate that a given question may require probing for appropriate responses, you should provide one or more useful probes next to the question in the questionnaire. This practice has two important advantages. First, you'll have more time to devise the best, most neutral probes. Second, all interviewers will use the same probes whenever they're needed. Thus, even if the probe isn't perfectly neutral, all respondents will be presented with the same stimulus. This is the same logical guideline discussed for question wording. Although a question should not be loaded or biased, it's essential that every respondent be presented with the same question, even if it's biased.

Coordination and Control

Most interview surveys require the assistance of several interviewers. In large-scale surveys, interviewers are hired and paid for their work. Student researchers might find themselves recruiting friends to help them interview. Whenever more than one interviewer is involved in a survey, their efforts must be carefully controlled. This control has two aspects: training interviewers and supervising them after they begin work.

The interviewers' training session should begin with the description of what the study is all about. Even though the interviewers may be involved only in the data-collection phase of the project, it will be useful to them to understand what will be done with the interviews they conduct and what purpose will be served. Morale and motivation are usually lower when interviewers don't know what's going on.

The training on how to interview should begin with a discussion of general guidelines and procedures, such as those discussed earlier in this section. Then the whole group should go through the questionnaire together—question by question. Don't simply ask if anyone has any questions about the first page of the questionnaire. Read the first question aloud, explain the purpose of the question, and then entertain any questions or comments the interviewers may have. Once all their questions and comments have been handled, go on to the next question in the questionnaire.

It's always a good idea to prepare specifications to accompany an interview questionnaire. *Specifications* are explanatory and clarifying comments about handling difficult or confusing situations that may occur with regard to particular questions in the questionnaire. When drafting the questionnaire, try to think of all the problem cases that might arise—the bizarre (or not so bizarre) circumstances that might make a question difficult to answer. The survey specifications should provide detailed guidelines on how to handle such situations. For example, even as simple a matter as age might present problems. Suppose a respondent says he or she will be 25 next week. The interviewer might not be sure whether to take the respondent's current age or the nearest one. The specifications for that question should explain what should be done. (Probably, you would specify that the age as of last birthday should be recorded in all cases.)

If you've prepared a set of specifications, review them with the interviewers when you go over the individual questions in the questionnaire. Make sure your interviewers fully understand the specifications and the reasons for them as well as the questions themselves.

This portion of the interviewer training is likely to generate many troublesome questions from your interviewers. They'll ask, "What should I do if . . . ?" In such cases, avoid giving a quick, offhand answer. If you have specifications, show how the solution to the problem could be determined from the specifications. If you do not have specifications, show how the preferred handling of the situation fits within the general logic of the question and the purpose of the study. Giving unexplained answers to such questions will only confuse the interviewers and cause them to take their work less seriously. If you don't know the answer to such a question when it is asked, admit it and ask for some time to decide on the best answer. Then think out

the situation carefully and be sure to give all the interviewers your answer, explaining your reasons.

Once you've gone through the whole questionnaire, conduct one or two demonstration interviews in front of everyone. Preferably, you should interview someone other than one of the interviewers. Realize that your interview will be a model for those you're training, so make it good. It would be best, moreover, if the demonstration interview were done as realistically as possible. Do not pause during the demonstration to point out how you've handled a complicated situation: Handle it, and then explain later. It is irrelevant if the person you're interviewing gives real answers or takes on some hypothetical identity for the purpose, as long as the answers are consistent.

After the demonstration interviews, pair off your interviewers and have them practice on each other. When they've completed the questionnaire, have them reverse roles and do it again. Interviewing is the best training for interviewing. As your interviewers practice on each other, wander around, listening in on the practice so you'll know how well they're doing. Once the practice is completed, the whole group should discuss their experiences and ask any other questions they may have.

The final stage of the training for interviewers should involve some "real" interviews. Have them conduct some interviews under the actual conditions that will pertain to the final survey. You may want to assign them people to interview, or perhaps they may be allowed to pick people themselves. Do not have them practice on people you've selected in your sample, however. After each interviewer has completed three to five interviews, have him or her check back with you. Look over the completed questionnaires for any evidence of misunderstanding. Again, answer any questions that the interviewers may have. Once you're convinced that a given interviewer knows what to do, assign some actual interviews, using the sample you've selected for the study.

It's essential to continue supervising the work of interviewers over the course of the study. You should check in with them after they conduct no more than 20 or 30 interviews. You might assign 20 interviews, have the interviewer bring back those questionnaires when they're completed, look them over, and assign another 20 or so. Although this may seem overly cautious, you must continually protect yourself against misunderstandings that may not be evident early in the study.

If you're the only interviewer in your study, these comments may not seem relevant. However, it would be wise, for example, to prepare specifications for potentially troublesome questions in your questionnaire. Otherwise, you run the risk of making ad hoc decisions during the course of the study that you'll later regret or forget. Also, the emphasis on practice applies equally to the one-person project and to the complex funded survey with a large interviewing staff.

TELEPHONE SURVEYS

For years telephone surveys had a rather bad reputation among professional researchers. Telephone surveys are limited by definition to people who have telephones. Years ago, this method produced a substantial social-class bias by excluding poor people from the surveys. This was vividly demonstrated by the *Literary Digest* fiasco of 1936. Recall that, even though voters were contacted by mail, the sample was partially selected from telephone subscribers, who were hardly typical in a nation just recovering from the Great Depression. By 1993, however, the U.S. Bureau of the Census (1996a: Table 1224) estimated that 93.4 percent of all housing units had telephones, so the earlier form of class bias has been substantially reduced.

A related sampling problem involved unlisted numbers. A survey sample selected from the pages of a local telephone directory would totally omit all those people—typically richer—who requested that their numbers not be published. This potential bias has been erased through a technique that has advanced telephone sampling substantially: random-digit dialing.

Telephone surveys have many advantages, which underlie the growing popularity of this method. Probably the greatest advantages are money and time, in that order. In a face-to-face

household interview, you may drive several miles to a respondent's home, find no one there, return to the research office, and drive back the next day—possibly finding no one there again. It's cheaper and quicker to let your fingers make the trips.

When interviewing by telephone, you can dress any way you please without affecting the answers respondents give. And sometimes respondents will be more honest in giving socially disapproved answers if they don't have to look you in the eye. Similarly, it may be possible to probe into more sensitive areas, though this isn't necessarily the case. People are, to some extent, more suspicious when they can't see the person asking them questions —perhaps a consequence of "surveys" aimed at selling magazine subscriptions and time-share condominiums.

Interviewers can communicate a lot about themselves over the phone, however, even though they can't be seen. For example, researchers worry about the impact of an interviewer's name (particularly if ethnicity is relevant to the study) and debate the ethics of having all interviewers use bland "stage names" such as Smith or Jones. (Female interviewers sometimes ask permission to do this, to avoid subsequent harassment from men they interview.)

Telephone surveys can allow greater control over data collection if several interviewers are engaged in the project. If all the interviewers are calling from the research office, they can get clarification from the person in charge whenever problems occur, as they inevitably do. Alone in the boondocks, an interviewer may have to wing it between weekly visits with the interviewing supervisor.

Finally, another important factor involved in the growing use of telephone surveys has to do with personal safety. Don Dillman (1978:4) describes the situation this way:

> Interviewers must be able to operate comfortably in a climate in which strangers are viewed with distrust and must successfully counter respondents' objections to being interviewed. Increasingly, interviewers must be willing to work at night to contact residents in many house-

holds. In some cases, this necessitates providing protection for interviewers working in potentially dangerous locations.

Concerns for safety thus work two ways to hamper face-to-face interviews. Potential respondents may refuse to be interviewed, fearing the stranger-interviewer. And the interviewers themselves may incur some risks. All this is made even worse by the possibility of the researchers being sued for huge sums if anything goes wrong.

There are problems involved in telephone interviewing, however. As I've already mentioned, the method is hampered by the proliferation of bogus "surveys" that are actually sales campaigns disguised as research. If you have any questions about any such call you receive, by the way, ask the interviewer directly whether you've been selected for a survey only or if a sales "opportunity" is involved. It's also a good idea, if you have any doubts, to get the interviewer's name, phone number, and company. Hang up if the caller refuses to provide any of these.

For the researcher, the ease with which people can hang up is another shortcoming of telephone surveys. Once you've been let inside someone's home for an interview, the respondent is unlikely to order you out of the house in midinterview. It's much easier to terminate a telephone interview abruptly, saying something like, "Whoops! Someone's at the door. I gotta go." or "OMIGOD! The pigs are eating my Volvo!" (That sort of thing is much harder to fake when the interviewer is sitting in your living room.)

Another potential problem for telephone interviewing is the prevalence of answering machines. A study conducted by Walker Research (1988) found that half of the owners of answering machines acknowledged using their machines to "screen" calls at least some of the time. Research by Peter Tuckel and Barry Feinberg (1991), however, showed that answering machines had not yet had a significant effect on the ability of telephone researchers to contact prospective respondents. Nevertheless, the researchers concluded that as answering machines continue to proliferate, "the

sociodemographic characteristics of owners will change." This fact makes it likely that "different behavior patterns associated with the utilization of the answering machine" will emerge (1991:216).

Computer-Assisted Telephone Interviewing (CATI)

In Chapter 14, we'll be looking at some of the ways computers have influenced the conduct of social research—particularly data processing and analysis. Computers are also changing the nature of telephone interviewing. One innovation is computer-assisted telephone interviewing (CATI). This method is increasingly used by academic, government, and commercial survey researchers. Though there are variations in practice, here's what CATI can look like.

Imagine an interviewer wearing a telephone headset, sitting in front of a computer monitor. The central computer has been programmed to select a telephone number at random and dials it. On the video screen is an introduction ("Hello, my name is . . .") and the first question to be asked ("Could you tell me how many people live at this address?").

When the respondent answers the phone, the interviewer says hello, introduces the study, and asks the first question displayed on the screen. When the respondent answers the question, the interviewer types that answer into the computer terminal—either the verbatim response to an open-ended question or the code category for the appropriate answer to a closed-ended question. The answer is immediately stored in the computer. The second question appears on the video screen, is asked, and the answer is entered into the computer. Thus, the interview continues.

In addition to the obvious advantages in terms of data collection, CATI automatically prepares the data for analysis; in fact, the researcher can begin analyzing the data before the interviewing is complete, thereby gaining an advanced view of how the analysis will turn out. Still another innovation that computer technology makes possible is described in the box entitled "Voice Capture.™"

VOICE CAPTURE™

by James E. Dannemiller
SMS Research, Honolulu

The development of various CATI techniques has been a boon to survey and marketing research, though mostly it has supported the collection, coding, and analysis of "data as usual." The Voice Capture™ technique developed by Survey Systems, however, offers quite unusual possibilities, which we are only beginning to explore.

In the course of a CATI-based telephone interview, the interviewer can trigger the computer to begin digitally recording the conversation with the respondent. Having determined that the respondent has recently changed his or her favorite TV news show, for example, the interviewer can ask, "Why did you change?" and begin recording the verbatim response. (Early in the interview, the interviewer has asked permission to record parts of the interview.)

Later on, coders can play back the responses and code them—much as they would do with the interviewer's typescript of the responses. This offers an easier and more accurate way of accomplishing a conventional task. But that's a tame use of the new capability.

It's also possible to incorporate such oral data as parts of a cross-tabulation during analysis. We may create a table of gender by age by reasons for switching TV news shows. Thus, we can hear, in turn, the responses of the young men, young women, middle-aged men, and so forth. In one such study we found the younger and older men tending to watch one TV news show, while the middle-aged men watched something else. Listening to the responses of the middle-aged men, one after another, we heard a common comment: "Well, now that I'm older . . . " This kind of aside might have been lost in the notes hastily typed by interviewers, but such comments stood out dramatically in the oral data. The middle-aged men seemed to be telling us they felt "maturity" required them to watch a particular show, while more years under their belts let them drift back to what they liked in the first place.

These kinds of data are especially compelling to clients, particularly in customer satisfaction studies. Rather than summarize what we feel a client's customers like and don't like, we can let the respondents speak directly to the client in their own words. It's like a focus group on demand. Going one step further, we have found that letting line employees (bank tellers, for example) listen to the responses has more impact than having their supervisors tell them what they are doing right or wrong.

As exciting as these experiences are, I have the strong feeling that we have scarcely begun to tap into the possibilities for such unconventional forms of data.

NEW TECHNOLOGIES AND SURVEY RESEARCH

As we have already seen in the case of computer-assisted telephone interviewing (CATI), many of the new technologies affecting people's lives also open new possibilities for survey research. For example, recent innovations in self-administered questionnaires make use of the computer. Among the techniques that are being tested are these (Nicholls, Baker, and Martin 1996):

CAPI (computer-assisted personal interviewing): Similar to CATI but used in face-to-face interviews rather than over the phone.

CASI (computer-assisted self-interviewing): A research worker brings a computer to the respondent's home, and the respondent reads questions on the computer screen and enters his or her own answers.

CSAQ (computerized self-administered questionnaire): The respondent receives the questionnaire via floppy disk, bulletin board, or other means and runs the software, which asks questions and accepts the respondent's answers. The respondent then returns the data file.

TDE (touchtone data entry): The respondent initiates the process by calling a number at the research organization. This prompts a series of computerized questions, which the respondent answers by pressing keys on the telephone keypad.

VR (voice recognition): Instead of asking the respondent to use the telephone keypad, as in TDE, this system accepts spoken responses.

Nicholls and colleagues report that such techniques are more efficient than conventional techniques, and they do not appear to result in a reduction of data quality.

Jeffery Walker (1994) has explored the possibility of conducting surveys by fax machine. Questionnaires are faxed to respondents, who are asked to fax their answers back. Of course, such surveys can only represent that part of the population that has fax machines. Walker reports that fax surveys don't achieve as high a response rate as do face-to-face interviews, but, because of the perceived urgency, they do produce higher response rates than do mail or telephone surveys. In one test case, all those who had ignored a mail questionnaire were sent a fax follow-up, and 83 percent responded.

I've already noted that, as a consumer of social research, you should be wary of "surveys" whose apparent purpose is to raise money for the sponsor. This practice has already invaded the realm of "fax surveys," evidenced by a fax entitled, "Should Hand Guns Be Outlawed?" Two fax numbers were provided for expressing either a "Yes" or "No" opinion. The smaller print noted, "Calls to these numbers cost $2.95 per minute, a small price for greater democracy. Calls take approx. 1 or 2 minutes." You can imagine where the $2.95 went.

The new technology of survey research includes the use of the Internet and the World Wide Web—two of the most far-reaching developments of the late 20th century. Some researchers feel that the Internet can also be used to conduct meaningful survey research.

An immediate objection that many social researchers make to online surveys concerns representativeness: Will the people who can be surveyed online be representative of meaningful populations, such as all U.S. adults, all voters, and so on? This is the criticism raised with regard to surveys via fax and, earlier, with regard to telephone surveys.

Camilo Wilson (1999), the founder of Cogix (http://www.cogix.com) points out that some populations are ideally suited to online surveys: specifically, those who visit a particular Web site. For example, Wilson indicates that market research for online companies *should* be conducted online, and his firm has developed software, ViewsFlash, for precisely that purpose. Although Web site surveys could easily collect data from all who visit a particular site, Wilson suggests that survey sampling techniques can provide sufficient consumer data without irritating thousands or millions of potential customers.

 To learn more about online surveys, you can go to http://www.cogix.com or use a search engine to find "online surveys."*

But how about general population surveys? As I write this, a debate is brewing within the survey research community. Humphrey Taylor and George Terhanian (1999:20) prompted part of the debate with an article, "Heady Days Are Here Again." Acknowledging the need for caution, they urged that online polling be given a fair hearing:

> One test of the credibility of any new data collection method hinges on its ability to reliably and accurately forecast voting behavior. For this

---

*Each time the Internet icon appears, you'll be given helpful leads for searching the World Wide Web.

reason, last fall we attempted to estimate the 1998 election outcomes for governor and US Senate in 14 states on four separate occasions using internet surveys.

The researchers compared their results with 52 telephone polls that addressed the same races. Online polling correctly picked 21 of the 22 winners, or 95 percent. However, simply picking the winner is not a sufficient test of effectiveness: How close did the polls come to the actual percentages received by the various candidates? Taylor and Terhanian report their online polls missed the actual vote by an average of 6.8 percentage points. The 52 telephone polls missed the same votes by an average of 6.2 percentage points.

Warren Mitofsky (1999) is a critic of online polling. In addition to disagreeing with the way Taylor and Terhanian calculated the ranges of error just reported, he has called for a sounder, theoretical basis on which to ground the new technique.

One key to online polling is the proper assessment and use of weights for different kinds of respondents—as was discussed in the context of quota sampling in Chapter 7. Taylor and Terhanian are aware of the criticisms of quota sampling, but their initial experiences with online polling suggest to them that the technique should be pursued. Indeed, they conclude by saying "This is an unstoppable train, and it is accelerating. Those who don't get on board run the risk of being left far behind" (1999:23).

Many of the cautions urged in relation to online surveys today are similar to those urged in relation to telephone surveys in the first edition of this book, in 1975. Whether online surveys will gain the respect and extensive use enjoyed by telephone surveys today remains to be seen. Students who consider using this technique should do so in full recognition of its potential shortcomings.

Researchers are amassing a body of experience with online surveying, yielding lessons for increasing success. For example, Survey Sampling, Inc. (2000) suggests the following do's and don'ts for conducting online surveys:

Do use consistent wording between the invitation and the survey. **Don't** use terms such as "unique ID number" in the invitation, then ask respondents to type their "password" when they get to the survey. Changing terminology can be confusing.

Do use plain, simple language.

Don't force the respondent to scroll down the screen for the URL for the study location.

Do offer to share selected results from the study with everyone who completes the survey. Respondents will often welcome information as a reward for taking the study, especially when they are young adults and teens.

Do plan the time of day and day of week to mail, depending on the subject of the study and type of respondent. Send the invitation late afternoon, evening, or weekend, when respondents are most likely to be reading mail at home, especially if the study requests respondents to check an item in the kitchen or other area in the home. If a parent-child questionnaire is planned, send the invitation late afternoon when children are home, not early in the day, when respondents can't complete the study because children are at school.

Do be aware of technical limitations. For example, WebTV users currently cannot

The World Wide Web is already seeing extensive use as a marketplace for surveys and other research techniques. As only a few illustrative examples, see the following:

- The Gallup Organization: http://www.gallup.com/
- Harris Poll Online: http://www.harrisinteractive.com/
- SMS Research: http://www.smshawaii.com/
- The Survey/Marketing Research e-Store: http://www.streamlinesurveys.com/Streamline/estore/index.htm
- Zogby International: http://www.zogby.com/

access surveys using Java. If respondents' systems need to be Java-enabled or require access to streaming video, alert panelists at the beginning of the study, not midway through.

Do test incentives, rewards, and prize drawings to determine the optimal offer for best response. Longer surveys usually require larger incentives.

Do limit studies to 15 minutes or less.

— Reprinted with permission.

COMPARISON OF THE DIFFERENT SURVEY METHODS

Now that we've seen several ways to collect survey data, let's take a moment to compare them directly.

Self-administered questionnaires are generally cheaper and quicker than face-to-face interview surveys. These considerations are likely to be important for an unfunded student wishing to undertake a survey for a term paper or thesis. Moreover, if you use the self-administered mail format, it costs no more to conduct a national survey than a local one of the same sample size. In contrast, a national interview survey (either face-to-face or by telephone) would cost far more than a local one. Also, mail surveys typically require a small staff: One person can conduct a reasonable mail survey alone, although you shouldn't underestimate the work involved. Further, respondents are sometimes reluctant to report controversial or deviant attitudes or behaviors in interviews but are willing to respond to an anonymous self-administered questionnaire.

Interview surveys also offer many advantages. For example, they generally produce fewer incomplete questionnaires. Although respondents may skip questions in a self-administered questionnaire, interviewers are trained not to do so. In CATI surveys, the computer offers a further check on this. Interview surveys, moreover, have typically achieved higher completion rates than have self-administered questionnaires.

Although self-administered questionnaires may be more effective for sensitive issues, interview surveys are definitely more effective for complicated ones. Prime examples include the enumeration of household members and the determination of whether a given address corresponds to more than one housing unit. Although the concept of housing unit has been refined and standardized by the Bureau of the Census and interviewers can be trained to deal with the concept, it's extremely difficult to communicate in a self-administered questionnaire. This advantage of interview surveys pertains generally to all complicated contingency questions.

With interviews, you can conduct a survey based on a sample of addresses or phone numbers rather than on names. An interviewer can arrive at an assigned address or call the assigned number, introduce the survey, and even—following instructions—choose the appropriate person at that address to respond to the survey. In contrast, self-administered questionnaires addressed to "occupant" receive a notoriously low response.

Finally, as we've seen, interviewers questioning respondents face-to-face can make important observations aside from responses to questions asked in the interview. In a household interview, they may note the characteristics of the neighborhood, the dwelling unit, and so forth. They may also note characteristics of the respondents or the quality of their interaction with the respondents—whether the respondent had difficulty communicating, was hostile, seemed to be lying, and so on.

The chief advantages of telephone surveys over those conducted face-to-face center primarily on time and money. Telephone interviews are much cheaper and can be mounted and executed quickly. Also, interviewers are safer when interviewing in high-crime areas. Moreover, the impact of the interviewers on responses is somewhat lessened when they can't be seen by the respondents. As only one indicator of the popularity of telephone interviewing, when Johnny Blair and his colleagues (1995) compiled a bibliography on sample designs for telephone interviews, they listed over 200 items.

Online surveys have many of the strengths and weaknesses of mail surveys. Once the available software has been further developed, they are likely to be substantially cheaper. An important weakness, however, lies in the difficulty of assuring that respondents to an online survey will be representative of some more general population.

Clearly, each survey method has its place in social research. Ultimately, you must balance the advantages and disadvantages of the different methods in relation to your research needs and your resources.

STRENGTHS AND WEAKNESSES OF SURVEY RESEARCH

Regardless of the specific method used, surveys —like other modes of observation in social research—have special strengths and weaknesses. You should keep these in mind when determining whether a survey is appropriate for your research goals.

Surveys are particularly useful in describing the characteristics of a large population. A carefully selected probability sample in combination with a standardized questionnaire offers the possibility of making refined descriptive assertions about a student body, a city, a nation, or any other large population. Surveys determine unemployment rates, voting intentions, and the like with uncanny accuracy. Although the examination of official documents—such as marriage, birth, or death records —can provide equal accuracy for a few topics, no other method of observation can provide this general capability.

Surveys—especially self-administered ones— make large samples feasible. Surveys of 2,000 respondents are not unusual. A large number of cases is very important for both descriptive and explanatory analyses, especially wherever several variables are to be analyzed simultaneously.

In one sense, surveys are flexible. Many questions may be asked on a given topic, giving you considerable flexibility in your analyses. Whereas an experimental design may require you to commit yourself in advance to a particular operational definition of a concept, surveys let you develop operational definitions from actual observations.

Finally, standardized questionnaires have an important strength in regard to measurement generally. Earlier chapters have discussed the ambiguous nature of most concepts: They have no ultimately real meanings. One person's religiosity is quite different from another's. Although you must be able to define concepts in those ways most relevant to your research goals, you may not find it easy to apply the same definitions uniformly to all subjects. The survey researcher is bound to this requirement by having to ask exactly the same questions of all subjects and having to impute the same intent to all respondents giving a particular response.

Survey research also has several weaknesses. First, the requirement of standardization often seems to result in the fitting of round pegs into square holes. Standardized questionnaire items often represent the least common denominator in assessing people's attitudes, orientations, circumstances, and experiences. By designing questions that will be at least minimally appropriate to all respondents, you may miss what is most appropriate to many respondents. In this sense, surveys often appear superficial in their coverage of complex topics. Although this problem can be partly offset by sophisticated analyses, it is inherent in survey research.

Similarly, survey research can seldom deal with the context of social life. Although questionnaires can provide information in this area, the survey researcher rarely develops a feel for the total life situation in which respondents are thinking and acting that, say, the participant observer can (see Chapter 10).

In many ways, surveys are inflexible. Studies involving direct observation can be modified as field conditions warrant, but surveys typically require that an initial study design remain unchanged throughout. As a field researcher, for example, you can become aware of an important new variable operating in the phenomenon you're studying and begin making careful observations of it. The survey researcher would probably be unaware of the new variable's importance and could do nothing about it in any event.

Finally, surveys are subject to the artificiality mentioned earlier in connection with experiments. Finding out that a person gives conservative answers to a questionnaire does not necessarily mean the person is conservative; finding out that a person gives prejudiced answers to a questionnaire does not necessarily mean the person is prejudiced. This shortcoming is especially salient in the realm of action. Surveys cannot measure social action; they can only collect self-reports of recalled past action or of prospective or hypothetical action.

The problem of artificiality has two aspects. First, the topic of study may not be amenable to measurement through questionnaires. Second, the act of studying that topic—an attitude, for example—may affect it. A survey respondent may have given no thought to whether the governor should be impeached until asked for his or her opinion by an interviewer. He or she may, at that point, form an opinion on the matter.

Survey research is generally weak on validity and strong on reliability. In comparison with field research, for example, the artificiality of the survey format puts a strain on validity. As an illustration, people's opinions on issues seldom take the form of strongly agreeing, agreeing, disagreeing, or strongly disagreeing with a specific statement. Their survey responses in such cases must be regarded as approximate indicators of what the researchers had in mind when they framed the questions. This comment, however, needs to be held in the context of earlier discussions of the ambiguity of validity itself. To say something is a valid or an invalid measure assumes the existence of a "real" definition of what's being measured, and many scholars now reject that assumption.

Reliability is a clearer matter. By presenting all subjects with a standardized stimulus, survey research goes a long way toward eliminating unreliability in observations made by the researcher. Moreover, careful wording of the questions can also reduce significantly the subject's own unreliability.

As with all methods of observation, a full awareness of the inherent or probable weaknesses of survey research can partially resolve them in some cases. Ultimately, though, researchers are on the safest ground when they can employ several research methods in studying a given topic.

SECONDARY ANALYSIS

As a mode of observation, survey research involves the following steps: (1) questionnaire construction, (2) sample selection, and (3) data collection, through either interviewing or self-administered questionnaires. As you've gathered, surveys are usually major undertakings. It's not unusual for a large-scale survey to take several months or even more than a year to progress from conceptualization to data in hand. (Smaller-scale surveys can, of course, be done more quickly.) Through a method called *secondary analysis,* however, researchers can pursue their particular social research interests—analyzing survey data from, say, a national sample of 2,000 respondents—while avoiding the enormous expenditure of time and money such a survey entails.

Secondary analysis is a form of research in which the data collected and processed by one researcher are reanalyzed—often for a different purpose—by another. Beginning in the 1960s, survey researchers became aware of the potential value that lay in archiving survey data for analysis by scholars who had nothing to do with the survey design and data collection. Even when one researcher had conducted a survey and analyzed the data, those same data could be further analyzed by others, who had slightly different interests. Thus, if you were interested in the relationship between political views and attitudes toward gender equality, you could examine that research question through the analysis of any data set that happened to contain questions relating to those two variables.

The initial data archives were very much like book libraries, with a couple of differences. First, instead of books, the data archives contained data sets: first as punched cards, then as magnetic tapes. Today they're typically contained on computer disks, CD-ROMs, or online servers. Second, whereas you're expected to return books to a conventional library, you can keep the data obtained from a data archive.

The best-known current example of secondary analysis is the General Social Survey (GSS). Every year or two, the federal government commissions the National Opinion Research Center (NORC) at the University of Chicago to conduct a major national survey to collect data on a large number of social science variables. These surveys are conducted precisely for the purpose of making data available to scholars at little or no cost.

Numerous other resources are available for identifying and acquiring survey data for secondary analysis. The Roper Center for Public Opinion Research at the University of Connecticut is one excellent resource. The center also publishes the journal *Public Perspective* on public opinion polling. Polling the Nations is an online repository for thousands of polls conducted in the United States and 70 other nations. A paid subscription allows users to obtain specific data results from studies they specify, rather than obtaining whole studies.

Outside the United States, the Netherlands Institute for Scientific Information Services allows users to track down European studies that contain variables of interest. You might also try the Central Archive for Social Science Research at the University of Cologne in Germany.

See the following sites for more on social science data sources:
- GSS: http://www.icpsr.umich.edu/gss/
- Roper Center for Public Opinion Research: http://www.ropercenter.uconn.edu/
- Polling the Nations: http://www.pollingthenations.com/
- Netherlands Institute for Scientific Information Services: http://www.niwi.knaw.nl/cgi-bin/nph-star_search.pl
- Central Archive for Social Science Research at the University of Cologne: http://www.za.uni-koeln.de/index-e.htm

---

The advantages of secondary analysis are obvious and enormous: It's cheaper and faster than doing original surveys, and, depending on who did the original survey, you may benefit from the work of topflight professionals. There are disadvantages,

A QUANDARY REVISITED

Professional survey research has been damaged in recent years by the actions of telemarketers who pretend they are conducting surveys. Potential respondents sometimes refuse to participate in a legitimate survey because they suspect that it's really a sales call.

Here are a few ways to determine the legitimacy of a survey.

1. Ask who is conducting the survey. The caller probably said something about it quickly at the outset, but ask him or her to repeat the information so you can write it down.
2. Ask for the telephone number of a supervisor or manager so you can call the people running the survey.
3. Ask whether the call involves a sales solicitation.

If the caller is reluctant to answer any of these questions, assume that the call is not a professional survey. You may respond as you deem appropriate. One possibility: Let them recite their entire sales pitch, and then ask, "Could you repeat that?"

---

however. The key problem involves the recurrent question of validity. When one researcher collects data for one particular purpose, you have no assurance that those data will be appropriate for your research interests. Typically, you'll find that the original researcher asked a question that "comes close" to measuring what you're interested in, but you'll wish the question had been asked just a little differently—or that another, related question had also been asked. Your question, then, is whether the question that was asked provides a valid measure of the variable you want to analyze. Nevertheless, secondary analysis can be immensely useful. Moreover, it illustrates once again the range of possibilities available in finding the answers to questions about social life. Although no single method unlocks all puzzles, there is no limit to the ways you can find out about things. And when you zero in on an issue from several independent directions, you gain that much more expertise.

Main Points

❑ Survey research, a popular social research method, is the administration of questionnaires to a sample of respondents selected from some population.

❑ Survey research is especially appropriate for making descriptive studies of large populations; survey data may be used for explanatory purposes as well.

❑ Questionnaires provide a method of collecting data by (1) asking people questions or (2) asking them to agree or disagree with statements representing different points of view. Questions may be open-ended (respondents supply their own answers) or closed-ended (they select from a list of provided answers).

❑ Items in a questionnaire should observe several guidelines: (1) The items must be clear and precise; (2) the items should ask only about one thing (i.e., double-barreled questions should be avoided); (3) respondents must be competent to answer the item; (4) respondents must be willing to answer the item; (5) questions should be relevant to the respondent; (6) items should ordinarily be short; (7) negative terms should be avoided so as not to confuse respondents; (8) the items should be worded to avoid biasing responses.

❑ The format of a questionnaire can influence the quality of data collected.

❑ A clear format for contingency questions is necessary to ensure that the respondents answer all the questions intended for them.

❑ The matrix question is an efficient format for presenting several items sharing the same response categories.

❑ The order of items in a questionnaire can influence the responses given.

❑ Clear instructions are important for getting appropriate responses in a questionnaire.

❑ Questionnaires should be pretested before being administered to the study sample.

❑ Questionnaires may be administered in three basic ways: through self-administered questionnaires, face-to-face interviews, or telephone surveys.

❑ It's generally advisable to plan follow-up mailings in the case of self-administered questionnaires, sending new questionnaires to those respondents who fail to respond to the initial appeal. Properly monitoring questionnaire returns will provide a good guide to when a follow-up mailing is appropriate.

❑ The essential characteristic of interviewers is that they be neutral; their presence in the data-collection process must not have any effect on the responses given to questionnaire items.

❑ Interviewers must be carefully trained to be familiar with the questionnaire, to follow the question wording and question order exactly, and to record responses exactly as they are given.

❑ Interviewers can use probes to elicit an elaboration on an incomplete or ambiguous response. Probes should be neutral. Ideally, all interviewers should use the same probes.

❑ Telephone surveys can be cheaper and more efficient than face-to-face interviews, and they can permit greater control over data collection. The development of computer-assisted telephone interviewing (CATI) techniques is especially promising.

❑ New technologies offer additional opportunities for social researchers. They include various kinds of computer-assisted data collection and analysis as well as the chance to conduct surveys by fax or over the Internet. The latter two methods, however, must be used with caution because respondents may not be representative of the intended population.

❏ The advantages of a self-administered questionnaire over an interview survey are economy, speed, lack of interviewer bias, and the possibility of anonymity and privacy to encourage candid responses on sensitive issues.

❏ The advantages of an interview survey over a self-administered questionnaire are fewer incomplete questionnaires and fewer misunderstood questions, generally higher return rates, and greater flexibility in terms of sampling and special observations.

❏ The principal advantages of telephone surveys over face-to-face interviews are the savings in cost and time. Telephone interviewers are also safer than in-person interviewers, and they may have a smaller effect on the interview itself.

❏ Online surveys have many of the strengths and weaknesses of mail surveys. Although they are cheaper to conduct, it can be difficult to ensure that the respondents represent a more general population.

❏ Survey research in general offers advantages in terms of economy, the amount of data that can be collected, and the chance to sample a large population. The standardization of the data collected represents another special strength of survey research.

❏ Survey research has the weaknesses of being somewhat artificial, potentially superficial, and relatively inflexible. It's difficult to use surveys to gain a full sense of social processes in their natural settings. In general, survey research is comparatively weak on validity and strong on reliability.

❏ Secondary analysis provides social researchers with an important option for "collecting" data cheaply and easily but at a potential cost in validity.

Key Terms

respondent	contingency question
questionnaire	response rate
open-ended questions	interview
closed-ended questions	probe
bias	secondary analysis

Review Questions

1. What closed-ended questions could you construct from each of the following open-ended questions?

    a. What was your family's total income last year?

    b. How do you feel about the space shuttle program?

    c. How important is religion in your life?

    d. What was your main reason for attending college?

    e. What do you feel is the biggest problem facing your community?

2. What are the main advantages and disadvantages of conducting surveys over the Internet?

3. A newspaper headline proclaims, "Most Americans oppose abortion, according to new survey." What methodological details do you want to know about the survey to help you interpret the results?

4. Look at your appearance right now. What aspects of your appearance might create a problem if you were interviewing a general cross section of the public?

Additional Readings

Babbie, Earl. 1990. *Survey Research Methods.* Belmont, CA: Wadsworth. A comprehensive overview of survey methods. (You thought I'd say it was lousy?) This textbook, although overlapping the present one somewhat, covers aspects of survey techniques omitted here.

Bradburn, Norman M., and Seymour Sudman. 1988. *Polls and Surveys: Understanding What They Tell Us.* San

Francisco: Jossey-Bass. These veteran survey researchers answer questions about their craft the general public commonly ask.

Dillman, Don A. 1978. *Mail and Telephone Surveys: The Total Design Method.* New York: Wiley. An excellent review of the methodological literature on mail and telephone surveys. Dillman makes many good suggestions for improving response rates.

Elder, Glen H., Jr., Eliza K. Pavalko, and Elizabeth C. Clipp. 1993. *Working with Archival Data: Studying Lives.* Newbury Park, CA: Sage. This book discusses the possibilities and techniques for using existing data archives in the United States, especially those providing longitudinal data.

Feick, Lawrence F. 1989. "Latent Class Analysis of Survey Questions That Include Don't Know Responses." *Public Opinion Quarterly* 53:525–47. "Don't know" can mean a variety of things, as this analysis indicates.

Fowler, Floyd J., Jr. 1995. *Improving Survey Questions: Design and Evaluation.* Thousand Oaks, CA: Sage. A comprehensive discussion of questionnaire construction, including a number of suggestions for pretesting questions. This book discusses the logic of obtaining information through survey questions and gives numerous guidelines for being effective. It also offers several examples of questions you might use.

Groves, Robert M. 1990. "Theories and Methods of Telephone Surveys." Pp. 221–40 in *Annual Review of Sociology* (vol. 16), edited by W. Richard Scott and Judith Blake. Palo Alto, CA: Annual Reviews. An attempt to place telephone surveys in the context of sociological and psychological theories and to address the various kinds of errors common to this research method.

Miller, Delbert. 1991. *Handbook of Research Design and Social Measurement.* Newbury Park, CA: Sage. A powerful reference work. This book, especially Part 6, cites and describes a wide variety of operational measures used in earlier social research. In several cases, the questionnaire formats used are presented. Though the quality of these illustrations is uneven, they provide excellent examples of possible variations.

Sheatsley, Paul F. 1983. "Questionnaire Construction and Item Writing." Pp. 195–230 in *Handbook of Survey Research,* edited by Peter H. Rossi, James D. Wright, and Andy B. Anderson. New York: Academic Press. An excellent examination of the topic by an expert in the field.

Smith, Eric R. A. N., and Peverill Squire. 1990. "The Effects of Prestige Names in Question Wording." *Public Opinion Quarterly* 54:97–116. Not only do prestigious names affect the overall responses given to survey questionnaires, they also affect such things as the correlation between education and the number of "don't know" answers.

Swafford, Michael. 1992. "Soviet Survey Research: The 1970's vs. the 1990's." *AAPOR News* 19 (3):3–4. The author contrasts the general repression of survey research during his first visit in 1973–74 with the renewed use of the method in more recent times. He notes, for example, that the Soviet government commissioned a national survey to determine public opinion on the possible reunification of Germany.

Tourangeau, Roger, Kenneth A. Rasinski, Norman Bradburn, and Roy D'Andrade. 1989. "Carryover Effects in Attitude Surveys." *Public Opinion Quarterly* 53:495–524. The authors asked six target questions in a telephone survey of 1,100 respondents, varying the questions immediately preceding the target questions. They found substantial differences.

Williams, Robin M., Jr. 1989. "The American Soldier: An Assessment, Several Wars Later." *Public Opinion Quarterly* 53:155–74. One of the classic studies in the history of survey research is reviewed by one of its authors.

Multimedia Resources

The Wadsworth Sociology Resource Center: Virtual Society
http://sociology.wadsworth.com/

Visit the companion Web site for the second edition of *The Basics of Social Research* to access a wide range of student resources. Begin by clicking on the Student Resources section of the book's Web site to access the following study tools:

- eBabbie Resource Center
- Planning a Research Project
- Doing Data Analysis
- Statistics Review
- Flash Cards
- Internet Links and Exercises
- InfoTrac College Edition: Exercises
- Quizzes

Visit the **eBabbie Resource Center** for an overview of each chapter and helpful online tutorials. Find information on budgeting and step-by-step examples of model research projects at **Planning a Research Project.** Learn how to use quantitative and qualitative data analysis programs at **Doing Data Analysis,** and brush up on your statistics at **Statistics Review.** You can also further your study by accessing **Internet Links and Exercises** related to chapter materials, **Flash Cards, Quizzes,** and many other learning tools.

InfoTrac College Edition
http://www.infotrac-college.com/ wadsworth/access.html
Access the latest news and journal articles with InfoTrac College Edition, an easy-to-use online data-

base of reliable, full-length articles from hundreds of top academic journals. Conduct an electronic search using the following search terms:

CATI
Mail survey
Opinion poll
Questionnaire
Respondent
Self-administered
Telephone poll
Telephone survey

10 QUALITATIVE FIELD RESEARCH

Paul Conklin/PhotoEdit/PictureQuest

What You'll Learn in This Chapter

Here you'll see that qualitative field research enables researchers to observe social life in its natural habitat: to go where the action is and watch. This type of research can produce a richer understanding of many social phenomena than can be achieved through other observational methods, provided that the researcher observes in a deliberate, well-planned, and active way.

In this chapter . . .

AN OPENING **QUANDARY**

To be sure, there are things to learn about doing experiments or surveys, but why does it take a whole chapter to talk about simply observing social life as it's happening? Don't researchers just show up and take notes?

INTRODUCTION

Several chapters ago, I suggested that you've been doing social research all your life. This idea should become even clearer as we turn to what probably seems like the most obvious method of making observations: qualitative field research. In a sense, we do field research whenever we observe or participate in social behavior and try to understand it, whether in a college classroom, in a doctor's waiting room, or on an airplane. Whenever we report our observations to others, we are reporting our field research efforts.

Such research is at once very old and very new in social science. Many of the techniques discussed in this chapter have been used by social researchers for centuries. Within the social sciences, anthropologists are especially associated with this method and have contributed to its development as a scientific technique. Moreover, something similar to this method is employed by many people who might not, strictly speaking, be regarded as social science researchers. Welfare department case workers are one example; newspaper reporters are another.

To take this last example further, consider that interviewing is a technique common to both journalism and sociology. A journalist uses the data to report a subject's attitude, belief, or experience—that's usually it. Sociologists, on the other hand, treat an interview as data that need to be analyzed in depth; their ultimate goal is to understand social life in the context of theory, using established analytical techniques. Although sociology and journalism use similar techniques, the two disciplines view and use data differently.

While many of the techniques involved in field research are "natural" activities, they are also skills to be learned and honed. This chapter discusses these skills in some detail, examining some of the major paradigms of field research and describing some specific techniques that make scientific field research more useful than the casual observation we all engage in.

As we'll see, there are many paradigms associated with field research, which comprises a wide range of studies. This range stems in part from differences among paradigms—specifically, the variety of theoretical approaches to basic questions such as "What is data?" "How should we collect data?" and "How should we analyze data?"

I use the term *qualitative field research* to distinguish this type of observation method from methods designed to produce data appropriate for quantitative (statistical) analysis. Thus, surveys provide data from which to calculate the percentage unemployed in a population, mean incomes, and so forth. Field research more typically yields qualitative data: observations not easily reduced to numbers. Thus, for example, a field researcher may note the "paternalistic demeanor" of leaders at a political rally or the "defensive evasions" of a public official at a public hearing without trying to express either the paternalism or the defensiveness as numerical quantities or degrees. Although field research can be used to collect quantitative data—for example, noting the number of interactions of various specified types within a field setting—typically, field research is qualitative.

Field observation also differs from some other models of observation in that it's not just a data-collecting activity. Frequently, perhaps typically, it's a theory-generating activity as well. As a field researcher, you'll seldom approach your task with

precisely defined hypotheses to be tested. More typically, you'll attempt to make sense out of an ongoing process that cannot be predicted in advance—making initial observations, developing tentative general conclusions that suggest particular types of further observations, making those observations and thereby revising your conclusions, and so forth. In short, the alternation of induction and deduction discussed in Part 1 of this book is perhaps nowhere more evident and essential than in good field research. For expository purposes, however, this chapter focuses primarily on some of the theoretical foundations of field research and on techniques of data collection. Chapter 13 discusses how to analyze qualitative data.

Keep in mind that the types of methods researchers use depend in part on the specific research questions they want to answer. For example, a question such as "How do women construct their everyday lives in order to perform their roles as mothers, partners, and breadwinners?" could be addressed by either in-depth interviews or direct observations—or both. The assessment of advertising campaigns might profit from focus group discussions. In most cases, researchers have many field research methods to choose from.

TOPICS APPROPRIATE TO FIELD RESEARCH

One of the key strengths of field research is how comprehensive a perspective it can give researchers. By going directly to the social phenomenon under study and observing it as completely as possible, researchers can develop a deeper and fuller understanding of it. As such, this mode of observation is especially, though not exclusively, appropriate to research topics and social studies that appear to defy simple quantification. Field researchers may recognize several nuances of attitude and behavior that might escape researchers using other methods.

Field research is especially appropriate to the study of those attitudes and behaviors best understood within their natural setting, as opposed to the somewhat artificial settings of experiments and surveys. For example, field research provides a superior method for studying the dynamics of religious conversion at a revival meeting, just as a statistical analysis of membership rolls would be a better way of discovering whether men or women were more likely to convert.

Finally, field research is well suited to the study of social processes over time. Thus, the field researcher might be in a position to examine the rumblings and final explosion of a riot as events actually occur rather than afterward in a reconstruction of the events.

Other good places to apply field research methods include campus demonstrations, courtroom proceedings, labor negotiations, public hearings, or similar events taking place within a relatively limited area and time. Several such observations must be combined in a more comprehensive examination over time and space.

In *Analyzing Social Settings* (1995: 101–13), John and Lyn Lofland discuss several elements of social life appropriate to field research.

1. *Practices:* Various kinds of behavior, such as talking or reading a book
2. *Episodes:* A variety of events such as divorce, crime, and illness
3. *Encounters:* Two or more people meeting and interacting
4. *Roles:* The analysis of the positions people occupy and the behavior associated with those positions: occupations, family roles, ethnic groups
5. *Relationships:* Behavior appropriate to pairs or sets of roles: mother-son relationships, friendships, and the like
6. *Groups:* Small groups, such as friendship cliques, athletic teams, and work groups.
7. *Organizations:* Formal organizations, such as hospitals or schools
8. *Settlements:* Small-scale "societies" such as villages, ghettos, and neighborhoods, as opposed to large societies such as nations, which are difficult to study
9. *Social worlds:* Ambiguous social entities with vague boundaries and populations, such as "the sports world" and "Wall Street"
10. *Lifestyles or subcultures:* How large numbers of people adjust to life in groups such as a "ruling class" or an "urban underclass"

In all these social settings, field research can reveal things that would not otherwise be apparent. Here's a concrete example.

One issue I'm particularly interested in (Babbie 1985) is the nature of responsibility for public matters: Who is responsible for maintaining the things that we share? Who's responsible for keeping public spaces—parks, malls, buildings, and so on—clean? Who's responsible for seeing that broken street signs get fixed? Or, if a strong wind knocks over garbage cans and rolls them around the street, who's responsible for getting them out of the street?

On the surface, the answer to these questions is pretty clear. We have formal and informal agreements in our society that assign responsibility for these activities. Government custodians are responsible for keeping public places clean. Transportation department employees are responsible for the street signs, and perhaps the police are responsible for the garbage cans rolling around on a windy day. And when these responsibilities are not fulfilled, we tend to look for someone to blame.

What fascinates me is the extent to which the assignment of responsibility for public things to specific individuals not only relieves others of the responsibility but actually prohibits them from taking responsibility. It's my notion that it has become unacceptable for someone like you or me to take personal responsibility for public matters that haven't been assigned to us.

Let me illustrate what I mean. If you were walking through a public park and you threw down a bunch of trash, you'd discover that your action was unacceptable to those around you. People would glare at you, grumble to each other; perhaps someone would say something to you about it. Whatever the form, you'd be subjected to definite, negative sanctions for littering. Now here's the irony. If you were walking through that same park, came across a bunch of trash that someone else had dropped, and cleaned it up, it's likely that your action would also be unacceptable to those around you. You'd probably face negative sanctions for cleaning it up.

When I first began discussing this pattern with students, most felt the notion was absurd. Although we would be negatively sanctioned for littering, cleaning up a public place would obviously bring positive sanctions: People would be pleased with us for doing it. Certainly, all my students said they would be pleased if someone cleaned up a public place. It seemed likely that everyone else would be pleased, too, if we asked them how they would react to someone's cleaning up litter in a public place or otherwise taking personal responsibility for fixing some social problem.

To settle the issue, I suggested that my students start fixing the public problems they came across in the course of their everyday activities. As they did so, I asked them to note the answers to two questions:

1. How did they feel while they were fixing a public problem they had not been assigned responsibility for?
2. How did others around them react?

My students picked up litter, fixed street signs, put knocked-over traffic cones back in place, cleaned and decorated communal lounges in their dorms, trimmed trees that blocked visibility at intersections, repaired public playground equipment, cleaned public restrooms, and took care of a hundred other public problems that weren't "their responsibility."

Most reported feeling very uncomfortable doing whatever they did. They felt foolish, goody-goody, conspicuous, and all the other feelings that usually keep us from performing these activities. In almost every case, their personal feelings of discomfort were increased by the reactions of those around them. One student was removing a damaged and long-unused newspaper box from the bus stop, where it had been a problem for months, when the police arrived, having been summoned by a neighbor. Another student decided to clean out a clogged storm drain on his street and found himself being yelled at by a neighbor who insisted that the mess should be left for the street cleaners. Everyone who picked up litter was sneered at, laughed at, and generally put down. One young man was picking up litter scattered around a trash can when a passerby sneered, "Clumsy!" It became clear to us that there are only three acceptable explanations for picking up litter in a public place:

1. You did it and got caught—somebody forced you to clean up your mess.
2. You did it and felt guilty.
3. You're stealing litter.

In the normal course of life in the United States, it's simply not acceptable for people to take responsibility for public things.

Clearly, we could not have discovered the nature and strength of agreements about taking personal responsibility for public things except through field research. Social norms suggest that taking responsibility is a good thing, sometimes referred to as good citizenship. Asking people what they thought about taking responsibility would have produced a solid consensus that it was good. Only going out into life, doing it, and watching what happened gave us an accurate picture.

As an interesting footnote to this story, my students and I found that whenever people could get past their initial reactions and discover that the students were simply taking responsibility for fixing things for the sake of having them work, the passersby tended to assist. Although there are some very strong agreements making it "unsafe" to take responsibility for public things, the willingness of one person to rise above those agreements seemed to make it safe for others to do so, and they did.

In summary, then, field research offers the advantage of probing social life in its natural habitat. Although some things can be studied adequately in questionnaires or in the laboratory, others cannot. And direct observation in the field lets researchers observe subtle communications and other events that might not be anticipated or measured otherwise.

SPECIAL CONSIDERATIONS IN QUALITATIVE FIELD RESEARCH

There are specific things to take into account in every research method, and qualitative field research is no exception. When you use field research methods, you're confronted with decisions about the role you'll play as an observer and your relations with the people you're observing. Let's examine some of the issues involved in these decisions.

The Various Roles of the Observer

In field research, observers can play any of several roles, including participating in what they want to observe (this was the situation of the students who fixed public things). In this chapter, I've used the term *field research* rather than the frequently used term *participant observation*, because field researchers need not always participate in what they're studying, though they usually will study it directly at the scene of the action. As Catherine Marshall and Gretchen Rossman (1995:60) point out:

> The researcher may plan a role that entails varying degrees of "participantness"—that is, the degree of actual participation in daily life. At one extreme is the full participant, who goes about ordinary life in a role or set of roles constructed in the setting. At the other extreme is the complete observer, who engages not at all in social interaction and may even shun involvement in the world being studied. And, of course, all possible complementary mixes along the continuum are available to the researcher.

The complete participant, in this sense, may be a genuine participant in what he or she is studying (for example, a participant in a campus demonstration) or may pretend to be a genuine participant. In any event, if acting as the complete participant, you let people see you only as a participant,

not as a researcher. For instance, if you're studying a group made up of uneducated and inarticulate people, it would not be appropriate for you to talk and act like a university professor or student.

This type of research introduces an ethical issue, one on which social researchers themselves are divided. Is it ethical to deceive the people you're studying in the hope that they will confide in you in ways that they will not confide in an identified researcher? Do the potential benefits to be gained from the research offset such considerations? Although many professional associations have addressed this issue, the norms to be followed remain somewhat ambiguous when applied to specific situations.

Related to this ethical consideration is a scientific one. No researcher deceives his or her subjects solely for the purpose of deception. Rather, it's done in the belief that the data will be more valid and reliable, that the subjects will be more natural and honest if they do not know the researcher is doing a research project. If the people being studied know they're being studied, they might modify their behavior in a variety of ways. First, they might expel the researcher. Second, they might modify their speech and behavior to appear more respectable than would otherwise be the case. Third, the social process itself might be radically changed. Students making plans to burn down the university administration building, for example, might give up the plan altogether once they learn that one of their group is a social scientist conducting a research project.

On the other side of the coin, if you're a complete participant, you may affect what you're studying. To play the role of participant, you must participate. Yet, your participation may importantly affect the social process you're studying. Suppose, for example, that you're asked for your ideas about what the group should do next. No matter what you say, you will affect the process in some fashion. If the group follows your suggestion, your influence on the process is obvious. If the group decides not to follow your suggestion, the process whereby the suggestion is rejected may affect what happens next. Finally, if you indicate that you just don't know what should be done next, you may be adding to a general feeling of uncertainty and in-decisiveness in the group.

Ultimately, anything the participant-observer does or does not do will have some effect on what's being observed; it's simply inevitable. More seriously, what you do or do not do may have an important effect on what happens. There is no complete protection against this effect, though sensitivity to the issue may provide a partial protection. (This influence, called the Hawthorne effect, was discussed more fully in Chapter 8.)

Because of these several considerations, ethical and scientific, the field researcher frequently chooses a different role from that of complete participant. You could participate fully with the group under study but make it clear that you were also undertaking research. As a member of the volleyball team, for example, you might use your position to launch a study in the sociology of sports, letting your teammates know what you're doing. There are dangers in this role also, however. The people being studied may shift much of their attention to the research project rather than focus on the natural social process, making the process being observed no longer typical. Or, conversely, you yourself may come to identify too much with the interests and viewpoints of the participants. You may begin to "go native" and lose much of your scientific detachment.

At the other extreme, the complete observer studies a social process without becoming a part of it in any way. Quite possibly, because of the researcher's unobtrusiveness, the subjects of study might not realize they're being studied. Sitting at a bus stop to observe jaywalking at a nearby intersection is one example. Although the complete observer is less likely to affect what's being studied and less likely to "go native" than the complete participant, she or he is also less likely to develop a full appreciation of what's being studied. Observations may be more sketchy and transitory.

Fred Davis (1973) characterizes the extreme roles that observers might play as "the Martian" and "the Convert." The latter involves delving deeper and deeper into the phenomenon under study, running the risk of "going native." We'll examine this risk further in the next section.

To appreciate the "Martian" approach, imagine that you were sent to observe some newfound life on Mars. Probably you would feel yourself inescapably separate from the Martians. Some social scientists adopt this degree of separation when observing cultures or social classes different from their own.

Marshall and Rossman (1995:60–61) also note that the researcher can vary the amount of time spent in the setting being observed: You can be a full-time presence on the scene or just show up now and then. Moreover, you can focus your attention on a limited aspect of the social setting or seek to observe all of it—framing an appropriate role to match your aims.

Different situations ultimately require different roles for the researcher. Unfortunately, there are no clear guidelines for making this choice—you must rely on your understanding of the situation and your own good judgment. In making your decision, however, you must be guided by both methodological and ethical considerations. Because these often conflict, your decision will frequently be difficult, and you may find sometimes that your role limits your study.

Relations to Subjects

Having introduced the different roles field researchers might play in connection with their observations, we now focus more specifically on how researchers may relate to the subjects of their study and to the subjects' points of view.

We've already noted the possibility of pretending to occupy social statuses we don't really occupy. Consider now how you would think and feel in such a situation.

Suppose you've decided to study a religious cult that has enrolled many people in your neighborhood. You might study the group by joining it or pretending to join it. Take a moment to ask yourself what the difference is between "really" joining and "pretending" to join. The main difference is whether or not you actually take on the beliefs, attitudes, and other points of view shared by the "real" members. If the cult members believe that Jesus will come next Thursday night to destroy the world and save the members of the cult, do you believe it or do you simply pretend to believe it?

Traditionally, social scientists have tended to emphasize the importance of "objectivity" in such matters. In this example, that injunction would be to avoid getting swept up in the beliefs of the group. Without denying the advantages associated with such objectivity, social scientists today also recognize the benefits gained by immersing themselves in the points of view they're studying, what Lofland and Lofland (1995:61) refer to as "insider understanding." Ultimately, you will not be able to fully understand the thoughts and actions of the cult members unless you can adopt their points of view as true—at least temporarily. To fully appreciate the phenomenon you've set out to study, you need to believe that Jesus is coming Thursday night.

Adopting an alien point of view is an uncomfortable prospect for most people. It can be hard enough merely to learn about views that seem strange to you; you may sometimes find it hard just to tolerate certain views. But to take them on as your own is ten times worse. Robert Bellah (1970, 1974) has offered the term *symbolic realism* to indicate the need for social researchers to treat the beliefs they study as worthy of respect rather than as objects of ridicule. If you seriously entertain this prospect, you may appreciate why William Shaffir and Robert Stebbins (1991:1) concluded that "fieldwork must certainly rank with the more disagreeable activities that humanity has fashioned for itself."

There is, of course, a danger in adopting the points of view of the people you're studying. When you abandon your objectivity in favor of adopting such views, you lose the possibility of seeing and understanding the phenomenon within frames of reference unavailable to your subjects. On the one hand, accepting the belief that the world will end Thursday night allows you to appreciate aspects of that belief available only to believers; stepping outside that view, however, makes it possible for you to consider some reasons why people might adopt such a view. You may discover that some did so as a consequence of personal trauma (such as unemployment or divorce) while others were

brought into the fold through their participation in particular social networks (for example, their whole bowling team joined the cult). Notice that the cult members might disagree with those "objective" explanations, and you might not come up with them to the extent that you had operated legitimately within the group's views.

The apparent dilemma here is that these postures offer important advantages but also seem mutually exclusive. In fact, it is possible to assume both postures. Sometimes you can simply shift viewpoints at will. When appropriate, you can fully assume the beliefs of the cult; later, you can step outside those beliefs (more accurately, you can step inside the viewpoints associated with social science). As you become more adept at this kind of research, you may come to hold contradictory viewpoints simultaneously, rather than switch back and forth.

During my study of trance channelers—people who allow spirits to occupy their bodies and speak through them—I found I could participate fully in channeling sessions without becoming alienated from conventional social science. Rather than "believing" in the reality of channeling, I found it possible to suspend beliefs in that realm: neither believing it to be genuine (like most of the other participants) nor disbelieving it (like most scientists). Put differently, I was open to either possibility. Notice how this differs from our normal need to "know" whether such things are legitimate or not.

Social researchers often refer to the concerns just discussed as a matter of *reflexivity,* in the sense of things acting on themselves. Thus, your own characteristics can affect what you see and how you interpret it. The issue is broader than that, however, and applies to the subjects as well as to the researcher. Imagine yourself interviewing a homeless person (1) on the street, (2) in a homeless shelter, or (3) in a social welfare office. The research setting could affect the person's responses. In other words, you might get different results because of where you conducted the interview. Moreover, you might act differently as a researcher in those different settings. If you reflect on this issue, you'll be able to identify other aspects of the research encounter that complicate the task of "simply observing what's so."

The problem we've just been discussing could be seen as psychological, occurring mostly inside the researchers' or subjects' heads. There is a corresponding problem at a social level, however. When you become deeply involved in the lives of the people you're studying, you're likely to be moved by their personal problems and crises. Imagine, for example, that one of the cult members becomes ill and needs a ride to the hospital. Should you provide transportation? Sure. Suppose someone wants to borrow money to buy a stereo. Should you loan it? Probably not. Suppose they need the money for food?

There are no black-and-white rules for resolving situations such as these, but you should realize that you will need to deal with them regardless of whether or not you reveal that you're a researcher. Such problems do not tend to arise in other types of research—surveys and experiments, for example—but they are part and parcel of field research.

This discussion of the field researcher's relations to subjects flies in the face of the conventional view of "scientific objectivity." Before concluding this section, let's take the issue one step further.

In the conventional view of science, there are implicit differences of power and status separating the researcher from the subjects of research. When we discussed experimental designs in Chapter 8, for example, it was obvious who was in charge: the experimenter. The experimenter organized things and told the subjects what to do. Often the experimenter was the only person who even knew what the research was really about. Something similar might be said about survey research. The person running the survey designs the questions, decides who will be selected for questioning, and is responsible for making sense out of the data collected.

Sociologists often look at these sorts of relationships as power or status relationships. In experimental and survey designs, the researcher clearly has more power and a higher status than do the people being studied. The researchers have a special knowledge that the subjects don't enjoy. They're not so crude as to say they're superior to their subjects, but there's a sense in which that's implicitly assumed. (Notice that there is a similar, implicit assumption about the writers and readers of textbooks.)

In field research, such assumptions can be problematic. When the early European anthropologists set out to study what were originally called "primitive" societies, there was no question but that the anthropologists knew best. Whereas the natives "believed" in witchcraft, for example, the anthropologists "knew" it wasn't really true. While the natives said some of their rituals would appease the gods, the anthropologists explained that the "real" functions of these rituals were the creation of social identity, the establishment of group solidarity, and so on.

The more social researchers have gone into the field to study their fellow humans face-to-face, however, the more they have become conscious of these implicit assumptions about researcher superiority, and the more they have considered alternatives. As we turn now to the various paradigms of field research, we'll see some of the ways in which that ongoing concern has worked itself out.

SOME QUALITATIVE FIELD RESEARCH PARADIGMS

Although I've described field research as simply going where the action is and observing it, there are actually many different approaches to this research method. This section examines several field research paradigms: naturalism, ethnomethodology, grounded theory, case studies and the extended case method, institutional ethnography, and participatory action research. Although this survey won't exhaust the variations on the method, it should give you a broad appreciation of the possibilities.

There aren't any specific methods attached to each of these paradigms. You could do ethnomethodology or institutional ethnography by analyzing court hearings or conducting group interviews, for example. The important distinctions of this section are *epistemological,* that is, having to do with what data mean, regardless of how they were collected.

Naturalism

Naturalism is an old tradition in qualitative research. The earliest field researchers operated on the positivist assumption that social reality was "out there," ready to be naturally observed and reported by the researcher as it "really is" (Gubrium and Holstein 1997). This tradition started in the 1930s and 1940s at the University of Chicago's sociology department, whose faculty and students fanned out across the city to observe and understand local neighborhoods and communities. The researchers of that era and their research approach are now often referred to as the "Chicago School."

One of the earliest and best-known studies that illustrates this research tradition is William Foote Whyte's ethnography of Cornerville, an Italian-American neighborhood, in his book *Street Corner Society* (1943). An **ethnography** is a study that focuses on detailed and accurate description rather than explanation. Like other naturalists, Whyte believed that in order to fully learn about social life on the streets, he needed to become more of an insider. He made contact with "Doc," his key informant, who appeared to be one of the street-gang leaders. Doc let Whyte enter his world, and Whyte got to participate in the activities of the people of Cornerville. His study offered something that surveys could not: a richly detailed picture of life among the Italian immigrants of Cornerville.

An important feature of Whyte's study is that he reported the reality of the people of Cornerville on their terms. The naturalist approach is based on telling "their" stories the way they "really are," not the way the ethnographer understands "them." The narratives collected by Whyte are taken at face value as the social "truth" of the Cornerville residents.

Forty years later, David A. Snow and Leon Anderson (1987) conducted exploratory field research into the lives of homeless people in Austin, Texas. Their main task was to understand how the homeless construct and negotiate their identity while knowing that the society they live in attaches a stigma to homelessness. Snow and Anderson believed that, to achieve this goal, the collection of data had to arise naturally. Like Whyte in *Street Corner Society,* they found some key informants whom they followed in their everyday journeys, such as at their day-labor pickup sites or under bridges. Snow and Anderson chose to memorize the conversations they participated in or the "talks" that homeless people had with each other. At the end of the day, the two researchers debriefed and wrote detailed field notes about all the "talks" they encountered. They also taped in-depth interviews with their key informants.

Snow and Anderson reported "hanging out" with homeless people over the course of 12 months for a total of 405 hours in 24 different settings. Out of these rich data, they identified three related patterns in homeless people's conversations. First, the homeless showed an attempt to "distance" themselves from other homeless people, from the low-status job they currently had, or from the Salvation Army they depended on. Second, they "embraced" their street-life identity, their group membership or a certain belief about why they are homeless. Third, they told "fictive stories" that always contrasted with their everyday life. For example, they would often say that they were making much more money than they really were, or even that they were "going to be rich."

Ethnomethodology

Ethnomethodology, which I introduced as a research paradigm in Chapter 2, is a very different approach to qualitative field research. It has its roots in the philosophical tradition of phenomenology, which can explain why ethnomethodologists are skeptical about the way people report their experience of reality (Gubrium and Holstein 1997). Alfred Schutz (1967, 1970), who introduced phenomenology, argued that reality was socially constructed rather than being "out there" for us to observe. People describe their world not "as it is" but "as they make sense of it." Thus, phenomenologists would argue that Whyte's street-corner men were describing their gang life as it made sense to them. Their reports, however, would not tell us how and why it made sense to them. For this reason, researchers cannot rely on their subjects' stories to depict social realities accurately.

Whereas traditional ethnographers believe in immersing themselves in a particular culture and reporting their informants' stories as if they represent reality, phenomenologists see a need to "make sense" out of the informants' perceptions of the world. Following in this tradition, some field researchers have tried to devise techniques that reveal how people make sense of their everyday world. As we saw in Chapter 2, the sociologist Harold Garfinkel suggested that researchers *break the rules* so that people's taken-for-granted expectations would become apparent. This is the technique that Garfinkel called "ethnomethodology."

Garfinkel became known for engaging his students to perform a series of what he called "breaching experiments" designed to break away from the ordinary (Heritage 1984). For instance, Garfinkel (1967) asked his students to do a "conversation clarification experiment." Students were told to engage in an ordinary conversation with an acquaintance or a friend and to ask for clarification about any of this person's statements. Through this technique, they uncovered elements of conversation that are normally taken for granted. Here are two examples of what Garfinkel's students reported (1967:42):

Case 1

The subject was telling the experimenter, a member of the subject's car pool, about having had a flat tire while going to work the previous day.

(S) I had a flat tire.

(E) What do you mean, you had a flat tire?

She appeared momentarily stunned. Then she answered in a hostile way: "What do you mean, 'What do you mean?' A flat tire is a flat tire. That is what I meant. Nothing special. What a crazy question."

Case 6

The victim waved his hand cheerily.

(S) How are you?

(E) How I am in regard of what? My health, my finances, my school work, my peace of mind, my . . . ?

(S) (Red in the face and suddenly out of control.) Look I was just trying to be polite. Frankly, I don't give a damn how you are.

By setting aside or "bracketing" their expectations from these everyday conversations, the experimenters made visible the subtleties of mundane interactions. For example, although "How are you?" has many possible meanings, none of us have any trouble knowing what it means in casual interactions, as the unsuspecting subject revealed in his final comment.

Ethnomethodologists, then, are not simply interested in subjects' perceptions of the world. In these cases, we could imagine that the subjects may have thought that the experimenters were rude, stupid, or arrogant. The conversation itself, not the informants, become the object of ethnomethodological studies. In general, in ethnomethodology the focus is on the "underlying patterns" of interactions that regulate our everyday lives.

Ethnomethodologists believe that researchers who use a naturalistic analysis "[lose] the ability to analyze the commonsense world and its culture if [they use] analytical tools and insights that are themselves part of the world or culture being studied" (Gubrium and Holstein 1997:43). Laurence Wieder provides an excellent example of how much a naturalistic approach differs from an ethnomethodological approach (Gubrium and Holstein 1997). In his study, *Language and Social Reality: The Case of Telling the Convict Code* (1988), Wieder started to approach convicts in a halfway house in a traditional ethnographic style: He was going to become an insider by befriending the inmates and by conducting participant observations. He took careful notes and recorded interactions among inmates and between inmates and staff. His first concern was to describe the life of the convicts of the halfway house the way it "really was" for them. Wieder's observations allowed him to report on a "convict code" that he thought was the source of the deviant behavior expressed by the inmates toward the staff. This code, which consisted of a series of rules such as "Don't kiss ass," "Don't snitch," and "Don't trust the staff," was followed by the inmates who interfered with the staff members' attempts to help them make the transition between prison and the community.

It became obvious to Wieder that the code was more than an explanation for the convicts' deviant behavior; it was a "method of moral persuasion and justification" (Wieder 1988:175). At this point he changed his naturalistic approach to an ethnomethodological one. Recall that whereas naturalistic field researchers aim to understand social life as the participants understand it, ethnomethodologists are more intent on identifying the methods through which understanding occurs. In the case of the convict code, Wieder came to see that convicts used the code to make sense of their own interactions with other convicts and with the staff. The ethnography of the halfway house thus shifted to an ethnography of the code. For instance, the convicts would say "You know I won't snitch," referring to the code as a way to justify their refusal to answer Wieder's question (p. 168). According to Wieder, the code "operated as a device for stopping or changing the topic of conversation" (p. 175). Even the staff would refer to the code to justify their reluctance to help the convicts. While the code was something that constrained behavior, it also functioned as a tool for the control of interactions.

Grounded Theory

Grounded theory originated from the collaboration of Barney Glaser and Anselm Strauss, sociologists who brought together two main traditions of research: positivism and interactionism. Essentially, **grounded theory** is the attempt to derive theories from an analysis of the patterns, themes, and common categories discovered in observational data. The first major presentation of this method can be found in Glaser and Strauss's book, *The Discovery of Grounded Theory* (1967). Grounded theory can be described as an approach that attempts to combine a naturalist approach with a positivist concern for a "systematic set of procedures" in doing qualitative research.

Strauss and Juliet Corbin (1990:44–46) have suggested that grounded theory allows the researcher to be scientific and creative at the same time, as long as the researcher follows three guidelines:

> **Periodically step back and ask:** What is going on here? Does what I think I see fit the reality of the data? The data themselves do not lie. . . .
>
> **Maintain an attitude of skepticism.** All theoretical explanations, categories, hypotheses, and questions about the data, whether they come directly or indirectly from the making of comparisons, the literature, or from experience, should be regarded as provisional. They always need to be checked out against the actual data, and never accepted as a fact. . . .
>
> **Follow the research procedures.** The data collection and analytical procedures are designed to give rigor to a study. At the same time they help you to break through biases, and lead you to examine at least some of your assumptions that might otherwise affect an unrealistic reading of the data.

Grounded theory emphasizes research procedures. In particular, systematic coding is important for achieving validity and reliability in the data analysis. Because of this somewhat positivistic view of data, grounded theorists are very open to the use of qualitative studies in conjunction with quantitative ones. Here are two examples of the implementation of this approach.

Studying Academic Change   Clifton F. Conrad's (1978) study of academic change in universities is an early example of the grounded theory approach. Conrad hoped to uncover the major sources of changes in academic curricula and at the same time understand the process of change. Using the grounded theory idea of *theoretical sampling*—whereby groups or institutions are selected on the basis of their theoretical relevance—Conrad chose four universities for the purpose of his study. In two, the main vehicle of change was the formal curriculum committee; in the other two, the vehicle of change was an ad hoc group.

Conrad explained, step by step, the advantage of using the grounded theory approach in building his theory of academic change. He described the process of systematically coding data in order to create categories that must "emerge" from the data and then assessing the fitness of these categories with each other. Going continuously from data to theory and theory to data allowed him to reassess the validity of his initial conclusions about academic change.

For instance, it first seemed that academic change was mainly caused by an administrator who was pushing for it. By reexamining the data and looking for more-plausible explanations, Conrad found the pressure of interest groups a more convincing source of change. The emergence of these interest groups actually allowed the administrator to become an agent of change.

Assessing how data from each of the two types of universities fit with the other helped refine theory building. This refinement process stands in contrast to a naturalist approach, in which the process of building theory would have stopped with Conrad's first interpretation.

Conrad concluded that changes in university curricula are based on the following process: Conflict and interest groups emerge because of internal and external social structural forces; they push for administrative intervention and recommendation to make changes in the current academic program; these changes are then made by the most powerful decision-making body.

Shopping Romania Much has been written about large-scale changes caused by the shift from socialism to capitalism in the former USSR and its Eastern-European allies. Patrick C. Jobes and his colleagues (1997) wanted to learn about the transition on a smaller scale among average Romanians. They focused on the task of shopping.

Noting that shopping is normally thought of as a routine, relatively rational activity, the researchers suggested that it could become a social problem in a radically changing economy. They used the grounded theory method to examine Romanian shopping as a social problem, looking for the ways in which ordinary people solved the problem.

Their first task was to learn something about how Romanians perceived and understood the task of shopping. The researchers—participants in

a social problems class—began by interviewing 40 shoppers and asking whether they had experienced problems in connection with shopping and what actions they had taken to cope with those problems.

Once the initial interviews were completed, the researchers reviewed their data, looking for categories of responses—the shoppers' most common problems and solutions. One of the most common problems was a lack of money. This led to the researchers' first working hypothesis: The "socio-economic position of shoppers would be associated with how they perceived problems and sought solutions" (1997:133). This and other hypotheses helped the researchers to focus their attention on more-specific variables in subsequent interviewing.

As they continued, they also sought to interview other types of shoppers. When they interviewed students, for example, they discovered that different types of shoppers were concerned with different kinds of goods, which in turn affected the problems faced and the solutions tried.

As additional hypotheses were developed in response to the continued interviewing, the researchers began to develop a more or less standardized set of questions to ask shoppers. Initially, all the questions were open-ended, but they eventually developed closed-ended items as well.

This study illustrates the key, inductive principles of grounded theory: data are collected in the absence of hypotheses. The initial data are used to determine the key variables as perceived by those being studied, and hypotheses about relationships among the variables are similarly derived from the data collected. Continuing data collection yields refined understanding and, in turn, sharpens the focus of data collection itself.

Case Studies and the Extended Case Method

Social researchers often speak of **case studies,** which focus attention on one or a few instances of some social phenomenon, such as a village, a family, or a juvenile gang. As Charles Ragin and Howard Becker (1992) point out, there is little consensus on what may constitute a "case" and the term is used broadly. The case being studied, for example, might be a period of time rather than a particular group of people. The limitation of attention to a particular instance of something is the essential characteristic of the case study.

The chief purpose of a case study may be descriptive, as when an anthropologist describes the culture of a preliterate tribe. Or the in-depth study of a particular case can yield explanatory insights, as when the community researchers Robert and Heley Lynd (1929, 1937) and W. Lloyd Warner (1949) sought to understand the structure and process of social stratification in small-town USA.

Case study researchers may seek only an idiographic understanding of the particular case under examination, or—as we've seen with grounded theory—case studies can form the basis for the development of more general, nomothetic theories.

Michael Burawoy and his colleagues (1991) have suggested a somewhat different relationship between case studies and theory. For them, the **extended case method** has the purpose of discovering flaws in, and then modifying, existing social theories. This approach differs importantly from some of the others already discussed.

Whereas the grounded theorists seek to enter the field with no preconceptions about what they'll find, Burawoy suggests just the opposite: to try "to lay out as coherently as possible what we expect to find in our site *before* entry" (Burawoy et al. 1991:9). Burawoy sees the extended case method as a way to rebuild or improve theory instead of approving or rejecting it. Thus, he looks for all the ways in which observations conflict with existing theories and what he calls "theoretical gaps and silences" (1991:10). This orientation to field research implies that knowing the literature beforehand is actually a must for Burawoy and his colleagues, whereas grounded theorists would worry that knowing what others have concluded might bias their observations and theories.

To illustrate the extended case method, I'll use two examples of studies by Burawoy's students.

Teacher-Student Negotiations Leslie Hurst (1991) set out to study the patterns of interaction between teachers and students of a junior high school. She went into the field armed with existing, contradictory theories about the "official" functions of the school. Some theories suggested that the purpose of schools was to promote social mobility, whereas others suggested that schools mainly reproduced the status quo in the form of a stratified division of labor. The official roles assigned to teachers and students could be interpreted in terms of either view.

Hurst was struck, however, by the contrast between these theories and the types of interactions she observed in the classroom. In her own experiences as a student, teachers had total rights over the mind, body, and soul of their pupils. She observed something very different at a school in a lower-middle-class neighborhood in Berkeley, California—Emerald Junior High School, where she volunteered as a tutor. She had access to several classrooms, the lunchroom, and the English Department's meetings. She wrote field notes based on the negotiation interactions between students and teachers. She explained the nature of the student-teacher negotiations she witnessed by focusing on the separation of functions among the school, the teacher, and the family.

In Hursts's observation, the school fulfilled the function of controlling its students' "bodies"—for example, by regulating their general movements and activities within the school. The students' "minds" were to be shaped by the teacher, whereas students' families were held responsible for their "souls"; that is, families were expected to socialize students regarding personal values, attitudes, sense of property, and sense of decorum. When students don't come to school with these values in hand, the teacher, according to Hurst, "must first negotiate with the students some compromise on how the students will conduct themselves and on what will be considered classroom decorum" (1991:185).

Hurst explained the constant bargaining between teachers and students is an expression of the separation between "the body," which is the school's concern, and "the soul" as family domain. The teachers, who had limited sanctioning power to control their students' minds in the classroom, were using forms of negotiations with students so that they could "control . . . the student's body and sense of property" (1991:185), or as Hurst defines it, "baby-sit" the student's body and soul.

Hurst says she differs from the traditional sociological perspectives as follows:

> I do not approach schools with a futuristic eye. I do not see the school in terms of training, socializing, or slotting people into future hierarchies. To approach schools in this manner is to miss the negotiated, chaotic aspects of the classroom and educational experience. A futurist perspective tends to impose an order and purpose on the school experience, missing its day-to-day reality. — (1991:186)

In summary, what emerges from Hurst's study is an attempt to improve the traditional sociological understanding of education by adding the idea that classroom, school, and family have separate functions, which in turn can explain the emergence of "negotiated order" in the classroom.

The Fight against AIDS   Katherine Fox (1991) set out to study an agency whose goal was to fight the AIDS epidemic by bringing condoms and bleach (for cleaning needles) to intravenous drug users. It's a good example of finding the limitations of well-used models of theoretical explanation in the realm of understanding deviance—specifically, the "treatment model" that predicted that drug users would come to the clinic and ask for treatment. Fox's interactions with outreach workers—most of whom were part of the community of drug addicts or former prostitutes—contradicted that model.

To begin, it was necessary to understand the drug users' subculture and use that knowledge to devise more realistic policies and programs. The target users had to be convinced, for example, that the program workers could be trusted, that they were really interested only in providing bleach and condoms. The target users needed to be sure they were not going to be arrested.

Fox's field research didn't stop with an examination of the drug users. She also studied the agency workers, discovering that the outreach program meant different things to the research directors and the outreach workers. Some of the volunteers who were actually providing the bleach and condoms were frustrated about the minor changes they felt they could make. Many thought the program was just a bandage on the AIDS and drug-abuse problems. Some resented having to take field notes. Directors, on the other hand, needed reports and field notes so that they could validate their research in the eyes of the federal and state agencies that financed the project. Fox's study showed how the AIDS research project developed the bureaucratic inertia typical of established organizations: Its goal became that of sustaining itself.

Both of these studies illustrate how the extended case method can operate. The researcher enters the field with full knowledge of existing theories but aims to uncover contradictions that require the modification of those theories.

Institutional Ethnography

Institutional ethnography is an approach originally developed by Dorothy Smith (1978) to better understand women's everyday experiences by discovering the power relations that shape those experiences. Today this methodology has been extended to the ideologies that shape the experiences of any oppressed subjects.

Smith and other sociologists believe that if researchers ask women or other members of subordinated groups about "how things work," they can discover the institutional practices that shape their realities (M. L. Campbell 1998; D. Smith 1978). The goal of such inquiry is to uncover forms of oppression that often are overlooked by more traditional types of research.

Dorothy Smith's methodology is similar to ethnomethodology in the sense that the subjects themselves are not the focus of the inquiry. The institutional ethnographer starts with the personal experiences of individuals but proceeds to uncover the institutional power relations that structure and govern those experiences. In this process, the researcher can reveal aspects of society that would have been missed by an inquiry that began with the official purposes of institutions.

This approach links the "microlevel" of everyday personal experiences with the "macrolevel" of institutions. As M. L. Campbell put it:

> Institutional ethnography, like other forms of ethnography, relies on interviewing, observations and document as data. Institutional ethnography departs from other ethnographic approaches by treating those data not as the topic or object of interest, but as "entry" into the social relations of the setting. The idea is to tap into people's expertise. — (1998:57)

Here are two examples of this approach.

Mothering, Schooling, and Child Development

Our first example of institutional ethnography is a study by Alison Griffith (1995), who collected data with Dorothy Smith on the relationship between mothering, schooling, and children's development. Griffith started by interviewing mothers from three cities of southern Ontario on their everyday work of creating a relationship between their families and the school. This was the starting point for other interviews with parents, teachers, school administrators, social workers, school psychologists, and central office administrators.

In her findings, Griffith explained how the discourse about mothering had shifted its focus over time from mother-child interactions to "child-centered" recommendations. She saw a distinct similarity in the discourse used by schools, the media (magazines and television programs), the state, and child development professionals.

Teachers and child development professionals saw the role of mothers in terms of a necessary collaboration between mothers and schools for the child's success not only in school but also in life. Because of unequal resources, all mothers do not participate in this discourse of "good" child development the same way. Griffith found that working-class mothers were perceived as weaker than middle-class mothers in the "stimulation" effort of schooling. Griffith argued that this child development discourse, embedded in the school institution, perpetuates the reproduction of class by making middle-class ideals for family-school relations the norm for everyone.

Compulsory Heterosexuality

The second illustration of institutional ethnography is taken from Didi Khayatt's (1995) study of the institutionalization of compulsory heterosexuality in schools and its effects on lesbian students. In 1990, Khayatt began her research by interviewing 12 Toronto lesbians, 15 to 24 years of age. Beginning with the young women's viewpoint, she then expanded her inquiry to other students, teachers, guidance counselors, and administrators.

Khayatt found that the school's administrative practices generated a *compulsory heterosexuality*, which produced a sense of marginality and vulnerability among lesbian students. For example, the school didn't punish harassment and name-calling against gay students. The issue of homosexuality was excluded from the curriculum lest it appear to students as an alternative to heterosexuality.

In both of the studies I've described, the inquiry began with the women's standpoint—mothers and lesbian students. However, instead of emphasizing the subjects' viewpoints, both analyses focused on the power relations that shaped these women's experiences and reality.

Participatory Action Research

Our final field research paradigm takes us further along in our earlier discussion of the status and power relationships linking researchers to the subjects of their research. Within the **participatory action research** paradigm (PAR), the researcher's function is to serve as a resource to those being studied—typically, disadvantaged groups—as an opportunity for them to act effectively in their own interest. The disadvantaged subjects define their problems, define the remedies desired, and take the lead in designing the research that will help them realize their aims.

This approach began in Third-World research development, but it spread quickly to Europe and North America (Gaventa 1991). It comes from a vivid critique of classical social science research. According to the PAR paradigm, traditional research is perceived as an "elitist model" (Whyte, Greenwood, and Lazes 1991) that reduces the "subjects" of research to "objects" of research. According to many advocates of this perspective, the distinction between the researcher and the researched should disappear. They argue that the subjects who will be affected by research should also be responsible for its design.

Implicit in this approach is the belief that research functions not only as a means of knowledge production but also as a "tool for the education and development of consciousness as well as mobilization for action" (Gaventa 1991:121–22). Advocates of participatory action research equate access to information with power and argue that this power has been kept in the hands of the dominant class, sex, ethnicity, or nation. Once people see themselves as researchers, they automatically regain power over knowledge.

Examples of this approach include community power structure research, corporate research, and "right-to-know" movements (Whyte, Greenwood, and Lazes 1991). Here are two examples of corporate research that used a PAR approach.

The Xerox Corporation A participatory action research project took place at the Xerox corporation at the instigation of leaders of both management and the union. Management's goal was to lower costs so that the company could thrive in an increasingly competitive market. The union suggested a somewhat broader scope: improving the quality of working life while lowering manufacturing costs and increasing productivity.

Company managers began by focusing attention on shop-level problems; they were less concerned with labor contracts or problematic managerial policies. At the time, management had a plan to start an "outsourcing" program that would lay off 180 workers, and the union had begun mobilizing to oppose the plan. Peter Lazes, a consultant hired by Xerox, spent the first month convincing management and the union to create a "cost study team" (CST) that included workers in the wire harness department.

Eight full-time workers were assigned to the CST for six months. Their task was to study the possibilities of making changes that would save the company $3.2 million and keep the 180 jobs. The team had access to all financial information

and was authorized to call on anyone within the company. This strategy allowed workers to make suggestions outside the realm usually available to them. According to Whyte and his colleagues, "reshaping the box enabled the CST to call upon management to explain and justify all staff services" (1991:27). Because of the changes suggested by the CST and implemented by management, the company saved the targeted $3.2 million.

Management was so pleased by this result that it expanded the wire harness CST project to three other departments that were threatened by competition. Once again, management was happy about the money saved by the teams of workers.

The Xerox case study is an interesting example of participatory action research because it shows how the production of knowledge does not always have to be an elitist enterprise. The "experts" do not necessarily have to be the professionals. According to Whyte and his colleagues, "at Xerox, participatory action research created and guided a powerful process of organizational learning—a process whereby leaders of labor and management learned from each other and from the consultant/facilitator, while he learned from them" (1991:30).

Mondragón-Cornell Project The Mondragón-Cornell PAR project is based on the Mondragón cooperative complex in the Basque area of Spain. The complex includes industrial-worker cooperatives, a consumer-worker cooperative, a cooperative bank, a cooperative research and development organization, and other supporting and linked structures (Whyte, Greenwood, and Lazes 1991:31).

The research project arose from three things: the need for organizational restructuring, the increase in national unemployment, and the large-scale expansion of the Mondragón complex. The project was proposed by the sociologist William Foote Whyte and jointly directed by Davydd Greenwood, the director of Cornell's Center for International Studies, and Jose Luis Gonzalez, a staff member in the Mondragón cooperative. Whyte was involved in the early stage of the project. Greenwood and Gonzalez played the roles of facilitators and consultants during the whole project, which was mainly financed with grants from the

Spain-U.S. Joint Committee on Educational and Cultural Exchange.

A main team of 15 cooperative members was established, and they set out to survey approximately 50 other members. From the first stage of the project, the research revolved around using the literature and theories about corporate culture and contrasting these concepts and theories with the everyday reality of the Mondragón cooperative. Early on, the team concluded that the cooperative needed to capitalize on the diversity of its members instead of implementing uniformity in organizational changes.

This approach to field research was particularly efficient in uncovering feelings that were shared by the cooperative members, such as commitment to the organization and an interest in rewards beyond a job and a secure salary. Collaboration with members of the cooperative who had diverse ages, experiences, formal positions, and educational backgrounds increased the level of commitment and a sense that they needed to find solutions in order to match their ideals with realities. Because "conceptualization and application went hand in hand" (Whyte, Greenwood, and Lazes 1991:39) in the analysis generated by the PAR project, it also encouraged the same process in finding solutions.

As you can see, the seemingly simple process of observing social action as it occurs has subtle though important variations. As we saw in Chapter 2, all our thoughts occur within, and are shaped by, paradigms, whether we're conscious of it or not. Qualitative field researchers have been unusually deliberate in framing a variety of paradigms to enrich the observation of social life.

CONDUCTING QUALITATIVE FIELD RESEARCH

So far in this chapter we've considered the kinds of topics appropriate to qualitative field research, special considerations in doing this kind of research, and a sampling of paradigms that direct different types of research efforts. Along the way we've seen some examples that illustrate field research in action. To round out the picture, we turn now to specific ideas and techniques for conducting field research, beginning with how researchers prepare for work in the field.

Preparing for the Field

Suppose for the moment that you've decided to undertake field research on a campus political organization. Let's assume further that you're not a member of that group, that you do not know a great deal about it, and that you will identify yourself to the participants as a researcher. This section will use this example and others to discuss some of the ways you might prepare yourself before undertaking direct observations.

As is true of all research methods, you would be well advised to begin with a search of the relevant literature, filling in your knowledge of the subject and learning what others have said about it (library research is discussed at length in Appendix A).

In the next phase of your research, you might wish to discuss the student political group with others who have already studied it or with anyone else likely to be familiar with it. In particular, you might find it useful to discuss the group with one or more informants (discussed in Chapter 7). Perhaps you have a friend who is a member, or you can meet someone who is. This aspect of your preparation is likely to be more effective if your relationship with the informant extends beyond your research role. In dealing with members of the group as informants, you should take care that your initial discussions do not compromise or limit later aspects of your research. Keep in mind that the impression you make on the informant, the role you establish for yourself, may carry over into your later effort. For example, creating the initial impression that you may be an undercover FBI agent is unlikely to facilitate later observations of the group.

You should also be wary about the information you get from informants. Although they may have more direct, personal knowledge of the subject under study than you do, what they "know" is probably a mixture of fact and point of view. Members of the political group in our example would be unlikely to provide completely unbiased information (so would members of opposing political groups).

Before making your first contact with the student group, then, you should already be quite familiar with it, and you should understand its general philosophical context.

There are many ways to establish your initial contact with the people you plan to study. How you do it will depend, in part, on the role you intend to play. Especially if you decide to take on the role of complete participant, you must find a way to develop an identity with the people to be studied. If you wish to study dishwashers in a restaurant, the most direct method would be to get a job as a dishwasher. In the case of the student political group, you might simply join the group.

Many of the social processes appropriate to field research are open enough to make your contact with the people to be studied rather simple and straightforward. If you wish to observe a mass demonstration, just be there. If you wish to observe patterns in jaywalking, hang around busy streets.

Whenever you wish to make more formal contact with the people, identifying yourself as a researcher, you must establish a rapport with them. You might contact a participant with whom you feel comfortable and gain that person's assistance. In studying a formal group, you might approach the groups' leaders, or you may find that one of your informants can introduce you.

While you'll probably have many options in making your initial contact with the group, realize that your choice can influence your subsequent observations. Suppose, for example, that you're studying a university and begin with high-level administrators. This choice is likely to have a couple of important consequences. First, your initial impressions of the university will be shaped to some extent by the administrators' views, which will be quite different from those of students or faculty. This initial impression may influence the way you observe and interpret events subsequently—especially if you're unaware of the influence.

Second, if the administrators approve of your research project and encourage students and faculty to cooperate with you, the latter groups will probably look on you as somehow aligned with the administration, which can affect what they say to you. For example, faculty members might be reluc-

tant to tell you about plans to organize through the teamsters' union.

In making a direct, formal contact with the people you want to study, you'll be required to give them some explanation of the purpose of your study. Here again, you face an ethical dilemma. Telling them the complete purpose of your research might eliminate their cooperation altogether or significantly affect their behavior. On the other hand, giving only what you believe would be an acceptable explanation may involve outright deception. Your decisions in this and other matters will probably be largely determined by the purpose of your study, the nature of what you're studying, the observations you wish to use, and similar factors, but ethical considerations must be taken into account as well.

Previous field research offers no fixed rule—methodological or ethical—to follow in this regard. Your appearance as a researcher, regardless of your stated purpose, may result in a warm welcome from people who are flattered that a scientist finds them important enough to study. Or, it may result in your being totally ostracized or worse. It probably wouldn't be a good idea, for example, to burst into a meeting of an organized crime syndicate and announce that you're writing a term paper on organized crime.

Qualitative Interviewing

In part, field research is a matter of going where the action is and simply watching and listening. As the baseball star Yogi Berra said, "You can observe a lot by watching"—provided that you're paying attention. At the same time, as I've already indicated, field research can involve more active inquiry. Sometimes it's appropriate to ask people questions and record their answers. Your on-the-spot observations of a full-blown riot will lack something if you don't know why people are rioting. Ask somebody.

We've already discussed interviewing in Chapter 9, and much of what was said there applies to qualitative field interviewing. The interviewing you'll do in connection with field observation, however, is different enough to demand a separate treatment. In surveys, questionnaires are rigidly structured; however, less structured interviews are more appropriate to field research. As Herbert and Riene Rubin (1995:43) describe the distinction: "Qualitative interviewing design is flexible, iterative, and continuous, rather than prepared in advance and locked in stone." They elaborate in this way:

> Design in qualitative interviewing is iterative. That means that each time you repeat the basic process of gathering information, analyzing it, winnowing it, and testing it, you come closer to a clear and convincing model of the phenomenon you are studying. . . .
>
> The continuous nature of qualitative interviewing means that the questioning is redesigned throughout the project. — (RUBIN AND RUBIN 1995:46, 47)

Unlike a survey, a **qualitative interview** is an interaction between an interviewer and a respondent in which the interviewer has a general plan of inquiry but not a specific set of questions that must be asked with particular words and in a particular order. At the same time, it is vital for the qualitative interviewer, like the survey interviewer, to be fully familiar with the questions to be asked. This allows the interview to proceed smoothly and naturally.

A qualitative interview is essentially a conversation in which the interviewer establishes a general direction for the conversation and pursues specific topics raised by the respondent. Ideally, the respondent does most of the talking. If you're talking more than 5 percent of the time, that's probably too much.

Steinar Kvale (1996:3–5) offers two metaphors for interviewing: the interviewer as a "miner" or as a "traveler." The first model assumes that the subject possesses specific information and that the interviewer's job is to dig it out. By contrast, in the second model, the interviewer

> wanders through the landscape and enters into conversations with the people encountered. The traveler explores the many domains of the country, as unknown territory or with maps, roaming freely around the territory. . . . The interviewer wanders along with the local inhabitants, asks

questions that lead the subjects to tell their own stories of their lived world.

Asking questions and noting answers is a natural human process, and it seems simple enough to add it to your bag of tricks as a field researcher. Be a little cautious, however. Wording questions is a tricky business. All too often, the way we ask questions subtly biases the answers we get. Sometimes we put our respondent under pressure to look good. Sometimes we put the question in a particular context that omits altogether the most relevant answers.

Suppose, for example, that you want to find out why a group of students is rioting and pillaging on campus. You might be tempted to focus your questioning on how students feel about the dean's recent ruling that requires students always to carry *The Basics of Social Research* with them on campus. (Makes sense to me.) Although you may collect a great deal of information about students' attitudes toward the infamous ruling, they may be rioting for some other reason. Perhaps most are simply joining in for the excitement. Properly done, field research interviewing enables you to find out.

Although you may set out to conduct interviews with a reasonably clear idea of what you want to ask, one of the special strengths of field research is its flexibility. In particular, the answers evoked by your initial questions should shape your subsequent ones. It doesn't work merely to ask preestablished questions and record the answers. Instead, you need to ask a question, listen carefully to the answer, interpret its meaning for your general inquiry, and then frame another question either to dig into the earlier answer or to redirect the person's attention to an area more relevant to your inquiry. In short, you need to be able to listen, think, and talk almost at the same time.

The discussion of probes in Chapter 9 provides a useful guide to getting answers in more depth without biasing later answers. More generally, field interviewers need the skills to be a good listener. Be more interested than interesting. Learn to say things like "How is that?" "In what ways?" "How do you mean that?" "What would be an example of that?" Learn to look and listen expectantly, and let the person you're interviewing fill in the silence.

At the same time, you can't afford to be a totally passive receiver. You'll go into your interviews with some general (or specific) questions you want answered and some topics you want addressed. At times you'll need the skill of subtly directing the flow of conversation.

There's something we can learn in this connection from the martial arts. The aikido master never resists an opponent's blow but instead accepts it, joins with it, and then subtly redirects it in a more appropriate direction. Field interviewing requires an analogous skill. Instead of trying to halt your respondent's line of discussion, learn to take what he or she has just said and branch that comment back in the direction appropriate to your purposes. Most people love to talk to anyone who's really interested. Stopping their line of conversation tells them that you are not interested; asking them to elaborate in a particular direction tells them that you are.

Consider this hypothetical example in which you're interested in why college students chose their majors.

YOU: What are you majoring in?

RESP: Engineering.

YOU: I see. How did you come to choose engineering?

RESP: I have an uncle who was voted the best engineer in Arizona in 1981.

YOU: Gee, that's great.

RESP: Yeah. He was the engineer in charge of developing the new civic center in Tucson. It was written up in most of the engineering journals.

YOU: I see. Did you talk to him about your becoming an engineer?

RESP: Yeah. He said that he got into engineering by accident. He needed a job when he graduated from high school, so he went to work as a laborer on a construction job. He spent eight years working his way up from the bottom, until he decided to go to college and come back nearer the top.

YOU: So is your main interest civil engineering, like your uncle, or are you more interested in some other branch of engineering?

RESP: Actually, I'm leaning more toward electrical engineering—computers, in particular. I

started messing around with microcomputers when I was in high school, and my long-term plan is . . .

Notice how the interview first begins to wander off into a story about the respondent's uncle. The first attempt to focus things back on the student's own choice of major ("Did you talk to your uncle . . . ?") fails. The second attempt ("So is your main interest . . . ?") succeeds. Now the student is providing the kind of information you're looking for. It's important for field researchers to develop the ability to "control" conversations in this fashion.

Herbert and Riene Rubin offer several ways to control a "guided conversation," such as the following:

> If you can limit the number of main topics, it is easier to maintain a conversational flow from one topic to another. Transitions should be smooth and logical. "We have been talking about mothers, now let's talk about fathers," sounds abrupt. A smoother transition might be, "You mentioned your mother did not care how you performed in school—was your father more involved?" The more abrupt the transition, the more it sounds like the interviewer has an agenda that he or she wants to get through, rather than wanting to hear what the interviewee has to say. — (1995:123)

Because field research interviewing is so much like normal conversation, researchers must keep reminding themselves that they are not having a normal conversation. In normal conversations, each of us wants to come across as an interesting, worthwhile person. If you watch yourself the next time you chat with someone you don't know too well, you'll probably find that much of your attention is spent on thinking up interesting things to say—contributions to the conversation that will make a good impression. Often, we don't really hear each other, because we're too busy thinking of what we'll say next. As an interviewer, the desire to appear interesting is counterproductive. The interviewer needs to make the other person seem interesting, by being interested—and is listening more than talking. (Do this in ordinary conversations, and people will actually regard you as a great conversationalist.)

John and Lyn Lofland (1995:56–57) suggest that investigators adopt the role of the "socially acceptable incompetent" when interviewing. That is, offer yourself as someone who does not understand the situation you find yourself in and must be helped to grasp even the most basic and obvious aspects of that situation: "A naturalistic investigator, almost by definition, is one who does not understand. She or he is 'ignorant' and needs to be 'taught.' This role of watcher and asker of questions is the quintessential *student* role" (Lofland and Lofland 1995:56).

Interviewing needs to be an integral part of the entire field research process. Later, I'll stress the need to review your observational notes every night—making sense out of what you've observed, getting a clearer feel for the situation you're studying, and finding out what you should pay more attention to in further observations. In the same fashion, you'll need to review your notes on interviews, recording especially effective questions and detecting all those questions you should have asked but didn't. Start asking such questions the next time you interview. If you have recorded the interviews, replay them as a useful preparation for future interviews.

Steinar Kvale (1996:88) details seven stages in the complete interviewing process:

1. *Thematizing:* clarifying the purpose of the interviews and the concepts to be explored
2. *Designing:* laying out the process through which you'll accomplish your purpose, including a consideration of the ethical dimension
3. *Interviewing:* doing the actual interviews
4. *Transcribing:* creating a written text of the interviews
5. *Analyzing:* determining the meaning of gathered materials in relation to the purpose of the study
6. *Verifying:* checking the reliability and validity of the materials
7. *Reporting:* telling others what you've learned

As with all other aspects of field research, interviewing improves with practice. Fortunately, it's something you can practice any time you want. Practice on your friends.

Focus Groups

While our discussions of field research so far have focused on studying people in the process of living their lives, researchers sometimes bring people into the laboratory for qualitative interviewing and observation. The focus group method, which is also called group interviewing, is essentially a qualitative method. It is based on structured, semi-structured, or unstructured interviews. It allows the researcher/interviewer to question systematically and simultaneously several individuals. Focus group techniques are typically used in marketing research, but not exclusively.

Imagine that you're thinking about introducing a new product. Let's suppose that you've invented a new computer that not only does word processing, spreadsheets, data analysis, and the like, but also contains a fax machine, AM/FM/TV tuner, CD player, dual-cassette unit, microwave oven, denture cleaner, and coffeemaker. To highlight its computing and coffee-making features, you're thinking of calling it "The Compulator." You figure the new computer will sell for about $28,000, and you want to know whether people are likely to buy it. Your prospects might be well served by focus groups.

In a **focus group,** typically 12 to 15 people are brought together in a room to engage in a guided discussion of some topic—in this case, the acceptability and salability of The Compulator. The subjects are selected on the basis of relevance to the topic under study. Given the likely cost of The Compulator, your focus group participants would probably be limited to upper-income groups, for example. Other, similar considerations might figure into the selection.

Participants in focus groups are not likely to be chosen through rigorous, probability sampling methods. This means that the participants do not statistically represent any meaningful population. However, the purpose of the study is to explore rather than to describe or explain in any definitive sense. Nevertheless, typically more than one focus group is convened in a given study because of the serious danger that a single group of 7 to 12 people will be too atypical to offer any generalizable insights.

William Gamson (1992) has used focus groups to examine how U.S. citizens frame their views of political issues. Having picked four issues—affirmative action, nuclear power, troubled industries, and the Arab-Israeli conflict—Gamson undertook a content analysis of press coverage to get an idea of the media context within which we think and talk about politics. Then the focus groups were convened for a firsthand observation of the process of people discussing issues with their friends.

Richard Krueger points to five advantages of focus groups:

1. The technique is a socially oriented research method capturing real-life data in a social environment.
2. It has flexibility.
3. It has high face validity.
4. It has speedy results.
5. It is low in cost. — (1988:47)

In addition to these advantages, group dynamics frequently bring out aspects of the topic that would not have been anticipated by the researcher and would not have emerged from interviews with individuals. In a side conversation, for example, a couple of the participants might start joking about the results of leaving out one letter from a product's name. This realization might save the manufacturer great embarrassment later on.

Krueger also notes some disadvantages of the focus group method, however:

1. Focus groups afford the researcher less control than individual interviews.
2. Data are difficult to analyze.
3. Moderators require special skills.
4. Difference between groups can be troublesome.
5. Groups are difficult to assemble.
6. The discussion must be conducted in a conducive environment. — (1988:44–45)

APPLYING THE RESULTS

While the focus group format lends itself to a wide range of research topics and purposes, it has become very popular in the field of market research, allowing researchers to explore public reactions to new and existing products, packaging, and advertising. Moreover, it taps into such attitudes in a social setting, with participants interacting with, and being influenced by, one another—mimicking the products' fate in real life.

In a focus group interview, more than in any other type of interview, the interviewer has to be a skilled moderator. Controlling the dynamic within the group is a major challenge. Letting one interviewee dominate the focus group interview reduces the likelihood that the other subjects will participate. This can generate the problem of group conformity or what Janis called "groupthink," which is the tendency for people in a group to conform with the opinions and decisions of the most outspoken members of the group. Interviewers need to be aware of this phenomenon and try to get everyone to participate fully on all the issues brought up in the interview. In addition, interviewers must resist bringing their own views into play by overdirecting the interview and the interviewees.

While focus group research differs from other forms of qualitative field research, it further illustrates the possibilities for doing social research face-to-face with those we wish to understand. In addition, David Morgan (1993) suggests that focus

 To see more on market research,* check out the Survey/Marketing eStore listing at http://www.streamlinesurveys.com/ Streamline/estore/focus.htm

---

*Each time the Internet icon appears, you'll be given helpful leads for searching the World Wide Web.

INTERVIEW TRANSCRIPT ANNOTATED WITH RESEARCHER MEMOS

Thursday August 26, 12:00–1:00

R: What is challenging for women directors on a daily experience, on a daily life?

J: Surviving.

R: OK. Could you develop a little bit on that? [I need to work on my interview schedule so that my interviewee answers with more elaboration without having to probe.]

J: Yeah, I mean it's all about trying to get, you know, in, trying to get the job, and try, you know, to do a great job so that you are invited back to the next thing. And particularly since they are so many, you know, difficulties in women directing. It makes it twice as hard to gain into this position where you do an incredible job, because . . . you can't just do an average job, you have to [347] do this job that just knocks your socks off all the time, and sometimes you don't get the opportunity to do that, because either you don't have a good producer or you have so many pressures that you can't see straight or your script is lousy, and you have to make a silk

purse out of sow's hair. You know, you have a lot of extra strikes against you than the average guy who has similar problems, because you are a woman and they look at it, and women are more visible than men . . . in unique positions. [It seems that Joy is talking about the particularities of the film industry. There are not that many opportunities and in order to keep working, she needs to build a certain reputation. It is only by continuing to direct that she can maintain or improve her reputation. She thinks that it is even harder for women but does not explain it.]

R: Hum . . . what about on the set did you experience, did it feel . . . did people make it clear that you were a woman, and you felt treated differently? [I am trying to get her to speak about more specific and more personal experiences without leading her answer]

J: Yeah, oh yeah, I mean . . . a lot of women have commiserated about, you know when you have to walk on the set for the first time, they're all used to working like a well-oiled machine and they say, "Oh, here is the

---

groups are an excellent device for generating questionnaire items for a subsequent survey.

Recording Observations

The greatest advantage of the field research method is the presence of an observing, thinking researcher at the scene of the action. Even tape recorders and cameras cannot capture all the relevant aspects of social processes. Consequently, in both direct observation and interviewing, it is vital to make full and accurate notes of what goes on. If possible, take notes on your observations while you observe. When that's not feasible, write down your notes as soon as possible afterward.

In your notes, include both your empirical observations and your interpretations of them. In other words, record what you "know" has hap-

pened and what you "think" has happened. Be sure to identify these different kinds of notes for what they are. For example, you might note that Person X spoke out in opposition to a proposal made by a group leader (an observation), that you *think* this represents an attempt by Person X to take over leadership of the group (an interpretation), and that you *think* you heard the leader comment to that effect in response to the opposition (a tentative observation).

Of course, you cannot hope to observe everything; nor can you record everything you do observe. Just as your observations will represent a sample of all possible observations, your notes will represent a sample of your observations. The idea, of course, is to record the most pertinent ones. The accompanying box, "Interview Transcript Annotated with Researcher Memos," provides an exwoman, something different" and sometimes they can be horrible, they can resist your directing and they can, they can sabotage you, by taking a long time to light, or to move sets, or to do something . . . and during that time you're wasting time, and that goes on a report, and the report goes to the front [368] office, and, you know, and so on and so on and so on and so forth. And people upstairs don't know what the circumstances are, and they are not about to fire a cinematographer that is on their show for ever and ever . . . nor do they want to know that this guy is a real bastard, and making your life a horror. They don't want to know that, so therefore, they go off, because she's a woman let's not hire any more women, since he has problems with women. You know, so, there is that aspect.

[I need to review the literature on institutional discrimination. It seems that the challenges that Joy is facing are not a matter of a particular individual. She is in a double bind situation where whether she complains or not, she will not be treated equal to men. Time seems to be one quantifiable measurement of how well she does her job and, as observed in other professions, the fact that she is a woman is perceived as a handicap. Review literature on women in high management position. I need to keep asking about the dynamics between my interviewees and the crew members on the set. The cinematographer has the highest status on the set under the director. Explore other interviews about reasons for conflict between them.]

[Methods (note to myself for the next interviews): try to avoid phone interviews unless specific request from the interviewee. It is difficult to assess how the interviewee feels with the questions. Need body language because I become more nervous about the interview process.]

Note: A number in brackets represents a word that was inaudible from the interview. It is the number that appeared on the transcribing machine, with each interview starting at count 0. The numbers help the researcher locate a passage quickly when he or she reviews the interview.

ample given by Sandrine Zerbib from an in-depth interview with a woman film director.

Some of the most important observations can be anticipated before you begin the study; others will become apparent as your observations progress. Sometimes you can make note taking easier by preparing standardized recording forms in advance. In a study of jaywalking, for example, you might anticipate the characteristics of pedestrians that are most likely to be useful for analysis—age, gender, social class, ethnicity, and so forth—and prepare a form in which observations of these variables can be recorded easily. Alternatively, you might develop a symbolic shorthand in advance to speed up recording. For studying audience participation at a mass meeting, you might want to construct a numbered grid representing the different sections of the meeting room; then you could record the location of participants easily, quickly, and accurately.

None of this advance preparation should limit your recording of unanticipated events and aspects of the situation. Quite the contrary, the speedy handling of anticipated observations can give you more freedom to observe the unanticipated.

You are already familiar with the process of taking notes, just as you already have at least informal experience with field research in general. Like good field research, however, good note taking requires careful and deliberate attention and involves specific skills. Some guidelines follow. (You can learn more from John and Lyn Lofland's *Analyzing Social Settings* [1995:91–96].)

First, don't trust your memory any more than you have to; it's untrustworthy. To illustrate this point, try this experiment. Recall the last three or four movies you saw that you really liked. Now, name five of the actors or actresses. Who had the longest hair? Who was the most likely to start conversations? Who was the most likely to make suggestions that others followed? Now, if you didn't have any trouble answering any of those questions, how sure are you of your answers? Would you be willing to bet a hundred dollars that a panel of impartial judges would observe what you recall?

Even if you pride yourself on having a photographic memory, it's a good idea to take notes either during the observation or as soon afterward as possible. If you take notes during observation, do it unobtrusively, because people are likely to behave differently if they see you taking down everything they say or do.

Second, it's usually a good idea to take notes in stages. In the first stage, you may need to take sketchy notes (words and phrases) in order to keep abreast of what's happening. Then go off by yourself and rewrite your notes in more detail. If you do this soon after the events you've observed, the sketchy notes should allow you to recall most of the details. The longer you delay, the less likely you'll be able to recall things accurately and fully.

I know this method sounds logical, but it takes self-discipline to put it into practice. Careful observation and note taking can be tiring, especially if it involves excitement or tension and if it extends over a long period. If you've just spent eight hours observing and making notes on how people have been coping with a disastrous flood, your first desire afterward will likely be to get some sleep, dry clothes, or a drink. You may need to take some inspiration from newspaper reporters who undergo the same sorts of hardships then write their stories to meet their deadlines.

Third, you'll inevitably wonder how much you should record. Is it really worth the effort to write out all the details you can recall right after the observation session? The general guideline is yes. Generally, in field research you can't be really sure of what's important and what's unimportant until you've had a chance to review and analyze a great volume of information, so you should record even things that don't seem important at the outset. They may turn out to be significant after all. Also,

the act of recording the details of something "unimportant" may jog your memory on something that is important.

Realize that most of your field notes will not be reflected in your final report on the project. Put more harshly, most of your notes will be "wasted." But take heart: Even the richest gold ore yields only about 30 grams of gold per metric ton, meaning that 99.997 percent of the ore is wasted. Yet, that 30 grams of gold can be hammered out to cover an area 18 feet square—the equivalent of about 685 book pages. So take a ton of notes, and plan to select and use only the gold.

Like other aspects of field research (and all research for that matter), proficiency comes with practice. The nice thing about field research is that you can begin practicing now and can continue practicing in almost any situation. You don't have to be engaged in an organized research project to practice observation and recording. You might start by volunteering to take the minutes at committee meetings, for example. Or just pick a sunny day on campus, find a shady spot, and try observing and recording some specific characteristics of the people who pass by. You can do the same thing at a shopping mall or a busy street corner. Remember that observing and recording are professional skills, and, like all worthwhile skills, they improve with practice.

STRENGTHS AND WEAKNESSES OF QUALITATIVE FIELD RESEARCH

Like all research methods, qualitative field research has distinctive strengths and weaknesses. As I've already indicated, field research is especially effective for studying subtle nuances in attitudes and behaviors and for examining social processes over time. As such, the chief strength of this method lies in the depth of understanding it permits. Whereas other research methods may be challenged as "superficial," this charge is seldom lodged against field research.

Flexibility is another advantage of field research. As discussed earlier, you may modify your field research design at any time. Moreover, you're always prepared to engage in field research, whenever the occasion should arise, whereas you could not as easily initiate a survey or an experiment.

Field research can be relatively inexpensive as well. Other social research methods may require expensive equipment or an expensive research staff, but field research typically can be undertaken by one researcher with a notebook and a pencil. This is not to say that field research is never expensive. The nature of the research project, for example, may require a large number of trained observers. Expensive recording equipment may be needed. Or you may wish to undertake participant observation of interactions in expensive Paris nightclubs.

Field research has several weaknesses as well. First, being qualitative rather than quantitative, it is not an appropriate means for arriving at statistical descriptions of a large population. Observing casual political discussions in Laundromats, for example, would not yield trustworthy estimates of the future voting behavior of the total electorate. Nevertheless, the study could provide important insights into how political attitudes are formed.

To assess field research further, let's focus on the issues of validity and reliability. Recall that validity and reliability are both qualities of measurements. Validity concerns whether measurements actually measure what they're supposed to rather than something else. Reliability, on the other hand, is a matter of dependability: If you made the same measurement again and again, would you get the same result? Let's see how field research stacks up in these respects.

Validity

Field research seems to provide measures with greater validity than do survey and experimental measurements, which are often criticized as superficial and not really valid. Let's review a couple of field research examples to see why this is so.

"Being there" is a powerful technique for gaining insights into the nature of human affairs in all their rich complexity. Listen, for example, to what this nurse reports about the impediments to patients' coping with cancer:

Common fears that may impede the coping process for the person with cancer can include the following:
—Fear of death—for the patient, and the implications his or her death will have for significant others.
—Fear of incapacitation—because cancer can be a chronic disease with acute episodes that may result in periodic stressful periods, the variability of the person's ability to cope and constantly adjust may require a dependency upon others for activities of daily living and may consequently become a burden.
—Fear of alienation—from significant others and health care givers, thereby creating helplessness and hopelessness.
—Fear of contagion—that cancer is transmissible and/or inherited.
—Fear of losing one's dignity—losing control of all bodily functions and being totally vulnerable. — (GARANT 1980:2167)

Observations and conceptualizations such as these are valuable in their own right. In addition, they can provide the basis for further research—both qualitative and quantitative.

Now listen to what Joseph Howell has to say about "toughness" as a fundamental ingredient of life on Clay Street, a white, working-class neighborhood in Washington, D.C.:

Most of the people on Clay Street saw themselves as fighters in both the figurative and literal sense. They considered themselves strong, independent people who would not let themselves be pushed around. For Bobbi, being a fighter meant battling the welfare department and cussing out social workers and doctors upon occasion. It meant spiking Barry's beer with sleeping pills and bashing him over the head with a broom. For Barry it meant telling off his boss and refusing to hang the door, an act that led to his being fired. It meant going through the ritual of a duel with Al. It meant pushing Bubba around and at times getting rough with Bobbi.

June and Sam had less to fight about, though if pressed they both hinted that they, too, would fight. Being a fighter led Ted into near conflict with Peg's brothers, Les into conflict with Lonnie, Arlene into conflict with Phyllis at the bowling alley, etc. — (1973:292)

Even without having heard the episodes Howell refers to in this passage, you have the distinct impression that Clay Street is a tough place to live in. That "toughness" comes through far more powerfully through these field observations than it would in a set of statistics on the median number of fistfights occurring during a specified period.

These examples point to the superior validity of field research, as compared with surveys and experiments. The kinds of comprehensive measurements available to the field researcher tap a depth of meaning in concepts such as common fears of cancer patients and "toughness" (or such as liberal and conservative) that are generally unavailable to surveys and experiments. Instead of specifying concepts, field researchers commonly give detailed illustrations.

Reliability

Field research has, however, a potential problem with reliability. Suppose you were to characterize your best friend's political orientations according to everything you know about him or her. Your assessment of your friend's politics would appear to have considerable validity; certainly it's unlikely to be superficial. We couldn't be sure, however, that another observer would characterize your friend's politics the same way you did, even with the same amount of observation.

In-depth, field research measurements are also often very personal. How I judge your friend's political orientation depends very much on my own, just as your judgment depends on your political orientation. Conceivably, then, you could describe your friend as middle-of-the-road, although I might feel that I've been observing a fire-breathing radical.

As I've suggested earlier, researchers who use qualitative techniques are conscious of this issue and take pains to address it. Individual researchers often sort out their own biases and points of view, and the communal nature of science means that

---

A QUANDARY REVISITED

Doesn't field research simply involve going where social life is happening and watching it unfold? As we've seen in the chapter, the answer is "yes and no."

On the one hand, field research is perhaps the most "natural" social research technique. At the same time, casual participant observation of social life involves many implicit assumptions, and subjective observations and analyses are wide open to error. Professional field research is conscious and deliberate, forcing researchers to question their assumptions, see alternatives, and choose carefully among those alternatives. For example, because the participant observer inevitably impacts what's being observed, field researchers need to make careful choices about their relationship to the subjects in their studies. Further, researchers have several paradigms or fundamentally different ways of "simply observing" what's in front of them. Certain field research methods involve special skills such as active listening and monitoring. Clearly, just showing up and watching does not make for good field research, though that's where it often begins.

---

their colleagues will help them in that regard. Nevertheless, it's prudent to be wary of purely descriptive measurements in field research—your own, or someone else's. If a researcher reports that the members of a club are fairly conservative, such a judgment is unavoidably linked to the researcher's own politics. You can be more trusting of *comparative* evaluations: identifying who is more conservative than who, for example. Even if you and I had different political orientations, we would probably agree pretty much in ranking the relative conservatism of the members of a group.

As we've seen, field research is a potentially powerful tool for social scientists, one that provides a useful balance against the strengths and weaknesses of experiments and surveys. The remaining chapters of Part 3 present additional modes of observation available to social researchers.

RESEARCH ETHICS IN QUALITATIVE FIELD RESEARCH

As I've noted repeatedly, all forms of social research raise ethical issues. By bringing researchers into direct and often intimate contact with their subjects, field research raises ethical concerns in a particularly dramatic way. Here are some of the issues mentioned by John and Lyn Lofland (1995:63):

- Is it ethical to talk to people when they do not know you will be recording their words?

- Is it ethical to get information for your own purposes from people you hate?

- Is it ethical to see a severe need for help and not respond to it directly?

- Is it ethical to be in a setting or situation but not commit yourself wholeheartedly to it?

- Is it ethical to develop a calculated stance toward other humans, that is, to be strategic in your relations?

- Is it ethical to take sides or to avoid taking sides in a factionalized situation?

- Is it ethical to "pay" people with trade-offs for access to their lives and minds?

- Is it ethical to "use" people as allies or informants in order to gain entree to other people or to elusive understandings?

Planning and conducting field research in a responsible way requires attending to these and other ethical concerns.

Main Points

- Field research involves the direct observation of social phenomena in their natural settings. Typically, field research is qualitative rather than quantitative.

- In field research, observation, data processing, and analysis are interwoven, cyclical processes.

- Field research is especially appropriate to topics and processes that are not easily quantifiable, that are best studied in natural settings, or that change over time. Among these topics are practices, episodes, encounters, roles, relationships, groups, organizations, settlements, social worlds, and lifestyles or subcultures.

- Among the special considerations involved in field research are the various possible roles of the observer and the researcher's relations with subjects. As a field researcher, you must decide whether to observe as an outsider or as a participant, whether or not to identify yourself as a researcher, and how to negotiate your relationships with subjects.

- Field research can be guided by any one of several paradigms, such as naturalism, ethnomethodology, grounded theory, case studies and the extended case method, institutional ethnography, and participatory action research.

- Preparing for the field involves doing background research, determining how to make contact with subjects, and resolving issues of what your relationship to your subjects will be.

- Field researchers often conduct in-depth interviews that are much less structured than those conducted in survey research. Qualitative interviewing is more of a guided conversation than a search for specific information. Effective interviewing involves skills of active listening and the ability to direct conversations unobtrusively.

- To create a focus group, researchers bring subjects together and observe their interactions as they explore a specific topic.

- Whenever possible, field observations should be recorded as they are made; otherwise, they should be recorded as soon afterward as possible.

❑ Among the advantages of field research are the depth of understanding it can provide, its flexibility, and (usually) its lack of costs.

❑ Compared with surveys and experiments, field research measurements generally have more validity but less reliability. Also, field research is generally not appropriate for arriving at statistical descriptions of large populations.

❑ Conducting field research responsibly involves confronting several ethical issues that arise from the researcher's direct contact with subjects.

Key Terms

naturalism	extended case method
ethnography	institutional ethnography
ethnomethodology	participatory action research
grounded theory	qualitative interview
case studies	focus group

Review Questions

1. Think of some group or activity you participate in or are very familiar with. In two or three paragraphs, describe how an outsider might effectively go about studying that group or activity. What should he or she read, what contacts should be made, and so on?

2. Choose any two of the paradigms discussed in this chapter. How might your hypothetical study from item 1 be conducted if you followed each? Compare and contrast the way these paradigms might work in the context of your study.

3. To explore the strengths and weaknesses of experiments, surveys, and field research, choose a general research area (e.g., prejudice, political orientation, education) and write brief descriptions of studies in that area that could be conducted using each of these three methods. In each case, why is the chosen method the most appropriate for the study you describe?

4. Return to the example you devised in response to item 1 above. What five ethical issues can you imagine having to confront if you were to undertake your study?

5. Using the Web, find a research report using the grounded theory method. Briefly, what are the study design and main findings?

Additional Readings

Denzin, Norman K., and Yvonna S. Lincoln, eds. 1994. *Handbook of Qualitative Research.* Thousand Oaks, CA: Sage. This handbook is an extensive collection of articles covering issues regarding the wide field of qualitative research. This book also exists in three volumes: Vol 1. *The Landscape of Qualitative Research: Theories and Issues;* Vol 2. *Strategies of Qualitative Inquiry;* and Vol. 3. *Interpreting Qualitative Materials.*

Emerson, Robert M., ed. 1988. *Contemporary Field Research.* Boston: Little, Brown. A diverse and interesting collection of articles on how field research contributes to understanding, the role of theory in such research, personal and relational issues that emerge, and ethical and political issues.

Gubrium, Jaber F., and James A. Holstein. 1997. *The New Language of Qualitative Method.* New York: Oxford University Press. This book provides the necessary foundations for understanding some of the main approaches or traditions in qualitative field research.

Johnson, Jeffrey C. 1990. *Selecting Ethnographic Informants.* Newbury Park, CA: Sage. The author discusses the various strategies that apply to the task of sampling in field research.

Kelle, Udo, ed. 1995. *Computer-Aided Qualitative Data Analysis: Theory, Methods, and Practice.* Thousand Oaks, CA: Sage. An international group of scholars report on their experiences with a variety of computer programs used in the analysis of qualitative data.

Kvale, Steinar. 1996. *InterViews: An Introduction to Qualitative Research Interviewing.* Thousand Oaks, CA: Sage. An in-depth presentation of in-depth interviewing. Besides presenting techniques, Kvale places interviewing in the context of postmodernism and other philosophical systems.

Lofland, John, and Lyn Lofland. 1995. *Analyzing Social Settings,* 3rd ed. Belmont, CA: Wadsworth. An unexcelled presentation of field research methods from beginning to end. This eminently readable book manages successfully to draw the links between the

logic of scientific inquiry and the nitty-gritty practical-
ities of observing, communicating, recording, filing,
reporting, and everything else involved in field research. In addition, the book contains a wealth of
references to field research illustrations.

Morgan, David L., ed. 1993. *Successful Focus Groups: Advancing the State of the Art.* Newbury Park, CA: Sage.
This collection of articles on the uses of focus groups
points to many aspects not normally considered.

Shaffir, William B., and Robert A. Stebbins, eds. 1991.
*Experiencing Fieldwork: An Inside View of Qualitative
Research.* Newbury Park, CA: Sage. Several field research practitioners discuss the nature of the craft
and recall experiences in the field. Here's an opportunity to gain a "feel" for the method as well as learn
some techniques.

Shostak, Arthur, ed. 1977. *Our Sociological Eye: Personal
Essays on Society and Culture.* Port Washington, NY:
Alfred. An orgy of social scientific introspection, this
delightful collection of first-person research accounts
offers concrete, inside views of the thinking process
in sociological research, especially field research.

Silverman, David. 1999. *Doing Qualitative Research: A
Practical Handbook.* Thousand Oaks, CA: Sage. This
book focuses on the process of collecting and interpreting qualitative data.

Strauss, Anselm, and Juliet Corbin. 1990. *Basics of Qualitative Research: Grounded Theory Procedures and
Techniques.* Newbury Park, CA: Sage. This is a very
important book to read before data collection and
during data analysis if you choose to take a grounded
theory approach.

Uwe, Flick. 1998. *An Introduction to Qualitative Research.*
Thousand Oaks, CA: Sage. This book provides a good
entrance to the large field of qualitative research.

Multimedia Resources

**The Wadsworth Sociology Resource
Center: Virtual Society**
http://sociology.wadsworth.com/
Visit the companion Web site for the second
edition of *The Basics of Social Research* to access a wide
range of student resources. Begin by clicking on the Student Resources section of the book's Web site to access
the following study tools:

- eBabbie Resource Center
- Planning a Research Project
- Doing Data Analysis
- Statistics Review
- Flash Cards
- Internet Links and Exercises
- InfoTrac College Edition: Exercises
- Quizzes

Visit the **eBabbie Resource Center** for an overview of
each chapter and helpful online tutorials. Find information on budgeting and step-by-step examples of model research projects at **Planning a Research Project.** Learn
how to use quantitative and qualitative data analysis programs at **Doing Data Analysis,** and brush up on your
statistics at **Statistics Review.** You can also further your
study by accessing **Internet Links and Exercises** related to chapter materials, **Flash Cards, Quizzes,** and
many other learning tools.

InfoTrac College Edition
http://www.infotrac-college.com/
wadsworth/access.html
Access the latest news and journal articles
with InfoTrac College Edition, an easy-to-use online database of reliable, full-length articles from hundreds of top
academic journals. Conduct an electronic search using
the following search terms:

Ethnomethodology
Field research
Grounded theory
In-depth interview
Naturalism
Participant observation
Participatory action research

11 UNOBTRUSIVE RESEARCH

Dion Ogust / The Image Works

What You'll Learn in This Chapter

This chapter will present overviews of three unobtrusive research methods: content analysis, the analysis of existing statistics, and historical/comparative analysis. Each of these methods allows researchers to study social life from afar, without influencing it in the process.

In this chapter . . .

INTRODUCTION

With the exception of the complete observer in field research, each of the modes of observation discussed so far requires the researcher to intrude to some degree on whatever he or she is studying. This is most obvious in the case of experiments,

---

AN OPENING **QUANDARY**

This chapter presents several research techniques that by definition have no impact on what is being studied. If the impact of the observer is such a problem in social research, why don't social scientists limit themselves to unobtrusive techniques?

---

followed closely by survey research. Even the field researcher, as we've seen, can change things in the process of studying them.

At least one previous example in this book, however, was totally exempt from that danger. Durkheim's analysis of suicide did nothing to affect suicides one way or the other (see Chapter 5). His study is an example of **unobtrusive research,** or methods of studying social behavior without affecting it. As you'll see, unobtrusive measures can be qualitative or quantitative.

This chapter examines three types of unobtrusive research methods: content analysis, analysis of existing statistics, and historical/comparative analysis. In content analysis, researchers examine a class of social artifacts that usually are written documents such as newspaper editorials. The Durkheim study is an example of the analysis of existing statistics. As you'll see, there are great masses of data all around you, awaiting your use in understanding social life. Finally, historical/comparative analysis, a form of research with a venerable history in the social sciences, is currently enjoying a resurgence of popularity. Like field research, historical/comparative analysis is a qualitative method, one in which the main resources for observation and analysis are historical records. The method's name includes the word *comparative* because social scientists—in contrast to historians who may simply describe a particular set of events—seek to discover common patterns that recur in different times and places.

To set the stage for our examination of these three research methods, I want to draw your attention to an excellent book that should sharpen your senses about the potential for unobtrusive measures in general. It is, among other things, the book from which I take the term *unobtrusive measures.*

In 1966, Eugene Webb and three colleagues published an ingenious little book on social research (revised in 1981) that has become a classic: *Unobtrusive Research.* It focuses, as you might have guessed, on the idea of unobtrusive or nonreactive research. Webb and his colleagues have played freely with the task of learning about human behavior by observing what people inadvertently leave behind them. Do you want to know what exhibits are the most popular at a museum? You could conduct a poll, but people might tell you what they thought you wanted to hear or what might make them look intellectual and serious. You could stand by different exhibits and count the viewers that came by, but people might come over to see what you were doing. Webb and his colleagues suggest you check the wear and tear on the floor in front of various exhibits. Those that have the most-worn tiles are probably the most popular. Want to know which exhibits are popular with little kids? Look for mucus on the glass cases. To get a sense of the most popular radio stations, you could arrange with an auto mechanic to check the radio dial settings for cars brought in for repair.

The possibilities are limitless. Like a detective investigating a crime, the social researcher looks for clues. If you stop to notice, you'll find that clues of social behavior are all around you. In a sense, everything you see represents the answer to some important social scientific question—all you have to do is think of the question.

Although problems of validity and reliability crop up in unobtrusive measures, a little ingenuity can either handle them or put them in perspective. I encourage you to look at Webb's book. It's enjoyable reading, and it can be a source of stimulation and insight for social inquiry through data that already exist. For now, let's turn our attention to three unobtrusive methods often employed by social scientists, beginning with content analysis.

CONTENT ANALYSIS

As I mentioned in the chapter introduction, **content analysis** is the study of recorded human communications. Among the forms suitable for study are books, magazines, Web pages, poems, newspapers, songs, paintings, speeches, letters, e-mail messages, bulletin board postings on the Internet, laws, and constitutions, as well as any components or collections thereof. Shulamit Reinharz (1992: 146–47) points out that feminist researchers have used content analysis to study "children's books, fairy tales, billboards, feminist nonfiction and fiction books, children's art work, fashion, fat-letter postcards, Girl Scout Handbooks, works of fine art, newspaper rhetoric, clinical records, research publications, introductory sociology textbooks, and citations, to mention only a few."

Topics Appropriate to Content Analysis

Content analysis is particularly well suited to the study of communications and to answering the classic question of communications research: "Who says what, to whom, why, how, and with what effect?" Are popular French novels more concerned with love than American ones? Was the popular British music of the 1960s more politically cynical than the popular German music during that period? Do political candidates who primarily address "bread and butter" issues get elected more often than those who address issues of high principle? Each of these questions addresses a social scientific research topic: The first might address national character, the second political orientations, and the third political process. Although such topics might be studied through observation of individual people, content analysis provides another approach.

An early example of content analysis is the work of Ida B. Wells. In 1891, Wells, whose parents had been slaves, wanted to test the widely held assumption that black men were being lynched in the South primarily for raping white women. As a research method, she examined newspaper articles on the 728 lynchings reported during the previous ten years. In only a third of the cases were the lynching victims even accused of rape, much less proven guilty. Primarily, they were charged with being insolent, not staying in "their place" (cited in Reinharz 1992:146).

More recently, the best-selling *Megatrends 2000* (Naisbitt and Aburdene 1990) used content analysis to determine the major trends in modern U.S. life. The authors regularly monitored thousands of local newspapers a month to discover local and regional trends for publication in a series of quarterly reports. Their book examines some of the trends they observed in the nation at large.

Some topics are more appropriately addressed by content analysis than by any other method of inquiry. Suppose that you're interested in violence on television. Maybe you suspect that the manufacturers of men's products are more likely to sponsor violent TV shows than are other kinds of sponsors. Content analysis would be the best way of finding out.

Briefly, here's what you would do. First, you'd develop operational definitions of the two key variables in your inquiry: *men's products* and *violence.* The section on coding, later in this chapter, will discuss some of the ways you could do that. Ultimately, you'd need a plan that would allow you to watch TV, classify sponsors, and rate the degree of violence on particular shows.

Next, you'd have to decide what to watch. Probably you'd decide (1) what stations to watch, (2) for what days or period, and (3) at what hours. Then, you'd stock up on beer and potato chips and start watching, classifying, and recording. Once you'd completed your observations, you'd be able to analyze the data you collected and determine whether men's product manufacturers sponsored more blood and gore than did other sponsors.

As a mode of observation, content analysis requires a thoughtful handling of the "what" that is being communicated. The analysis of data collected in this mode, as in others, addresses the "why" and "with what effect."

Sampling in Content Analysis

In the study of communications, as in the study of people, you often can't observe directly all that you would like to explore. In your study of TV violence and sponsorship, I advise against attempting to watch everything that's broadcast. It wouldn't be possible, and your brain would probably short-circuit before you got close to discovering that for yourself. Usually, then, it's appropriate to sample. Let's begin by revisiting the idea of units of analysis. We'll then review some of the sampling techniques that might be applied to them in content analysis.

Units of Analysis As I discussed in Chapter 4, determining appropriate units of analysis—the individual units that we make descriptive and explanatory statements about—can be a complicated task. For example, if we wish to compute the average family income, the individual family is the unit of analysis. But we'll have to ask individual members of families how much money they make. Thus, individuals will be the units of observation, even though the individual family remains the unit of analysis. Similarly, we may wish to compare crime rates of different cities in terms of their size, geographical region, racial composition, and other differences. Even though the characteristics of these cities are partly a function of the behaviors and characteristics of their individual residents, the cities would ultimately be the units of analysis.

The complexity of this issue is often more apparent in content analysis than in other research methods, especially when the units of observation differ from the units of analysis. A few examples should clarify this distinction.

Let's suppose we want to find out whether criminal law or civil law makes the most distinctions between men and women. In this instance, individual laws would be both the units of observation and the units of analysis. We might select a sample of a state's criminal and civil laws and then categorize each law by whether it makes a distinction between men and women. In this fashion, we could determine whether criminal or civil law distinguishes by gender the most.

Somewhat differently, we might wish to determine whether states that enact laws distinguishing between different racial groups are more likely than other states to enact laws distinguishing between men and women. Although the examination of this question would also involve coding individual acts of legislation, the unit of analysis in this case is the individual state, not the law.

Or, changing topics radically, let's suppose we're interested in representationalism in painting. If we wish to compare the relative popularity of representational and nonrepresentational paintings, the individual paintings will be our units of analysis. If, on the other hand, we wish to discover whether representationalism in painting is more characteristic of wealthy or impoverished painters, of educated or uneducated painters, of capitalist or socialist painters, the individual painters will be our units of analysis.

It's essential that this issue be clear, because sample selection depends largely on what the unit of analysis is. If individual writers are the units of analysis, the sample design should select all or a sample of the writers appropriate to the research question. If books are the units of analysis, we should select a sample of books, regardless of their authors. Bruce Berg (1989:112–13) points out that even if you plan to analyze some body of textual materials, the units of analysis might be words, themes, characters, paragraphs, items (such as a book or letter), concepts, semantics, or combinations of these. Figure 11-1 illustrates some of those possibilities.

I'm not suggesting that sampling should be based solely on the units of analysis. Indeed, we may often subsample—select samples of subcategories—for each individual unit of analysis. Thus, if writers are the units of analysis, we might (1) select a sample of writers from the total population of writers, (2) select a sample of books written by each writer selected, and (3) select portions of each selected book for observation and coding.

Finally, let's look at a trickier example: the study of TV violence and sponsors. What's the unit of analysis for the research question "Are the manufacturers of men's products more likely to sponsor

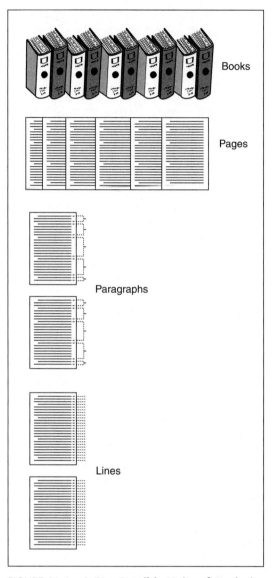

FIGURE 11-1 **A Few Possible Units of Analysis for Content Analysis**

violent shows than are other sponsors?" Is it the TV show? The sponsor? The instance of violence?

In the simplest study design, it would be none of these. Though you might structure your inquiry in various ways, the most straightforward design would be based on the commercial as the unit of analysis. You would use two kinds of observational

Sponsor	Men's Product?			Number of Instances of Violence	
	Yes	No	?	Before	After
Grunt Aftershave	✓			*6*	*4*
Brute Jock Straps	✓			*6*	*4*
Bald-No-More Lotion	✓			*4*	*3*
Grunt Aftershave	✓			*3*	*0*
Snowflake Toothpaste		✓		*3*	*0*
Godliness Cleanser		✓		*3*	*0*
Big Thumb Hammers			✓	*0*	*1*
Snowflake Toothpaste		✓		*1*	*0*
Big Thumb Hammers			✓	*1*	*0*
Buttercup Bras		✓		*0*	*0*

FIGURE 11-2 **Example of Recording Sheet for TV Violence**

units: the commercial and the program (the show that gets squeezed in between commercials). You'd want to observe both units. You would classify commercials by whether they advertised men's products and the programs by their violence. The program classifications would be transferred to the commercials occurring near them. Figure 11-2 provides an example of the kind of record you might keep.

Notice that in the research design illustrated in Figure 11-2, all the commercials occurring in the same program break are bracketed and get the same scores. Also, the number of violent instances recorded as following one commercial is the same as the number preceding the next break. This simple design allows us to classify each commer-

cial by its sponsorship and the degree of violence associated with it. Thus, for example, the first Grunt Aftershave commercial is coded as being a men's product and as having 10 instances of violence associated with it. The Buttercup Bra commercial is coded as not being a men's product and as having no violent instances associated with it.

In the illustration, we have four men's product commercials with an average of 7.5 violent instances each. The four commercials classified as definitely not men's products have an average of 1.75, and the two that might or might not be considered men's products have an average of 1 violent instance each. If this pattern of differences persisted across a much larger number of observations, we'd probably conclude that manufacturers of men's products are more likely to sponsor TV violence than are other sponsors.

The point of this illustration is to demonstrate how units of analysis figure into data collection and analysis. You need to be clear about your unit of analysis before planning your sampling strategy, but in this case you can't sample commercials. Unless you have access to the stations' broadcasting logs, you won't know when the commercials are going to occur. Moreover, you need to observe the programming as well as the commercials. As a result, you must set up a sampling design that will include everything you need to observe.

In designing the sample, you would need to establish the universe to be sampled from. In this case, what TV stations will you observe? What will be the period of the study—number of days? And during which hours of each day will you observe? Then, how many commercials do you want to observe and code for analysis? Watch television for a while and find out how many commercials occur each hour; then you can figure out how many hours of observation you'll need.

Now you're ready to design the sample selection. As a practical matter, you wouldn't have to sample among the different stations if you had assistants—each of you could watch a different channel during the same time period. But let's suppose you're working alone. Your final sampling frame, from which a sample will be selected and watched, might look something like this:

Jan. 7, Channel 2, 7–9 P.M.
Jan. 7, Channel 4, 7–9 P.M.
Jan. 7, Channel 9, 7–9 P.M.
Jan. 7, Channel 2, 9–11 P.M.
Jan. 7, Channel 4, 9–11 P.M.
Jan. 7, Channel 9, 9–11 P.M.
Jan. 8, Channel 2, 7–9 P.M.
Jan. 8, Channel 4, 7–9 P.M.
Jan. 8, Channel 9, 7–9 P.M.
Jan. 8, Channel 2, 9–11 P.M.
Jan. 8, Channel 4, 9–11 P.M.
Jan. 8, Channel 9, 9–11 P.M.
Jan. 9, Channel 2, 7–9 P.M.
Jan. 9, Channel 4, 7–9 P.M.
etc.

Notice that I've made several decisions for you in the illustration. First, I've assumed that channels 2, 4, and 9 are the ones appropriate to your study. I've assumed that you found the 7 to 11 P.M. prime-time hours to be the most relevant and that two-hour periods would do the job. I picked January 7 out of a hat for a starting date. In practice, of course, all these decisions should be based on your careful consideration of what would be appropriate to your particular study.

Once you have become clear about your units of analysis and the observations appropriate to those units and have created a sampling frame like the one I've illustrated, sampling is simple and straightforward. The alternative procedures available to you are the same ones described in Chapter 7: random, systematic, stratified, and so on.

Sampling Techniques As we've seen, in the content analysis of written prose, sampling may occur at any or all of several levels, including the contexts relevant to the works. Other forms of communication may also be sampled at any of the conceptual levels appropriate to them.

In content analysis, we could employ any of the conventional sampling techniques discussed in Chapter 7. We might select a random or systematic sample of French and U.S. novelists, of laws passed in the state of Mississippi, or of Shakespearean soliloquies. We might select (with a random start) every 23rd paragraph in Tolstoy's *War and Peace.* Or, we might number all the songs recorded by the Beatles and select a random sample of 25.

Stratified sampling is also appropriate to content analysis. To analyze the editorial policies of U.S. newspapers, for example, we might first group all newspapers by the region of the country or size of the community in which they are published, frequency of publication, or average circulation. We might then select a stratified random or systematic sample of newspapers for analysis. Having done so, we might select a sample of editorials from each selected newspaper, perhaps stratified chronologically.

Cluster sampling is equally appropriate to content analysis. Indeed, if individual editorials were to be the unit of analysis in the previous example, then the selection of newspapers at the first stage of sampling would be a cluster sample. In an analysis of political speeches, we might begin by selecting a sample of politicians; each politician would represent a cluster of political speeches. The TV commercial study described previously is another example of cluster sampling.

Again, sampling need not end when we reach the unit of analysis. If novels are the unit of analysis in a study, we might select a sample of novelists, subsamples of novels written by each selected author, and a sample of paragraphs within each novel. We would then analyze the content of the paragraphs for the purpose of describing the novels themselves. (*Note:* researchers speak of samples within samples as "subsamples.")

Let's turn now to the coding or classification of the material being observed. Part 4 discusses the manipulation of such classifications to draw descriptive and explanatory conclusions.

Coding in Content Analysis

Content analysis is essentially a **coding** operation. *Coding* is the process of transforming raw data into a standardized form. In content analysis, communications—oral, written, or other—are coded or classified according to some conceptual framework. Thus, for example, newspaper editorials may be coded as liberal or conservative. Radio broadcasts can be coded as propagandistic or not, novels as romantic or not, paintings as representational or not, and political speeches as containing character assassinations or not. Recall that terms such as these are subject to many interpretations, and the researcher must specify definitions clearly.

Coding in content analysis involves the logic of conceptualization and operationalization as discussed in Chapter 5. As in other research methods, you must refine your conceptual framework and develop specific methods for observing in relation to that framework.

Manifest and Latent Content In the earlier discussions of field research, we found that the researcher faces a fundamental choice between depth and specificity of understanding. Often, this represents a choice between validity and reliability, respectively. Typically, field researchers opt for depth, preferring to base their judgments on a broad range of observations and information, even at the risk that another observer might reach a different judgment of the same situation. Survey research—through the use of standardized questionnaires—represents the other extreme: total specificity, even though the specific measures of variables may not be fully satisfactory as valid reflections of those variables. The content analyst has some choice in this matter, however.

Coding the **manifest content**—the visible, surface content—of a communication is analogous to using a standardized questionnaire. To determine, for example, how erotic certain novels are, you might simply count the number of times the word *love* appears in each novel or the average number of appearances per page. Or, you might use a list of such words as *love, kiss, hug,* and *caress,* each of which might serve as an indicator of the erotic nature of the novel. This method would have the advantage of ease and reliability in coding and of letting the reader of the research report know precisely how eroticism was measured. It would have a disadvantage, on the other hand, in terms of validity. Surely the phrase *erotic novel* conveys a richer and deeper meaning than the number of times the word *love* is used.

Alternatively, you may code the **latent content** of the communication: its underlying meaning. In the present example, you might read an entire novel or a sample of paragraphs or pages and make an overall assessment of how erotic the novel was. Although your total assessment might very well be influenced by the appearance of words such as *love* and *kiss,* it would not depend fully on their frequency.

Clearly, this second method seems better designed for tapping the underlying meaning of communications, but its advantage comes at a cost to reliability and specificity. Especially if more than one person is coding the novel, somewhat different definitions or standards may be employed. A passage that one coder regards as erotic may not seem erotic to another. Even if you do all the coding yourself, there's no guarantee that your definitions and standards will remain constant throughout the enterprise. Moreover, the reader of your research report will likely be uncertain about the definitions you've employed.

Wherever possible, the best solution to this dilemma is to use both methods. For example, Carol Auster was interested in changes in the socialization of young women in Girl Scouts. To explore this, she undertook a content analysis of the Girl Scout manuals as revised over time. In particular, Auster was interested in the view that women should be limited to homemaking. Her analysis of the manifest content suggested a change: "I found that while 23% of the badges in 1913 centered on home life, this was true of only 13% of the badges in 1963 and 7% of the badges in 1980" (1985:361).

An analysis of the latent content also pointed to an emancipation of Girl Scouts, similar to that occurring in U.S. society at large. The change of uniform was one indicator: "The shift from skirts to pants may reflect an acknowledgement of the more physically active role of women as well as the variety of physical images available to modern women" (Auster 1985:362). Supporting evidence was found in the appearance of badges such as "Science Sleuth," "Aerospace," and "Ms. Fix-It."

Conceptualization and the Creation of Code Categories   For all research methods, conceptualization and operationalization typically involve the interaction of theoretical concerns and empirical observations. If, for example, you believe some newspaper editorials to be liberal and others to be conservative, ask yourself why you think so. Read some editorials, asking yourself which ones are liberal and which ones are conservative. Was the political orientation of a particular editorial most clearly indicated by its manifest content or by its tone? Was your decision based on the use of certain terms (for example, *leftist, fascist,* and so on) or on the support or opposition given to a particular issue or political personality?

Both inductive and deductive methods should be used in this activity. If you're testing theoretical propositions, your theories should suggest empirical indicators of concepts. If you begin with specific empirical observations, you should attempt to derive general principles relating to them and then apply those principles to the other empirical observations.

Bruce Berg (1989:111) places code development in the context of grounded theory and likens it to solving a puzzle:

> Coding and other fundamental procedures associated with grounded theory development are certainly hard work and must be taken seriously, but just as many people enjoy finishing a complicated jigsaw puzzle, many researchers find great satisfaction in coding and analysis. As researchers . . . begin to see the puzzle pieces come together to form a more complete picture, the process can be downright thrilling.

Throughout this activity, remember that the operational definition of any variable is composed of the attributes included in it. Such attributes, moreover, should be mutually exclusive and exhaustive. A newspaper editorial, for example, should not be described as both liberal and conservative, though you should probably allow for some to be middle-of-the-road. It may be sufficient for your purposes to code novels as erotic or nonerotic, but you may also want to consider that some could be anti-erotic. Paintings might be classified as representational or not, if that satisfied your research purpose, or you might wish to classify them as impressionistic, abstract, allegorical, and so forth.

Realize further that different levels of measurement may be used in content analysis. You may, for example, use the nominal categories of liberal and conservative for characterizing newspaper editorials, or you might wish to use a more refined ordinal ranking, ranging from extremely liberal to extremely conservative. Bear in mind, however, that the level of measurement implicit in your coding methods—nominal, ordinal, interval, or ratio—does not necessarily reflect the nature of your variables. If the word *love* appeared 100 times in Novel A and 50 times in Novel B, you would be justified in saying that the word *love* appeared twice as often in Novel A, but not that Novel A was twice as erotic as Novel B. Similarly, agreeing with twice as many anti-Semitic statements in a questionnaire does not necessarily make one twice as anti-Semitic.

Counting and Record Keeping If you plan to evaluate your content analysis data quantitatively, your coding operation must be amenable to data processing. This means, first, that the end product of your coding must be numerical. If you're counting the frequency of certain words, phrases, or other manifest content, the coding is necessarily numerical. But even if you're coding latent content on the basis of overall judgments, it will be necessary to represent your coding decision numerically: 1 = very liberal, 2 = moderately liberal, 3 = moderately conservative, and so on.

Second, your record keeping must clearly distinguish between your units of analysis and your units of observation, especially if these two are different. The initial coding, of course, must relate to the units of observation. If novelists are the units of analysis, for example, and you wish to characterize them through a content analysis of their novels, your primary records will represent novels as the units of observation. You may then combine your scoring of individual novels to characterize each novelist, the unit of analysis.

Third, when counting, it will normally be important to record the base from which the counting is done. It would probably be useless to know the number of realistic paintings produced by a given painter without knowing the number he or she had painted all together; the painter would be regarded as realistic if a high percentage of paintings were of that genre. Similarly, it would tell us little that the word *love* appeared 87 times in a novel if we didn't know about how many words there were in the entire novel. The issue of observational base is most easily resolved if every observation is coded in terms of one of the attributes making up a variable. Rather than simply counting the number of liberal editorials in a given collection, for example, code each editorial by its political orientation, even if it must be coded "no apparent orientation."

Let's suppose we want to describe and explain the editorial policies of different newspapers. Figure 11-3 presents part of a tally sheet that might result from the coding of newspaper editorials. Note that newspapers are the units of analysis. Each newspaper has been assigned an identification number to facilitate mechanized processing.

The second column has a space for the number of editorials coded for each newspaper. This will be an important piece of information, since we want to be able to say, for example, "Of all the editorials, 22 percent were pro–United Nations," not just "There were eight pro–United Nations editorials."

One column in Figure 11-3 is for assigning a subjective overall assessment of the newspapers' editorial policies. (Such assessments might later be compared with the several objective measures.) Other columns provide space for recording numbers of editorials reflecting specific editorial positions. In a real content analysis, there would be spaces for recording other editorial positions plus noneditorial information about each newspaper, such as the region in which it is published, its circulation, and so forth.

Qualitative Data Analysis Not all content analysis results in counting. Sometimes a qualitative assessment of the materials is most appropriate, as in Carol Auster's examination of changes in Girl Scout uniforms and handbook language.

Bruce Berg (1989:123–25) discusses "negative case testing" as a technique for qualitative hypothesis testing. First, in the grounded theory tradition, you begin with an examination of the data, which may yield a general hypothesis. Let's say that you're examining the leadership of a new community association by reviewing the minutes of meetings to see who made motions that were subsequently passed. Your initial examination of the data suggests that the wealthier members are the most likely to assume this leadership role.

The second stage in the analysis is to search your data to find all the cases that would contradict the initial hypothesis. In this instance, you would look for poorer members who made successful motions and wealthy members who never did. Third, you must review each of the disconfirming cases and either (1) give up the hypothesis or (2) see how it needs to be fine-tuned.

Let's say that in your analysis of disconfirming cases, you notice that each of the unwealthy leaders has a graduate degree, while each of the wealthy nonleaders has very little formal education. You may revise your hypothesis to consider

Newspaper ID	Number of editorials evaluated	SUBJECTIVE EVALUATION 1. Very liberal 2. Moderately liberal 3. Middle-of-road 4. Moderately conservative 5. Very conservative	Number of "isolationist" editorials	Number of "pro-UN" editorials	Number of "anti-UN" editorials
001	37	2	0	8	0
002	26	5	10	0	6
003	44	4	2	1	2
004	22	3	1	2	3
005	30	1	0	6	0

FIGURE 11-3 **Sample Tally Sheet (Partial)**

both education and wealth as routes to leadership in the association. Perhaps you'll discover some threshold for leadership (a white-collar job, a level of income, and a college degree) beyond which those with the most money, education, or both are the most active leaders.

This process is an example of what Barney Glaser and Anselm Strauss (1967) called *analytic induction*. It's inductive in that it primarily begins with observations, and it's analytic because it goes beyond description to find patterns and relationships among variables.

There are, of course, dangers in this form of analysis, as in all others. The chief risk here is that you'll misclassify observations so as to support your emerging hypothesis. You may erroneously conclude that a nonleader didn't graduate from college or you may decide that the job of factory foreman is "close enough" to being white-collar.

Berg (1989:124) offers techniques for avoiding these errors:

1. If there are sufficient cases, select some at random from each category in order to avoid merely picking those that best support the hypothesis.
2. Give at least three examples in support of every assertion you make about the data.
3. Have your analytic interpretations carefully reviewed by others uninvolved in the research project to see whether they agree.
4. Report whatever inconsistencies you do discover—any cases that simply do not fit your hypotheses. Realize that few social patterns are 100 percent consistent, so you may have discovered something important even if it doesn't apply to absolutely all of social life.

APPLYING THE RESULTS

Content analysis has many practical applications. Consider, for example, the periodic allegations that the mass media have particular biases: Conservatives often claim that the media are liberal, and liberals often claim the opposite. Notice how the tools of content analysis permit you to select media materials for analysis, define indicators of liberal or conservative slant, and then replace simple allegations with research findings.

Or consider charges that tobacco ads are aimed at youth and minority groups. Is it really true? What elements in ads could be regarded as indicators of such targeting? How would you select a sample of advertisements for study? Again, you could conduct research to address the issue on factual grounds.

However, you should be honest with your readers in that regard.

An Illustration of Content Analysis

Several studies have indicated that women are stereotyped on television. R. Stephen Craig (1992) took this line of inquiry one step further to examine the portrayal of both men and women during different periods of television programming.

To study gender stereotyping in television commercials, Craig selected a sample of 2,209 network commercials during several periods between January 6 and 14, 1990.

> The weekday day part (in this sample, Monday–Friday, 2–4 P.M.) consisted exclusively of soap operas and was chosen for its high percentage of women viewers. The weekend day part (two consecutive Saturday and Sunday afternoons during sports telecasts) was selected for its high percentage of men viewers. Evening "prime

TABLE 11-1   Percentages of Adult Primary Visual Characters by Sex Appearing in Commercials in Three Day Parts

	Daytime	Evening	Weekend
Adult male	40	52	80
Adult female	60	48	20

Source: R. Stephen Craig, "The Effect of Television Day Part on Gender Portrayals in Television Commercials: A Content Analysis," *Sex Roles* 26, nos. 5/6 (1992): 204.

> time" (Monday–Friday, 9–11 P.M.) was chosen as a basis for comparison with past studies and the other day parts. — (1992:199)

Each of the commercials was coded in several ways. "Characters" were coded as:

All male adults
All female adults
All adults, mixed gender
Male adults with children or teens (no women)
Female adults with children or teens (no men)
Mixture of ages and genders

In addition, Craig's coders noted which character was on the screen longest during the commercial—the "primary visual character"—as well as the roles played by the characters (such as spouse, celebrity, parent), the type of product advertised (such as body product, alcohol), the setting (such as kitchen, school, business), and the voice-over narrator.

Table 11-1 indicates the differences in the times when men and women appeared in commercials. Women were more common during the daytime (with its soap operas), men predominated during the weekend commercials (with its sports programming), and men and women were equally represented during evening prime time.

Craig found other differences in the ways men and women were portrayed.

> Further analysis indicated that male primary characters were proportionately more likely than females to be portrayed as celebrities and professionals in every day part, while women were proportionately more likely to be portrayed as interviewer/demonstrators, parent/spouses,

or sex object/models in every day part. . . .
Women were proportionately more likely to appear as sex object/models during the weekend than during the day. — (1992:204)

The research also showed that different products were advertised during different time periods. As you might have imagined, almost all the daytime commercials dealt with body, food, or home products. These products accounted for only one in three on the weekends. Instead, weekend commercials stressed automotive products (29 percent), business products or services (27 percent), or alcohol (10 percent). There were virtually no alcohol ads during evenings and daytime.

As you might suspect, women were most likely to be portrayed in home settings, men most likely to be shown away from home. Other findings dealt with the different roles played by men and women.

The women who appeared in weekend ads were almost never portrayed without men and seldom as the commercial's primary character. They were generally seen in roles subservient to men (e.g., hotel receptionist, secretary, or stewardess), or as sex objects or models in which their only function seemed to be to lend an aspect of eroticism to the ad. — (CRAIG 1992:208)

Although some of Craig's findings may seem unsurprising, remember that "common knowledge" does not always correspond with reality. It's always worthwhile to check out widely held assumptions. And even when we think we know about a given situation, it's often useful to know more specific details such as those provided by a content analysis like this one.

Strengths and Weaknesses of Content Analysis

Probably the greatest advantage of content analysis is its economy in terms of both time and money. A single college student could undertake a content analysis, whereas undertaking a survey, for example, might not be feasible. There is no requirement for a large research staff; no special equipment is required. As long as you have access to the

material to be coded, you can undertake content analysis.

Safety is another advantage of content analysis. If you discover you've botched up a survey or an experiment, you may be forced to repeat the whole research project with all its attendant costs in time and money. If you botch up your field research, it may be impossible to redo the project; the event under study may no longer exist. In content analysis, it is usually easier to repeat a portion of the study than it is in other research methods. You might be required, moreover, to recode only a portion of your data rather than all of it.

A third advantage of content analysis is that it permits you to study processes occurring over a long time. You might focus on the imagery of African Americans conveyed in U.S. novels of 1850 to 1860, for example, or you might examine changing imagery from 1850 to the present.

Finally, content analysis has the advantage of all unobtrusive measures, namely, that the content analyst seldom has any effect on the subject being studied. Because the novels have already been written, the paintings already painted, the speeches already presented, content analyses can have no effect on them.

Content analysis has disadvantages as well. For one thing, it's limited to the examination of recorded communications. Such communications may be oral, written, or graphic, but they must be recorded in some fashion to permit analysis.

As we've seen, content analysis has both advantages and disadvantages in terms of validity and reliability. Problems of validity are likely unless you happen to be studying communication processes per se.

On the other side of the ledger, the concreteness of materials studied in content analysis strengthens the likelihood of reliability. You can always code and recode and even recode again if you want, making certain that the coding is consistent. In field research, by contrast, there's probably nothing you can do after the fact to ensure greater reliability in observation and categorization.

Let's move from content analysis now and turn to a related research method: the analysis of existing data. Although numbers rather than communications are the substance analyzed in this case, I think you'll see the similarity to content analysis.

ANALYZING EXISTING STATISTICS

Frequently you can or must undertake social scientific inquiry through the use of official or quasi-official statistics. This differs from secondary analysis, in which you obtain a copy of someone else's data and undertake your own statistical analysis. In this section, we're going to look at ways of using the data analyses that others have already done.

This method is particularly significant because existing statistics should always be considered at least a supplemental source of data. If you were planning a survey of political attitudes, for example, you would do well to examine and present your findings within a context of voting patterns, rates of voter turnout, or similar statistics relevant to your research interest. Or, if you were doing evaluation research on an experimental morale-building program on an assembly line, probably statistics on absenteeism, sick leave, and so on would be interesting and revealing in connection with the data your own research would generate. Existing statistics, then, can often provide a historical or conceptual context within which to locate your original research.

Existing statistics can also provide the main data for a social scientific inquiry. An excellent example is the classic study mentioned at the beginning of this chapter, Emile Durkheim's *Suicide* ([1897] 1951). Let's take a closer look at Durkheim's work before considering some of the special problems this method presents.

Durkheim's Study of Suicide

Why do people kill themselves? Undoubtedly, every suicide case has a unique history and explanation, yet all such cases could no doubt be grouped according to certain common causes: financial failure, trouble in love, disgrace, and other kinds of personal problems. The French sociologist Emile Durkheim had a slightly different question in mind

when he addressed the matter of suicide, however. He wanted to discover the environmental conditions that encouraged or discouraged it, especially social conditions.

The more Durkheim examined the available records, the more patterns of differences became apparent to him. All of these patterns interested him. One of the first things to attract his attention was the relative stability of suicide rates. Looking at several countries, he found suicide rates to be about the same year after year. He also discovered that a disproportionate number of suicides occurred in summer, leading him to hypothesize that temperature might have something to do with suicide. If this were the case, suicide rates should be higher in the southern European countries than in the temperate ones. However, Durkheim discovered that the highest rates were found in countries in the central latitudes, so temperature couldn't be the answer.

He explored the role of age (35 was the most common suicide age), gender (men outnumbered women around four to one), and numerous other factors. Eventually, a general pattern emerged from different sources.

In terms of the stability of suicide rates over time, for instance, Durkheim found the pattern was not totally stable. There were spurts in the rates during times of political turmoil, which occurred in several European countries around 1848. This observation led him to hypothesize that suicide might have something to do with "breaches in social equilibrium." Put differently, social stability and integration seemed to be a protection against suicide.

This general hypothesis was substantiated and specified through Durkheim's analysis of a different set of data. The countries of Europe had radically different suicide rates. The rate in Saxony, for example, was about ten times that of Italy, and the relative ranking of various countries persisted over time. As Durkheim considered other differences among the various countries, he eventually noticed a striking pattern: Predominantly Protestant countries had consistently higher suicide rates than did Catholic ones. The predominantly Protestant countries had 190 suicides per million population; mixed Protestant-Catholic countries, 96; and predominantly Catholic countries, 58 (Durkheim [1897] 1951:152).

Although suicide rates thus seemed to be related to religion, Durkheim reasoned that some other factor, such as level of economic and cultural development, might explain the observed differences among countries. If religion had a genuine effect on suicide, then the religious difference would have to be found *within* given countries as well. To test this idea, Durkheim first noted that the German state of Bavaria had both the most Catholics and the lowest suicide rates in that country, whereas heavily Protestant Prussia had a much higher suicide rate. Not content to stop there, however, Durkheim examined the provinces composing each of those states.

Table 11-2 shows what he found. As you can see, in both Bavaria and Prussia, provinces with the highest proportion of Protestants also had the highest suicide rates. Increasingly, Durkheim became confident that religion played a significant role in the matter of suicide.

Returning eventually to a more general theoretical level, Durkheim combined the religious findings with the earlier observation about increased suicide rates during times of political turmoil. Put most simply, Durkheim suggested that many suicides are a product of *anomie,* "normlessness," or a general sense of social instability and disintegration. During times of political strife, people may feel that the old ways of society are collapsing. They become demoralized and depressed, and suicide is one answer to the severe discomfort. Seen from the other direction, social integration and solidarity—reflected in personal feelings of being part of a coherent, enduring social whole—offer protection against depression and suicide. That was where the religious difference fit in. Catholicism, as a far more structured and integrated religious system, gave people a greater sense of coherence and stability than did the more loosely structured Protestantism.

From these theories, Durkheim created the concept of *anomic suicide.* More important, as you know, he added the concept of *anomie* to the lexicon of the social sciences.

TABLE 11-2 **Suicide Rates in Various German Provinces, Arranged in Terms of Religious Affiliation**

Religious Character of Province	Suicides per Million Inhabitants
Bavarian Provinces (1867–1875) *	
Less than 50% Catholic	
Rhenish Palatinate	167
Central Franconia	207
Upper Franconia	204
Average	192
50% to 90% Catholic	
Lower Franconia	157
Swabia	118
Average	135
More than 90% Catholic	
Upper Palatinate	64
Upper Bavaria	114
Lower Bavaria	19
Average	75
Prussian Provinces (1883–1890)	
More than 90% Protestant	
Saxony	309.4
Schleswig	312.9
Pomerania	171.5
Average	264.6x
68% to 89% Protestant	
Hanover	212.3
Hesse	200.3
Brandenburg and Berlin	296.3
East Prussia	171.3
Average	220.0
40% to 50% Protestant	
West Prussia	123.9
Silesia	260.2
Westphalia	107.5
Average	163.6
28% to 32% Protestant	
Posen	96.4
Rhineland	100.3
Hohenzollern	90.1
Average	95.6

**Note:* The population below 15 years has been omitted.

Source: Adapted from Emile Durkheim, *Suicide* (Glencoe, IL: Free Press, [1897] 1951), 153.

This account of Durkheim's classic study is greatly simplified, of course. Anyone studying social research would profit from studying the original. For our purposes, though, Durkheim's approach provides a good illustration of the possibilities for research contained in the masses of data regularly gathered and reported by government agencies.

Units of Analysis

The unit of analysis involved in the analysis of existing statistics is often *not* the individual. Durkheim, for example, was required to work with political-geographical units: countries, regions, states, and cities. The same situation would probably appear if you were to undertake a study of crime rates, accident rates, disease, and so forth. By their nature, most existing statistics are aggregated: They describe groups.

The aggregate nature of existing statistics can present a problem, though not an insurmountable one. As we saw, for example, Durkheim wanted to determine whether Protestants or Catholics were more likely to commit suicide. The difficulty was that none of the records available to him indicated the religion of those people who committed suicide. Ultimately, then, it was not possible for him to say whether Protestants committed suicide more often than Catholics, though he inferred as much. Because Protestant countries, regions, and states had higher suicides than did Catholic countries, regions, and states, he drew the obvious conclusion.

There's danger in drawing this kind of conclusion, however. It's always possible that patterns of behavior at a group level do not reflect corresponding patterns on an individual level. Such errors are said to be due to an ecological fallacy, which was discussed in Chapter 4. In the case of Durkheim's study, it was altogether possible, for example, that it was Catholics who committed suicide in the predominantly Protestant areas. Perhaps Catholics in predominantly Protestant areas were so badly persecuted that they were led into despair and suicide. In that case it would be possible for Protestant countries to have high suicide rates without any Protestants committing suicide.

Durkheim avoided the danger of the ecological fallacy in two ways. First, his general conclusions were based as much on rigorous, theoretical deductions as on the empirical facts. The correspondence between theory and fact made a counter-explanation, such as the one I just made up, less

likely. Second, by extensively retesting his conclusions in a variety of ways, Durkheim further strengthened the likelihood that they were correct. Suicide rates were higher in Protestant countries than in Catholic ones; higher in Protestant regions of Catholic countries than in Catholic regions of Protestant countries; and so forth. The replication of findings added to the weight of evidence supporting his conclusions.

Problems of Validity

Whenever we base our research on an analysis of data that already exist, we're obviously limited to what exists. Often, the existing data don't cover exactly what we're interested in, and our measurements may not be altogether valid.

Two characteristics of science are used to handle the problem of validity in the analysis of existing statistics: *logical reasoning* and *replication*. Durkheim's strategy provides an example of logical reasoning. Although he could not determine the religion of people who committed suicide, he reasoned that most of the suicides in a predominantly Protestant region would be Protestants.

Replication can be a general solution to problems of validity in social research. Recall the earlier discussion of the interchangeability of indicators (Chapter 5). Crying in sad movies isn't necessarily a valid measure of compassion; nor is putting little birds back in their nests or giving money to charity. None of these things, taken alone, would prove that one group (women, say) were more compassionate than another (men). But if women appeared more compassionate than men by all these measures, that would create a weight of evidence in support of the conclusion. In the analysis of existing statistics, a little ingenuity and reasoning can usually turn up several independent tests of a given hypothesis. If all the tests seem to confirm the hypothesis, then the weight of evidence supports the validity of the measure.

Problems of Reliability

The analysis of existing statistics depends heavily on the quality of the statistics themselves: Are they accurate reports of what they claim to report? This

IS AMERICA #1?

On September 19, 1999, ABC-TV broadcast a special show, hosted by John Stossel, to examine where the United States stood in the ranking of the world's societies. As the show unfolded, it became clear that the USA was doing okay—arguably #1—and that the key to our success was primarily due to our laissez-faire capitalist system. To make the latter point more strongly, Stossel pointed to other success stories that also owed their success to laissez-faire capitalism.

According to Stossel, Hong Kong stood out among the world's nations as the leader of free-market economics. As evidence of Hong Kong's success, Stossel reported that it had "the only government in the world that makes a surplus, a big surplus." What do you think about that conclusion? Is it convincing to you?

Here's what the media watchdog, Fairness and Accuracy in Reporting (FAIR), had to say about Stossel's assertion:

> As anyone who pays attention to Washington politics knows, the U.S. government has been running a federal budget surplus for more than a year; it amounted to $70 billion last year. Other countries with budget surpluses last year included the United Kingdom, Can-

can be a substantial problem, because the weighty tables of government statistics, for example, are sometimes grossly inaccurate.

Consider research into crime. Because a great deal of this research depends on official crime statistics, this body of data has come under critical evaluation. The results have not been too encouraging. As an illustration, suppose you were interested in tracing long-term trends in marijuana use in the United States. Official statistics on the numbers of people arrested for selling or possessing it would seem to be a reasonable measure of use, right? Not necessarily.

To begin, you face a hefty problem of validity. Before the passage of the Marihuana Tax Act in 1937, "grass" was legal in the United States, so arrest records would not give you a valid measure of use. But even if you limited your inquiry to the post-1937 era, you would still have problems of reliability, stemming from the nature of law enforcement and crime record keeping.

Law enforcement, for example, is subject to various pressures. A public outcry against marijuana, led perhaps by a vocal citizens' group, often results in a police "crackdown on drug trafficking"—especially if it occurs during an election or budget year. A sensational story in the press can have a similar effect. In addition, the volume of other business facing police affects marijuana arrests.

In tracing the pattern of drug arrests in Chicago between 1942 and 1970, Lois DeFleur (1975) has demonstrated that the official records present a far less accurate history of drug use than of police practices and political pressure on police. On a different level of analysis, Donald Black (1970) and others have analyzed the factors influencing whether an offender is actually arrested by police or let off with a warning. Ultimately, official crime statistics are influenced by whether specific offenders are well or poorly dressed, whether they are polite or abusive to police officers, and so forth. When we consider unreported crimes, sometimes estimated to be as much as ten times the number of crimes known to police, the reliability of crime statistics gets even shakier.

These comments concern crime statistics at a local level. Often it's useful to analyze national crime statistics, such as those reported in the FBI's annual *Uniform Crime Reports*. Additional problems are introduced at the national level. Different local jurisdictions define crimes differently. Also, participation in the FBI program is voluntary, so the data are incomplete.

Finally, the process of record keeping affects

ada, Australia, Denmark, Finland, Iceland, Ireland, New Zealand, Norway and Sweden.

Stossel went on to contrast Hong Kong (a capitalist success story) with the alternative to a free economy: "stagnation, and often poverty. Consider China, now mired in Third World poverty. They were once the leader of the world." Again, FAIR suggests a different assessment:

Actually, China's economy is anything but "stagnant." As the Treasury Department's Lawrence Summers said in a speech last year, "China has been the fastest growing economy in history since [economic] reform began

in 1980." While China has adopted some aspects of market economics, a large proportion of its business firms are still owned by the government.

In the media and elsewhere, you'll often find assertions of fact that appear to be based on statistical analyses. However, it's usually a good idea to check the facts.

 Source: Fairness and Accuracy in Reporting, "Action Alert: ABC News Gives up on Accuracy?" September 28, 1999.* Accessed at http://www.fair.org/activism/stossel-america.html

the records that are kept and reported. Whenever a law-enforcement unit improves its record-keeping system—computerizes it, for example—the apparent crime rates increase dramatically. This can happen even if the number of crimes committed, reported, and investigated does not increase.

Researchers' first protection against the problems of reliability in the analysis of existing statistics is awareness—knowing that the problem may exist. Investigating the nature of the data collection and tabulation may enable you to assess the nature and degree of unreliability so you can judge its potential impact on your research interest. If you also use logical reasoning and replication, you can usually cope with the problem.

The box "Is America #1?" provides an example of what you might discover by carefully examining the use of existing statistics.

Sources of Existing Statistics

It would take a whole book just to list the sources of data available for analysis. In this section, I'll mention a few sources and point you in the direction of finding others relevant to your research interest.

Undoubtedly, the single most valuable book you can buy is the annual *Statistical Abstract of the United States,* published by the United States Department of Commerce. Unquestionably the best source of data about the United States, it includes statistics on the individual states and (less extensively) cities as well as on the nation as a whole. Where else can you learn the number of work stoppages in the country year by year, the residential property taxes of major cities, the number of water pollution discharges reported around the country, the number of business proprietorships in the nation, and hundreds of other such handy bits of information? To make things even better, Hoover's Business Press offers the same book in soft cover for less cost. The commercial version, entitled *The American Almanac,* shouldn't be confused with other almanacs that are less reliable and less useful for social scientific research. Better yet, you can buy the *Statistical Abstract* on a CD-ROM, making the search for and transfer of data quite easy. You can also access it through the World Wide Web.

Suppose you were interested in the issue of income discrimination by gender. You could examine this rather easily through the *Statistical Abstract*

---

*Each time the Internet icon appears, you'll be given helpful leads for searching the World Wide Web.

Here are a few of the many data sources you can find on the World Wide Web:

- Bureau of the Census: http://www.census.gov/
- The Bureau of Labor Statistics: http://stats.bls.gov/blshome.html
- The Bureau of Transportation Statistics: http://www.bts.gov/
- The Central Intelligence Agency: http://www.cia.gov/
- The Department of Education: http://www.ed.gov/
- The Federal Bureau of Investigation: http://www.fbi.gov/
- The State and Local Governments: http://www.piperinfo.com/~piper/state/states.html
- The *Statistical Abstract*: http://www.census.gov/statab/www/
- The U.S. Government Printing Office: http://www.access.gpo.gov
- The World Bank: http://www.worldbank.org/

data. Here, for example, is a look at gender, education, and income (adapted from No. 758 of the online data for 1999 at http://www.census.gov/prod/99pubs/99statab/sec14.pdf):

Average Earnings of Year-Round, Full-Time Workers

	Men	Women	Ratio of Women/Men Earnings
All workers	43,709	29,261	0.67
Less than 9th grade	22,746	14,957	0.66
9th–12th grades	27,638	18,594	0.67
HS graduates	32,611	22,656	0.69
Some college	39,367	26,562	0.67
Associate degree	40,465	29,776	0.74
Bachelors or more	66,393	41,626	0.63

These data point to a consistent difference between the incomes of men and women, even when both groups have achieved the same levels of edu-

cation. Other variables could explain the differences, however; we'll return to this issue in Chapter 14.

Federal agencies—the Departments of Labor, Agriculture, Transportation, and so forth—publish countless data series. To find out what's available, go to your library, find the government documents section, and spend a few hours browsing through the shelves. You'll come away with a clear sense of the wealth of data available to your insight and ingenuity. You can also visit these departments' Web sites or the U.S. Government Printing Office site and look around.

World statistics are available through the United Nations. Its *Demographic Yearbook* presents annual vital statistics (births, deaths, and other data relevant to population) for the individual nations of the world. Other publications report a variety of other kinds of data. Again, a trip to your library, along with a Web search, is the best introduction to what's available.

The amount of data provided by nongovernment agencies is as staggering as the amount your taxes buy. Chambers of commerce often publish data reports on business, as do private consumer groups. Ralph Nader has information on automobile safety, and Common Cause covers politics and government. And, as mentioned earlier, George Gallup publishes reference volumes on public opinion as tapped by Gallup Polls since 1935.

Organizations such as the Population Reference Bureau publish a variety of demographic data, U.S. and international, that a secondary analyst could use. Their *World Population Data Sheet* and *Population Bulletin* are resources heavily used by social scientists. Social indicator data can be found in the journal *SINET: A Quarterly Review of Social Reports and Research on Social Indicators, Social Trends, and the Quality of Life.*

The sources I've listed are only a tiny fraction of the thousands that are available. With so much data already collected, the lack of funds to support expensive data collection is no reason for not doing good and useful social research.

The availability of existing statistics makes it possible to create some fairly sophisticated measures. The accompanying box, "Suffering Around the World," (pp. 330–331) describes an analysis published by the Population Crisis Committee based on the kinds of data available in government practice.

HISTORICAL/COMPARATIVE ANALYSIS

Historical/comparative research differs substantially from the methods discussed so far, though it overlaps somewhat with field research, content analysis, and the analysis of existing statistics. It involves the use of historical methods by sociologists, political scientists, and other social scientists.

The discussion of longitudinal research designs in Chapter 4 notwithstanding, our examination of research methods so far has focused primarily on studies anchored in one point in time and in one locale, whether a small group or a nation. Although accurately portraying the main thrust of contemporary social scientific research, this focus conceals the fact that social scientists are also interested in tracing the development of social forms over time and comparing those developmental processes across cultures. After describing some major instances of historical and comparative research in the past, this section discusses the key elements of this method.

Examples of Historical/Comparative Analysis

August Comte, who coined the term *sociologie,* saw that new discipline as the final stage in a historical development of ideas. With his broadest brush, he painted an evolutionary picture that took humans from a reliance on religion to metaphysics to science. With a finer brush, he portrayed science as evolving from the development of biology and the other natural sciences to the development of psychology and, finally, to the development of scientific sociology.

A great many later social scientists have also turned their attention to broad historical processes. Several have examined the historical pro-

gression of social forms from the simple to the complex, from rural-agrarian to urban-industrial societies. The U.S. anthropologist Lewis Morgan, for example, saw a progression from "savagery" to "barbarism" to "civilization" (1870). Robert Redfield, another anthropologist, has more recently written of a shift from "folk society" to "urban society" (1941). Emile Durkheim saw social evolution largely as a process of ever-greater division of labor ([1893] 1964). In a more specific analysis, Karl Marx examined economic systems progressing historically from primitive to feudal to capitalistic forms ([1867] 1967). All history, he wrote in this context, was a history of class struggle—the "haves" struggling to maintain their advantages and the "have-nots" struggling for a better lot in life. Looking beyond capitalism, Marx saw the development of socialism and finally communism.

Not all historical studies in the social sciences have had this evolutionary flavor, however. Some social scientific readings of the historical record, in fact, point to grand cycles rather than to linear progressions. No scholar better represents this view than Pitirim A. Sorokin. A participant in the Russian Revolution of 1917, Sorokin served as secretary to Prime Minister Kerensky. Both Kerensky and Sorokin fell from favor, however, and Sorokin began his second career—as a U.S. sociologist.

Whereas Comte read history as a progression from religion to science, Sorokin (1937–1940) suggested that societies alternate cyclically between two points of view, which he called "ideational" and "sensate." Sorokin's sensate point of view defines reality in terms of sense experiences. The ideational, by contrast, places a greater emphasis on spiritual and religious factors. Sorokin's reading of the historical record further indicated that the passage between the ideational and sensate was through a third point of view, which he called the "idealistic." This third view combined elements of the sensate and ideational in an integrated, rational view of the world.

These examples indicate some of the topics historical/comparative researchers have examined. To get a better sense of what historical/comparative research entails, let's look at a few examples in somewhat more detail.

SUFFERING AROUND THE WORLD

In 1992, the Population Crisis Committee, a nonprofit organization committed to combating the population explosion, undertook to analyze the relative degree of suffering in nations around the world. Every country with a population of one million or more was evaluated in terms of the following ten indicators—with a score of 10 on any indicator representing the highest level of adversity:

Life expectancy
Daily per capita calorie supply
Percentage of the population with access to
    clean drinking water
Proportion of infant immunization
Rate of secondary school enrollment
Gross national product
Inflation
Number of telephones per 1,000 people
Political freedom
Civil rights

Here's how the world's nations ranked in terms of these indicators. Remember, high scores are signs of overall suffering.

Extreme Human Suffering:

93—Mozambique
92—Somalia
89—Afghanistan, Haiti, Sudan
88—Zaire
87—Laos
86—Guinea, Angola
85—Ethiopia, Uganda
84—Cambodia, Sierra Leone
82—Chad, Guinea-Bissau
81—Ghana, Burma
79—Malawi
77—Cameroon, Mauritania
76—Rwanda, Vietnam, Liberia
75—Burundi, Kenya, Madagascar, Yemen

High Human Suffering:

74—Ivory Coast
73—Bhutan, Burkina Faso, Central African Republic
71—Tanzania, Togo
70—Lesotho, Mali, Niger, Nigeria
69—Guatemala, Nepal

Weber and the Role of Ideas In his analysis of economic history, Karl Marx put forward a view of economic determinism. That is, he postulated that economic factors determined the nature of all other aspects of society. For example, Marx's analysis showed that a function of European churches was to justify and support the capitalist status quo—religion was a tool of the powerful in maintaining their dominance over the powerless. "Religion is the sigh of the oppressed creature," Marx wrote in a famous passage, "the sentiment of a heartless world, and the soul of soulless conditions. It is the opium of the people" (Bottomore and Rubel [1843] 1956:27).

Max Weber, a German sociologist, disagreed. Without denying that economic factors could and did affect other aspects of society, Weber argued that economic determinism did not explain everything. Indeed, Weber said, economic forms could come from noneconomic ideas. In his research in the sociology of religion, Weber examined the extent to which religious institutions were the source of social behavior rather than mere reflections of economic conditions. His most noted statement of this side of the issue is found in *The Protestant Ethic and the Spirit of Capitalism* ([1905] 1958). Here's a brief overview of Weber's thesis.

John Calvin (1509–1564), a French theologian, was an important figure in the Protestant reformation of Christianity. Calvin taught that the ultimate salvation or damnation of every individual had already been decided by God; this idea is called *predestination*. Calvin also suggested that God communicated his decisions to people by making them either successful or unsuccessful during their earthly existence. God gave each person an earthly "calling"—an occupation or profession—and manifested their success or failure through that medium. Ironically, this point of view led Calvin's followers to seek proof of their coming salvation by

SUFFERING AROUND THE WORLD (continued)

68—Bangladesh, Bolivia, Zambia
67—Pakistan
66—Nicaragua, Papua-New Guinea, Senegal, Swaziland, Zimbabwe
65—Iraq
64—Gambia, Congo, El Salvador, Indonesia, Syria
63—Comores, India, Paraguay, Peru
62—Benin, Honduras
61—Lebanon, China, Guyana, South Africa
59—Egypt, Morocco
58—Ecuador, Sri Lanka
57—Botswana
56—Iran
55—Suriname
54—Algeria, Thailand
53—Dominican Republic, Mexico, Tunisia, Turkey
51—Libya, Colombia, Venezuela
50—Brazil, Oman, Philippines

Moderate Human Suffering:

49—Solomon Islands
47—Albania
45—Vanuatu
44—Jamaica, Romania, Saudi Arabia, Seychelles, Yugoslavia (former)
43—Mongolia
41—Jordan

40—Malaysia, Mauritius
39—Argentina
38—Cuba, Panama
37—Chile, Uruguay, North Korea
34—Costa Rica, South Korea, United Arab Emirates
33—Poland
32—Bulgaria, Hungary, Qatar
31—Soviet Union (former)
29—Bahrain, Hong Kong, Trinidad and Tobago
28—Kuwait, Singapore
25—Czechoslovakia, Portugal, Taiwan

Minimal Human Suffering:

21—Israel
19—Greece
16—United Kingdom
12—Italy
11—Barbados, Ireland, Spain, Sweden
8—Finland, New Zealand
7—France, Iceland, Japan, Luxembourg
6—Austria, Germany
5—United States
4—Australia, Norway
3—Canada, Switzerland
2—Belgium, Netherlands
1—Denmark

working hard, saving their money, and generally striving for economic success.

In Weber's analysis, Calvinism provided an important stimulus for the development of capitalism. Rather than "wasting" their money on worldly comforts, the Calvinists reinvested it in their economic enterprises, thus providing the capital necessary for the development of capitalism. In arriving at this interpretation of the origins of capitalism, Weber researched the official doctrines of the early Protestant churches, studied the preaching of Calvin and other church leaders, and examined other relevant historical documents.

In three other studies, Weber conducted detailed historical analyses of Judaism ([1934] 1952) and the religions of China ([1934] 1951) and India ([1934] 1958). Among other things, Weber wanted to know why capitalism had not developed in the ancient societies of China, India, and Israel. In none of the three religions did he find any teaching

that would have supported the accumulation and reinvestment of capital—strengthening his conclusion about the role of Protestantism in that regard.

Japanese Religion and Capitalism Weber's thesis regarding Protestantism and capitalism has become a classic in the social sciences. Not surprisingly, other scholars have attempted to test it in other historical situations. No analysis has been more interesting, however, than Robert Bellah's examination of the growth of capitalism in Japan during the late nineteenth and early twentieth centuries, entitled *Tokugawa Religion* (1957).

As both an undergraduate and a graduate student, Bellah had developed interests in Weber and in Japanese society. Given these two interests, it was perhaps inevitable that he would, in 1951, first conceive his Ph.D. thesis topic as "nothing less than an 'Essay on the Economic Ethic of Japan' to be a companion to Weber's studies of China, India, and Judaism: *The Economic Ethic of the World Religions*" (recalled in Bellah 1967:168). Originally, Bellah sketched his research design as follows:

> Problems would have to be specific and limited—no general history would be attempted—since time span is several centuries. Field work in Japan on the actual economic ethic practiced by persons in various situations, with, if possible, controlled matched samples from the U.S. (questionnaires, interviews, etc.). — (1967:168)

Bellah's original plan, then, called for surveys of contemporary Japanese and Americans. However, he did not receive the financial support necessary for the study as originally envisioned. So instead, he immersed himself in the historical records of Japanese religion, seeking the roots of the rise of capitalism in Japan.

Over the course of several years' research, Bellah uncovered numerous leads. In a 1952 term paper on the subject, Bellah felt he had found the answer in the samurai code of Bushido and in the Confucianism practiced by the samurai class:

> Here I think we find a real development of this worldly asceticism, at least equaling anything found in Europe. Further, in this class the idea of duty in occupation involved achievement without traditionalistic limits, but to the limits of one's capacities, whether in the role of bureaucrat, doctor, teacher, scholar, or other role open to the Samurai. — (QUOTED IN BELLAH 1967:171)

The samurai, however, made up only a portion of Japanese society. So Bellah kept looking at the religions among Japanese generally. His understanding of the Japanese language was not yet very good, but he wanted to read religious texts in the original. Under these constraints and experiencing increased time pressure, Bellah decided to concentrate his attention on a single group: Shingaku, a religious movement among merchants in the eighteenth and nineteenth centuries. He found that Shingaku had two influences on the development of capitalism. It offered an attitude toward work similar to the Calvinist notion of a "calling," and it had the effect of making business a more accept-

able calling for the Japanese. Previously, commerce had had a very low standing in Japan.

In other aspects of his analysis, Bellah examined the religious and political roles of the Emperor and the economic impact of periodically appearing emperor cults. Ultimately, Bellah's research pointed to the variety of religious and philosophical factors that laid the groundwork for capitalism in Japan. It seems unlikely that he would have achieved anything approaching that depth of understanding if he had been able to pursue his original plan to interview matched samples of U.S. and Japanese citizens.

These examples of historical/comparative research should have given you some sense of the potential power of the method. Let's turn now to an examination of the sources and techniques used in this method.

Sources of Historical/Comparative Data

As we saw in the case of existing statistics, there is no end of data available for analysis in historical research. To begin, historians may already have reported on whatever it is you want to examine, and their analyses can give you an initial grounding in the subject, a jumping-off point for more in-depth research.

Most likely you'll ultimately want to go beyond others' conclusions and examine some "raw data" to draw your own conclusions. These data vary, of course, according to the topic under study. In Bellah's study of Tokugawa religion, raw data included the sermons of Shingaku teachers. When W. I. Thomas and Florian Znaniecki (1918) studied the adjustment process for Polish peasants coming to the United States early in this century, they examined letters written by the immigrants to their families in Poland. (They obtained the letters through newspaper advertisements.) Other researchers have analyzed old diaries. Such personal documents only scratch the surface, however. In discussing procedures for studying the history of family life, Ellen Rothman points to the following sources:

> In addition to personal sources, there are public records which are also revealing of family his-

tory. Newspapers are especially rich in evidence on the educational, legal, and recreational aspects of family life in the past as seen from a local point of view. Magazines reflect more general patterns of family life; students often find them interesting to explore for data on perceptions and expectations of mainstream family values. Magazines offer several different kinds of sources at once: visual materials (illustrations and advertisements), commentary (editorial and advice columns), and fiction. Popular periodicals are particularly rich in the last two. Advice on many questions of concern to families—from the proper way to discipline children to the economics of wallpaper—fills magazine columns from the early nineteenth century to the present. Stories that suggest common experiences or perceptions of family life appear with the same continuity. — (1981:53)

Organizations generally document themselves, so if you're studying the development of some organization—as Bellah studied Shingaku, for example—you should examine its official documents: charters, policy statements, speeches by leaders, and so on. Once when I was studying the rise of a contemporary Japanese religious group—Sokagakkai—I discovered not only weekly newspapers and magazines published by the group but also a published collection of all the speeches given by the original leaders. With these sources, I could trace changes in recruitment patterns over time. At the outset, followers were enjoined to enroll all the world. Later, the emphasis shifted specifically to Japan. Once a sizable Japanese membership had been established, an emphasis on enrolling all the world returned (Babbie 1966).

Often, official government documents provide the data needed for analysis. To better appreciate the history of race relations in the United States, A. Leon Higginbotham, Jr. (1978) examined some 200 years of laws and court cases involving race. Himself the first African American appointed a federal judge, Higginbotham found that the law, rather than protecting African Americans, embodied bigotry and oppression. In the earliest court cases, there was considerable ambiguity over whether African Americans were indentured servants or, in fact, slaves. Later court cases and laws clarified the matter—holding African Americans to be something less than human.

The sources of data for historical analysis are too extensive to cover even in outline here, though the few examples we've looked at should suggest some ideas. Whatever resources you use, however, a couple of cautions are in order.

As we saw in the case of existing statistics, you can't trust the accuracy of records—official or unofficial, primary or secondary. Your protection lies in replication: In the case of historical research, that means corroboration. If several sources point to the same set of "facts," your confidence in them might reasonably increase.

At the same time, you need always be wary of bias in your data sources. If all your data on the development of a political movement are taken from the movement itself, you're unlikely to gain a well-rounded view of it. The diaries of well-to-do gentry of the Middle Ages may not give you an accurate view of life in general during those times. Where possible, obtain data from a variety of sources representing different points of view. Here's what Bellah said regarding his analysis of Shingaku:

> One could argue that there would be a bias in what was selected for notice by Western scholars. However, the fact that there was material from Western scholars with varied interests from a number of countries and over a period of nearly a century reduced the probability of bias. — (BELLAH 1967:179)

The issues raised by Bellah are important ones. As Ron Aminzade and Barbara Laslett indicate in the box "Reading and Evaluating Documents," there is an art in knowing how to regard such documents and what to make of them.

Incidentally, the critical review that Aminzade and Laslett urge for the reading of historical documents is useful in many areas of your life besides the pursuit of historical/comparative research. Consider applying some of their questions to presidential press conferences, advertising, or (gasp) college textbooks. None of these offers a direct view of reality; all have human authors and human subjects.

READING AND EVALUATING DOCUMENTS

by Ron Aminzade and Barbara Laslett
University of Minnesota

The purpose of the following comments is to give you some sense of the kind of interpretive work historians do and the critical approach they take toward their sources. It should help you to appreciate some of the skills historians develop in their efforts to reconstruct the past from residues, to assess the evidentiary status of different types of documents, and to determine the range of permissible inferences and interpretations. Here are some of the questions historians ask about documents:

1. Who composed the documents? Why were they written? Why have they survived all these years? What methods were used to acquire the information contained in the documents?
2. What are some of the biases in the documents and how might you go about checking or correcting them? How inclusive or representative is the sample of individuals, events, and so on, contained in the document? What were the institutional constraints and the general organizational routines under which the document was prepared? To what extent does the document provide more of an index of institutional activity than of the phenomenon being studied? What is the time lapse between the observation of the events documented and the witnesses' documentation of them? How confidential or public was the document meant to be? What role did etiquette, convention, and custom play in the presentation of the material contained within the document? If you relied solely upon the evidence contained in these documents, how might your vision of the past be distorted? What other kinds of documents might you look at for evidence on the same issues?
3. What are the key categories and concepts used by the writer of the document to organize the information presented? What selectivities or silences result from these categories of thought?
4. What sorts of theoretical issues and debates do these documents cast light on? What kinds of historical and/or sociological questions do they help to answer? What sorts of valid inferences can one make from the information contained in these documents? What sorts of generalizations can one make on the basis of the information contained in these documents?

Analytical Techniques

The analysis of historical/comparative data is another large subject that I can't cover exhaustively here. Moreover, because historical/comparative research is usually a qualitative method, there are no easily listed steps to follow in the analysis of historical data. Nevertheless, a few comments are in order.

Max Weber used the German term *verstehen*—"understanding"—in reference to an essential quality of social research. He meant that the researcher must be able to take on, mentally, the circumstances, views, and feelings of those being studied, so that the researcher can interpret their actions appropriately. Certainly this concept applies to historical/comparative research. It is the researcher's imaginative understanding that breathes life and meaning into the evidence being analyzed.

The historical/comparative researcher must find patterns among the voluminous details describing the subject matter of study. Often, this takes the form of what Weber called *ideal types:*

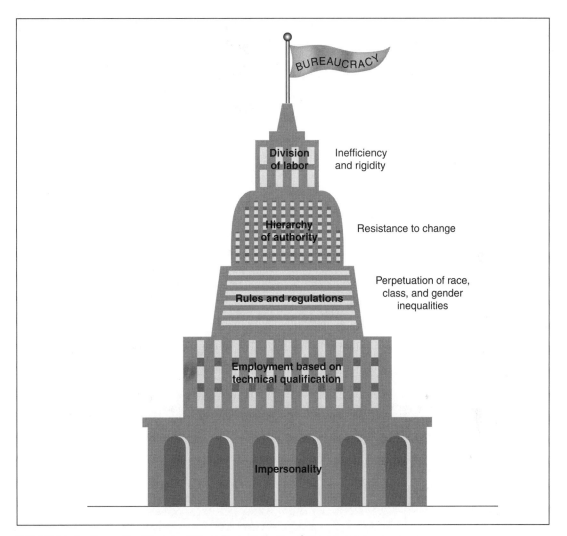

Division
of labor   Inefficiency
and rigidity

Hierarchy
of authority   Resistance to change

Rules and regulations   Perpetuation of race,
class, and gender
inequalities

Employment based on
technical qualification

Impersonality

FIGURE 11-4 **Some Positive and Negative Aspects of Bureaucracy**

Source: Diana Kendall, *Sociology in Our Times,* 3rd ed. (Belmont, CA: Wadsworth, ©2001). Used by permission.

conceptual models composed of the essential characteristics of social phenomena. Thus, for example, Weber himself did considerable research on bureaucracy. Having observed numerous actual bureaucracies, Weber ([1925] 1946) detailed those qualities essential to bureaucracies in general: jurisdictional areas, hierarchically structured authority, written files, and so on. Weber did not merely list those characteristics common to all the actual bureaucracies he observed. Rather, to create a theoretical model of the "perfect" (ideal type) bureau-

cracy, he needed to understand fully the essentials of bureaucratic operation. Figure 11-4 offers a more recent, graphic portrayal of some positive and negative aspects of bureaucracy as a general social phenomenon.

Often, historical/comparative research is informed by a particular theoretical paradigm. Thus, Marxist scholars may undertake historical analyses of particular situations—such as the history of Latinos and Latinas in the United States—to determine whether they can be understood in terms of the Marxist version of conflict theory. Sometimes historical/comparative researchers attempt to replicate prior studies in new situations—for example, Bellah's study of Tokugawa religion in the context of Weber's studies of religion and economics.

While historical/comparative research is often regarded as a qualitative rather than quantitative technique, this is by no means necessary. Historical analysts sometimes use time-series data to monitor changing conditions over time, such as data on population, crime rates, unemployment, and infant mortality rates. The analysis of such data sometimes requires sophistication, however. For example, Larry Isaac and Larry Griffin (1989) discuss the uses of a variation on regression in determining the meaningful breaking points in historical processes as well as for specifying the periods within which certain relationships occur among variables. Criticizing the tendency to regard history as a steadily unfolding process, the authors focus their attention on the statistical relationship between unionization and the frequency of strikes, demonstrating that the relationship has shifted importantly over time.

Isaac and Griffin raise several important issues regarding the relationship among theory, research methods, and the "historical facts" they address. Their analysis, once again, warns against the naive assumption that history as documented necessarily coincides with what actually happened.

A QUANDARY **REVISITED**

Unobtrusive research techniques allow researchers to avoid having an effect on what is being studied. Given that, why don't social scientists limit themselves to these techniques?

As we've seen, each of the unobtrusive techniques presented in this chapter has shortcomings of its own. The most general is that we may not be able to find existing statistics, recorded communications, or historical records that provide valid and reliable data relevant to the topic we wish to study. Other techniques, such as experiments, surveys, and field research, allow us to generate original data to fill such voids.

The ideal approach is to use multiple techniques, including unobtrusive ones. The use of multiple approaches to research can substantiate our findings when they agree, and they can broaden our understanding of the subject matter when they do not.

Main Points

- Unobtrusive measures are ways of studying social behavior without affecting it in the process.

- Content analysis is a social research method appropriate for studying human communications. Researchers can use it to study not only communication processes but other aspects of social behavior as well.

- Common units of analysis in content analysis include units of communication—words, paragraphs, books, and so forth. Standard proba-

bility sampling techniques are sometimes appropriate in content analysis.

- Content analysis involves coding—transforming raw data into categories based on some conceptual scheme. Coding may attend to both manifest and latent content. The determination of latent content requires judgments on the part of the researcher.

- Both quantitative and qualitative techniques are appropriate for interpreting content analysis data.

- The advantages of content analysis include economy, safety, and the ability to study processes occurring over a long time. Its disadvantages are that it is limited to recorded communications and can raise issues of reliability and validity.

❑ A variety of government and nongovernment agencies provide aggregate statistical data for studying aspects of social life.

❑ Problems of validity in the analysis of existing statistics can often be handled through logical reasoning and replication.

❑ Existing statistics often have problems of reliability, so they must be used with caution.

❑ Social scientists use historical/comparative methods to discover patterns in the histories of different cultures.

❑ Although often regarded as a qualitative method, historical/comparative research can make use of quantitative techniques.

Key Terms

unobtrusive research	manifest content
content analysis	latent content
coding	historical/comparative research

Review Questions

1. Is the Republican or the Democratic party the more supportive of free speech? In two or three paragraphs, outline a content analysis design to answer this question. Be sure to specify units of analysis, sampling methods, and the relevant measurements.

2. Social scientists often contrast the sense of "community" in villages, small towns, and neighborhoods with life in large, urban societies. What, in your opinion, are the essential qualities of an ideal type of community?

3. How might you compare lifestyles in different societies around the world, using pictures on the World Wide Web?. What cultural features could you look for? How would you identify differences and similarities?

4. How "old" is the college or university you're attending? When you decide on an age, specify *what* is that old. Is it people, buildings, or something else? Cite the sources you might use in arriving at your conclusion. Discuss any ambiguities that might exist in determining the age of your college or university.

Additional Readings

Baker, Vern, and Charles Lambert. 1990. "The National Collegiate Athletic Association and the Governance of Higher Education." *Sociological Quarterly* 31 (3): 403–21. A historical analysis of the factors producing and shaping the NCAA.

Berg, Bruce L. 1998. *Qualitative Research Methods for the Social Sciences.* 3rd ed. Boston: Allyn & Bacon. Contains excellent materials on unobtrusive measures, including a chapter on content analysis. While focusing on qualitative research, Berg shows the logical links between qualitative and quantitative approaches.

Evans, William. 1996. "Computer-Supported Content Analysis: Trends, Tools, and Techniques." *Social Science Computer Review* 14 (3): 269–79. Here's a review of current computer software for content analysis, such as CETA, DICTION, INTEXT, MCCA, MECA, TEXTPACK, VBPro, and WORDLINK.

Øyen, Else, ed. 1990. *Comparative Methodology: Theory and Practice in International Social Research.* Newbury Park, CA: Sage. Here are a variety of viewpoints on different aspects of comparative research. Appropriately, the contributors are from many different countries.

U.S. Bureau of the Census. 1996. *Statistical Abstract of the United States, 1996, National Data Book and Guide to Sources.* Washington, DC: U.S. Government Printing Office. This is absolutely the best book bargain available (present company excluded). Although the hundreds of pages of tables of statistics are not exciting bedtime reading—the plot is a little thin—it is an absolutely essential resource volume for every social scientist. This document is now also available on CD-ROM.

Webb, Eugene T., Donald T. Campbell, Richard D. Schwartz, Lee Sechrest, and Janet Belew Grove. 1981. *Nonreactive Measures in the Social Sciences.* Boston: Houghton Mifflin. A compendium of unobtrusive measures. Includes physical traces, a variety of archival sources, and observations. Good discussion of the ethics involved and the limitations of such measures.

Weber, Robert Philip. 1990. *Basic Content Analysis.* Newbury Park, CA: Sage. Here's an excellent beginner's book for the design and execution of content analysis. Both general issues and specific techniques are presented.

Multimedia Resources

The Wadsworth Sociology Resource Center: Virtual Society
http://sociology.wadsworth.com/

Visit the companion Web site for the second edition of *The Basics of Social Research* to access a wide range of student resources. Begin by clicking on the Student Resources section of the book's Web site to access the following study tools:

- eBabbie Resource Center
- Planning a Research Project
- Doing Data Analysis
- Statistics Review
- Flash Cards
- Internet Links and Exercises
- InfoTrac College Edition: Exercises
- Quizzes

Visit the **eBabbie Resource Center** for an overview of each chapter and helpful online tutorials. Find information on budgeting and step-by-step examples of model research projects at **Planning a Research Project.** Learn how to use quantitative and qualitative data analysis programs at **Doing Data Analysis,** and brush up on your statistics at **Statistics Review.** You can also further your study by accessing **Internet Links and Exercises** related to chapter materials, **Flash Cards, Quizzes,** and many other learning tools.

InfoTrac College Edition
http://www.infotrac-college.com/ wadsworth/access.html

Access the latest news and journal articles with InfoTrac College Edition, an easy-to-use online database of reliable, full-length articles from hundreds of top academic journals. Conduct an electronic search using the following search terms:

Content analysis
Emile Durkheim
Historical/comparative
Karl Marx
Latent content
Manifest content
Max Weber
Unobtrusive

12 EVALUATION RESEARCH

Wojnarowicz / The Image Works

What You'll Learn in This Chapter

Now you're going to see one of the most rapidly growing uses of social research: the evaluation of social interventions. You'll come away from this chapter able to judge whether social programs have succeeded or failed.

In this chapter . . .

AN OPENING **QUANDARY**

Why is there so much continuing debate over issues that straightforward social research would likely resolve? For example, people still debate whether the threat of the death penalty successfully deters murderers. Can't that outcome be tested once and for all?

INTRODUCTION

You may not be familiar with *Twende na Wakati* ("Let's Go with the Times"), but it's the most popular radio show in Tanzania. It's a soap opera. The main character, Mkwaju, is a truck driver with some pretty traditional ideas about gender roles and sex. By contrast, Fundi Mitindo, a tailor, and his wife, Mama Waridi, have more modern ideas regarding the roles of men and women particularly regarding the issues of overpopulation and family planning.

Twende na Wakati was the creation of Population Communications International (PCI) and other organizations working in conjunction with the Tanzanian government in response to two problems facing that country: (1) a population growth rate over twice that of the rest of the world and (2) an AIDS epidemic particularly heavy along the international truck route, where more than a fourth of the truck drivers and over half the commercial sex workers were found to be HIV positive in 1991. The prevalence of contraceptive use was 11 percent (Rogers et al. 1996:5–6).

The purpose of the soap opera was to bring about a change in knowledge, attitudes, and practices (KAP) relating to contraception and family planning. Rather than instituting a conventional educational campaign, PCI felt it would be more effective to illustrate the message through entertainment.

Between 1993 and 1995, 108 episodes of *Twende na Wakati* were aired, aiming at the 67 percent of Tanzanians who listen to the radio. Eighty-four percent of the radio listeners reported listening to the PCI soap opera, making it the most popular show in the country. Ninety percent of the show's listeners recognized Mkwaju, the sexist truck driver, and only three percent regarded him as a positive role model. Over two-thirds identified Mama Waridi, a businesswoman, and her tailor husband as positive role models.

Surveys conducted to measure the impact of the show indicated it had affected knowledge, attitudes, and behavior. For example, 49 percent of the married women who listened to the show said they now practiced family planning, compared with only 19 percent of the nonlisteners. There were other impacts:

> Some 72 percent of the listeners in 1994 said that they adopted an HIV/AIDS prevention behavior because of listening to *"Twende na Wakati,"* and this percentage increased to 82 percent in our 1995 survey. Seventy-seven percent of these individuals adopted monogamy, 16 percent began using condoms, and 6 percent stopped sharing razors and/or needles. — (ROGERS ET AL. 1996:21)

We can judge the effectiveness of the soap opera because of a particular form of social science. *Evaluation research*—sometimes called *program evaluation*—refers to a research purpose rather than a specific research method. This purpose is to evaluate the impact of social interventions such as new teaching methods, innovations in parole, and a host of others. Many methods—surveys, experiments, and so on—can be used in evaluation research.

Evaluation research is probably as old as social research itself. Whenever people have instituted a social reform for a specific purpose, they have paid attention to its actual consequences, even if they have not always done so in a conscious, deliberate,

---

APPLYING THE RESULTS

The research evaluating the soap operas produced in Tanzania serves many practical functions. To begin, it tells the producers whether they have been successful in delivering each of their messages. These data can help them fine-tune their presentations and make it easier to promote similar programs. Soap operas promoting small families, safe sex, and the liberation of women have been produced in several other countries in Africa as well as Asia and Latin America, and the list is still growing.

---

or sophisticated fashion. In recent years, however, the field of evaluation research has become an increasingly popular and active research specialty, as reflected in textbooks, courses, and projects. Moreover, the growth of evaluation research points to a more general trend in the social sciences. As a researcher, you'll likely be asked to conduct evaluations of your own.

In part, the growth of evaluation research reflects social scientists' increasing desire to make a difference in the world. At the same time, we can't discount the influence of (1) an increase in federal requirements that program evaluations must accompany the implementation of new programs and (2) the availability of research funds to fulfill those requirements. In any case, it seems clear that social scientists will be bringing their skills into the real world more than ever before.

This chapter looks at some of the key elements in this form of social research. After considering the kinds of topics commonly subjected to evaluation, we'll move through some of its main operational aspects: measurement, study design, and execution. As you'll see, formulating questions is as important as answering them. Because it occurs within real life, evaluation research has special problems, some of which we'll examine. Besides logistical problems, special ethical issues arise from evaluation research generally and in its specific, technical procedures. As you review reports of program evaluations, you should be especially sensitive to these problems.

Evaluation is a form of applied research—that is, it's intended to have some real-world effect. It will be useful, therefore, to consider whether and how it's actually applied. As you'll see, the obvious implications of an evaluation research project do not necessarily affect real life. They may become the focus of ideological, rather than scientific, debates. They may simply be denied out of hand, for political or other reasons. Perhaps most typically, they may simply be ignored and forgotten, left to collect dust in bookcases across the land.

This chapter concludes with a look at a particular resource for large-scale evaluation—social indicators research. This type of research is also a rapidly growing specialty. Essentially, it involves the creation of aggregated indicators of the "health" of society, similar to the economic indicators that give diagnoses and prognoses of economies.

TOPICS APPROPRIATE TO EVALUATION RESEARCH

Evaluation research is appropriate whenever some social intervention occurs or is planned. A *social intervention* is an action taken within a social context for the purpose of producing some intended result. In its simplest sense, **evaluation research** is a process of determining whether a social intervention has produced the intended result.

The topics appropriate to evaluation research are limitless. When the federal government abolished the selective service system, military researchers began paying special attention to the impact on enlistments. As individual states have liberalized their marijuana laws, researchers have sought to learn the consequences, both for marijuana use and for other forms of social behavior. Do no-fault divorce reforms increase the number of divorces, and do related social problems increase or decrease? Has no-fault automobile insurance really brought down insurance policy premiums?

Some years ago, a project evaluating the nation's driver education programs, conducted by the National Highway and Transportation Safety Administration (NHTSA), stirred up a controversy. Philip Hilts reported on the study's findings:

> For years the auto insurance industry has given large insurance discounts for children who take drivers' education courses, because statistics show that they have fewer accidents.
>
> The preliminary results of a new major study, however, indicate that drivers' education does not prevent or reduce the incidence of traffic accidents at all. — (HILTS 1981:4)

Based on an analysis of 17,500 young people in DeKalb County, Georgia (including Atlanta), the preliminary findings indicate that students who took drivers' education had just as many accidents and traffic violations as those who didn't. The study also seemed to reveal some subtle aspects of driver training.

First, it suggested that the apparent impact of driver education was largely a matter of self-selection. The kind of students who took drivers' education were less likely to have accidents and traffic violations—with or without driver training. Students with high grades, for example, were more likely to sign up for driver training, and they were also less likely to have accidents.

More startling, however, was the suggestion that driver training courses may have actually increased traffic accidents! The existence of drivers' education may encourage some students to get their licenses earlier than if there were no such courses. In a study of ten Connecticut towns that discontinued driver training, it was found that about three-fourths of those who probably would have been licensed through their classes delayed getting licenses until they were 18 or older (Hilts 1981:4).

As you might imagine, these results were not well received by those most closely associated with driver training. This matter was complicated, moreover, by the fact that the NHTSA study was also evaluating a new, more intensive training program—and the preliminary results showed that the new program was effective.

Here's a very different example of evaluation research. Rudolf Andorka, a Hungarian sociologist, has been particularly interested in his country's shift to a market economy. Even before the dramatic events in Eastern Europe in 1989, Andorka and his colleagues had been monitoring the nation's "second economy"—jobs pursued outside the socialist economy. Their surveys followed the rise and fall of such jobs and examined their impact within Hungarian society. One conclusion was that "the second economy, which earlier probably tended to diminish income inequalities or at least improved the standard of living of the poorest part of the population, in the 1980s increasingly contributed to the growth of inequalities" (Andorka 1990:111).

As you can see, the questions appropriate to evaluation research are of great practical significance: Jobs, programs, and investments as well as beliefs and values are at stake. Let's now examine how these questions are answered—how evaluations are conducted.

FORMULATING THE PROBLEM: ISSUES OF MEASUREMENT

Several years ago, I headed an institutional research office that conducted research directly relevant to the operation of the university. Often, we were asked to evaluate new programs in the curriculum. The following description shows the problem that arose in that context, and it points to one of the key barriers to good evaluation research.

Faculty members would appear at my office to say they'd been told by the university administration to arrange for an evaluation of the new program they had permission to try. This points to a common problem: Often the people whose programs are being evaluated aren't thrilled at the prospect. For them, an independent evaluation threatens the survival of the programs and perhaps even their jobs.

The main problem I want to introduce, however, has to do with the purpose of the intervention to be evaluated. The question "What is the intended result of the new program?" often produced

a vague response such as "Students will get an in-depth and genuine understanding of mathematics, instead of simply memorizing methods of calculations." Fabulous! And how could we measure that "in-depth and genuine understanding"? Often, I was told that the program aimed at producing something that could not be measured by conventional aptitude and achievement tests. No problem there; that's to be expected when we're innovating and being unconventional. What would be an unconventional measure of the intended result? Sometimes this discussion came down to an assertion that the effects of the program would be "unmeasurable."

There's the common rub in evaluation research: measuring the "unmeasurable." Evaluation research is a matter of finding out whether something is there or not there, whether something happened or didn't happen. To conduct evaluation research, we must be able to operationalize, observe, and recognize the presence or absence of what is under study.

Often, outcomes can be derived from published program documents. Thus, when Edward Howard and Darlene Norman (1981) evaluated the performance of the Vigo County Public Library in Indiana, they began with the statement of purpose previously adopted by the library's Board of Trustees.

> To acquire by purchase or gift, and by recording and production, relevant and potentially useful information that is produced by, about, or for the citizens of the community;
>
> To organize this information for efficient delivery and convenient access, furnish the equipment necessary for its use, and provide assistance in its utilization; and
>
> To effect maximum use of this information toward making the community a better place in which to live through aiding the search for understanding by its citizens. — (1981:306)

As the researchers said, "Everything that VCPL does can be tested against the Statement of Purpose." They then set about creating operational measures for each of the purposes.

While "official" purposes of interventions are often the key to designing an evaluation, this may not always be sufficient. Anna-Marie Madison (1992), for example, warns that programs designed to help disadvantaged minorities do not always reflect what the proposed recipients of the aid may need and desire:

> The cultural biases inherent in how middle-class white researchers interpret the experiences of low-income minorities may lead to erroneous assumptions and faulty propositions concerning causal relationships, to invalid social theory, and consequently to invalid program theory. Descriptive theories derived from faulty premises, which have been legitimized in the literature as existing knowledge, may have negative consequences for program participants. — (1992:38)

In setting up an evaluation, then, researchers must pay careful attention to issues of measurement. Let's take a closer look at the types of measurements that evaluation researchers must deal with.

Specifying Outcomes

As I've already suggested, a key variable for evaluation researchers to measure is the outcome, or what is called the *response variable*. If a social program is intended to accomplish something, we must be able to measure that something. If we want to reduce prejudice, we need to be able to measure prejudice. If we want to increase marital harmony, we need to be able to measure that.

It is essential to achieve agreements on definitions in advance:

> The most difficult situation arises when there is disagreement as to standards. For example, many parties may disagree as to what defines *serious* drug abuse—is it defined best as 15% or more of students using drugs weekly, 5% or more using hard drugs such as cocaine or PCP monthly, students beginning to use drugs as young as seventh grade, or some combination of the dimensions of rate of use, nature of use, and age of user? . . . Applied researchers should, to the degree possible, attempt to achieve con-

sensus from research consumers in advance of the study (e.g., through advisory groups) or at least ensure that their studies are able to produce data relevant to the standards posited by all potentially interested parties. — (HEDRICK, BICKMAN, AND ROG 1993:27)

In some cases you may find that the definitions of a problem and a sufficient solution are defined by law or agency regulations; if so, you must be aware of such specifications and accommodate them. Moreover, whatever the agreed-upon definitions, you must also achieve agreement on how the measurements will be made. Because there are different possible methods for estimating the percentage of students "using drugs weekly," for example, you'd have to be sure that all the parties involved understood and accepted the method(s) you'd chosen.

In the case of the Tanzanian soap opera, there were several outcome measures. In part, the purpose of the program was to improve knowledge about both family planning and AIDS. Thus, for example, one show debunked the belief that the AIDS virus was spread by mosquitoes and could be avoided by the use of insect repellant. Studies of listeners showed a reduction in that belief (Rogers et al. 1996:21).

PCI also wanted to change Tanzanian attitudes toward family size, gender roles, HIV/AIDS, and other related topics; the research indicated that the show had affected these as well. Finally, the program aimed at affecting behavior. We've already seen that radio listeners reported changing their behavior with regard to AIDS prevention. They reported a greater use of family planning as well. However, because there's always the possibility of a gap between what people say they do and what they actually do, the researchers sought independent data to confirm their conclusions.

Tanzania's national AIDS-control program had been offering condoms free of charge to citizens. In the areas covered by the soap opera, the number of condoms given out increased sixfold between 1992 and 1994. This far exceeded the increase of 1.4 times in the control area, where broadcasters did not carry the soap opera.

Measuring Experimental Contexts

Measuring the dependent variables directly involved in the experimental program is only a beginning. As Henry Riecken and Robert Boruch (1974:120–21) point out, it's often appropriate and important to measure those aspects of the context of an experiment that researchers think might affect the experiment. Though external to the experiment itself, some variables may affect it.

Suppose, for example, that you were conducting an evaluation of a program aimed at training unskilled people for employment. The primary outcome measure would be their success at gaining employment after completing the program. You would, of course, observe and calculate the subjects' employment rate, but you should also determine what has happened to the employment/unemployment rates of society at large during the evaluation. A general slump in the job market should be taken into account in assessing what might otherwise seem a pretty low employment rate for subjects. Or, if all the experimental subjects get jobs following the program, you should consider any general increase in available jobs. Combining complementary measures with proper control-group designs should allow you to pinpoint the effects of the program you're evaluating.

Specifying Interventions

Besides making measurements relevant to the outcomes of a program, researchers must measure the program intervention—the experimental stimulus. In part, this measurement will be handled by the assignment of subjects to experimental and control groups, if that's the research design. Assigning a person to the experimental group is the same as scoring that person yes on the stimulus, and assignment to the control group represents a score of no. In practice, however, it's seldom that simple.

Let's stick with the job-training example. Some people will participate in the program; others will not. But imagine for a moment what job-training programs are probably like. Some subjects will participate fully; others will miss a lot of sessions or

fool around when they are present. So we may need measures of the extent or quality of participation in the program. If the program is effective, you should find that those who participated fully have higher employment rates than do those who participated less.

Other factors may further confound the administration of the experimental stimulus. Suppose we're evaluating a new form of psychotherapy that's designed to cure sexual impotence. Several therapists administer it to subjects composing an experimental group. We plan to compare the recovery rate of the experimental group with that of a control group, which receives some other therapy or none at all. It may be useful to include the names of the therapists treating specific subjects in the experimental group, because some may be more effective than others. If this turns out to be the case, we must find out why the treatment worked better for some therapists than for others. What we learn will further develop our understanding of the therapy itself.

Specifying the Population

In evaluating an intervention, it's important to define the population of subjects for whom the program is appropriate. Ideally, all or a sample of appropriate subjects will then be assigned to experimental and control groups as warranted by the study design. Defining the population, however, can itself involve specifying measurements. If we're evaluating a new form of psychotherapy, it's probably appropriate for people with mental problems. But how will "mental problems" be defined and measured? The job-training program mentioned previously is probably intended for people who are having trouble finding work, but what counts as "having trouble"?

Beyond defining the relevant population, then, the researcher should make fairly precise measurements on the variables considered in the definition. For example, even though the randomization of subjects in the psychotherapy study would ensure an equal distribution of those with mild and severe mental problems into the experimental and control groups, we'd need to keep track of the relative severity of different subjects' problems in case the therapy turns out to be effective for only those with mild disorders. Similarly, we should measure such demographic variables as gender, age, race, and so forth in case the therapy works only for women, the elderly, or some other group.

New versus Existing Measures

In providing for the measurement of these different kinds of variables, the researcher must continually choose whether to create new measures or to use ones already devised by others. If a study addresses something that's never been measured before, the choice is easy. If it addresses something that others have tried to measure, the researcher will need to evaluate the relative worth of various existing measurement devices in terms of her or his specific research situations and purpose. Recall that this is a general issue in social research that applies well beyond evaluation research. Let's examine briefly the advantages of creating new measures versus using existing ones.

Creating measurements specifically for a study can offer greater relevance and validity. If the psychotherapy we're evaluating aims at a specific aspect of recovery, we can create measures that pinpoint that aspect. We might not be able to find any standardized psychological measures that hit that aspect right on the head. However, creating our own measure will cost us the advantages to be gained from using preexisting measures. Creating good measures takes time and energy, both of which could be saved by adopting an existing technique. Of greater scientific significance, measures that have been used frequently by other researchers carry a body of possible comparisons that might be important to our evaluation. If the experimental therapy raises scores by an average of ten points on a standardized test, we'll be in a position to compare that therapy with others that had been evaluated using the same measure. Finally, measures with a long history of use usually have known degrees of validity and reliability, but newly created measures will require pretesting or will be used with considerable uncertainty.

Operationalizing Success/Failure

Potentially, one of the most taxing aspects of evaluation research is determining whether the program under review succeeded or failed. The purpose of the foreign language program mentioned earlier may be to help students better learn the language, but how much better is enough? The purpose of the conjugal visit program at a prison may be to raise morale, but how high does morale need to be raised to justify the program?

As you may anticipate, there are almost never clear-cut answers to questions like these. This dilemma has surely been the source of what is generally called *cost/benefit analysis.* How much does the program cost in relation to what it returns in benefits? If the benefits outweigh the cost, keep the program going. If the reverse, junk it. That's simple enough, and it seems to apply in straightforward economic situations: If it cost you $20 to produce something and you can sell it for only $18, there's no way you can make up the difference in volume.

Unfortunately, the situations usually faced by evaluation researchers are seldom amenable to straightforward economic accounting. The foreign language program may cost the school district $100 per student, and it may raise students' performances on tests by an average of 15 points. Because the test scores can't be converted into dollars, there's no obvious ground for weighing the costs and benefits.

Sometimes, as a practical matter, the criteria of success and failure can be handled through competition among programs. If a different foreign language program costs only $50 per student and produces an increase of 20 points in test scores, it would undoubtedly be considered more successful than the first program—assuming that test scores were seen as an appropriate measure of the purpose of both programs and the less expensive program had no negative, unintended consequences.

Ultimately, the criteria of success and failure are often a matter of agreement. The people responsible for the program may commit themselves in advance to a particular outcome that will be regarded as an indication of success. If that's the case, all you need to do is make absolutely certain that the research design will measure the specified outcome. I mention this obvious requirement simply because researchers sometimes fail to meet it, and there's little or nothing more embarrassing than that.

In summary, researchers must take measurement quite seriously in evaluation research, carefully determining all the variables to be measured and getting appropriate measures for each. As I've implied, however, such decisions are typically not purely scientific ones. Evaluation researchers often must work out their measurement strategy with the people responsible for the program being evaluated. It usually doesn't make sense to determine whether a program achieves Outcome X when its purpose is to achieve Outcome Y. (Realize, however, that evaluation designs sometimes have the purpose of testing for unintended consequences.)

There is a political aspect to these choices, also. Because evaluation research often affects other people's professional interests—their pet program may be halted, or they may be fired or lose professional standing—the results of evaluation research are often argued about.

Let's turn now to some of the evaluation designs commonly employed by researchers.

TYPES OF EVALUATION RESEARCH DESIGNS

As I noted at the start of this chapter, evaluation research is not itself a method, but rather one application of social research methods. As such, it can involve any of several research designs. Here we'll consider three main types of research design that are appropriate for evaluations: experimental designs, quasi-experimental designs, and qualitative evaluations.

Experimental Designs

Many of the experimental designs introduced in Chapter 8 can be used in evaluation research. By way of illustration, let's see how the classical experimental model might be applied to our evaluation of the new psychotherapeutic treatment for sexual impotence.

In designing our evaluation, we should begin by identifying a population of patients appropriate to the therapy. This identification might be made by researchers experimenting with the new therapy. Let's say we're dealing with a clinic that already has 100 patients being treated for sexual impotence. We might take that group and the clinic's definition of sexual impotence as a starting point, and we should maintain any existing assessments of the severity of the problem for each specific patient.

For purposes of the evaluation research, however, we would need to develop a more specific measure of impotence. Maybe it would involve whether patients have sexual intercourse at all (within a specified time), how often they have intercourse, or whether and how often they reach orgasm. Alternatively, the outcome measure might be based on the assessments of independent therapists not involved in the therapy who interview the patients later. In any event, we would need to agree on the measures to be used.

In the simplest design, we would assign the 100 patients randomly to experimental and control groups; the former would receive the new therapy, and the latter would be taken out of therapy altogether during the experiment. Because ethical practice would probably prevent withdrawing therapy altogether from the control group, however, it's more likely that the control group would continue to receive their conventional therapy.

Having assigned subjects to the experimental and control groups, we would need to agree on the length of the experiment. Perhaps the designers of the new therapy feel it ought to be effective within two months, and an agreement could be reached. The duration of the study doesn't need to be rigid, however. One purpose of the experiment and evaluation might be to determine how long it actually takes for the new therapy to be effective. Conceivably, then, an agreement could be struck to measure recovery rates weekly, say, and let the ultimate length of the experiment rest on a continual review of the results.

Let's suppose the new therapy involves showing pornographic movies to patients. We'd need to specify that stimulus. How often would patients see the movies, and how long would each session be? Would they see the movies in private or in groups? Should therapists be present? Perhaps we should observe the patients while the movies are being shown and include our observations among the measurements of the experimental stimulus. Do some patients watch the movies eagerly but others look away from the screen? We'd have to ask these kinds of questions and create specific measurements to address them.

Having thus designed the study, all we have to do is "roll 'em." The study is set in motion, the observations are made and recorded, and the mass of data is accumulated for analysis. Once the study has run its course, we can determine whether the new therapy had its intended—or perhaps some unintended—consequences. We can tell whether the movies were most effective for patients with mild problems or severe ones, whether they worked for young subjects but not older ones, and so forth.

This simple illustration should show you how the standard experimental designs presented in Chapter 8 can be used in evaluation research. Many, perhaps most, of the evaluations reported in the research literature don't look exactly like this illustration, however. Because it's nested in real life, evaluation research often calls for quasi-experimental designs. Let's see what this means.

Quasi-Experimental Designs

Quasi experiments are distinguished from "true" experiments primarily by the lack of random assignment of subjects to an experimental and a control group. In evaluation research, it's often impossible to achieve such an assignment of subjects. Rather than forgo evaluation altogether, researchers sometimes create and execute research designs that give some evaluation of the program in question. This section describes some of these designs.

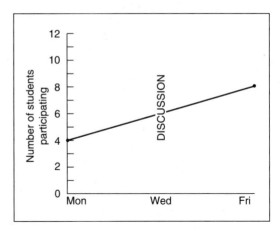

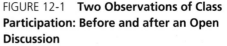

FIGURE 12-1 **Two Observations of Class Participation: Before and after an Open Discussion**

Time-Series Designs To illustrate the **time-series design**—studies that involve measurements taken over time—I'll begin by asking you to assess the meaning of some hypothetical data. Suppose I come to you with what I say is an effective technique for getting students to participate in classroom sessions of a course I'm teaching. To prove my assertion, I tell you that on Monday only four students asked questions or made a comment in class; on Wednesday I devoted the class time to an open discussion of a controversial issue raging on campus; and on Friday, when we returned to the subject matter of the course, eight students asked questions or made comments. In other words, I contend, the discussion of a controversial issue on Wednesday has doubled classroom participation. This simple set of data is presented graphically in Figure 12-1.

Have I persuaded you that the open discussion on Wednesday has had the consequence I claim for it? Probably you'd object that my data don't prove the case. Two observations (Monday and Friday) aren't really enough to prove anything. Ideally I should have had two classes, with students assigned randomly to each, held an open discussion in only one, and then compared the two on Friday. But I don't have two classes with random assignment of students. Instead, I've been keeping a

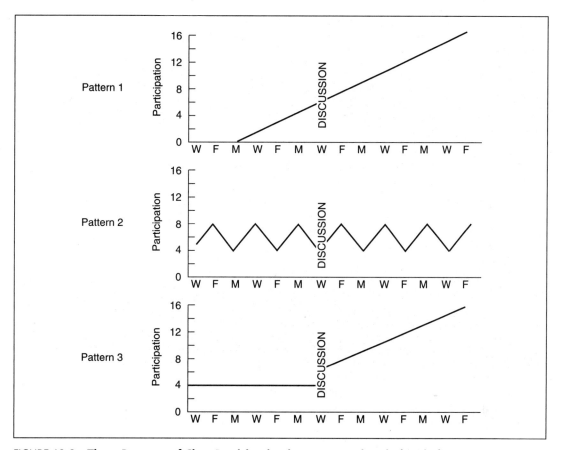

FIGURE 12-2 **Three Patterns of Class Participation in a Longer Historical Period**

record of class participation throughout the semester for the one class. This record allows you to conduct a time-series evaluation.

Figure 12-2 presents three possible patterns of class participation over time, both before and after the open discussion on Wednesday. Which of these patterns would give you some confidence that the discussion had the impact I contend it had?

If the time-series results looked like the first pattern in Figure 12-2, you'd probably conclude that the process of greater class participation had begun on the Wednesday before the discussion and had continued, unaffected, after the day devoted to the discussion. The long-term data suggest that the trend would have occurred even without the discussion on Wednesday. The first pattern, then, contradicts my assertion that the special discussion increased class participation.

The second pattern contradicts my assertion by indicating that class participation has been bouncing up and down in a regular pattern throughout the semester. Sometimes it increases from one class to the next, and sometimes it decreases; the open discussion on that Wednesday simply came at a time when the level of participation was about to increase. More to the point, we note that class participation decreased again at the class following the alleged postdiscussion increase.

Only the third pattern in Figure 12-2 supports my contention that the open discussion mattered. As depicted there, the level of discussion before that Wednesday had been a steady four students per class. Not only did the level of participation double following the day of discussion, but it continued to increase afterward. Although these data do not protect us against the possible influence of some extraneous factor (I might also have mentioned that participation would figure into students' grades), they do exclude the possibility that the increase results from a process of maturation (indicated in the first pattern) or from regular fluctuations (indicated in the second).

Nonequivalent Control Groups The time-series design just described involves only an "experimental" group; it doesn't provide the value to be gained from having a control group. Sometimes, when researchers can't create experimental and control groups by random assignment from a common pool, they can find an existing "control" group that appears similar to the experimental group. Such a group is called a **nonequivalent control group.** If an innovative foreign language program is being tried in one class in a large high school, for example, you may be able to find another foreign language class in the same school that has a very similar student population: one that has about the same composition in terms of grade in school, gender, ethnicity, IQ, and so forth. The second class, then, could provide a point of comparison. At the end of the semester, both classes could be given the same foreign language test, and you could compare performances.

Here's how two junior high schools were selected for purposes of evaluating a program aimed at discouraging tobacco, alcohol, and drug use:

> The pairing of the two schools and their assignment to "experimental" and "control" conditions was not random. The local Lung Association had identified the school where we delivered the program as one in which administrators were seeking a solution to admitted problems of smoking, alcohol, and drug abuse. The "control" school was chosen as a convenient and nearby demographic match where administrators were willing to allow our surveying and breath-testing procedures. The principal of that

school considered the existing program of health education to be effective and believed that the onset of smoking was relatively uncommon among his students. The communities served by the two schools were very similar. The rate of parental smoking reported by the students was just above 40 percent in both schools. — (MCALISTER ET AL. 1980:720)

In the initial set of observations, the experimental and control groups reported virtually the same (low) frequency of smoking. Over the 21 months of the study, smoking increased in both groups, but it increased less in the experimental group than in the control group, suggesting that the program had an impact on students' behavior.

Multiple Time-Series Designs Sometimes the evaluation of processes occurring outside of "pure" experimental controls can be made easier by the use of more than one time-series analysis. **Multiple time-series designs** are an improved version of the nonequivalent control group design just described. Carol Weiss has presented a useful example:

> An interesting example of multiple time series was the evaluation of the Connecticut crackdown on highway speeding. Evaluators collected reports of traffic fatalities for several periods before and after the new program went into effect. They found that fatalities went down after the crackdown, but since the series had had an unstable up-and-down pattern for many years, it was not certain that the drop was due to the program. They then compared the statistics with time-series data from four neighboring states where there had been no changes in traffic enforcement. Those states registered no equivalent drop in fatalities. The comparison lent credence to the conclusion that the crackdown had had some effect. — (1972:69)

Although this study design is not as good as one in which subjects are assigned randomly, it's nonetheless an improvement over assessing the experimental group's performance without any comparison. That's what makes these designs quasi experiments instead of just fooling around. The key in assessing this aspect of evaluation studies is comparability, as the following example illustrates.

A growing concern in the poor countries of the world, rural development has captured the attention and support of many rich countries. Through national foreign assistance programs and through international agencies such as the World Bank, the developed countries are in the process of sharing their technological knowledge and skills with the developing countries. Such programs have had mixed results, however. Often, modern techniques do not produce the intended results when applied in traditional societies.

Rajesh Tandon and L. Dave Brown (1981) undertook an experiment in which technological training would be accompanied by instruction in village organization. They felt it was important for poor farmers to learn how to organize and exert collective influence within their villages—getting needed action from government officials, for example. Only then would their new technological skills bear fruit.

Both intervention and evaluation were attached to an ongoing program in which 25 villages had been selected for technological training. Two poor farmers from each village had been trained in new agricultural technologies. Then they had been sent home to share their new knowledge with their fellow villagers and to organize other farmers into "peer groups" who would assist in spreading that knowledge. Two years later, the authors randomly selected two of the 25 villages (subsequently called Group A and Group B) for special training and 11 others as controls. A careful comparison of demographic characteristics showed the experimental and control groups to be strikingly similar to each other, suggesting they were sufficiently comparable for the study.

The peer groups from the two experimental villages were brought together for special training in organization building. The participants were given some information about organizing and making demands on the government; they were also given opportunities to act out dramas similar to the

situations they faced at home. The training took three days.

The outcome variables considered by the evaluation all had to do with the extent to which members of the peer groups initiated group activities designed to improve their situation. Six types were studied. "Active initiative," for example, was defined as "active effort to influence persons or events affecting group members versus passive response or withdrawal" (Tandon and Brown 1981: 180). The data for evaluation came from the journals that the peer group leaders had been keeping since their initial technological training. The researchers read through the journals and counted the number of initiatives taken by members of the peer groups. Two researchers coded the journals independently and compared their work to test the reliability of the coding process.

Figure 12-3 compares the number of active initiatives by members of the two experimental groups with those coming from the control groups. Similar results were found for the other outcome measures.

Notice two things about the graph. First, there's a dramatic difference in the number of initiatives by the two experimental groups as compared with the eleven controls. This seems to confirm the effectiveness of the special training program. Second, notice that the number of initiatives also increased among the control groups. The researchers explain this latter pattern as a result of contagion. Because all the villages were near each other, the lessons learned by peer group members in the experimental groups were communicated in part to members of the control villages.

This example illustrates the strengths of multiple time-series designs in situations where true experiments are inappropriate to the program being evaluated.

Qualitative Evaluations

I've laid out the steps involved in tightly structured, mostly quantitative evaluation research, but evaluations can also be less structured and more qualitative. For example, Pauline Bart and Patricia

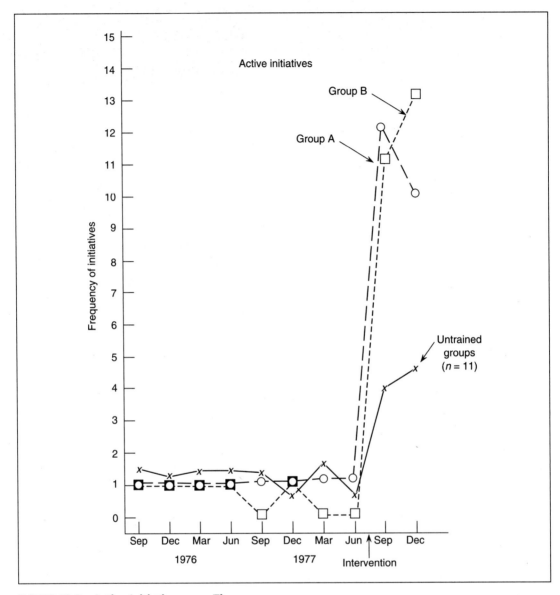

FIGURE 12-3 **Active Initiatives over Time**

Source: Rajesh Tandon and L. Dave Brown, "Organization-Building for Rural Development: An Experiment in India," *Journal of Applied Behavior Science,* April–June 1981, p. 182.

O'Brien (1985) wanted to evaluate different ways to stop rape, so they undertook in-depth interviews with both rape victims and women who had successfully fended off rape attempts.

Sometimes even structured, quantitative evaluations can yield unexpected, qualitative results. Paul Steel is a social researcher specializing in the

evaluation of programs aimed at pregnant drug users. One program he evaluated involved counseling by public health nurses, who warned pregnant drug users that continuing to use drugs would likely result in underweight babies whose skulls would be an average of 10 percent smaller than normal. In his in-depth interviews with program participants, however, he discovered that the program omitted one important piece of information: that undersized babies were a bad thing. Many of the young women Steel interviewed thought that smaller babies would mean easier deliveries.

In another program, a local district attorney had instituted what would generally be regarded as a progressive, enlightened program. If a pregnant drug user were arrested, she could avoid prosecution if she would (1) agree to stop using drugs and (2) successfully complete a drug rehabilitation program. Again, in-depth interviews suggested that the program did not always operate on the ground the way it did in principle. Specifically, Steel discovered that whenever a young woman was arrested for drug use, the other inmates would advise her to get pregnant as soon as she was released on bail. That way, she could avoid prosecution (personal communication, November 22, 1993).

The most effective evaluation research is one that combines qualitative and quantitative components. While making statistical comparisons is useful, so is gaining an in-depth understanding of the processes producing the observed results—or preventing the expected results from appearing.

The evaluation of the Tanzanian soap opera, presented earlier in this chapter, employed several research techniques. I've already mentioned the listener surveys and data obtained from clinics. In addition, the researchers conducted numerous focus groups to probe more deeply into the impact the shows had on listeners. Also, content analyses were done on the soap opera episodes themselves and on the many letters received from listeners. Both quantitative and qualitative analyses were undertaken (Swalehe et al. 1995).

THE SOCIAL CONTEXT

Many of the comments in previous sections have hinted at the possibility of problems in the actual execution of evaluation research projects. Of course, all forms of research can run into problems, but evaluation research has a special propensity for it. This section looks at some of the logistical problems and special ethical issues in evaluation research. It concludes with some observations about using evaluation research results.

Logistical Problems

In a military context, *logistics* refers to moving supplies around—making sure people have food, guns, and tent pegs when they need them. Here, I use it to refer to getting subjects to do what they're supposed to do, getting research instruments distributed and returned, and other seemingly unchallenging tasks. These tasks are more challenging than you might guess!

Motivating Sailors When Kent Crawford, Edmund Thomas, and Jeffrey Fink (1980) set out to find a way to motivate "low performers" in the U.S. Navy, they found out just how many problems can occur. The purpose of the research was to test a three-pronged program for motivating sailors who were chronically poor performers and often in trouble aboard ship. First, a workshop was to be held for supervisory personnel, training them in effective leadership of low performers. Second, a few supervisors would be selected and trained as special counselors and role models—people the low performers could turn to for advice or just as sounding boards. Finally, the low performers themselves would participate in workshops aimed at training them to be more motivated and effective in their work and in their lives. The project was to be conducted aboard a particular ship, with a control group selected from sailors on four other ships.

To begin, the researchers reported that the supervisory personnel were not exactly thrilled with the program.

> Not surprisingly, there was considerable resistance on the part of some supervisors toward dealing with these issues. In fact, their reluctance to assume ownership of the problem was reflected by "blaming" any of several factors that can contribute to their personnel problem. The recruiting system, recruit training, parents, and society at large were named as influencing low performance—factors that were well beyond the control of the supervisors. — (CRAWFORD, THOMAS, AND FINK 1980:488)

Eventually, the reluctant supervisors came around and "this initial reluctance gave way to guarded optimism and later to enthusiasm" (1980:489).

The low performers themselves were even more of a problem, however. The research design called for pre- and posttesting of attitudes and personalities, so that changes brought about by the program could be measured and evaluated.

> Unfortunately, all of the LPs (Low Performers) were strongly opposed to taking these so-called personality tests and it was therefore concluded that the data collected under these circumstances would be of questionable validity. Ethical concerns also dictated that we not force "testing" on the LPs. — (CRAWFORD, THOMAS, AND FINK 1980:490)

As a consequence, the researchers had to rely on interviews with the low performers and on the judgments of supervisors for their measures of attitude change. The subjects continued to present problems, however.

Initially, the ship's command ordered 15 low performers to participate in the experiment. Of the 15, however, one went into the hospital, another was assigned duties that prevented participation, and a third went "over the hill" (absent without leave). Thus, the experiment began with 12 subjects. But before it was completed, three more subjects completed their enlistments and left the Navy, and another was thrown out for disciplinary reasons. The experiment concluded, then, with 8 subjects. Although the evaluation pointed to positive results, the very small number of subjects warranted caution in any generalizations from the experiment.

The special, logistical problems of evaluation research grow out of the fact that it occurs within the context of real life. Although evaluation research is modeled after the experiment—which suggests that the researchers have control over what happens—it takes place within frequently uncontrollable daily life. Of course, the participant-observer in field research doesn't have control over what's observed either, but that method doesn't strive for control. Given the objectives of evaluation research, lack of control can create real dilemmas for the researcher.

Administrative Control As suggested in the previous example, the logistical details of an evaluation project often fall to program administrators. Let's suppose you're evaluating the effects of a "conjugal visit" program on the morale of married prisoners. The program allows inmates periodic visits from their spouses during which they can have sexual relations. On the fourth day of the program, a male prisoner dresses up in his wife's clothes and escapes. Although you might be tempted to assume that his morale was greatly improved by escaping, that turn of events would complicate your study design in many ways. Perhaps the warden will terminate the program altogether, and where's your evaluation then? Or, if the warden is brave, he or she may review the files of all those prisoners you selected randomly for the experimental group and veto the "bad risks." There goes the comparability of your experimental and control groups. As an alternative, stricter security measures may be introduced to prevent further escapes, and the security measures may have a dampening effect on morale. So the experimental stimulus has changed in the middle of your research project. Some of the data will reflect the original stimulus; other data will reflect the modification. Although you'll probably be able to sort it all out, your carefully designed study has become a logistical snake pit.

Or suppose you've been engaged to evaluate the effect of race-relations lectures on prejudice in the army. You've carefully studied the soldiers available to you for study, and you've randomly assigned some to attend the lectures and others to stay away. The rosters have been circulated weeks in advance, and at the appointed day and hour, the lectures begin. Everything seems to be going smoothly until you begin processing the files: The names don't match. Checking around, you discover that military field exercises, KP duty, and a variety of emergencies required some of the experimental subjects to be elsewhere at the time of the lectures. That's bad enough, but then you learn that helpful commanding officers sent others to fill in for the missing soldiers. And whom do you suppose they picked to fill in? Soldiers who didn't have anything else to do or who couldn't be trusted to do anything important. You might learn this bit of in-

TESTING SOAP OPERAS IN TANZANIA

by William N. Ryerson
Executive Vice-President
Population Communications International

Twende na Wakati ("Let's Go With the Times") has been broadcast on Radio Tanzania since mid-1993 with support from the United Nations Population Fund. The program was designed to encourage family planning use and AIDS prevention measures.

There were many different elements to the research. One was a nationwide, random-sample survey given prior to the first airing of the soap opera in June 1993 and then annually after that. Many interviewers faced particularly interesting challenges. For example, one interviewer, Fridolan Banzi, had never been in or on water in his life and couldn't swim. He arranged for a small boat to take him through the rough waters of Lake Victoria so he could carry out his interviews at a village that had no access by road. He repeated this nerve-wracking trip each year afterward in order to measure the change in that village.

Another interviewer, Mr. Tende, was invited to participate in a village feast that the villagers held to welcome him and to indicate their enthusiasm about having been selected for the study. They served him barbequed rats. Though they weren't part of his normal diet, he ate them anyway to be polite and to ensure that the research interviews could be carried out in that village.

Still another interviewer, Mrs. Masanja, was working in a village in the Pwani region along the coast of the Indian Ocean when cholera broke out in that village. She wisely chose to abandon the interviews there, which reduced the 1993 sample size by one ward.

The unsung heroes of this research, the Tanzanian interviewers, deserve a great deal of credit for carrying out this important work under difficult circumstances.

formation a week or so before the deadline for submitting your final report on the impact of the race-relations lectures.

These are some of the logistical problems confronting evaluation researchers. You need to be familiar with the problems to understand why some research procedures may not measure up to the design of the classical experiment. As you read reports of evaluation research, however, you'll find that—my earlier comments notwithstanding—it is possible to carry out controlled social research in conjunction with real-life experiments.

The accompanying box, "Testing Soap Operas in Tanzania," describes some of the logistical problems involved in the research discussed at the outset of this chapter.

Just as evaluation research has special logistical problems, it also raises special ethical concerns. Because ethical problems can affect the scientific quality of the research, we should look at them briefly.

Some Ethical Issues

Ethics and evaluation are intertwined in many ways. Sometimes the social interventions being evaluated raise ethical issues. Evaluating the impact of busing schoolchildren to achieve educational integration will throw the researchers directly into the political, ideological, and ethical issues of busing itself. It's not possible to evaluate a sex education program in elementary schools without becoming involved in the heated issues surrounding sex education itself, and the researcher will find it difficult to remain impartial. The evaluation study design will *require* that some children receive sex education—in fact, you may very well be the one who decides which children

THE IMPACT OF "THREE-STRIKES" LAWS

SACRAMENTO (AP)—The author of California's five-year-old "three strikes" law says it's prevented more than a million crimes and has saved $21.7 billion.

Secretary of State Bill Jones offered his interpretation of the "three strikes" results to a Doris Tate Crimes Victim Bureau conference on Friday in Sacramento. — *BayInsider*, March 1, 1999

The 1990s saw the passage of "three-strikes" laws at the federal level and in numerous states. The intention was to reduce crime rates by locking up "career criminals." Under the 1994 California law, for example, having a past felony conviction would double your punishment when convicted of your second felony, and the third felony conviction would bring a mandatory sentence of 25 years to life. Over the years, only California has enforced such laws with any vigor.

Those who supported the passage of "three-strikes" legislation, such as Bill Jones, quoted above, have been quick to link the dramatic drop in crime rates during the 1990s to the new policy of getting tough with career criminals. While acknowledging that "three-strikes" may not be the only cause of the drop in crime, Jones added, "If you can have a 51 percent reduction in the homicide rate in five years, I would guarantee you three strikes is a big part of the reason."

In spite of the politicians' guarantees, other observers have looked for additional evidence to support the impact of "three-strikes" laws. Some critics of these laws, for example, have noted that crime rates have been dropping dramatically across the country, not only in California but also in states that have no "three-strikes" laws and in those where the courts have not enforced the "three-strikes" laws that exist. In fact, crime rates have dropped in those California counties that have tended to ignore that state's law. Moreover, the drop in California crime rates began before the "three-strikes" law went into effect.

In 1994, Peter Greenwood and his colleagues at the Rand Corporation estimated that implementation of the law would cost California's criminal justice system approximately $5.5 billion more per year, especially in prison costs as

---

do. (From a scientific standpoint, you *should* be in charge of selection.) This means that when parents become outraged that their child is being taught about sex, you'll be directly responsible.

Now let's look on the "bright" side. Maybe the experimental program is of great value to those participating in it. Let's say that the new industrial safety program being evaluated reduces injuries dramatically. What about the control-group members who were deprived of the program by the research design? The evaluators' actions could be an important part of the reason that a control-group subject suffered an injury.

Sometimes the name of evaluation research has actually been a mask for unethical behavior. In Chapter 9 I discussed push polls, which pretend to evaluate the impact of various political campaign accusations but intend to spread malicious misinformation. That's not the worst example, however.

In 1932, researchers in Tuskegee, Alabama, began a program of providing free treatment for syphilis to poor, black men suffering from the disease. Over the years that followed, several hundred men participated in the program. What they didn't know was that they were not actually receiving any treatment at all; the physicians conducting the study merely wanted to observe the natural progress of the disease. Even after penicillin was found to be an effective cure, the researchers still withheld the treatment. While there is unanimous agreement today as to the unethical nature of the study, this was not the case at the time. Even when the study began being reported in research publications, the researchers refused to acknowledge they had done anything wrong. When professional complaints were finally lodged with the U.S. Center for Disease Control in 1965, there was no reply (J. Jones 1981).

"career criminals" were sentenced to longer terms. While the Rand group did not deny that the "three-strikes" legislation would have *some* impact on crime—those serving long terms in prison can't commit crimes on the streets—a followup study (Greenwood, Rydell, and Model 1996) suggested it was an inefficient way of attacking crime. They estimated that a million dollars spent on "three-strikes" would prevent 60 crimes, whereas the same amount spent on programs encouraging high school students to stay in school and graduate would prevent 258 crimes.

Criminologists have long recognized that most crimes are committed by young men. Focusing attention on older "career criminals" has little or no affect on the youthful offenders. In fact, "three-strikes" sentences disproportionately fall on those approaching the end of their criminal careers by virtue of growing older.

In a more general critique, John Irwin and James Austin (1997) suggest that people in the United States tend to overuse prisons as a solution to crime, ignoring other, more effective, solutions. Often, imprisonment causes problems more serious than those it was intended to remedy.

As with many other social interventions, however, much of the support for "three-strikes" laws in California and elsewhere have mostly to do with public emotions about crime and the political implications of such emotions. Thus, evaluation research on these laws may eventually bring about changes, but its impact is likely to be much slower than you might logically expect.

Sources: "State Saved $21.7 Billion with Five-Year-Old 'Three Strikes' Law," *BayInsider,* March 1, 1999 [Online]: http://www.bayinsider.com/news/1999/03/01/three_strikes.html; Peter W. Greenwood et al., *Three Strikes and You're Out: Estimated Benefits and Costs of California's New Mandatory-Sentencing Law* (Santa Monica, CA: Rand Corporation, 1994); Peter W. Greenwood, C. Peter Rydell, and Karyn Model, *Diverting Children from a Life of Crime: Measuring Costs and Benefits* (Santa Monica, CA: Rand Corporation, 1996); John Irwin and James Austin, *It's About Time: America's Imprisonment Binge* (Belmont, CA: Wadsworth 1997).

My purpose in these comments has not been to cast a shadow on evaluation research. Rather, I want to bring home the real-life consequences of the evaluation researcher's actions. Ultimately, all social research has ethical components (see Chapter 3).

Use of Research Results

One more facts-of-life aspect of evaluation research concerns how evaluations are used. Because the purpose of evaluation research is to determine the success or failure of social interventions, you might think it reasonable that a program would automatically be continued or terminated based on the results of the research.

Reality isn't that simple and reasonable, however. Other factors intrude on the assessment of evaluation research results, sometimes blatantly and sometimes subtly. As president, Richard Nixon appointed a blue-ribbon national commission to study the consequences of pornography. After a diligent, multifaceted evaluation, the commission reported that pornography didn't appear to have any of the negative social consequences often attributed to it. Exposure to pornographic materials, for example, didn't increase the likelihood of sex crimes. You might have expected liberalized legislation to follow from the research. Instead, the president said the commission was wrong.

Less dramatic examples of the failure to follow the implications of evaluation research could be listed endlessly. Undoubtedly every evaluation researcher can point to studies he or she conducted —studies providing clear research results and obvious policy implications—that were ignored, as the accompanying box, "The Impact of 'Three-Strikes' Laws," illustrates.

There are three important reasons why the implications of the evaluation research results are not always put into practice. First, the implications may not always be presented in a way that the nonresearchers can understand. Second, evaluation results sometimes contradict deeply held beliefs. That was certainly true in the case of the pornography commission. If everybody *knows* that pornography is bad, that it causes all manner of sexual deviance, then it's likely that research results to the contrary will have little immediate impact. By the same token, people thought Copernicus was crazy when he said the earth revolved around the sun. Anybody could tell the earth was standing still. The third barrier to the use of evaluation results is vested interests. If I've devised a new rehabilitation program that I'm convinced will keep ex-convicts from returning to prison, and if people have taken to calling it "the Babbie Plan," how do you think I'm going to feel when your evaluation suggests the program doesn't work? I might apologize for misleading people, fold up my tent, and go into another line of work. But more likely, I'd call your research worthless and begin intense lobbying with the appropriate authorities to have my program continue.

In the earlier example of the evaluation of drivers' education, Philip Hilts reported some of the reactions to the researchers' preliminary results:

> Ray Burneson, traffic safety specialist with the National Safety Council, criticized the study, saying that it was a product of a group (NHTSA) run by people who believe "that you can't do anything to train drivers. You can only improve medical facilities and build stronger cars for when the accidents happen. . . . This knocks the whole philosophy of education." — (1981:4)

By its nature, evaluation research takes place in the midst of real life, affecting it and being affected by it. Here's another example, well known to social researchers.

Rape Reform Legislation   For years, many social scientists and other observers have noted certain problems with the prosecution of rape cases. All too often, it is felt, the victim ends up suffering almost as much on the witness stand as in the rape itself. Frequently, she is portrayed by the defense lawyers as having encouraged the sex act and being of shady moral character; other personal attacks are intended to deflect responsibility from the accused rapist.

Criticisms such as these have resulted in a variety of state-level legislation aimed at remedying the problems. Cassie Spohn and Julie Horney (1990) were interested in tracking the impact of such legislation. The researchers summarize the ways in which new laws were intended to make a difference:

> The most changes are: (1) redefining rape and replacing the single crime of rape with a series of graded offenses defined by the presence or absence of aggravating conditions; (2) changing the consent standard by eliminating the requirement that the victim physically resist her attacker; (3) eliminating the requirement that the victim's testimony be corroborated; and (4) placing restrictions on the introduction of evidence of the victim's prior sexual conduct. — (1990:2)

It was generally expected that such legislation would encourage women to report being raped and would increase convictions when the cases were brought to court. To examine the latter expectation, the researchers focused on the period from 1970 to 1985 in Cook County, Illinois: "Our data file includes 4,628 rape cases, 405 deviate sexual assault cases, 745 aggravated criminal sexual assault cases, and 37 criminal sexual assault cases." (1990:4) Table 12-1 shows some of what they discovered.

Spohn and Horney summarized these findings as follows:

> The only significant effects revealed by our analyses were increases in the average maximum prison sentences; there was an increase of almost 48 months for rape and of almost 36 months for sex offenses. Because plots of the data indicated an increase in the average sentence before the reform took effect, we modeled the series with the intervention moved back one year earlier than the actual reform date. The size

TABLE 12-1   **Analysis of Rape Cases before and after Legislation**

	Rape	
	Before (N = 2252)	After (N = 2369)
Outcome of case		
Convicted of original charge	45.8%	45.4%
Convicted of another charge	20.6	19.4
Not convicted	33.6	35.1
Median prison sentence in months		
For those convicted of original charge	96.0	144.0
For those convicted of another charge	36.0	36.0

of the effect was even larger and still significant, indicating that the effect should not be attributed to the legal reform. — (1990:10)

Notice in the table that there was virtually no change in the percentages of cases ending in conviction for rape or some other charge (e.g., assault). Hence the change in laws didn't have any effect on the likelihood of conviction. As the researchers note, the one change that *is* evident—an increase in the length of sentences—cannot be attributed to the reform legislation itself.

In addition to the analysis of existing statistics, Spohn and Horney interviewed judges and lawyers to determine what they felt about the impact of the laws. Their responses were somewhat more encouraging.

> Judges, prosecutors and defense attorneys in Chicago stressed that rape cases are taken more seriously and rape victims treated more humanely as a result of the legal changes. These educative effects clearly are important and should please advocates of rape reform legislation. — (1990:17)

Thus, the study found other effects besides the qualitative results the researchers looked for. This study demonstrates the importance of following up on social interventions to determine whether, in what ways, and to what degree they accomplish their intended results.

Preventing Wife Battering   In a somewhat similar study, researchers in Indianapolis focused their attention on the problem of wife battering, with a special concern for whether prosecuting the batterers can lead to subsequent violence. David Ford and Mary Jean Regoli (1992) set about studying the consequences of various options for prosecution allowed within the "Indianapolis Prosecution Experiment" (IPE).

Wife-battering cases can follow a variety of patterns, as Ford and Regoli summarize:

> After a violent attack on a woman, someone may or may not call the police to the scene. If the police are at the scene, they are expected to investigate for evidence to support probable cause for a warrantless arrest. If it exists, they may arrest at their discretion. Upon making such an on-scene arrest, officers fill out a probable cause affidavit and slate the suspect into court for an initial hearing. When the police are not called, or if they are called but do not arrest, a victim may initiate charges on her own by going to the prosecutor's office and swearing out a probable cause affidavit with her allegation against the man. Following a judge's approval, the alleged batterer may either be summoned to court or be arrested on a warrant and taken to court for his initial hearing. — (1992:184)

What if a wife brings charges against her husband and then reconsiders later on? Many courts have a policy of prohibiting such actions, in the belief that they are serving the interests of the victim by forcing the case to be pursued to completion. In the IPE, however, some victims are offered the possibility of dropping the charges if they so choose later in the process. In addition, the court offers several other options. Because wife battering is largely a function of sexism, stress, and an inability to deal with anger, some of the innovative possibilities in the IPE involve educational classes with anger-control counseling.

If the defendant admits his guilt and is willing to participate in an anger-control counseling program, the judge may postpone the trial for that purpose and can later dismiss the charges if the defendant successfully completes the program. Alternatively, the defendant may be tried and, if found guilty, be granted probation provided he participates in the anger-control program. Finally, the defendant can be tried and, if found guilty, given a conventional punishment such as imprisonment.

Which of these possibilities most effectively prevents subsequent wife battering? That's the question Ford and Regoli addressed. Here are some of their findings.

First, their research shows that men who are brought to court for a hearing are less likely to continue beating their wives, no matter what the outcome of the hearing. Simply being brought into the criminal justice system has an impact.

Second, women who have the right to drop charges later on are less likely to be abused subsequently than those who do not have that right. In particular, the combined policies of arresting defendants by warrant and allowing victims to drop charges provides victims with greater security from subsequent violence than do any of the other prosecution policies (Ford and Regoli 1992).

However, giving victims the right to drop charges has a somewhat strange impact. Women who exercise that right are more likely to be abused later than those who insist on the prosecution proceeding to completion. The researchers interpret this as showing that future violence can be decreased when victims have a sense of control supported by a clear and consistent alliance with criminal justice agencies.

> A decisive system response to any violation of conditions for pretrial release, including of course new violence, should serve notice that

the victim-system alliance is strong. It tells the defendant that the victim is serious in her resolve to end the violence and that the system is unwavering in its support of her interest in securing protection. — (FORD AND REGOLI 1992:204)

The effectiveness of anger-control counseling cannot be assessed simply. Policies aimed at getting defendants into anger-control counseling seem to be relatively ineffective in preventing new violence. The researchers note, however, that the policy effects should not be confused with actual counseling outcomes. Some defendants scheduled for treatment never received it. Considerably more information on implementing counseling is needed for a proper evaluation.

Moreover, the researchers caution that the results of their research point to general patterns, and that individual battered wives must choose courses of action appropriate to their particular situations and should not act blindly on the basis of the overall patterns. The research is probably more useful in what it says about ways of structuring the criminal justice system (giving victims the right to drop charges, for example) than in guiding the actions of individual victims.

Finally, the IPE offers an example of a common problem in evaluation research. Often, actual practices differ from what might be expected in principle. For example, the researchers considered the impact of different alternatives for bringing suspects into court: Specifically, the court can issue either a summons ordering the husband to appear in court or a warrant to have the husband arrested. The researchers were concerned that having the husband arrested might actually add to his anger over the situation. They were somewhat puzzled, therefore, to find no difference in the anger of husbands summoned or arrested.

The solution of the puzzle lay in the discrepancy between principle and practice:

> Although a warrant arrest should in principle be at least as punishing as on-scene arrest, in practice it may differ little from a summons. A man usually knows about a warrant for his arrest and often elects to turn himself in at his conve-

nience, or he is contacted by the warrant service agency and invited to turn himself in. Thus, he may not experience the obvious punishment of, say, being arrested, handcuffed, and taken away from a workplace. — (FORD 1989:9–10)

In summary, many factors besides the scientific quality of evaluation research affect how its results are used. And, as we saw earlier, factors outside the evaluator's control can affect the quality of the study itself. But this "messiness" is balanced by the potential contributions that evaluation research can make toward the betterment of human life.

SOCIAL INDICATORS RESEARCH

I want to conclude this chapter with a type of research that combines what you've learned about evaluation research and about the analysis of existing data. A rapidly growing field in social research involves the development and monitoring of **social indicators,** aggregated statistics that reflect the social condition of a society or social subgroup. Researchers use social indicators to monitor aspects of social life in much the way that economists use indexes such as gross national product (GNP) per capita as an indicator of a nation's economic development.

Suppose we wanted to compare the relative health conditions in different societies, One strategy would be to compare their death rates (number of deaths per 1,000 population). Or, more specifically, we could look at infant mortality: the number of infants who die during their first year of life among every 1,000 births. Depending on the particular aspect of health conditions we were interested in, we could devise any number of other measures: physicians per capita, hospital beds per capita, days of hospitalization per capita, and so forth. Notice that intersocietal comparisons are facilitated by calculating per capita rates (dividing by the size of the population).

Before we go further, recall from Chapter 11 the problems involved in using existing statistics. In a word, they're often unreliable, reflecting their modes of collection, storage, and calculation. With this in mind, we'll look at some of the ways we can use social indicators for evaluation research on a large scale.

The Death Penalty and Deterrence

Does the death penalty deter capital crimes such as murder? This question is hotly debated every time a state considers eliminating or reinstating capital punishment and every time someone is executed. Those supporting capital punishment often argue that the threat of execution will keep potential murderers from killing people. Opponents of capital punishment often argue that it has no effect in that regard. Social indicators can help shed some light on the question.

If capital punishment actually deters people from committing murder, then we should expect to find murder rates lower in those states that have the death penalty than in those that do not. The relevant comparisons in this instance are not only possible, they've been compiled and published. Table 12-2 presents data compiled by William Bailey (1975) that directly contradict the view that the death penalty deters murderers. In both 1967 and 1968, those states with capital punishment had dramatically *higher* murder rates than did those without capital punishment. Some people criticized the interpretation of Bailey's data, saying that most states have not used the death penalty in recent years, even when they had it on the books. That could explain why it hasn't seemed to work as a deterrent. Further analysis, however, contradicts this explanation. When Bailey compared those states that hadn't used the death penalty with those that had, he found no real difference in murder rates.

The patterns uncovered by Bailey have persisted. In 1999, for example, the average murder rate in states with the death penalty was 5.5 per 100,000 population, and in states without the death penalty, it was 3.6 per 100,000 (Death Penalty Information Center 1999).

Another counterexplanation is possible, however. It could be the case that the interpretation given Bailey's data was *backward.* Maybe the existence of the death penalty as an option was a

TABLE 12-2   **Average Rate per 100,000 Population of First- and Second-Degree Murders for Capital-Punishment and Non-Capital-Punishment States, 1967 and 1968**

	Non-Capital-Punishment States		Capital-Punishment States	
	1967	*1968*	*1967*	*1968*
First-degree murder	.18	.21	1.47	1.58
Second-degree murder	.30	.43	1.92	1.03
Total murders	.48	.64	1.38	1.59

Source: Adapted from William C. Bailey, "Murder and Capital Punishment," in William J. Chambliss, ed., *Criminal Law in Action.* Copyright © 1975 by John Wiley & Sons, Inc. Used by permission.

consequence of high murder rates: Those states with high rates instituted it; those with low rates didn't institute it or repealed it if they had it on the books. It could be the case, then, that instituting the death penalty would bring murder rates down, and repealing it would increase murders and still produce—in a broad aggregate—the data presented in Table 12-2. Not so, however. Analyses over time do not show an increase in murder rates when a state repeals the death penalty nor a decrease in murders when one is instituted. A more recent examination by Bailey and Ruth D. Peterson (1994) confirmed the earlier findings and also indicated that law enforcement officials doubted the deterrent effect.

For more on the death penalty, see the Death Penalty Information Center at http://www.deathpenaltyinfo.org/deter .html*

Notice from the preceding discussion that it's possible to use social indicators data for comparison across groups either at one time or across some period of time. Often, doing both sheds the most light on the subject.

At present, work on the use of social indicators is proceeding on two fronts. On the one hand, researchers are developing ever more-refined indicators—finding which indicators of a general variable are the most useful in monitoring social life. At the same time, research is being devoted to discovering the relationships among variables within whole societies.

To find out more about social indicators, search for "social indicators" on the Web or check out the Sociometrics Corporation: http://www.socio.com/

Computer Simulation

One of the most exciting prospects for social indicators research lies in the area of computer simulation. As researchers begin compiling mathematical equations describing the relationships that link social variables to one another (for example, the relationship between growth in population and the number of automobiles), those equations can be stored and linked to one another in a computer. With a sufficient number of adequately accurate equations on tap, researchers one day will be able to test the implications of specific social changes by computer rather than in real life.

Suppose a state contemplated doubling the size of its tourism industry, for example. We could enter that proposal into a computer simulation model and receive in seconds or minutes a description of all the direct and indirect consequences of the increase in tourism. We could know what new public facilities would be required, which public agencies such as police and fire departments would have to be increased and by how much, what the labor force would look like, what kind of training would be required to provide it, how much new in-

---

*Each time the Internet icon appears, you'll be given helpful leads for searching the World Wide Web.

come and tax revenue would be produced, and so forth, through all the intended and unintended consequences of the action. Depending on the results, the public planners might say, "Suppose we increased the industry only by half," and have a new printout of consequences immediately.

An excellent illustration of computer simulation linking social and physical variables is to be found in the research of Donella and Dennis Meadows and their colleagues at Dartmouth and the Massachusetts Institute of Technology. They've taken as input data some known and estimated reserves of various nonreplaceable natural resources (for example, oil, coal, iron), past patterns of population and economic growth, and the relationships between growth and use of resources. Using a complex computer simulation model, they've been able to project, among other things, the probable number of years various resources will last in the face of alternative usage patterns in the future. Going beyond the initially gloomy projections, such models also make it possible to chart out less gloomy futures, specifying the actions required to achieve them. Clearly, the value of computer simulation is not limited to evaluation research, though it can serve an important function in that regard.

This potentiality points to the special value of evaluation research in general. Throughout human history, we've been tinkering with our social arrangements, seeking better results. Evaluation research provides a means for us to learn right away whether a particular tinkering really makes things better. Social indicators allow us to make that determination on a broad scale; coupling them with computer simulation opens up the possibility of knowing how much we would like a particular intervention without having to experience its risks.

A QUANDARY **REVISITED**

The purpose of evaluation research is to determine whether social interventions or programs have had their desired effects. No matter how much research is done, however, debates tend to persist.

As we've seen in this chapter, the political and ideological viewpoints that inform positions on such issues often are deeply ingrained and withstand disconfirming information. Moreover, evaluation research occurs in the "real world," where people's self-interests are often affected by such assessments. Research that reports a program is ineffective threatens the jobs of those employed by the program, not to mention the reputations of those who created it.

Evaluation research typically examines studies with multiple variables. This means that researchers can easily argue about which variable caused an apparent effect. They may even argue that some other, previously unthought of variable is responsible (recall the discussion of internal and external invalidity in Chapter 8). Further, many of the variables evaluated tend to be somewhat ambiguous: What is "happiness," "motivation," and so forth? In short, there is usually ample "wiggle room" for people who disagree with the findings of evaluation research.

Main Points

❑ Evaluation research is a form of applied research that studies the effects of social interventions.

❑ A careful formulation of the problem, including relevant measurements and criteria of suc-

cess or failure, is essential in evaluation research. In particular, evaluators must carefully specify outcomes, measure experimental contexts, specify the intervention being studied and the population targeted by the intervention, decide whether to use existing measures or devise new ones, and assess the potential cost-effectiveness of an intervention.

❑ Evaluation researchers typically use experimental or quasi-experimental designs. Examples of quasi-experimental designs include time-series studies and the use of nonequivalent control groups.

❑ Evaluators can also use qualitative methods of data collection. Both quantitative and qualitative data analyses can be appropriate in evaluation research, sometimes in the same study.

❑ Evaluation research entails special logistical and ethical problems because it's embedded in the day-to-day events of real life.

❑ The implications of evaluation research won't necessarily be put into practice, especially if they conflict with official points of view.

❑ Social indicators can provide an understanding of broad social processes.

❑ Computer-simulation models hold the promise of allowing researchers to study the possible results of social interventions without having to incur those results in real life.

Key Terms

evaluation research	nonequivalent control group
quasi experiments	multiple time-series designs
time-series design	social indicators

Review Questions

1. Review the evaluation of the Navy low-performer program discussed in this chapter. How would you redesign the program and the evaluation to avoid the problems that appeared in the actual study?

2. Take a minute to think of the many ways your society has changed during your own lifetime. How would you specify those changes as social indicators that could be used in monitoring the quality of life in your society?

3. Identify at least three deliberate social interventions, such as lowering the voting age to

18. For each, how would you (1) specify the perceived problem and (2) describe the kind of research that would evaluate whether the intervention was successful in solving the perceived problem?

4. Think of something at your college that you feel could be improved. Now think of something that could be done to solve the problem you've identified. Pursue this line of thought until you've developed a clear operational definition of how you would know when the problem had been solved through some intervention. What future measurement would represent success?

Additional Readings

Berg, Richard, and Peter H. Rossi. 1998. *Thinking about Program Evaluation.* Thousand Oaks, CA: Sage. Great book if you're looking for gaining good foundations in evaluation research while enjoying a wide range of examples.

Bickman, Leonard, and Debra J. Rog, eds. 1998. *Handbook of Applied Social Research Methods.* Thousand Oaks, CA: Sage. The two editors of this book have provided examples that illustrate all stages in evaluation, from planning through the collection and analysis of data. In addition, they cover ethical issues in the particular context of an evaluation research.

Chen, Huey-Tsyh. 1990. *Theory-Driven Evaluations.* Newbury Park, CA: Sage. Chen argues that evaluation research must be firmly based in theory if it is to be meaningful and useful.

Cunningham, J. Barton. 1993. *Action Research and Organizational Development.* Westport, CT: Praeger. This book urges researchers to bridge the gap between theory and action, becoming engaged participants in the evolution of organizational life and using social research to monitor problems and solutions.

Hedrick, Terry E., Leonard Bickman, and Debra J. Rog. 1993. *Applied Research Design: A Practical Guide.* Newbury Park, CA: Sage. This introduction to evaluation research is, as its subtitle claims, a practical guide, dealing straight-out with the compromises that must usually be made in research design and execution.

Rossi, Peter H., and Howard E. Freeman. 1996. *Evaluation: A Systematic Approach.* Newbury Park, CA: Sage.

This thorough examination of evaluation research is an excellent resource. In addition to discussing the key concepts of evaluation research, the authors provide numerous examples that might help you guide your own designs.

Multimedia Resources

The Wadsworth Sociology Resource Center: Virtual Society
http://sociology.wadsworth.com/

Visit the companion Web site for the second edition of *The Basics of Social Research* to access a wide range of student resources. Begin by clicking on the Student Resources section of the book's Web site to access the following study tools:

- eBabbie Resource Center
- Planning a Research Project
- Doing Data Analysis
- Statistics Review
- Flash Cards
- Internet Links and Exercises
- InfoTrac College Edition: Exercises
- Quizzes

Visit the **eBabbie Resource Center** for an overview of each chapter and helpful online tutorials. Find information on budgeting and step-by-step examples of model research projects at **Planning a Research Project.** Learn how to use quantitative and qualitative data analysis programs at **Doing Data Analysis,** and brush up on your statistics at **Statistics Review.** You can also further your study by accessing **Internet Links and Exercises** related to chapter materials, **Flash Cards, Quizzes,** and many other learning tools.

InfoTrac College Edition
http://www.infotrac-college.com/ wadsworth/access.html

Access the latest news and journal articles with InfoTrac College Edition, an easy-to-use online database of reliable, full-length articles from hundreds of top academic journals. Conduct an electronic search using the following search terms:

Capital-punishment evaluation
Computer social simulation
Evaluation research
Program assessment
Program evaluation
Quasi experiment
Social indicators research

David Falconer / Folio, Inc.

Part Four

ANALYSIS OF DATA

In this part of the book, we'll discuss the analysis of qualitative and quantitative data, and we'll examine the steps that separate observation from the final reporting of findings.

In Chapter 1, I made a fundamental distinction between qualitative and quantitative data. In the subsequent discussions, we've seen that many of the fundamental concerns in social research apply equally to both types of data. The analysis of qualitative and quantitative data, however, is quite different and will be discussed separately.

I hope the chapters that make up this part of the book will give you some of the tools and sharpen the insights needed to produce sophisticated data analyses, whether qualitative or quantitative.

Chapter 13 examines ways to analyze qualitative data. We'll look at some of the basic steps involved, and then we'll take an in-depth look at some of the computer software now available to qualitative analysts.

The logic of quantitative data analysis is presented in Chapter 14. We'll begin by examining the methods of analyzing and presenting the data related to a single variable. Then we'll turn to the relationship between two variables and learn how to construct and read simple percentage tables. The chapter ends with a preview of multivariate analysis.

13 QUALITATIVE DATA ANALYSIS

Jim Whitmer/Stock Boston

What You'll Learn in This Chapter

Here you'll see that qualitative data analysis is the nonnumerical
assessment of observations made through participant observation,
content analysis, in-depth interviews, and other qualitative research
techniques. Although qualitative analysis is an art as much as a
science, it has its own logic and techniques, some of which are
enhanced by special computer programs.

In this chapter . . .

INTRODUCTION

Chapter 14 will deal with the *quantitative* analysis of social research data, sometimes called *statistical analysis.* Recent decades of social science research have tended to focus on quantitative data analysis techniques. This focus, however, sometimes conceals another approach to making sense of social

AN OPENING **QUANDARY**

Why do researchers sometimes use qualitative analyses when it might have been possible to use statistics? Isn't a statistical, or quantitative, analysis a more "scientific" way to study poverty, discrimination, and so forth?

observations: **qualitative analysis**—methods for examining social research data without converting them to a numerical format. This approach predates quantitative analysis. It remains a useful approach to data analysis and is even enjoying a resurgence of interest among social scientists.

Although statistical analyses may intimidate some students, the steps involved can sometimes be learned by rote. That is, with practice, the rote exercise of quantitative skills can produce an ever-more sophisticated understanding of the logic that lies behind those techniques.

It is much more difficult to teach qualitative analysis as a series of rote procedures. In this case, understanding must precede practice. In this chapter, we begin with the links between research and theory in qualitative analysis. Then we examine some procedures that have proven useful in pursuing the theoretical aims. After considering some simple manual techniques, we'll take some computer programs out for a spin.

LINKING THEORY AND ANALYSIS

As suggested in Chapter 10 and elsewhere in this book, qualitative research methods involve a continuing interplay between data collection and theory. As a result, I've already talked about qualitative data analysis in earlier discussions of field research and content analysis. In quantitative research, it's sometimes easy to get caught up in the logistics of data collection and in the statistical analysis of data, thereby losing sight of theory for a time. This is less likely in qualitative research, where data collection, analysis, and theory are more intimately intertwined.

In the discussions to follow, we'll use the image of theory offered by Anselm Strauss and Juliet Corbin (1994:278) as consisting of "*plausible* relationships proposed among *concepts* and *sets of concepts.*" They stress "plausible" to indicate that theories represent our best understanding of how life operates. The more our research confirms a particular set of relationships among particular concepts, however, the more confident we become that our understanding corresponds to social reality.

While qualitative research is sometimes undertaken for purely descriptive purposes—such as the anthropologist's ethnography detailing ways of life in a previously unknown tribe—the rest of this chapter focuses primarily on the search for explanatory patterns. As we'll see, sometimes the patterns occur over time, and sometimes they take the form of causal relations among variables. Let's look at some of the ways qualitative researchers uncover such patterns.

Discovering Patterns

John and Lyn Lofland (1995:127–45) suggest six different ways of looking for patterns in a particular research topic. Let's suppose you're interested in analyzing child abuse in a certain neighborhood. Here are some questions you might ask yourself to make sense out of your data.

1. Frequencies: How often does child abuse occur among families in the neighborhood under study? (Realize that there may be a difference between the frequency and what people are willing to tell you.)
2. Magnitudes: What are the levels of abuse? How brutal are they?
3. Structures: What are the different types of abuse: physical, mental, sexual? Are they related in any particular manner?

4. Processes: Is there any order among the elements of structure? Do abusers begin with mental abuse and move on to physical and sexual abuse, or does the order of elements vary?
5. Causes: What are the causes of child abuse? Is it more common in particular social classes or among different religious or ethnic groups? Does it occur more often during good times or bad?
6. Consequences: How does child abuse affect the victims, in both the short and the long term? What changes does it cause in the abusers?

For the most part, in examining your data you'll look for patterns appearing across several observations that typically represent different cases under study. A. Michael Huberman and Matthew B. Miles (1994) offer two strategies for cross-case analysis: variable-oriented and case-oriented analyses. **Variable-oriented analysis** is similar to a model we've already discussed from time to time in this book. If we were trying to predict the decision to attend college, Huberman and Miles suggest, we might consider variables such as "gender, socioeconomic status, parental expectations, school performance, peer support, and decision to attend college" (1994:435). Thus, we would determine whether men or women were more likely to attend college. The focus of our analysis would be on interrelations among variables, and the people observed would be primarily the carriers of those variables.

Variable-oriented analysis may remind you of the discussion in Chapter 1 that introduced the idea of nomothetic explanation. The aim here is to achieve a partial, overall explanation using a relatively few number of variables. The political pollster who attempts to explain voting intentions on the basis of two or three key variables is using this approach. There is no pretense that the researcher can predict every individual's behavior nor even explain any one person's motivations in full. Sometimes, though, it's useful to have even a partial explanation of overall orientations and actions.

You may also recall Chapter 1's introduction of idiographic explanation, wherein we attempt to understand a particular case fully. In the voting example, we would attempt to learn everything we could about all the factors that came into play in determining one person's decision on how to vote. This orientation lies at the base of what Huberman and Miles call a **case-oriented analysis.**

> In a case-oriented analysis, we would look more closely into a particular case, say, Case 005, who is female, middle-class, has parents with high expectations, and so on. These are, however, "thin" measures. To do a genuine case analysis, we need to look at a full history of Case 005; Nynke van der Molen, whose mother trained as a social worker but is bitter over the fact that she never worked outside the home, and whose father wants Nynke to work in the family florist shop. Chronology is also important: two years ago, Nynke's closest friend decided to go to college, just before Nynke began work in a stable and just before Nynke's mother showed her a scrapbook from social work school. Nynke then decided to enroll in veterinary studies.
> — (1994:436)

This abbreviated commentary should give some idea of the detail involved in this type of analysis. Of course, an entire analysis would be more extensive and pursue issues in greater depth. This full, idiographic examination, however, tells us nothing about people in general. It offers nothing in the way of a theory about why people choose to attend college.

Even so, in addition to understanding one person in great depth, the researcher sees the critical elements of the subject's experiences as instances of more general social concepts or variables. For example, Nynke's mother's social work training can also be seen as "mother's education." Her friend's decision can be seen as "peer influence." More specifically, these could be seen as independent variables having an impact on the dependent variable of attending college.

Of course, one case does not a theory make— hence Huberman and Miles's reference to **cross-case analysis,** in which the researcher turns to other subjects, looking into the full details of their lives as well but paying special note to the variables that seemed important in the first case. How much and what kind of education did other subjects' mothers have? Is there any evidence of close friends attending college?

Some subsequent cases will closely parallel the first one in the apparent impact of particular variables. Other cases will bear no resemblance to the first. These latter cases may require the identification of other important variables, which may invite the researcher to explore why some cases seem to reflect one pattern while others reflect another.

Grounded Theory Method

The cross-case method just described should sound somewhat familiar. In the discussion of grounded theory in Chapter 10, we saw how qualitative researchers sometimes attempt to establish theories on a purely inductive basis. This approach begins with observations rather than hypotheses and seeks to discover patterns and develop theories from the ground up, with no preconceptions, though some research may build and elaborate on earlier grounded theories.

Grounded theory was first developed by the sociologists Barney Glaser and Anselm Strauss (1967) in an attempt to come to grips with their clinical research in medical sociology. Since then, it has evolved as a method, with the cofounders taking it in slightly different directions. The following discussion will deal with the basic concepts and procedures of the **Grounded Theory Method (GTM).**

> You can hear Barney Glaser discuss grounded theory on the Web* at http://www.groundedtheory.com/vidseries1.html

In addition to the fundamental, inductive tenet of building theory from data, GTM employs the **constant comparative method.** As Glaser and

---

*Each time the Internet icon appears, you'll be given helpful leads for searching the World Wide Web.

Strauss originally described this method, it involved four stages (1967:105–13):

1. "Comparing incidents applicable to each category." As Glaser and Strauss researched the reactions of nurses to the possible death of patients in their care, the researchers found that the nurses were assessing the "social loss" attendant upon a patient's death. Once this concept arose in the analysis of one case, they looked for evidence of the same phenomenon in other cases. When they found the concept arising in the cases of several nurses, they compared the different incidents. This process is similar to conceptualization as described in Chapter 5—specifying the nature and dimensions of the many concepts arising from the data.

2. "Integrating categories and their properties." Here the researcher begins to note relationships among concepts. In the assessment of social loss, for example, Glaser and Strauss found that nurses took special notice of a patient's age, education, and family responsibilities. For these relationships to emerge, however, it was necessary for the researchers to have noticed all these concepts.

3. "Delimiting the theory." Eventually, as the patterns of relationships among concepts become clearer, the researcher can ignore some of the concepts initially noted but evidently irrelevant to the inquiry. In addition to the number of categories being reduced, the theory itself may become simpler. In the examination of social loss, for example, Glaser and Strauss found that the assessment processes could be generalized beyond nurses and dying patients: They seemed to apply to the ways all staff dealt with all patients (dying or not).

4. "Writing theory." Finally, the researcher must put his or her findings into words to be shared with others. As you may have already experienced for yourself, the act of communicating your understanding of a topic actually modifies and even improves your own grasp of it. In GTM, the writing stage is regarded as a part of the research process. A later section of this chapter (on memoing) elaborates on this point.

This brief overview should give you an idea of how grounded theory proceeds. The many techniques associated with GTM can be found both in print and on the Web. One of the key publications is Anselm Strauss and Juliet Corbin's *Basics of Qualitative Research* (1990), which elaborates and extends many of the concepts and techniques found in the original Glaser/Strauss volume.

 You might want to explore Gaelle T. Morin's "Grounded Theory Methodology on the Web" at http://www.geocities .com/ResearchTriangle/Lab/1491/ gtm-i-gtmweb.html

GTM is only one analytical approach to qualitative data. In the remainder of this section, we'll take a look at some other specialized techniques.

Semiotics

Semiotics is commonly defined as the "science of signs" and has to do with symbols and meanings. It's commonly associated with content analysis, which was discussed in Chapter 11, though it can be applied in a variety of research contexts.

Peter K. Manning and Betsy Cullum-Swan (1994:466) offer some sense of the applicability of semiotics, as follows: "Although semiotics is based on language, language is but one of the many sign systems of varying degrees of unity, applicability, and complexity. Morse code, etiquette, mathematics, music, and even highway signs are examples of semiotic systems."

There is no meaning inherent in any sign, however. Meanings reside in minds. So, a particular sign means something to a particular person. However, the agreements we have about the meanings associated with particular signs make semiotics a *social* science. As Manning and Cullum-Swan point out:

SIGN	MEANING
1. Poinsettia	a. Good luck
2. Horseshoe	b. First prize
3. Blue ribbon	c. Christmas
4. "Say cheese"	d. Acting
5. "Break a leg"	e. Smile for a picture

FIGURE 13-1   **Matching Signs and Their Meanings**

For example, a lily is an expression linked conventionally to death, Easter, and resurrection as a content. Smoke is linked to cigarettes and to cancer, and Marilyn Monroe to sex. Each of these connections is social and arbitrary, so that many kinds of links exist between expression and content. — (1994:466)

To explore this contention, see if you can link the signs with their meanings in Figure 13-1. I'm confident enough that you know all the "correct" associations that there's no need for me to give the answers. (OK, you should have said 1c, 2a, 3b, 4e, 5d.) The point is this: What do any of these signs have to do with their "meanings"? Draft an email message to a Martian social scientist explaining the logic at work here. (You might want to include some "emoticons" like :) —another example of semiotics.)

While there is no doubt a story behind each of the linkages in Figure 13-1, the meanings you and I "know" today are socially constructed. Semiotic analysis involves a search for the meanings intentionally or unintentionally attached to signs.

Consider the sign shown in Figure 13-2, from a hotel lobby in Portland, Oregon. What's being communicated by the rather ambiguous sign? The first sentence seems to be saying that the hotel is up to date with the current move away from tobacco in the United States. Guests who want a smoke-free environment need look no farther: This is a healthy place to stay. At the same time, says the second sentence, the hotel would not like to be seen as inhospitable to smokers. There's room for everyone under this roof. No one needs to feel excluded. This sign is more easily understood within a marketing paradigm than one of logic.

The "signs" examined in semiotics, of course, are not limited to this kind of sign. Most are quite different, in fact. *Signs* are any things that are assigned special meanings. They can include such things as logos, animals, people, and consumer products. Sometimes the symbolism is a bit subtle. A classic analysis can be found in Erving Goffman's *Gender Advertisements* (1979). Goffman focused on advertising pictures found in magazines and newspapers. The overt purpose of the ads, of course, was to sell specific products. But what else was communicated, Goffman asked. What in particular did the ads say about men and women?

Analyzing pictures containing both men and women, Goffman was struck by the fact that men were almost always bigger and taller than the women accompanying them. (In many cases, in fact, the picture managed to convey the distinct impression that the women were merely accompanying the men.) While the most obvious explanation is that men are, on average, heavier and taller than women, Goffman suggested the pattern had a different meaning: that size and placement implied *status*. Those larger and taller presumably had higher social standing—more power and authority (1979:28). Goffman suggested that the ads communicated that men were more important than women.

In the spirit of Freud's comment that "sometimes a cigar is just a cigar" (he was a smoker), how would you decide whether the ads simply reflected the biological differences in the average sizes of men and women or whether they sent a message about social status? In part, Goffman's conclusion was based on an analysis of the exceptional cases: those in which the women appeared taller than the men. In these cases, the men were typically of a lower social status—the chef beside the society matron, for example. This confirmed Goffman's main point that size and height indicated social status.

The same conclusion was to be drawn from pictures with men of different heights. Those of higher status were taller, whether it was the gentleman speaking to a waiter or the boss guiding the work of his younger assistants. Where actual height was unclear, Goffman noted the placement of heads in

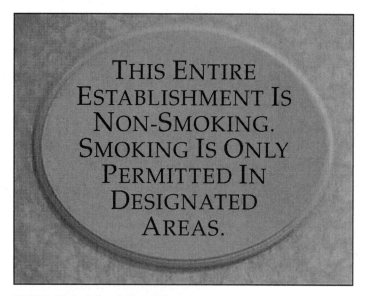

FIGURE 13-2 **Mixed Signals?**

the picture. The assistants were crouching down while the boss leaned over them. The servant's head was bowed so it was lower than that of the master.

The latent message conveyed by the ads, then, was that the higher a person's head appeared in the ad, the more important that person was. And in the great majority of ads containing men and women, the former were clearly portrayed as more important. The subliminal message in the ads, whether intended or not, was that men are more powerful and enjoy a higher status than do women.

Goffman examined several differences in the portrayal of men and women besides physical size. As another example, men were typically portrayed in active roles, women in passive ones. The (male) doctor examined the child while the (female) nurse or mother looked on, often admiringly. A man guided a woman's tennis stroke (all the while keeping his head higher than hers). A man gripped the reins of his galloping horse, while a woman rode behind him with her arms wrapped around his waist. A woman held the football, while a man kicked it. A man took a photo, which contained only women.

Goffman suggested that such pictorial patterns subtly perpetuated a host of gender stereotypes.

Even as people spoke publicly about gender equality, these advertising photos established a quiet backdrop of men and women in the "proper roles."

Conversation Analysis

Ethnomethodology, as you'll recall, aims to uncover the implicit assumptions and structures in social life. **Conversation analysis (CA)** seeks to pursue that aim through an extremely close scrutiny of the way we converse with one another. In the examination of ethnomethodology in Chapter 10, you saw some examples of conversation analysis. Here we'll look a little more deeply into that technique.

David Silverman (1993, 125f), reviewing the work of other CA theorists and researchers, speaks of three fundamental assumptions. First, conversation is a socially structured activity. Like other social structures, it has established rules of behavior. For example, we're expected to take turns, with only one person speaking at a time. In telephone conversations, the person answering the call is expected to speak first (e.g., "Hello"). You can verify the existence of this rule, incidentally, by picking up the phone without speaking. You may recall that this is the sort of thing ethnomethodologists tend to do.

Second, Silverman points out that conversations must be understood contextually. The same utterance will have totally different meanings in different contexts. For example, notice how the meaning of "Same to you!" varies if preceded by "I don't like your looks" or by "Have a nice day."

Third, CA aims to understand the structure and meaning of conversation through excruciatingly accurate transcripts of conversations. Not only are the exact words recorded, but all the uhs, ers, bad grammar, and pauses are also noted. Pauses, in fact, are recorded to the nearest tenth of a second.

The practical uses of this type of analysis are many. Ann Marie Kinnel and Douglas Maynard (1996), for example, analyzed conversations between staff and clients at an HIV testing clinic to examine how information about safe sex was communicated. Among other things, they found that the staff tended to provide standard information rather than try to speak directly to a client's specific circumstances. Moreover, they seemed reluctant to give direct advice about safe sex, settling for information alone.

These discussions should give you some sense of the variety of qualitative analysis methods available to researchers. Now let's look at some of the data-processing and data-analysis techniques commonly used in qualitative research.

QUALITATIVE DATA PROCESSING

Let me begin this section with a warning. The activity we are about to examine is as much art as science. At the very least, there are no cut-and-dried steps that guarantee success.

It's a lot like learning how to paint with watercolors or compose a symphony. Education in such activities is certainly possible, and university courses are offered in both. Each has its own conventions and techniques as well as tips you may find useful as you set out to create art or music. However, instruction can carry you only so far. The final product must come from you. Much the same can be said of qualitative data processing.

This section presents some ideas relating to the coding of qualitative data, writing memos, and mapping concepts graphically. Although far from a "how-to" manual, these ideas give a useful starting point for finding order in qualitative data.

Coding

Whether you've engaged in participant observation, in-depth interviewing, collecting biographical narratives, doing content analysis, or some other form of qualitative research, you will now be in the possession of a growing mass of data—most typically in the form of textual materials. Now what do you do?

The key process in the analysis of qualitative social research data is *coding*—classifying or categorizing individual pieces of data—coupled with some kind of retrieval system. Together, these procedures allow you to retrieve materials you may later be interested in.

Let's say you're chronicling the growth of a social movement. You recall writing up some notes about the details of the movement's earliest beginnings. Now you need that information. If all your notes have been catalogued by topic, retrieving those you need should be straightforward. As a simple format for coding and retrieval, you might have created a set of file folders labeled with various topics, such as "History." Data retrieval in this case means pulling out the "History" folder and rifling through the notes contained therein until you find what you need.

As you'll see later in this chapter, there are now some sophisticated computer programs that allow for a faster, more certain, and more precise retrieval process. Rather than looking through a "History" file, you can go directly to notes dealing with the "Earliest History" or the "Founding" of the movement.

Coding has another, even more important purpose. As discussed earlier, the aim of data analysis is the discovery of patterns among the data, patterns that point to a theoretical understanding of social life. The coding and relating of concepts is key to this process and requires a more refined system than a set of manila folders. In this section, we'll assume that you'll be doing your coding manually. The concluding section of the chapter will illustrate the use of computer programs for qualitative data analysis.

Coding Units As you may recall from the earlier discussion of content analysis, for statistical analysis it's important to identify a standardized unit of analysis prior to coding. If you were comparing American and French novels, for example, you might evaluate and code sentences, paragraphs, chapters, or whole books. It would be important, however, to code the same units for each novel analyzed. This uniformity is necessary in a quantitative analysis, as it allows us to report something like "23 percent of the paragraphs contained metaphors." This is only possible if we've coded the same unit—paragraphs—in each of the novels.

Coding data for a qualitative analysis, however, is quite different. The *concept* is the organizing principle for qualitative coding. Here the units of text appropriate for coding will vary within a given document. Thus, in a study of organizations, "Size" might require only a few words per coding unit, whereas "Mission" might take a few pages. Or, a lengthy description of a heated stockholders meeting might be coded as "Internal Dissent."

Realize also that a given code category may be applied to textual materials of quite different lengths. For example, some references to the organization's mission may be brief, others lengthy. Whereas standardization is a key principle in quantitative analysis, this is not the case in qualitative analysis.

Coding as a Physical Act Before continuing with the logic of coding, let's take a moment to see what it actually looks like. John and Lyn Lofland (1995: 188) offer this description of manual filing:

> Prior to the widespread availability of personal computers beginning in the late 1980s, coding frequently took the specific physical form of *filing*. The researcher established an expanding set of file folders with code names on the tabs and physically placed either the item of data itself or a note that located it in the appropriate

file folder. . . . Before photocopying was easily available and cheap, some fieldworkers typed their fieldnotes with carbon paper, wrote codes in the margins of the copies of the notes, and cut them up with scissors. They then placed the resulting slips of paper in corresponding file folders.

As the Loflands point out, personal computers have greatly simplified this task. However, the image of slips of paper that contain text and are put in folders representing code categories is useful for understanding the process of coding. In the next section, when I suggest that we code a textual passage with a certain code, imagine that we have the passage typed on a slip of paper and that we place it in a file folder bearing the name of the code. Whenever we assign two codes to a passage, imagine placing duplicate copies of the passage in two different folders representing the two codes.

Creating Codes So, what should your code categories be? Glaser and Strauss (1967:101f) allow for the possibility of coding data for the purpose of testing hypotheses that have been generated by prior theory. In that case, then, the codes would be suggested by the theory, in the form of variables.

In this section, however, we're going to focus on the more common process of **open coding.** Strauss and Corbin (1990:62) define it as follows:

> Open coding is the part of analysis that pertains specifically to the naming and categorizing of phenomena through close examination of data. Without this first basic analytical step, the rest of the analysis and communication that follows could not take place. During open coding the data are broken down into discrete parts, closely examined, compared for similarities and differences, and questions are asked about the phenomena as reflected in the data. Through this process, one's own and others' assumptions about phenomena are questioned or explored, leading to new discoveries.

Here's a concrete example to illustrate how you might proceed. Suppose you're interested in the religious bases for homophobia. You've interviewed some people opposed to homosexuality who cite a religious basis for their feelings. Specifically, they refer you to these passages in the Book of Leviticus (Revised Standard Version):

18:22 You shall not lie with a male as with a woman; it is an abomination.

20:13 If a man lies with a male as with a woman, both of them have committed an abomination; they shall be put to death, their blood is upon them.

Although the point of view expressed here seems unambiguous, you might decide to examine it in more depth. Perhaps a qualitative analysis of Leviticus can yield a fuller understanding of where these injunctions against homosexuality fit into the larger context of Judeo-Christian morality.

Let's start our analysis by examining the two passages just quoted. We might begin by coding each passage with the label "Homosexuality." This is clearly a key concept in our analysis. Whenever we focus on the issue of homosexuality in our analysis of Leviticus, we want to consider these two passages.

Because homosexuality is such a key concept, let's look more closely into what it means within the data under study. We first notice the way *homosexuality* is identified: a man lying with a man "as with a woman." While we can imagine a lawyer seeking admission to heaven saying, "But here's my point; if we didn't actually lie down . . ." it seems safe to assume the passage refers to having sex, though it is not clear what specific acts might or might not be included.

Notice, however, that the injunctions appear to concern *male* homosexuality only; lesbianism is not mentioned. In our analysis, then, each of these passages might also be coded "Male Homosexuality." This illustrates two more aspects of coding: (1) Each unit can have more than one code and (2) hierarchical codes (one included within another) can be used. Now each passage has two codes assigned to it.

An even more general code might be introduced at this point: "Prohibited Behavior." This is important for two reasons. First, homosexuality is not inherently wrong, from an analytical stand-

point. The purpose of the study is to examine the way it's made wrong by the religious texts in question. Second, our study of Leviticus may turn up other behaviors that are prohibited.

There are at least two more critical concepts in the passages: "Abomination" and "Put to Death." Notice that while these are clearly related to "Prohibited Behavior," they are hardly the same. Parking without putting money in the meter is prohibited, but few would call it an abomination and fewer still would demand the death penalty for that transgression. Let's assign these two new codes to our first two passages.

At this point, we want to branch out from the two key passages and examine the rest of Leviticus. We therefore examine and code each of the remaining chapters and verses. In our subsequent analyses, we'll use the codes we have already and add new ones as appropriate. When we do add new codes, it will be important to review the passages already coded to see whether the new codes apply to any of them.

Here are the passages we decide to code "Abomination." (I've boldfaced the abominations.)

7:18 If any of the flesh of the sacrifice of **his peace offering is eaten on the third day,** he who offers it shall not be accepted, neither shall it be credited to him; it shall be an abomination, and he who eats of it shall bear his iniquity.

7:21 And if any one **touches an unclean thing,** whether the uncleanness of man or an unclean beast or any unclean abomination, **and then eats of the flesh of the sacrifice** of the LORD's peace offerings, that person shall be cut off from his people.

11:10 But **anything in the seas or the rivers that has not fins and scales,** of the swarming creatures in the waters and of the living creatures that are in the waters, is an abomination to you.

11:11 They shall remain an abomination to you; **of their flesh you shall not eat, and their carcasses you shall have in abomination.**

11:12 **Everything in the waters that has not fins and scales** is an abomination to you.

11:13 And these you shall have in abomination among the birds, they **shall not be eaten,** they are an abomination: the **eagle,** the **vulture,** the **osprey,**

11:14 the **kite,** the **falcon** according to its kind,

11:15 every **raven** according to its kind,

11:16 the **ostrich,** the **nighthawk,** the **sea gull,** the **hawk** according to its kind,

11:17 the **owl,** the **cormorant,** the **ibis,**

11:18 the **water hen,** the **pelican,** the **carrion vulture,**

11:19 the **stork,** the **heron** according to its kind, the **hoopoe,** and the **bat.**

11:20 **All winged insects that go upon all fours** are an abomination to you.

11:41 **Every swarming thing** that swarms upon the earth is an abomination; it shall not be eaten.

11:42 Whatever goes on its belly, and whatever goes on all fours, or whatever has many feet, all the **swarming things** that swarm upon the earth, you shall not eat; for they are an abomination.

11:43 You shall not make yourselves abominable with **any swarming thing that swarms;** and you shall not defile yourselves with them, lest you become unclean.

18:22 You shall not **lie with a male as with a woman;** it is an abomination.

19:6 It shall be eaten the same day you offer it, or on the morrow; and anything left over until the third day shall be burned with fire.

19:7 **If it is eaten at all on the third day,** it is an abomination; it will not be accepted,

19:8 and every one who eats it shall bear his iniquity, because he has profaned a holy thing of the LORD; and that person shall be cut off from his people.

20:13 **If a man lies with a male as with a woman,** both of them have committed an abomination; they shall be put to death, their blood is upon them.

20:25 You shall therefore make a distinction between the clean beast and the unclean, and between the unclean bird and the clean; **you shall not make yourselves abominable by beast or by bird or by anything with which the ground** teems, which I have set apart for you to hold unclean.

Male homosexuality, then, isn't the only abomination identified in Leviticus. As you compare these passages, looking for similarities and differences, it will become apparent that most of the abominations have to do with dietary rules—specifically those potential foods deemed "unclean." Other abominations flow from the mishandling of ritual sacrifices. "Dietary Rules" and "Ritual Sacrifices" thus represent additional concepts and codes to be used in our analysis.

Earlier, I mentioned the death penalty as another concept to be explored in our analysis. When we take this avenue, we discover that many behaviors besides male homosexuality warrant the death penalty. Among them are these:

20:2 Giving your children to Molech (human sacrifice)

20:9 Cursing your father or mother

20:10 Adultery with your neighbor's wife

20:11 Adultery with your father's wife

20:12 Adultery with your daughter-in-law

20:14 Taking a wife and her mother also

20:15 Men having sex with animals (the animals are to be killed, also)

20:16 Women having sex with animals

20:27 Being a medium or wizard

24:16 Blaspheming the name of the Lord

24:17 Killing a man

As you can see, the death penalty is broadly applied in Levicitus: everything from swearing to murder, including male homosexuality somewhere in between.

An extended analysis of prohibited behavior, short of abomination and death, also turns up a lengthy list. Among them are slander, vengeance, grudges, cursing the deaf, and putting stumbling blocks in front of blind people. In chapter 19, verse 19, Leviticus quotes God as ordering, "You shall not let your cattle breed with a different kind; you shall not sow your field with two kinds of seed; nor shall there come upon you a garment of cloth made of two kinds of stuff." Shortly thereafter, he adds, "You shall not eat any flesh with the blood in it. You

APPLYING THE RESULTS

A simple qualitative analysis such as this sheds new light on a key civil rights issues in the United States today. People are harassed, discriminated against, and even killed because of their sexual orientation. When GSS respondents are asked for their opinions about homosexuality, "Always wrong" is the most frequent response selected, followed by "Never wrong," with small minorities choosing more moderate, mixed views.

Anti-gay sentiments and actions are often justified on religious grounds, specifically the passages in Leviticus cited in this chapter. But the longer list of abominations in Leviticus was used in the TV series *West Wing* to debunk the homophobic preaching of a radio talk-jock.

shall not practice augury or witchcraft. You shall not round off the hair on your temples or mar the edges of your beard." Tattoos were prohibited, though Leviticus is silent on body piercing. References to all of these practices would be coded "Prohibited Acts" and perhaps given additional codes as well (recall "Dietary Rules").

I hope this brief glimpse into a possible analysis will give you some idea of the process by which codes are generated and applied. You should also have begun to see how such coding would allow you to better understand the messages being put forward in a text and to retrieve data appropriately as you need them.

Memoing

In the Grounded Theory Method, the coding process involves more than simply categorizing chunks of text. As you code data, you should also be using the technique of **memoing**—writing memos or notes to yourself and others involved in the project. Some of what you write during analy-

sis may end up in your final report; much of it will at least stimulate what you write.

In GTM, these memos have a special significance. Strauss and Corbin (1990:197f) distinguish three kinds of memos: code notes, theoretical notes, and operational notes.

Code notes identify the code labels and their meanings. This is particularly important because, as in all social science research, most of the terms we use with technical meanings also have meanings in everyday language. It's essential, therefore, to write down a clear account of what you mean by the codes used in your analysis. In the Leviticus analysis, for example, you would want a code note regarding the meaning of *abomination* and how you've used that code in your analysis of text.

Theoretical notes cover a variety of topics: reflections of the dimensions and deeper meanings of concepts, relationships among concepts, theoretical propositions, and so on. All of us occasionally ruminate over the nature of something, try to think it out, to make sense out of it. In qualitative data analysis, it's vital to write down these thoughts, even those you'll later discard as useless. They will vary greatly in length, though you should limit each to a single main thought so that you can sort and organize them all later. In the Leviticus analysis, one theoretical note might discuss the way that most of the injunctions implicitly address the behavior of *men,* with women being mostly incidental.

Operational notes deal primarily with methodological issues. Some will draw attention to data-collection circumstances that may be relevant to understanding the data later on. Others will consist of notes directing future data collection.

Writing these memos occurs throughout the data-collection and analysis process. Thoughts demanding memos will come to you as you reread notes or transcripts, code chunks of text, or discuss the project with others. It's a good idea to get in the habit of writing out your memos as soon as possible after the thoughts come to you.

John and Lyn Lofland (1995:193f) spe?¹ memoing somewhat differently, descri⁺¹ that come closer to the final writing sta. *mental memo* is a detailed analytic rendering of some relatively specific matter. Depending on the scale of the project, the worker may write from one to several dozen or more of these. Built out of selective codes and codings, these are the most basic prose cannon fodder, as it were, of the project.

— (1995:194)

The *sorting memo* is based on several elemental memos and presents key themes in the analysis. Whereas we create elemental memos as they come to mind, with no particular rhyme nor reason, we write sorting memos as an attempt to discover or create reason among the data being analyzed. A sorting memo will bring together a set of related elemental memos. A given project may see the creation of several sorting memos dealing with different aspects of the project.

Finally, the *integrating memo* ties together the several sorting memos to bring order to the whole project. It tells a coherent and comprehensive story, casting it in a theoretical context. In any real project, however, there are many different ways of bringing about this kind of closure. Hence, the data analysis may result in several integrating memos.

Notice that whereas we often think of writing as a linear process, starting at the beginning and moving through to the conclusion, memoing is very different. It might be characterized as a process of creating chaos and then finding order within it.

To explore this process further, refer to the works cited in this discussion and at the end of the chapter. You'll also find a good deal of information on the Web. Ultimately, the best education in this process comes from practice. Even if you don't have a research project underway, you can practice now on class notes. Or start a journal and code it.

 For Barney Glaser's rules on memoing, go to http://www.vlsm.org/gnm/gnm-gtm3.html

Concept Mapping

It should be clear by now that qualitative-data analysts spend a lot of time committing thoughts to paper (or to a computer file), but this process is not

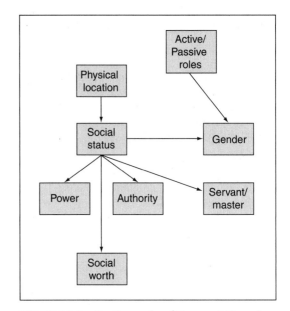

FIGURE 13-3　**An Example of Concept Mapping**

limited to text alone. Often, we can think out relationships among concepts more clearly by putting the concepts in a graphical format, a process called **concept mapping.** Some researchers find it useful to put all their major concepts on a single sheet of paper, while others spread their thoughts across several sheets of paper, blackboards, magnetic boards, computer pages, or other media. Figure 13-3 shows how we might think out some of the concepts of Goffman's examination of gender and advertising. (This image was created through the use of Inspiration, a concept-mapping computer program.)

Incidentally, many of the topics discussed in this section have useful applications in quantitative as well as qualitative analyses. Certainly, concept mapping is appropriate in both types of analysis. The several types of memos would also be useful in both. And the discussion of coding readily applies to the coding of open-ended questionnaire responses for the purpose of quantification and statistical analysis. (We'll look at coding again in the next chapter, on quantifying data.)

Having noted the overlap of qualitative and quantitative techniques, it seems fitting now to address an instrument that is primarily associated

sex	homosex	death	Verse	Passage
X	X	X	20:13	If a man lies with a male as with a woman, both of them have committed an abomination; they shall be put to death, their blood is upon them.
X		X	20:12	If a man lies with his daughter-in-law, both of them shall be put to death; they have committed incest, their blood is upon them.
X		X	20:15	If a man lies with a beast, he shall be put to death; and you shall kill the beast.
		X	20:09	For every one who curses his father or his mother shall be put to death; he has cursed his father or his mother, his blood is upon him.
		X	20:02	Any man of the people of Israel, or of the strangers that sojourn in Israel, who gives any of his children to Molech shall be put to death.
X	X		18:22	You shall not lie with a male as with a woman; it is an abomination.

FIGURE 13-4 **Using a Spreadsheet for Qualitative Analysis**

with quantitative research but that is proving very valuable for qualitative analysts as well—the personal computer.

COMPUTER PROGRAMS FOR QUALITATIVE DATA

The advent of computers—mainframe and personal—has been a boon to quantitative research, allowing the rapid calculation of extremely complex statistics. The importance of the computer for qualitative research has been somewhat more slowly appreciated. Some qualitative researchers were quick to adapt the basic capacities of computers to nonnumerical tasks, but it took a bit longer for programmers to address the specific needs of qualitative research. Today, however, several powerful programs are available.

Let's start this section with a brief overview of some of the ways you can use basic computer tools in qualitative research. Perhaps only those who can recall hours spent with carbon paper and White-out can fully appreciate the glory of computers as a note-taking device. "Easier editing" and "easier duplication" simply don't capture the scope of the advance.

Moving beyond the basic recording and storage of data, simple word-processing programs can be used for some data analysis. The "find" or "search" command will take you to passages containing key words. Or, going one step further, you can type code words alongside passages in your notes so that you can search for those keywords later.

Database and spreadsheet programs can also be used for processing and analyzing qualitative data. Figure 13-4 is a simple illustration of how some of the verses from Leviticus might be manipulated within a spreadsheet. The three columns to the left represent three of the concepts we've discussed. An "x" means that the passage to the right contains that concept. As shown, the passages are sorted in such a way as to gather all those dealing with punishment by death. Another simple "sort" command would gather all those dealing with sex, with homosexuality, or any of the other concepts coded.

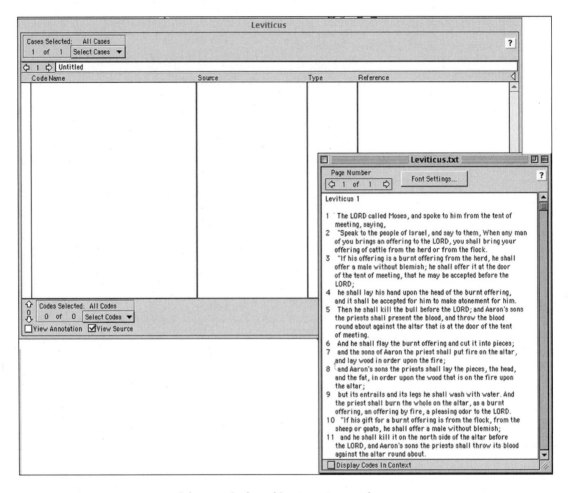

FIGURE 13-5   **How Text Materials Are Displayed in HyperResearch**

This brief illustration should give you some idea of the possibilities for using readily available programs as tools in qualitative data analysis. Happily, there are now a large number of programs created specifically for that purpose.

Leviticus as Seen through HyperResearch

Let's take a closer look at how qualitative data analysis programs operate. We'll consider one of the programs just mentioned, HyperResearch. Although each of the programs has somewhat different features and different approaches to analysis,

HyperResearch is a popular program for teaching qualitative social research, and it offers a fairly representative view of the genre. We'll begin with a brief examination of Leviticus, and then we'll examine a project focused on understanding the experiences of women film directors.

While the text materials to be coded can be typed directly into HyperResearch, usually materials already in existence—such as field notes or, in this case, the verses of Leviticus—are *imported* into the program. Menu-based commands do this easily, though the text must be in a *plaintext* format (i.e., without word processing or other formatting).

Figure 13-5 shows how the text is displayed

Here's an excellent list of qualitative data analysis tools. It was prepared by sociologists at the University of Surrey, England at http://www.soc.surrey.ac.uk/sru/SRU1.html. This Web site also provides a brief description of each of the programs listed, along with the price and contact, if available.

> The Ethnograph
> HyperQual
> HyperResearch
> HyperSoft
> NUD*IST
> QUALPRO
> QUALOG
> Textbase Alpha
> SONAR
> Atlas-ti

within HyperResearch. For the illustrations in this section, I have used the Macintosh version of the program, but the Windows version is similar.

In one window, we see the beginning of the text materials being analyzed: the Book of Leviticus. Notice the window behind it. This window in HyperResearch is key because this is where we store the coding of passages.

To begin our analysis, let's code the two anti-homosexual passages. The first step is to create a code category. Clicking "Codes/Edit Codes . . . " will open the "Code List Editor" window. Clicking "Add . . . " will open another window where you can create a code, as Figure 13-6 shows.

Once we click "OK," the new code will be added to the list (it's the first one in this example) and we can enter a definition of the code if we wish.

Now that we've created the code, we can use it as follows. Let's find the first antihomosexual passage in Leviticus 18:22. We simply scroll through the window containing the text and locate the passage. Then we drag the cursor across it to select it. The "Codes/Encode" option opens up our list of codes, allowing us to select how we want the passage coded. In this case, we highlight "AntiGay" and click the "Select" button as indicated in Figure 13-7.

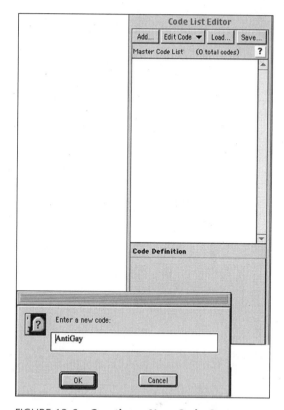

FIGURE 13-6 **Creating a New Code Category**

HyperResearch will also code the document automatically, based on instructions you specify. Here's a simple example. Since we want to examine those things considered abominations in Leviticus, we can instruct HyperResearch to code every passage where the word appears.

 You can learn more about HyperResearch at the Web site for this book and at the HyperResearch Web site: http://research-ware.com/

We begin with the "Codes/Autocode . . . " menu command. Using the tabs in the new window, we are asked to select the data file to be coded and the code to be used. This procedure lets us create a new code if the one we want is not in the list. When we specify the phrase to search for, the program

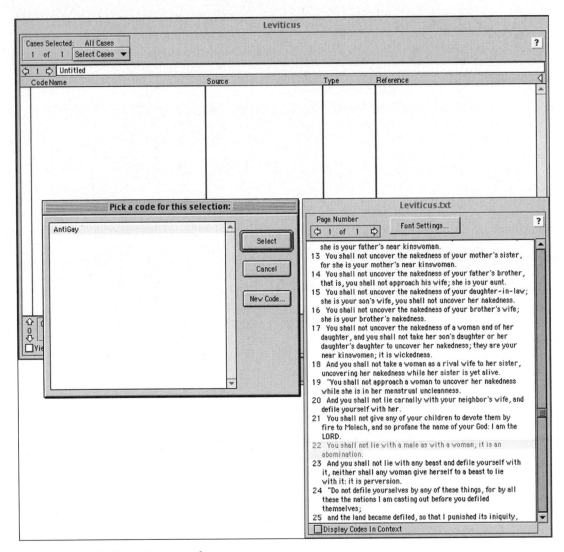

FIGURE 13-7 **Coding a Passage of Text**

lets us specify a chunk of text to be included in the coded passage. In Figure 13-8, the program has been instructed to include 50 characters on either side of the word *abomination.*

Clicking the "Autocode" button causes the program to scan the document and code those passages that contain the term *abomination.* All the coding, whether manual or automatic, is stored in the first window seen upon launching the program (Figure 13-5). Notice that each entry contains the code name, the document that was coded, the type of document (text, images, and other types of data), and the location in the document. Clicking one of the entries will take you to the passage in question, as shown in Figure 13-9.

This example should have introduced you to how you can use computers to analyze qualitative data. HyperResearch also supports much more complex analyses: locating passages on the basis of several codes, for example.

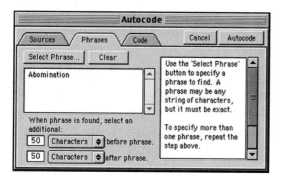

FIGURE 13-8 **Autocoding "Abomination"**

Sandrine Zerbib: Using NVivo to Understand Women Film Directors

Now let's probe more deeply into the possibilities of computerized qualitative data analysis. Sandrine Zerbib is a French sociologist interested in understanding the special difficulties faced by women breaking into the male-dominated world of film direction. To address this issue, she interviewed 30 women directors in depth. Having compiled hours of recorded interviews, she turned to a popular program, NVivo (a product of NUD*IST), as a vehicle for analysis. In the rest of this section, she describes her experiences with the ongoing process of qualitative data analysis.

For those of you who feel uncomfortable using new programs or computer programs in general, NVivo should work well. It is visually clear and intuitive, and it requires mostly dragging (moving text or objects using a mouse).

To learn more about the tools in this software package, let's look at a project file I created using NVivo. Figure 13-10 shows the opened browser window containing my interview with Berta, one of the 30 film directors I interviewed. The "Coding Strip" view allows you to visualize "nodes" (i.e., codes) associated with the text. Parts of the same passage were coded with more than one node, which explains the overlapping of strips. The program automatically colors the coding strips in ways intended to help you differentiate between nodes. The "Coder" window is opened on the right. You

can click on the + symbol to visualize "child" nodes (or subnodes) of a particular "tree" node (main node). As you can see in this figure, when you click on the "first job in the industry" node located in the coder window, the passages associated with that node are automatically highlighted.

In Figure 13-11, a new interview has been imported as a "Rich Text Format" into the project. You can either import a file with a particular formatting of headings and font styles or make these formatting changes while using NVivo. Using styles and other text-formatting tools can help you refine your coding system. For instance, it can be very useful to create styles that format the interviewer's and the interviewee's narratives differently; this way, you can see at a glance which type of narrative you're reading. Of course, taking the time to format makes more sense in some cases than in others. For instance, I found it extremely useful to use formatting in projects that comprise a large number of interviews.

As you explore possibilities, you'll find certain formatting choices more helpful than others. You'll want to consider such things as how much a change in font or typeface expands or contracts your text, and how easy or hard a given format is to read. For example, you might increase the size of or use boldface for the interviewer's text rather than the interviewee's text, because the interviewer speaks far less than the interviewee and you do not want your text to be too long. In Figure 13-11 I chose to italicize my part of the interview.

Another important feature of NVivo is that it allows you to attach passages to nodes easily. You can select passages based on content in units as small as a single word, then drag them where you want them. In Figure 13-11, I highlighted part of a paragraph, opened the "career experience" node, and simply dragged it to the child node "early artist" located in the coder within the "career experience" category.

Another helpful tool is the "attribute" function of NVivo. By clicking on the colorful cube icon, you can open the attribute browser any time for any of your interviews or other texts. This allows you to begin a content analysis by creating attributes (or

Leviticus

Cases Selected: All Cases
1 of 1 Select Cases ▼

?

◁ 1 ▷ Untitled

Code Name	Source	Type	Reference
AntiGay	Leviticus.txt	TEXT	84585,84662
Abomination	Leviticus.txt	TEXT	25307,25417
Abomination	Leviticus.txt	TEXT	25797,25907
Abomination	Leviticus.txt	TEXT	42488,42598
Abomination	Leviticus.txt	TEXT	42641,42751
Abomination	Leviticus.txt	TEXT	42724,42834
Abomination	Leviticus.txt	TEXT	42849,42959
Abomination	Leviticus.txt	TEXT	43303,43413
Abomination	Leviticus.txt	TEXT	43732,43842
Abomination	Leviticus.txt	TEXT	46333,46443
Abomination	Leviticus.txt	TEXT	46562,46672
Abomination	Leviticus.txt	TEXT	84601,84711
Abomination	Leviticus.txt	TEXT	85113,85223
Abomination	Leviticus.txt	TEXT	85212,85322
Abomination	Leviticus.txt	TEXT	85458,85568
Abomination	Leviticus.txt	TEXT	86491,86601
Abomination	Leviticus.txt	TEXT	92699,92809

Codes Selected: By Name
17 of 17 Select Codes ▼
☐ View Annotation ☑ View Source

Leviticus.txt

Page Number   Font Settings...
◁ 1 of 1 ▷

?

shall be eaten,
17 but what remains of the flesh of the sacrifice on the third
    day shall be burned with fire.
18 If any of the flesh of the sacrifice of his peace offering is
    eaten on the third day, he who offers it shall not be
    accepted, neither shall it be credited to him; it shall be an
    abomination, and he who eats of it shall bear his iniquity.
19 "Flesh that touches any unclean thing shall not be eaten; it
    shall be burned with fire. All who are clean may eat flesh,
20 but the person who eats of the flesh of the sacrifice of the
    LORD's peace offerings while an uncleanness is on him, that
    person shall be cut off from his people.
21 And if any one touches an unclean thing, whether the
    uncleanness of man or an unclean beast or any unclean
    abomination, and then eats of the flesh of the sacrifice of
    the LORD's peace offerings, that person shall be cut off from
    his people."
22 The LORD said to Moses,
23 "Say to the people of Israel, You shall eat no fat, of ox, or
    sheep, or goat.
24 The fat of an animal that dies of itself, and the fat of one
    that is torn by beasts, may be put to any other use, but on
    no account shall you eat it.
25 For every person who eats of the fat of an animal of which an
    offering by fire is made to the LORD shall be cut off from
    his people.
26 Moreover you shall eat no blood whatever, whether of fowl or
    of animal, in any of your dwellings.
27 Whoever eats any blood, that person shall be cut off from his
    people."
28 The LORD said to Moses,

☐ Display Codes In Context

FIGURE 13-9   **Locating a Coded Passage**

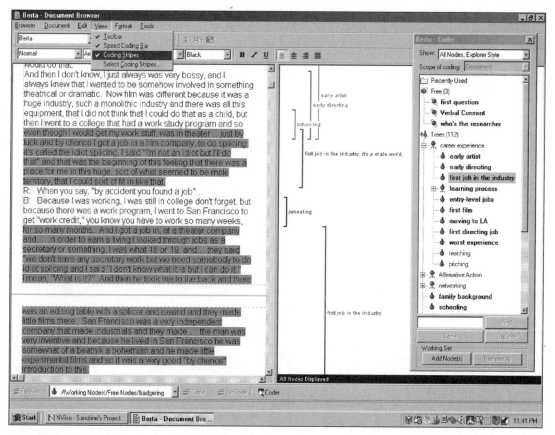

FIGURE 13-10   **Viewing the Interview with Berta**

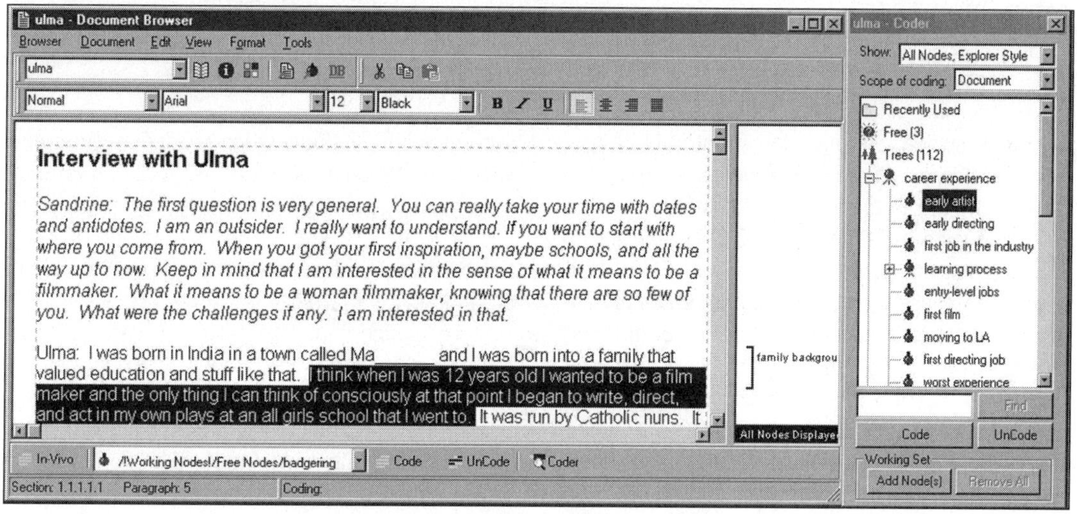

FIGURE 13-11   **Example of Formatted Text and Attaching Passages to Nodes**

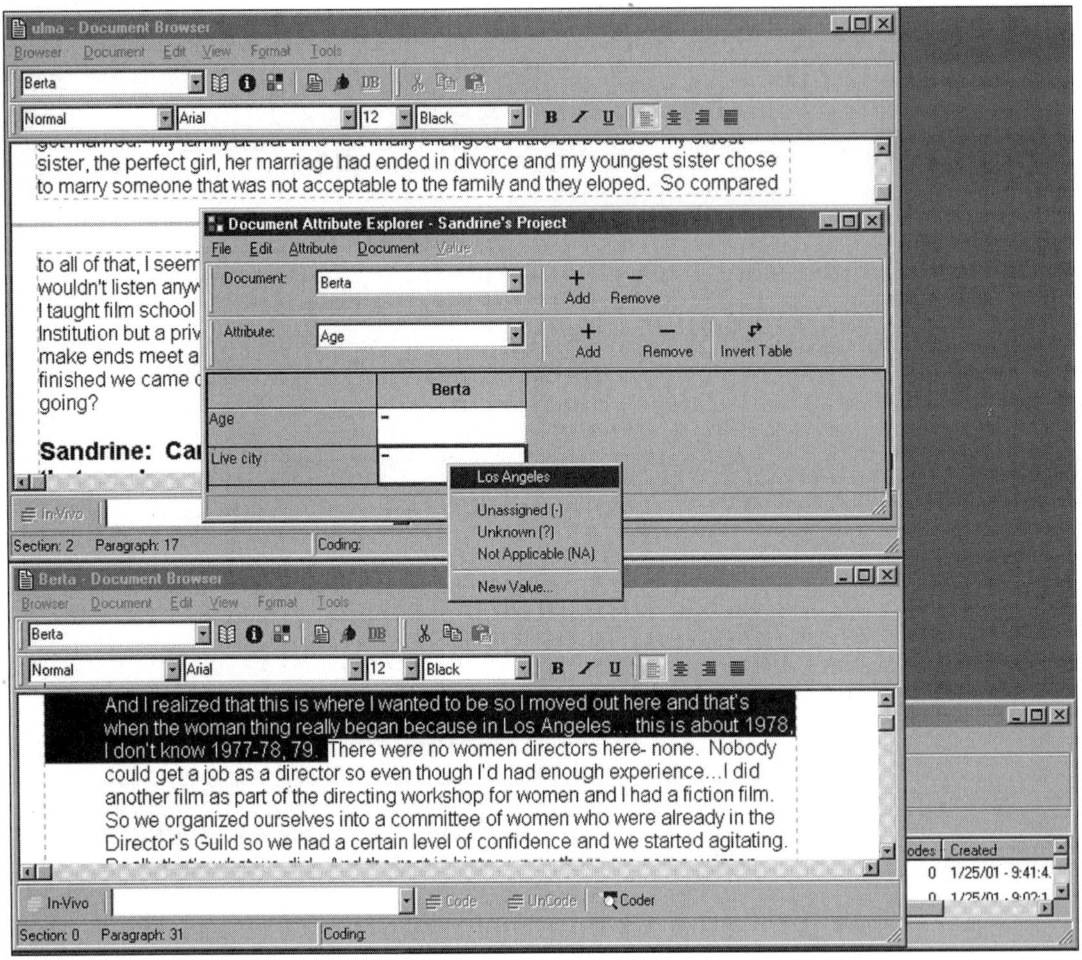

FIGURE 13-12 **Using the "Attribute" Function**

variables) such as age, sex, date of interview, number of children, and so on. There is no rule as to how many attributes you should create, but you need to weigh the time spent on choosing attributes against the potential usefulness of those attributes. Because values previously created are automatically presented as possible choices, it's easy to keep your names and definitions of attributes consistent as you move from text to text. For instance, in Figure 13-12, I had typed "Los Angeles" under the category "Live city" (defined as "Where does the interviewee live?") while coding my interview with Ulma. The value "Los Angeles" was later automatically available when I coded my interview with Berta. You can see that using "LA" in one instance and "Los Angeles" in another might have caused problems. The "attribute" function of NVivo can also help you generate an interview profile or a quantitative analysis.

To organize your attributes, you can create either "free" nodes or "tree" nodes. *Free nodes* are independent nodes, which means that you can't organize them in any type of hierarchy or structure. They are generally those nodes that cannot be related to others. *Tree nodes,* as their name suggests, can be organized into a hierarchy. You can create

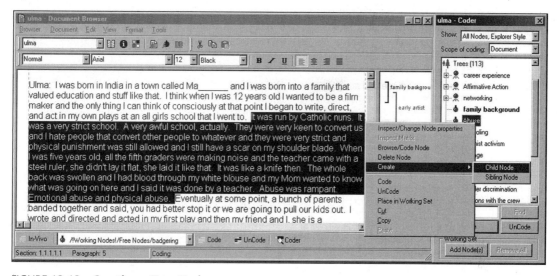

FIGURE 13-13 **Creating a Tree Node**

"child" nodes and "sibling" nodes in relation to them. Tree nodes are the most helpful for analysis purposes. To create a tree node, click on "trees" next to the green trees symbol and then click on the right button of your mouse. A window like the one in Figure 13-13 will appear. Next, click on "create" and choose either "child node" or "sibling node." In Figure 13-13, I had created a node called "abuse" and needed to create a child node called "school abuse," because my interviewee was telling me about the abuse she had experienced at school. I created a child node and then typed "school abuse" over the default "tree node" just created under the "abuse" node. If Ulma reported abuse experienced at home, I could have created another child node called "domestic abuse" under "abuse" or simply a sibling node to "school abuse." Again, I can select any passage from the text and drag it to any number of nodes.

When creating new nodes, be sure to attach a description to each. It's easy to forget what you originally meant if you don't write it down. You can define each node with the "properties" command. With this feature, you can also keep track of the time you created a particular node and who cre-

ated it in case you're working with other coders. In Figure 13-14, I added the description "Mental and physical abuse inflicted by school authority" to the "school abuse" node. You can modify the properties of the nodes you create and the documents you import at any time. Be aware, however, that NVivo is not conceived to merge project files. If you plan to have several people coding the same data, each will have to work on the same NVivo project file at different times instead of simultaneously, that is, they will not be able to combine separate files at the end. If this is an issue, N5 (also from NUD*IST) is an alternative program that allows files to be merged.

Coding can be tedious and time consuming. However, the analysis it allows may be priceless. You can use NVivo for generating reports for all or specific nodes or texts. For instance, in Figure 13-15 I inquired about all interviewee narratives that were coded under the "sabotage" node. As you can see, all passages are extracted under the "sabotage" node browser. Each interview name is specified, as well as the size of each passage. From this window, you could turn on the coding strip view and see how each passage is associated with nodes

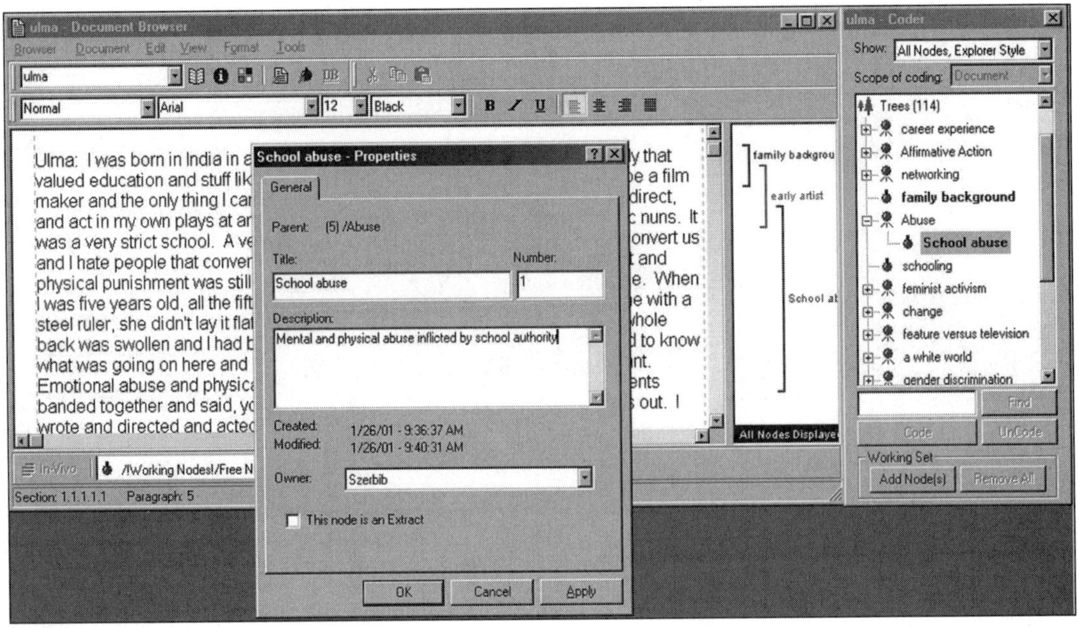

FIGURE 13-14 **Adding a Description**

other than "sabotage." You could also do further coding from this window by simply dragging selected text to a node. Finally, you could generate a report by attribute; for example, you could get all passages that have to do with "gender discrimination" for women who live in Los Angeles or New York and compare them.

There are many analytical tools besides NVivo that are available to you for qualitative analysis. Discover more by downloading free demos from Scolari's Web site at www.scolari.com

THE QUALITATIVE ANALYSIS OF QUANTITATIVE DATA

While it is important and appropriate to distinguish between qualitative and quantitative research and to discuss them separately, they are by no means incompatible or in competition. You need to operate in both modes to explore your full potential as a social researcher.

A QUANDARY REVISITED

Quantification requires a simplification of data through a loss of detail. Sometimes those details are critical to understanding the "whole picture." You've experienced this if you've ever found yourself being categorized by someone else. Let's say you express some political opinion and someone who knows your major says, "Well, of course, you're a sociology [or other] major," as though they now "know" a long list of things about you—some true, some false—that will now shape their "understanding" of the opinion you expressed. You may have experienced being similarly categorized in terms of your religion, your race, where you come from, or your gender. A similar loss can occur in the quantification of data, where a limited number of categories takes the place of varied details. Qualitative analysis, while coding and categorizing, aims at staying closer to the original details.

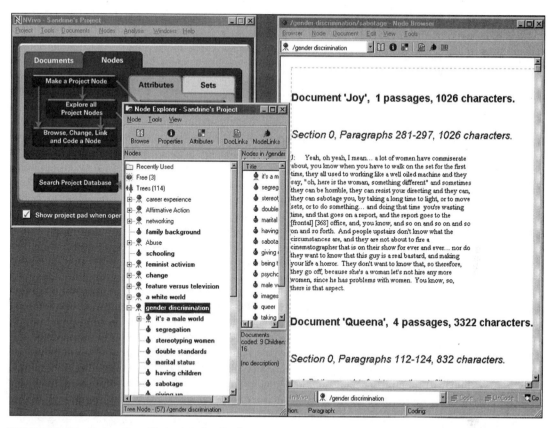

FIGURE 13-15 **Extracting Materials by Node**

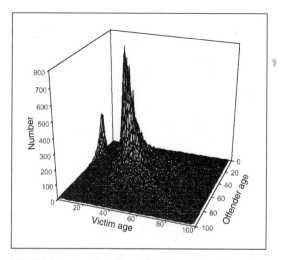

FIGURE 13-16 **Number of One-on-One Homicides by Age of Victim and Age of Offender, Raw Data** *Source:* From Maltz (1998)

Chapter 14 explores some ways in which quantitative analyses can strengthen qualitative studies. Before we move on, however, let's look at an example of how quantitative data demand qualitative assessment.

Figure 13-16 presents FBI data on the hour and day of crimes committed in the United States. These data are often presented in a tabular form, but notice how clearly the patterns of crime appear in this three-dimensional graph. Even though the graph is based on statistical data, it conveys its meaning quite clearly. While summarizing it in the form of equations may be useful for certain purposes, it would add nothing to the clarity of the picture itself. Thus the qualitative assessment of the graph clarifies the quantitative data in a way that no other representation could. Here's a case where a picture is truly worth a thousand words.

Main Points

- Qualitative analysis is the nonnumerical examination and interpretation of observations.

- Qualitative analysis involves a continual interplay between theory and analysis. In analyzing qualitative data, we seek to discover patterns such as changes over time or possible causal links between variables.

- Examples of approaches to the discovery and explanation of such patterns are Grounded Theory Method (GTM), semiotics, and conversation analysis.

- The processing of qualitative data is as much art as science. Three key tools for preparing data for analysis are coding, memoing, and concept mapping.

- In contrast to the standardized units used in coding for statistical analyses, the units to be coded in qualitative analyses may vary within a document. Although codes may be derived from the theory being explored, more often researchers use open coding, in which codes are suggested by the researchers' examination and questioning of the data.

- Memoing is appropriate at several stages of data processing to capture code meanings, theoretical ideas, preliminary conclusions, and other thoughts that will be useful during analysis.

- Concept mapping uses diagrams to explore relationships in the data graphically.

- Several computer programs, such as NVivo, are specifically designed to assist researchers in the analysis of qualitative data.

- Researchers need both qualitative and quantitative analysis for the fullest understanding of social science data.

Key Terms

qualitative analysis
variable-oriented analysis
case-oriented analysis

semiotics
conversation analysis
  (CA)

cross-case analysis
Grounded Theory
  Method (GTM)
constant comparative
  method

open coding
memoing
concept mapping

Review Questions

1. Review Goffman's examination of gender advertising, and collect and analyze a set of advertising photos from magazines or newspapers. What is the relationship between gender and status in the materials you found?

2. Review the discussion of homosexuality in the Book of Leviticus. How might the examination be structured as a cross-case analysis?

3. Imagine you were conducting a cross-case analysis of revolutionary documents such as the Declaration of Independence and the Declaration of the Rights of Man and of the Citizen (from the French Revolution). What key concepts might you code in the following sentence?

> When in the Course of human events, it becomes necessary for one people to dissolve the political bands which have connected them with another, and to assume among the Powers of the earth, the separate and equal station to which the Laws of Nature and of Nature's God entitle them, a decent respect to the opinions of mankind requires that they should declare the causes which impel them to the separation.

4. What would be a good code note and theoretical note for item 3?

Additional Readings

Denzin, Norman K., and Yvonna S. Lincoln. 1994. *Handbook of Qualitative Research.* Thousand Oaks, CA: Sage. Here's a rich resource covering many aspects of qualitative research, in both theory and practice.

Glaser, Barney G., and Anselm L. Strauss. 1967. *The Discovery of Grounded Theory: Strategies for Qualitative Research.* Chicago: Aldine. This is the classic statement of grounded theory with practical suggestions that are still useful today.

Hutchby, Ian, and Robin Wooffitt. 1998. *Conversation Analysis: Principles, Practices and Applications.* Cambridge, England: Polity Press. An excellent overview of the conversation analysis method. The book examines the theory behind the technique, how to use it, and some possible applications.

Strauss, Anselm, and Juliet Corbin. 1990. *Basics of Qualitative Research: Grounded Theory, Procedures and Techniques.* Thousand Oaks, CA: Sage. This updated statement of grounded theory offers special guidance on coding and memoing.

Multimedia Resources

The Wadsworth Sociology Resource Center: Virtual Society
http://sociology.wadsworth.com/

Visit the companion Web site for the second edition of *The Basics of Social Research* to access a wide range of student resources. Begin by clicking on the Student Resources section of the book's Web site to access the following study tools:

- eBabbie Resource Center
- Planning a Research Project
- Doing Data Analysis
- Statistics Review
- Flash Cards

- Internet Links and Exercises
- InfoTrac College Edition: Exercises
- Quizzes

Visit the **eBabbie Resource Center** for an overview of each chapter and helpful online tutorials. Find information on budgeting and step-by-step examples of model research projects at **Planning a Research Project.** Learn how to use quantitative and qualitative data analysis programs at **Doing Data Analysis,** and brush up on your statistics at **Statistics Review.** You can also further your study by accessing **Internet Links and Exercises** related to chapter materials, **Flash Cards, Quizzes,** and many other learning tools.

InfoTrac College Edition
http://www.infotrac-college.com/
wadsworth/access.html

Access the latest news and journal articles with InfoTrac College Edition, an easy-to-use online database of reliable, full-length articles from hundreds of top academic journals. Conduct an electronic search using the following search terms:

Concept mapping
Conversation analysis
Grounded theory
Qualitative data analysis
Semiotics

14 QUANTITATIVE DATA ANALYSIS

David Falconer/Folio, Inc.

What You'll Learn in This Chapter

Often, social data are converted to numerical form for statistical analyses. In this chapter, we'll begin with the process of quantifying data, then turn to analysis. Quantitative analysis may be descriptive or explanatory; it may involve, one, two, or several variables. We begin our examination of how quantitative analyses are done with some simple but powerful ways of manipulating data in order to attain research conclusions.

In this chapter . . .

AN OPENING **QUANDARY**

In Chapter 13, we saw several inherent shortcomings in quantitative data. These shortcomings primarily centered on standardization and superficiality in the face of a social reality that is varied and deep. Can anything meaningful be learned from data that sacrifice meaningful detail in order to permit numerical manipulations?

INTRODUCTION

In Chapter 13, we saw some of the logic and techniques by which social researchers analyze the qualitative data they have collected. This chapter will examine **quantitative analysis,** or the techniques by which researchers convert data to a numerical form and subject it to statistical analyses.

To begin we'll look at *quantification*—the process of converting data to a numerical format. This involves converting social science data into a *machine-readable form*—a form that can be read and manipulated by computers and similar machines used in quantitative analysis.

The rest of the chapter will present the logic and some of the techniques of quantitative data analysis—starting with the simplest case, univariate analysis, which involves one variable, then discussing bivariate analysis, which involves two variables. We'll end with a brief introduction to multivariate analysis, or the examination of several variables simultaneously, such as *age, education,* and *prejudice.*

Before we can do any sort of analysis, we need to quantify our data. Let's turn now to the basic steps involved in converting data into machine-readable forms amenable to computer processing and analysis.

QUANTIFICATION OF DATA

Today, quantitative analysis is almost always done by computer programs such as SPSS and Micro-Case. For those programs to work their magic, they must be able to read the data you've collected in your research. If you've conducted a survey, for example, some of your data are inherently numerical: age or income, for example. While the writing and check marks on a questionnaire are qualitative in nature, a scribbled age is easily converted to quantitative data.

Other data are also easily quantified: transforming male and female into "1" and "2" is hardly rocket science. Researchers can also easily assign numerical representations to such variables as *religious affiliation, political party,* and *region of the country.*

Some data are more challenging, however. If a survey respondent tells you that he or she thinks the biggest problem facing Woodbury, Vermont, today is "the disintegrating ozone layer," the computer can't process that response numerically. You must translate by *coding* the responses. We have already discussed coding in connection with content analysis (Chapter 11) and again in connection with qualitative data analysis (Chapter 13). Now we look at coding specifically for quantitative analysis.

To conduct a quantitative analysis, researchers often must engage in a coding process after the data have been collected. For example, open-ended questionnaire items result in nonnumerical responses, which must be coded before analysis. As with content analysis, the task is to reduce a wide variety of idiosyncratic items of information to a more limited set of attributes composing a variable. Suppose, for example, that a survey researcher asks respondents, "What is your occupation?" The responses to such a question will vary considerably. Although it will be possible to assign each reported occupation reported a separate numerical code, this procedure will not facilitate analysis, which typically depends on several subjects having the same attribute.

The variable *occupation* has many preestablished coding schemes. One such scheme distinguishes professional and managerial occupations, clerical occupations, semiskilled occupations, and so forth. Another scheme distinguishes different sectors of the economy: manufacturing, health, education, commerce, and so forth. Still others combine both. Using an established coding scheme gives you the advantage of being able to compare your research results with those of other studies.

> To learn more about preestablished coding schemes, visit the Bureau of Labor Statistics to learn about their Standard Occupational Classification (http://stats .bls.gov/soc/soc_majo.htm).*

The occupational coding scheme you choose should be appropriate to the theoretical concepts being examined in your study. For some studies, coding all occupations as either white-collar or blue-collar might be sufficient. For others, self-employed and not self-employed might be sufficient. Or a peace researcher might wish to know only whether the occupation depended on the defense establishment or not.

> **SPSS** In SPSS,** categories can be easily combined using the "Transform/Recode" menu option. It is usually wise to create a new variable ("Into Different Variables"). That way, you'll have the new set of categories without losing the original details. In this SPSS procedure, you (1) select the variable to recode, (2) give a name to the new variable, and (3) indicate which codes on the old variable will be included in each of the codes of the new variable.

---

*Each time the Internet icon appears, you'll be given helpful leads for searching the World Wide Web.
**Each time the SPSS icon appears, it indicates that the topic under discussion could be pursued through the use of this software program.

Although the coding scheme should be tailored to meet particular requirements of the analysis, one general guideline should be kept in mind. If the data are coded to maintain a great deal of detail, code categories can always be combined during an analysis that does not require such detail. If the data are coded into relatively few, gross categories, however, there's no way during analysis to recreate the original detail. To keep your options open, it's a good idea to code your data in greater detail than you plan to use in the analysis.

Developing Code Categories

There are two basic approaches to the coding process. First, you may begin with a relatively well-developed coding scheme, derived from your research purpose. Thus, as suggested previously, the peace researcher might code occupations in terms of their relationship to the defense establishment. Or, you may want to use an existing coding scheme so that you can compare your findings with those of previous research.

The alternative method is to generate codes from your data, as discussed in Chapter 13. Let's say we've asked students in a self-administered campus survey to say what they believe is the biggest problem facing their college today. Here are a few of the answers they might have written in.

Tuition is too high
Not enough parking spaces
Faculty don't know what they are doing
Advisors are never available
Not enough classes offered
Cockroaches in the dorms
Too many requirements
Cafeteria food is infected
Books cost too much
Not enough financial aid

Take a minute to review these responses and see whether you can identify some categories represented. Realize that there is no right answer; there are several coding schemes that might be generated from these answers.

TABLE 14-1 **Student Responses That Can Be Coded "Financial Concerns"**

	Financial Concerns
Tuition is too high	x
Not enough parking spaces	
Faculty don't know what they are doing	
Advisors are never available	
Not enough classes offered	
Cockroaches in the dorms	
Too many requirements	
Cafeteria food is infected	
Books cost too much	x
Not enough financial aid	x

Let's start with the first response: "Tuition is too high." What general areas of concern does that response reflect? One obvious possibility is "Financial Concerns." Are there other responses that would fit into that category? Table 14-1 shows which of the questionnaire responses could fit into that category.

In more general terms, the first answer can also be seen as reflecting nonacademic concerns. This categorization would be relevant if your research interest included the distinction between academic and nonacademic concerns. If that were the case, the responses might be coded as shown in Table 14-2.

Notice that I didn't code the response "Books cost too much" in Table 14-2, because this concern could be seen as representing both of the categories. Books are part of the academic program, but their cost is not. This signals the need to refine the coding scheme we are developing. Depending on our research purpose, we might be especially interested in identifying any problems that had an academic element; hence we'd code this one "Academic." Just as reasonably, however, we might be more interested in identifying nonacademic problems and would code the response accordingly. Or, as another alternative, we might create a separate category for responses that involved both academic and nonacademic matters.

TABLE 14-2 **Student Concerns Coded as "Academic" and "Nonacademic"**

	Academic	Nonacademic
Tuition is too high		x
Not enough parking spaces		x
Faculty don't know what they are doing	x	
Advisors are never available	x	
Not enough classes offered	x	
Cockroaches in the dorms		x
Too many requirements	x	
Cafeteria food is infected		x
Books cost too much		
Not enough financial aid		x

TABLE 14-3 **Nonacademic Concerns Coded as "Administrative" or "Facilities"**

	Academic	Administrative	Facilities
Tuition is too high		x	
Not enough parking spaces			x
Faculty don't know what they are doing	x		
Advisors are never available	x		
Not enough classes offered	x		
Cockroaches in the dorms			x
Too many requirements	x		
Cafeteria food is infected			x
Books cost too much	x		
Not enough financial aid		x	

As yet another alternative, we might want to separate nonacademic concerns into those involving administrative matters and those dealing with campus facilities. Table 14-3 shows how the first ten responses would be coded in that event.

As these few examples illustrate, there are many possible schemes for coding a set of data. Your choices should match your research purposes and reflect the logic that emerges from the data themselves. Often, you'll find yourself modifying the code categories as the coding process proceeds. Whenever you change the list of categories, however, you must review the data already coded to see whether changes are in order.

Like the set of attributes composing a variable, and like the response categories in a closed-ended questionnaire item, code categories should be both exhaustive and mutually exclusive. Every piece of information being coded should fit into one and only one category. Problems arise whenever a given response appears to fit equally into more than one code category or whenever it fits into no category: Both signal a mismatch between your data and your coding scheme.

If you're fortunate enough to have assistance in the coding process, you'll need to train your coders in the definitions of code categories and show them how to use those categories properly. To do so, explain the meaning of the code categories and give several examples of each. To make sure your coders fully understand what you have in mind, code several cases ahead of time. Then ask your

---

POLVIEWS

We hear a lot of talk these days about liberals and conservatives. I'm going to show you a seven-point scale on which the political views that people might hold are arranged from extremely liberal—point 1—to extremely conservative—point 7. Where would you place yourself on this scale?

1. Extremely liberal
2. Liberal
3. Slightly liberal
4. Moderate, middle of the road
5. Slightly conservative
6. Conservative
7. Extremely conservative
8. Don't know
9. No answer

ATTEND

How often do you attend religious services?

0. Never
1. Less than once a year
2. About once or twice a year
3. Several times a year
4. About once a month
5. 2–3 times a month
6. Nearly every week
7. Every week
8. Several times a week
9. Don't know, No answer

FIGURE 14-1 **A Partial Codebook**

coders to code the same cases without knowing how you coded them. Finally, compare your coders' work with your own. Any discrepancies will indicate an imperfect communication of your coding scheme to your coders. Even with perfect agreement between you and your coders, however, it's best to check the coding of at least a portion of the cases throughout the coding process.

If you're not fortunate enough to have assistance in coding, you should still obtain some verification of your own reliability as a coder. Nobody's perfect, especially a researcher hot on the trail of a finding. Suppose that you're studying an emerging cult and that you have the impression that people who do not have a regular family will be the most likely to regard the new cult as a family substitute. The danger is that whenever you discover a subject who reports no family, you'll unconsciously try to find some evidence in the subject's comments that the cult is a substitute for family. If at all possible, then, get someone else to code some of your cases to see whether that person makes the same assignments you made.

Codebook Construction

The end product of the coding process is the conversion of data items into numerical codes. These codes represent attributes composing variables, which, in turn, are assigned locations within a data file. A **codebook** is a document that describes the locations of variables and lists the assignments of codes to the attributes composing those variables.

A codebook serves two essential functions. First, it is the primary guide used in the coding process. Second, it is your guide for locating variables and interpreting codes in your data file during analysis. If you decide to correlate two variables as a part of your analysis of your data, the codebook tells you where to find the variables and what the codes represent.

Figure 14-1 is a partial codebook created from two variables from the General Social Survey. Though there is no one right format for a codebook, this example presents some of the common elements.

Notice first that each variable is identified by an abbreviated variable name: *POLVIEWS, ATTEND*. We can determine the church attendance of respondents, for example, by referencing *ATTEND*. This example uses the format established by the General Social Survey, which has been carried over into SPSS. Other data sets and/or analysis programs might format variables differently. Some use numerical codes in place of abbreviated names, for example. You must, however, have some identifier that will allow you to locate and use the variable in question.

Next, every codebook should contain the full definition of the variable. In the case of a questionnaire, the definition consists of the exact word-ings of the questions asked, because, as we've seen, the wording of questions strongly influences the answers returned. In the case of *POLVIEWS,* you know that respondents were handed a card containing the several political categories and asked to pick the one that best fit them.

The codebook also indicates the attributes com-posing each variable. In *POLVIEWS,* for example, respondents could characterize their political ori-entations as "Extremely liberal," "Liberal," "Slightly liberal," and so forth.

Finally, notice that each attribute also has a nu-meric label. Thus, in *POLVIEWS,* "Extremely liberal" is code category 1. These numeric codes are used in various manipulations of the data. For example, you might decide to combine categories 1 through 3 (all the "liberal" responses). It's easier to do this with code numbers than with lengthy names.

> You can visit the GSS codebook online at http://www.icpsr.umich.edu/GSS99/codebook.htm. If you know the symbolic name (e.g., POLVIEWS), you can locate it in the Mnemonic listing. Otherwise, you can browse the "Index by Subject" to find all the different questions that have been asked regarding a particular topic.

Data Entry

In addition to transforming data into quantitative form, researchers interested in qualitative analysis also need to convert data into a machine-readable format, so that computers can read and manipu-late the data. There are many ways of accomplish-ing this step, depending on the original form of your data and also the computer program you will use for analyzing the data. I'll simply introduce you to the process here. If you find yourself undertaking this task, you should be able to tailor your work to the particular data source and program you are using.

If your data have been collected by question-naire, you might do your coding on the question-naire itself. Then, data entry specialists (including

yourself) could enter the data into, say, an SPSS data matrix or into an Excel spreadsheet and later imported into SPSS.

Sometimes, social researchers use optical scan sheets for data collection. These sheets can be fed into machines that will convert the black marks into data, which can be imported into the analysis program. This procedure will only work with sub-jects who are comfortable using such sheets, and it will usually be limited to closed-ended questions.

Sometimes, data entry occurs in the process of data collection. In Computer Assisted Telephone Interviewing, for example, the interviewer keys responses directly into the computer, where the data are compiled for analysis (see Chapter 9). Even more effortlessly, online surveys can be con-structed so that the respondents enter their own answers directly into the accumulating database, without the need for an intervening interviewer or data entry person.

Once data have been fully quantified and en-tered into the computer, researchers can begin quantitative analysis. Let's look at a the three cases mentioned at the start of this chapter: univariate, bivariate, and multivariate analyses.

UNIVARIATE ANALYSIS

The simplest form of quantitative analysis, **uni-variate analysis,** involves describing a case in terms of a single variable—specifically, the distri-bution of attributes that comprise it. For example, if *gender* was measured, we would look at how many of the subjects were men and how many were women.

Distributions

The most basic format for presenting univariate data is to report all individual cases, that is, to list the attribute for each case under study in terms of the variable in question. Let's take as an ex-ample the General Social Survey (GSS) data on at-tendance at religious services, *ATTEND.* Table 14-4 presents the results of an SPSS analysis of this variable.

TABLE 14-4 **GSS Attendance at Religious Services, 1973–1993**

Attend   How Often R Attends Religious Services

Value Label	Value	Frequency	Percent	Valid Percent	Cum Percent
NEVER	0	224	14.9	15.1	15.1
LT ONCE A YEAR	1	139	9.3	9.4	24.5
ONCE A YEAR	2	180	12.0	12.1	36.6
SEVRL TIMES A YR	3	194	12.9	13.1	49.7
ONCE A MONTH	4	84	5.6	5.7	55.4
2-3X A MONTH	5	136	9.1	9.2	64.5
NRLY EVERY WEEK	6	114	7.6	7.7	72.2
EVERY WEEK	7	294	19.6	19.8	92.0
MORE THN ONCE WK	8	118	7.9	8.0	100.0
DK,NA	9	17	1.1	Missing	
	Total	1,500	100.0	100.0	

Valid cases   1,483   Missing cases   17

Let's examine the table, piece by piece. First, if you look near the bottom of the table, you'll see that the sample being analyzed has a total of 1,500 cases. In the last row above the totals, you'll see that 17 of the 1,500 respondents either said they didn't know (DK) or gave no answer (NA) in response to this question. So our assessment of U.S. attendance at religious services during the two decades from 1973 to 1993 will be based on the 1,483 respondents who answered the question.

Go back to the top of the table now. You'll see that 224 people said they never went to religious services. This number in and of itself tells us nothing about religious practices. If the data we're examining were based on 3,000 respondents instead of 1,500, we could assume that about 448 people would have said they never attended religious services. Neither the number 224 nor the number 448, in itself, gives us an idea of whether the "average American" goes to church a little or a lot.

By analogy, suppose your best friend tells you that she drank a six-pack of beer. Is that a little beer or a lot? The answer, of course, depends on whether she consumed the beer in a month, a week, a day, or an hour. In the case of religious participation, similarly, we need some basis for assessing the number that represents the people who never attend church.

One way to assess the number is to calculate the percentage of all respondents who said they never go to church. If you were to divide 224 by the 1,483 who gave some answer, you would get 15.1 percent, which appears in the table as the "valid percent." Now we can say that 15 percent, or roughly one U.S. adult in seven, reports never attending religious services.

This result is more meaningful, but does it suggest that people in the United States are generally nonreligious? A further look at Table 14-4 shows that the response category most often chosen was "every week," with 19.8 percent of the respondents giving that answer. Add to that the 8 percent who report attending religious services more than once a week, and we find that over a fourth (27.8 percent) of U.S. adults say they attend religious services at least once a week. As you can see, each new comparison gives a more complete picture of the data.

A description of the number of times that the various attributes of a variable are observed in a sample is called a **frequency distribution.** Sometimes it's easiest to see a frequency distribution in a graph. Figure 14-2 was created by SPSS from the GSS data on *ATTEND.* The vertical scale on the left side of the graph indicates the percentages selecting each of the answers displayed along the

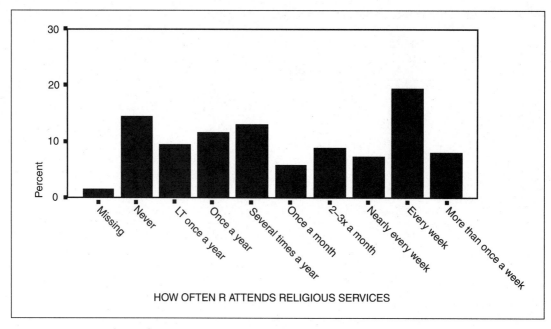

FIGURE 14-2   **Bar Chart of GSS ATTEND, 1973–1993**

horizontal axis of the graph. Take a minute to notice how the percentages in Table 14-4 correspond with the heights of the bars in Figure 14-2.

Central Tendency

Beyond simply reporting the overall distribution of values, sometimes called the *marginal frequencies* or just the *marginals,* you may choose to present your data in the form of an **average** or measure of *central tendency.* You're already familiar with the concept of central tendency from the many kinds of averages you use in everyday life to express the "typical" value of a variable. For instance, in baseball a batting average of .300 says that a batter gets a hit three out of every ten opportunities—on average. Over the course of a season, a hitter might go through extended periods without getting any hits at all and go through other periods when he or she gets a bunch of hits all at once. Over time, though, the central tendency of the batter's performance can be expressed as getting three hits in every ten chances. Similarly, your grade point average expresses the "typical" value of all your grades taken together, even though some of them might

be A's, others B's, and one or two might be C's (I know you never get anything lower than a C).

Averages like these are more properly called the arithmetic **mean** (the result of dividing the sum of the values by the total number of cases). The mean is only one way to measure central tendency or "typical" values. Two other options are the **mode** (the most frequently occurring attribute) and the **median** (the middle attribute in the ranked distribution of observed attributes). Here's how the three averages would be calculated from a set of data.

Suppose you're conducting an experiment that involves teenagers as subjects. They range in age from 13 to 19, as indicated in the following table:

Age	Number
13	3
14	4
15	6
16	8
17	4
18	3
19	3

Now that you've seen the actual ages of the 31 subjects, how old would you say they are in general, or "on average"? Let's look at three different ways you might answer that question.

The easiest average to calculate is the mode, the most frequent value. As you can see, there were more 16-year-olds (eight of them) than any other age, so the modal age is 16, as indicated in Figure 14-3. Technically, the modal age is the category "16," which may include some people who are closer to 17 than 16 but who haven't yet reached that birthday.

Figure 14-3 also demonstrates the calculation of the mean. There are three steps: (1) multiply each age by the number of subjects who have that age, (2) total the results of all those multiplications, and (3) divide that total by the number of subjects.

In the case of age, a special adjustment is needed. As indicated in the discussion of the mode, those who call themselves "13" actually range from exactly 13 years old to those just short of 14. It is reasonable to assume, moreover, that as a group the "13-year-olds" in the country are evenly distributed within that one-year span, making their average age 13.5 years. This is true for each of the age groups. Hence, it is appropriate to add 0.5 years to the final calculation, making the mean age 16.37, as indicated in Figure 14-3.

The third measure of central tendency, the median, represents the "middle" value: Half are above it, half below. If we had the precise ages of each subject (for example, 17 years and 124 days), we'd be able to arrange all 31 subjects in order by age, and the median for the whole group would be the age of the middle subject.

As you can see, however, we do not know precise ages; our data constitute "grouped data" in this regard. For example, three people who are not precisely the same age have been grouped in the category "13-year-olds."

Figure 14-3 illustrates the logic of calculating a median for grouped data. Because there are 31 subjects altogether, the "middle" subject would be subject number 16 if they were arranged by age—15 teenagers would be younger and 15 older. Look at the bottom portion of Figure 14-3, and you'll see that the middle person is one of the eight 16-year-

olds. In the enlarged view of that group, we see that number 16 is the third from the left.

Because we do not know the precise ages of the subjects in this group, the statistical convention here is to assume they are evenly spread along the width of the group. In this instance, the possible ages of the subjects go from 16 years and no days to 16 years and 364 days. Strictly speaking, the range, then, is 364/365 days. As a practical matter, it's sufficient to call it one year.

If the eight subjects in this group were evenly spread from one limit to the other, they would be one-eighth of a year apart from each other—a 0.125–year interval. Look at the illustration and you'll see that if we place the first subject half the interval from the lower limit and add a full interval to the age of each successive subject, the final one is half an interval from the upper limit.

What we've done is calculate, hypothetically, the precise ages of the eight subjects—assuming their ages were spread out evenly. Having done this, we merely note the age of the middle subject—16.31—and that is the median age for the group.

Whenever the total number of subjects is an even number, of course, there is no middle case. To get the median, you merely calculate the mean of the two values on either side of the midpoint in the ranked data. Suppose, for example, that there was one more 19-year-old in our sample, giving us a total of 32 cases. The midpoint would then fall between subjects 16 and 17. The median would therefore be calculated as $(16.31 + 16.44)/2 = 16.38$.

As you can see in Figure 14-3, the three measures of central tendency produce three different values for our set of data, which is often (but not necessarily) the case. Which measure, then, best represents the "typical" value? More generally, which measure of central tendency should we prefer? The answer depends on the nature of your data and the purpose of your analysis. For example, whenever means are presented, you should be aware that they are susceptible to extreme values—a few very large or very small numbers. As only one example, the (mean) average person in Redmond, Washington, has a net worth in excess

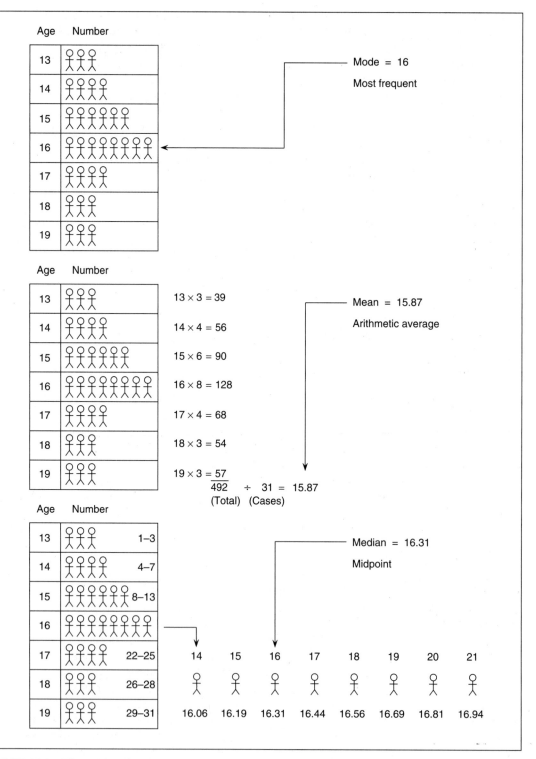

FIGURE 14-3 **Three "Averages"**

of a million dollars. If you were to visit Redmond, however, you would not find that the "average" resident lives up to your idea of a millionaire. The very high mean reflects the influence of one extreme case among Redmond's 40,000 residents—Bill Gates of Microsoft, who has a net worth (at the time this is being written) of tens of billions of dollars. Clearly, the median wealth would give you a more accurate picture of the residents of Redmond as a whole.

This example should illustrate the need to choose carefully among the various measures of central tendency. A course or textbook in statistics will give you a fuller understanding of the variety of situations in which each is appropriate.

Dispersion

Averages offer readers the advantage of reducing the raw data to the most manageable form: A single number (or attribute) can represent all the detailed data collected in regard to the variable. This advantage comes at a cost, of course, because the reader cannot reconstruct the original data from an average. Summaries of the dispersion of responses can somewhat alleviate this disadvantage.

Dispersion refers to the way values are distributed around some central value, such as an average. The simplest measure of dispersion is the range: the distance separating the highest from the lowest value. Thus, besides reporting that our subjects have a mean age of 15.87, we might also indicate that their ages ranged from 13 to 19.

A more sophisticated measure of dispersion is the **standard deviation.** This measure was briefly mentioned in Chapter 7 as the standard error of a sampling distribution. Essentially, the standard deviation is an index of the amount of variability in a set of data. A higher standard deviation means that the data are more dispersed; a lower standard deviation means that they are more bunched together. Figure 14-4 illustrates the basic idea. Notice that the professional golfer not only has a lower mean score but is also more consistent—repre-

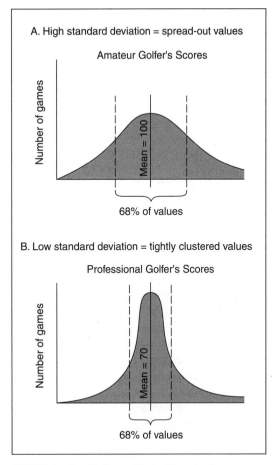

FIGURE 14-4 **High and Low Standard Deviations**

sented by the smaller standard deviation. The duffer, on the other hand, has a higher average but is also less consistent: sometimes doing much better, sometimes much worse.

There are many other measures of dispersion. In reporting intelligence test scores, for example, researchers might determine the *interquartile range,* the range of scores for the middle 50 percent of subjects. If the top one-fourth had scores ranging from 120 to 150, and if the bottom one-fourth had scores ranging from 60 to 90, the report might say that the interquartile range was from 90 to 120 (or 30 points) with a mean score of, let's say, 102.

Continuous and Discrete Variables

The preceding calculations are not appropriate for all variables. To understand this point, we must distinguish between two types of variables: continuous and discrete. A **continuous variable** (or ratio variable) increases steadily in tiny fractions. An example is age, which increases steadily with each increment of time. A **discrete variable** jumps from category to category without intervening steps. Examples include *gender, military rank,* or *year in college* (you go from being a sophomore to a junior in one step).

In analyzing a discrete variable—a nominal or ordinal variable, for example—some of the techniques discussed previously do not apply. Strictly speaking, modes should be calculated for nominal data, medians for interval data, and means for ratio data, not for nominal data (see Chapter 5). If the variable in question is *gender,* for example, raw numbers (23 of the cross-dressing outlaw bikers in our sample are women) or percentages (7 percent are women) can be appropriate and useful analyses, but neither a median nor a mean would make any sense. Calculating the mode would be legitimate, though not very revealing, since it would only tell us "most were men." However, the mode for data on religious affiliation might be more interesting, as in "most people in the United States are Protestant."

Detail versus Manageability

In presenting univariate and other data, you'll be constrained by two goals. On the one hand, you should attempt to provide your reader with the fullest degree of detail regarding those data. On the other hand, the data should be presented in a manageable form. As these two goals often directly counter each other, you'll find yourself continually seeking the best compromise between them. One useful solution is to report a given set of data in more than one form. In the case of age, for example, you might report the distribution of ungrouped ages *plus* the mean age and standard deviation.

As you can see from this introductory discussion of univariate analysis, this seemingly simple matter can be rather complex. In any event, the lessons of this section pave the way for a consideration of subgroup comparisons and bivariate analyses.

SUBGROUP COMPARISONS

Univariate analyses describe the units of analysis of a study and, if they are a sample drawn from some larger population, allow us to make descriptive inferences about the larger population. Bivariate and multivariate analyses are aimed primarily at explanation. Before turning to explanation, however, we should consider the case of subgroup description.

Often it's appropriate to describe subsets of cases, subjects, or respondents. Table 14-5, for example, presents income data for men and women separately. In addition, the table presents the ratio

TABLE 14-5 **Median Earnings of Year-Round, Full-Time Civilian Workers with Earnings by Gender: 1967–1977**

Year	Women	Men	Ratio of Women's to Men's Earnings
1977	$8,618	$14,626	.59
1976	8,622	14,323	.60
1975	8,449	14,175	.60
1974	8,565	14,578	.59
1973	8,639	15,254	.57
1972	8,551	14,778	.58
1971	8,369	14,064	.61
1970	8,307	13,993	.59
1969	8,227	13,976	.59
1968	7,763	13,349	.58
1967	7,503	13,021	.58

Source: Adapted from U.S. Bureau of the Census, Current Population Reports, Series P-23, No. 100, *Statistical Portrait of Women in the United States: 1978* (Washington, DC: U.S. Government Printing Office, 1978).

TABLE 14-6 **Ratio of Women-to-Men Earnings among Full-Time Wage and Salary Workers: 1985–1998**

	Ratio of Women's to Men's Median Weekly Earnings
1985	.68
1990	.72
1995	.75
1998	.76

Source: U.S. Bureau of the Census, *Statistical Abstract of the United States* (Washington, DC: U.S. Government Printing Office, 1999): 445.

of women's median income to that of men, showing that women in the labor force earned a little over half what men earned in the indicated period.

In some situations, the researcher presents subgroup comparisons purely for descriptive purposes. More often, the purpose of subgroup descriptions is comparative: Women earn less than men. In the present case, it is assumed that there is something about being a woman that resulted in their lower incomes. When we compare incomes of African Americans and whites, a similar assumption is made. In such cases, the analysis is based on an assumption of causality: one variable causing another, as in *gender* causing *income.*

When the data shown in Table 14-5 were published by the Census Bureau in 1978, they added legitimacy to the growing outcry over the discrimination of women in the U.S. economy. Both politics and research have focused on the issue ever since, though the discrepancy in male and female earnings has hardly been resolved.

More-recent statistics reflect some improvement but still show a distinct difference. At the end of the 1990s, the average full-time, year-round female worker earned 76 cents for each dollar earned by her male counterpart. See Table 14-6 for the recent wage ratios.

Before moving on to the logic of bivariate, causal analysis, let's consider another example of subgroup comparisons—one that will let us address some table-formatting issues.

APPLYING THE RESULTS

Data like those in Tables 14-5 and 14-6 provide a powerful arguing point and incentive to action for those committed to gender equality. As we'll see later in this chapter, several factors that may legitimately explain why women earn less than men: Women are more likely to work part-time, they are likely to have less seniority (because of child-bearing and child-rearing absences), and so forth. However, we'll also see that quantitative analyses of data such as these allow us to test such explanations. In this case, women earn less even when researchers account for such legitimate explanations.

"Collapsing" Response Categories

"Textbook examples" of tables are often simpler than you'll typically find in published research reports or in your own analyses of data, so this section and the next one address two common problems and suggest solutions.

Let's begin by turning to the data reported in Table 14-7, which reports data collected in a multinational poll conducted by the *New York Times,* CBS News, and the *Herald Tribune* in 1985, concerning attitudes about the United Nations. The question reported in Table 14-7 deals with general attitudes about the way the UN was handling its job.

Here's the question: How do people in the five nations reported in Table 14-7 compare in their support for the kind of job the UN was doing? As you review the table, you may find there are simply so many numbers that it's hard to see any meaningful pattern.

Part of the problem with Table 14-7 lies in the relatively small percentages of respondents selecting the two extreme response categories: the UN is doing a very good or a very poor job. Furthermore, although it might be tempting to read only

TABLE 14-7 **Attitudes toward the United Nations: "How is the UN doing in solving the problems it has had to face?"**

	West Germany	Britain	France	Japan	United States
Very good job	2%	7%	2%	1%	5%
Good job	46	39	45	11	46
Poor job	21	28	22	43	27
Very poor job	6	9	3	5	13
Don't know	26	17	28	41	10

Source: "5-Nation Survey Finds Hope for U.N.," *New York Times,* June 26, 1985, p. 6.

the second line of the table—those saying "good job"—that would be improper. Looking at only the second row, we would conclude that West Germany and the United States were the most positive (46 percent) about the UN's performance, followed closely by France (45 percent), with Britain (39 percent) less positive than any of those three and Japan (11 percent) the least positive of all.

This procedure is inappropriate in that it ignores all those respondents who gave the most positive answer of all: "very good job." In a situation like this, you should combine or "collapse" the two ends of the range of variation. In this instance, combine "very good" with "good" and "very poor" with "poor." If you were to do this in the analysis of your own data, it would be wise to add the raw frequencies together and recompute percentages for the combined categories, but in analyzing a published table such as this one, you can simply add the percentages as illustrated by the results shown in Table 14-8.

With the collapsed categories illustrated in Table 14-8, we can now rather easily read across the several national percentages of people who said the UN was doing at least a good job. Now the United States appears the most positive; Germany, Britain, and France are only slightly less positive and are nearly indistinguishable from one another; and Japan stands alone in its quite low assessment of the UN's performance. Although the conclusions to be drawn now do not differ radically from what we might have concluded from simply reading the second line of Table 14-7, we should note that Britain now appears relatively more supportive.

Here's the risk I'd like to spare you. Suppose you had hastily read the second row of Table 14-7 and

noted that the British had a somewhat lower assessment of the job the UN was doing than was true of people in the United States, West Germany, and France. You might feel obliged to think up an explanation for why that was so—possibly creating an ingenious psychohistorical theory about the painful decline of the once powerful and dignified British Empire. Then, once you had touted your "theory" about it, someone else might point out that a proper reading of the data would show the British were actually not really less positive than the other three nations. This is not a hypothetical risk. Errors like these happen frequently, but they can be avoided by collapsing answer categories where appropriate.

Handling "Don't Knows"

Tables 14-7 and 14-8 illustrate another common problem in the analysis of survey data. It's usually a good idea to give people the option of saying "don't know" or "no opinion" when asking for their opinions on issues. But what do you do with those answers when you analyze the data?

Notice there is a good deal of variation in the national percentages saying "don't know" in this instance, ranging from only 10 percent in the United States to 41 percent in Japan. The presence of substantial percentages saying they don't know can confuse the results of a tables like these. For example, were the Japanese so much less likely to say the UN was doing a good job simply because so many didn't express any opinion?

Here's an easy way to recalculate percentages, with the "don't knows" excluded. Look at the first column of percentages in Table 14-8: West Ger-

TABLE 14-8 **Collapsing Extreme Categories**

	West Germany	Britain	France	Japan	United States
Good job or better	48%	46%	47%	12%	51%
Poor job or worse	27	37	25	48	40
Don't know	26	17	28	41	10

TABLE 14-9 **Omitting the "Don't Knows"**

	West Germany	Britain	France	Japan	United States
Good job or better	65%	55%	65%	20%	57%
Poor job or worse	35%	45%	35%	81%	44%

many's answers to the question about the UN's performance. Notice that 26 percent of the respondents said they didn't know. This means that those who said "good" or "bad" job—taken together—represent only 74 percent (100 minus 26) of the whole. If we divide the 48 percent saying "good job or better" by .74 (the proportion giving any opinion), we can say that 65 percent "of those with an opinion" said the UN was doing a good or very good job (48%/.74 = 65%).

Table 14-9 presents the whole table with the "don't knows" excluded. Notice that these new data offer a somewhat different interpretation than do the previous tables. Specifically, it would now appear that France and West Germany were the most positive in their assessments of the UN, with the United States and Britain a bit lower. Although Japan still stands out as lowest in this regard, it has moved from 12 percent to 20 percent positive.

At this point, having seen three versions of the data, you may be asking yourself: Which is the right one? The answer depends on your purpose in analyzing and interpreting the data. For example, if it is not essential for you to distinguish "very good" from "good," it makes sense to combine them, because it's easier to read the table.

Whether to include or exclude the "don't knows" is harder to decide in the abstract. It may be a very important finding that such a large percentage of the Japanese had no opinion—if you wanted to find out whether people were familiar with the

work of the UN, for example. On the other hand, if you wanted to know how people might vote on an issue, it might be more appropriate to exclude the "don't knows" on the assumption that they wouldn't vote or that ultimately they would be likely to divide their votes between the two sides of the issue.

In any event, the *truth* contained within your data is that a certain percentage said they didn't know and the remainder divided their opinions in whatever manner they did. Often, it's appropriate to report your data in both forms—with and without the "don't knows"—so your readers can also draw their own conclusions.

Numerical Descriptions in Qualitative Research

Although this chapter deals primarily with quantitative research, the discussions are also relevant to qualitative studies. The findings of in-depth, qualitative studies often can be verified by some numerical testing. Thus, for example, when David Silverman wanted to compare the cancer treatments received by patients in private clinics with those in Britain's National Health Service, he primarily chose in-depth analyses of the interactions between doctors and patients:

> My method of analysis was largely qualitative and . . . I used extracts of what doctors and patients had said as well as offering a brief

ethnography of the setting and of certain behavioural data. In addition, however, I constructed a coding form which enabled me to collate a number of crude measures of doctor and patient interactions. — (SILVERMAN 1993:163)

Not only did the numerical data fine-tune Silverman's impressions based on his qualitative observations, but his in-depth understanding of the situation allowed him to craft an ever more appropriate quantitative analysis. Listen to the interaction between qualitative and quantitative approaches in this lengthy discussion:

> My overall impression was that private consultations lasted considerably longer than those held in the NHS clinics. When examined, the data indeed did show that the former were almost twice as long as the latter (20 minutes as against 11 minutes) and that the difference was statistically highly significant. However, I recalled that, for special reasons, one of the NHS clinics had abnormally short consultations. I felt a fairer comparison of consultations in the two sectors should exclude this clinic and should only compare consultations taken by a single doctor in both sectors. This subsample of cases revealed that the difference in length between NHS and private consultations was now reduced to an average of under 3 minutes. This was still statistically significant, although the significance was reduced. Finally, however, if I compared only *new* patients seen by the same doctor, NHS patients got 4 minutes more on the average— 34 minutes as against 30 minutes in the private clinic. — (SILVERMAN 1993:163–64)

This example further demonstrates the special power that can be gained from a combination of approaches in social research. The combination of qualitative and quantitative analyses can be especially potent.

BIVARIATE ANALYSIS

In contrast to univariate analysis, subgroup comparisons involve two variables. In this respect subgroup comparisons constitute a kind of **bivariate**

TABLE 14-10 **Church Attendance Reported by Men and Women in 1996**

	Men	**Women**
Weekly	25%	34%
Less often	75	66
100% =	(901)	(1134)

Source: General Social Survey, National Opinion Research Center.

analysis—that is, the analysis of two variables simultaneously. However, as with univariate analysis, the purpose of subgroup comparisons is largely descriptive. Most bivariate analysis in social research adds another element: determining relationships between the variables themselves. Thus, univariate analysis and subgroup comparisons focus on describing the people (or other units of analysis) under study, whereas bivariate analysis focuses on the variables and their empirical relationships.

Table 14-10 could be regarded as an instance of subgroup comparison: It independently describes the church attendance of men and women, as reported in the 1996 General Social Survey. It shows—comparatively and descriptively—that the women under study attended church more often than did the men. However, the same table, seen as an explanatory bivariate analysis, tells a somewhat different story. It suggests that the variable *gender* has an effect on the variable *church attendance.* That is, we can view the behavior as a dependent variable that is partially determined by the independent variable, *gender.*

Explanatory bivariate analyses, then, involve the "variable language" introduced in Chapter 1. In a subtle shift of focus, we are no longer talking about men and women as different subgroups but about *gender* as a variable: one that has an influence on other variables. The theoretical interpretation of Table 14-10 might be taken from Charles Glock's Comfort Hypothesis as discussed in Chapter 2:

1. Women are still treated as second-class citizens in U.S. society.
2. People denied status gratification in the secular society may turn to religion as an alternative source of status.

3. Hence, women should be more religious than men.

The data presented in Table 14-10 confirm this reasoning. Thirty-four percent of the women attend church weekly, as compared with 25 percent of the men.

Adding the logic of causal relationships among variables has an important implication for the construction and reading of percentage tables. One of the chief bugaboos for new data analysts is deciding on the appropriate "direction of percentaging" for any given table. In Table 14-10, for example, I've divided the group of subjects into two subgroups—men and women—and then described the behavior of each subgroup. That is the correct method for constructing this table. Notice, however, that we could—however inappropriately—construct the table differently. We could first divide the subjects into different degrees of church attendance and then describe each of those subgroups in terms of the percentage of men and women in each. This method would make no sense in terms of explanation, however. Table 14-10 suggests that your gender will affect your frequency of church attendance. Had we used the other method of construction, the table would suggest that your church attendance affects whether you are a man or a woman—which makes no sense. Your behavior cannot determine your gender.

A related problem complicates the lives of new-data analysts. How do you read a percentage table? There is a temptation to read Table 14-10 as follows: "Of the women, only 34 percent attended church weekly, and 66 percent said they attended less often; therefore, being a woman makes you less likely to attend church frequently." This is, of course, an incorrect reading of the table. Any conclusion that *gender*—as a variable—has an effect on church attendance must hinge on a comparison between men and women. Specifically, we compare the 34 percent with the 25 percent and note that women are more likely than men to attend church weekly. The comparison of subgroups, then, is essential in reading an explanatory bivariate table.

In constructing and presenting Table 14-10, I have used a convention called *percentage down.*

This term means that you can add the percentages down each column to total 100 percent. You read this form of table across a row. For the row labeled "weekly," what percentage of the men attend weekly? What percentage of the women attend weekly?

The direction of percentaging in tables is arbitrary, and some researchers prefer to percentage across. They would organize Table 14-10 so that "men" and "women" were shown on the left side of the table, identifying the two rows, and "weekly" and "less often" would appear at the top to identify the columns. The actual numbers in the table would be moved around accordingly, and each row of percentages would total 100 percent. In that case, you would read the table down a column, still asking what percentage of men and women attended frequently. The logic and the conclusion would be the same in either case; only the form would differ.

In reading a table that someone else has constructed, therefore, you need to find out in which direction it has been percentaged. Usually this will be labeled or be clear from the logic of the variables being analyzed. As a last resort, however, you should add the percentages in each column and each row. If each of the columns totals 100 percent, the table has been percentaged down. If the rows total 100 percent each, it has been percentaged across. The rule, then, is as follows:

1. If the table is percentaged down, read across.
2. If the table is percentaged across, read down.

Percentaging a Table

Figure 14-5 reviews the logic by which we create percentage tables from two variables. I've used as variables *gender* and *attitudes toward equality for men and women.*

Here's another example. Suppose we're interested in learning something about newspaper editorial policies regarding the legalization of marijuana. We undertake a content analysis of editorials on this subject that have appeared during a given year in a sample of daily newspapers across the nation. Each editorial has been classified as

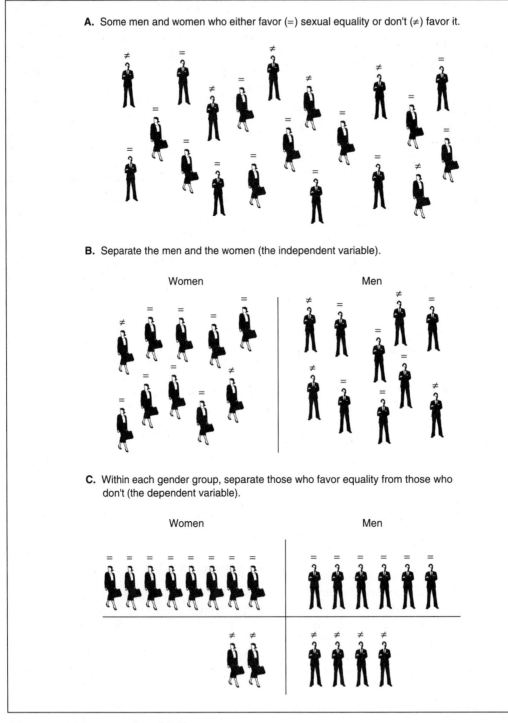

A. Some men and women who either favor (=) sexual equality or don't (≠) favor it.

B. Separate the men and the women (the independent variable).

Women   Men

C. Within each gender group, separate those who favor equality from those who don't (the dependent variable).

Women   Men

FIGURE 14-5   **Percentaging a Table**

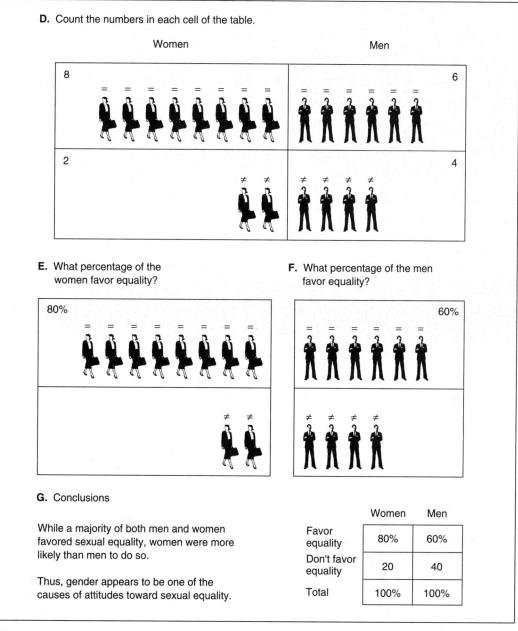

D. Count the numbers in each cell of the table.

Women   Men

8   6

2   4

E. What percentage of the women favor equality?

F. What percentage of the men favor equality?

80%   60%

G. Conclusions

While a majority of both men and women favored sexual equality, women were more likely than men to do so.

Thus, gender appears to be one of the causes of attitudes toward sexual equality.

	Women	Men
Favor equality	80%	60%
Don't favor equality	20	40
Total	100%	100%

FIGURE 14-5   **(continued)**

favorable, neutral, or unfavorable toward the legalization of marijuana. Perhaps we wish to examine the relationship between editorial policies and the types of communities in which the newspapers are published, thinking that rural newspapers might be more conservative in this regard than urban ones.

Thus, each newspaper (hence, each editorial) has been classified in terms of the population of the community in which it is published.

Table 14-11 presents some hypothetical data describing the editorial policies of rural and urban newspapers. Note that the unit of analysis in this

TABLE 14-11 Hypothetical Data Regarding Newspaper Editorials on the Legalization of Marijuana

Editorial Policy Toward Legalizing Marijuana	Community Size	
	Under 100,000	Over 100,000
Favorable	11%	32%
Neutral	29	40
Unfavorable	60	28
100% =	(127)	(438)

example is the individual editorial. Table 14-11 tells us that there were 127 editorials about marijuana in our sample of newspapers published in communities with populations under 100,000. (*Note:* This cutting point is chosen for simplicity of illustration and does not mean that *rural* refers to a community of less than 100,000 in any absolute sense.) Of these, 11 percent (14 editorials) were favorable toward legalization of marijuana, 29 percent were neutral, and 60 percent were unfavorable. Of the 438 editorials that appeared in our sample of newspapers published in communities of more than 100,000 residents, 32 percent (140 editorials) were favorable toward legalizing marijuana, 40 percent were neutral, and 28 percent were unfavorable.

When we compare the editorial policies of rural and urban newspapers in our imaginary study, we find—as expected—that rural newspapers are less favorable toward the legalization of marijuana than are urban newspapers. We determine this by noting that a larger percentage (32 percent) of the urban editorials were favorable than the percentage of rural ones (11 percent). We might note as well that more rural than urban editorials were unfavorable (60 percent compared with 28 percent). Note that this table assumes that the size of a community might affect its newspapers' editorial policies on this issue, rather than that editorial policy might affect the size of communities.

Constructing and Reading Bivariate Tables

Let's now review the steps involved in the construction of explanatory bivariate tables:

1. The cases are divided into groups according to the attributes of the independent variable.
2. Each of these subgroups is then described in terms of attributes of the dependent variable.
3. Finally, the table is read by comparing the independent variable subgroups with one another in terms of a given attribute of the dependent variable.

Let's repeat the analysis of gender and attitude on sexual equality following these steps. For the reasons outlined previously, *gender* is the independent variable; *attitude toward sexual equality* constitutes the dependent variable. Thus, we proceed as follows:

1. The cases are divided into men and women.
2. Each gender subgrouping is described in terms of approval or disapproval of sexual equality.
3. Men and women are compared in terms of the percentages approving of sexual equality.

In the example of editorial policies regarding the legalization of marijuana, *size of community* is the independent variable, and a *newspaper's editorial policy* the dependent variable. The table would be constructed as follows:

1. Divide the editorials into subgroups according to the sizes of the communities in which the newspapers are published.
2. Describe each subgroup of editorials in terms of the percentages favorable, neutral, or unfavorable toward the legalization of marijuana.
3. Compare the two subgroups in terms of the percentages favorable toward the legalization of marijuana.

Bivariate analyses typically have an explanatory causal purpose. These two hypothetical examples have hinted at the nature of causation as social scientists use it.

Tables such as the ones we've been examining are commonly called **contingency tables:** Values of the dependent variable are contingent on (depend on) values of the independent variable. Although contingency tables are common in social science, their format has never been standardized. As a result, you'll find a variety of formats in research literature. As long as a table is easy to read and interpret, there's probably no reason to strive for standardization. However, there are several guidelines that you should follow in the presentation of most tabular data.

1. A table should have a heading or a title that succinctly describes what is contained in the table.
2. The original content of the variables should be clearly presented—in the table itself if at all possible or in the text with a paraphrase in the table. This information is especially critical when a variable is derived from responses to an attitudinal question, because the meaning of the responses will depend largely on the wording of the question.
3. The attributes of each variable should be clearly indicated. Though complex categories will have to be abbreviated, their meaning should be clear in the table and, of course, the full description should be reported in the text.
4. When percentages are reported in the table, the base on which they are computed should be indicated. It's redundant to present all the raw numbers for each category, because these could be reconstructed from the percentages and the bases. Moreover, the presentation of both numbers and percentages often confuses a table and makes it more difficult to read.
5. If any cases are omitted from the table because of missing data ("no answer," for example), their numbers should be indicated in the table.

INTRODUCTION TO MULTIVARIATE ANALYSIS

The logic of **multivariate analysis,** or the analysis of more than two variables simultaneously, can be seen as an extension of bivariate analysis.

Specifically, we can construct multivariate tables on the basis of a more complicated subgroup description by following essentially the same steps outlined for bivariate tables. Instead of one independent variable and one dependent variable, however, we'll have more than one independent variable. Instead of explaining the dependent variable on the basis of a single independent variable, we'll seek an explanation through the use of more than one independent variable.

Let's return to the example of church attendance. Suppose we believe that age would also affect such behavior (Glock's Comfort Hypothesis suggests that older people are more religious than younger people). As the first step in table construction, we would divide the total sample into subgroups based on the attributes of both independent variables simultaneously: younger men, older men, younger women, and older women. Then the several subgroups would be described in terms of the dependent variable, *church attendance,* and comparisons would be made. Table 14-12, from an analysis of the 1973–1993 General Social Survey data, is the result.

Table 14-12 has been percentaged down and therefore should be read across. The interpretation of this table warrants several conclusions:

1. Among both men and women, older people attend church more often than do younger people. Among women, 32 percent of those under 40 and 48 percent of those 40 and older attend church weekly. Among men, the respective figures are 24 and 33 percent.

TABLE 14-12 **Multivariate Relationship: Church Attendance, Gender, and Age**

"How often do you attend religious services?"				
	Under 40		40 and Older	
	Men	Women	Men	Women
About weekly*	24%	32%	33%	48%
Less often	76	68	67	52
100% =	(325)	(383)	(323)	(452)

*About weekly = "More than once a week," "Weekly," and "Nearly every week."
Source: A random sample of GSS respondents in 1973, 1983, and 1993.

2. Within each age group, women attend more frequently than men. Among those respondents under 40, 32 percent of the women attend weekly, compared with 24 percent of the men. Among those 40 and over, 48 percent of the women and 33 percent of the men attend weekly.

3. As measured in the table, gender and age appear to have comparable effects on attendance at religious services.

4. Age and gender have independent effects on church attendance. Within a given attribute of one independent variable, different attributes of the second still affect behaviors.

5. Similarly, the two independent variables have a cumulative effect on behaviors. Older women attend the most often (48 percent), and younger men attend the least often (24 percent).

Before we conclude this section, it will be useful to note an alternative format for presenting such data. Several of the tables presented in this chapter are somewhat inefficient. When the dependent variable, *church attendance,* is dichotomous (having exactly two attributes), knowing one attribute permits the reader to reconstruct the other easily. Thus, if we know that 32 percent of the women under 40 attend church weekly, then we know automatically that 68 percent attend less often. So reporting the percentages who attend less often is unnecessary.

On the basis of this recognition, Table 14-12 could be presented in the alternative format of Table 14-13. In Table 14-13, the percentages of people saying they attend church about weekly are reported in the cells representing the intersections of the two independent variables. The numbers presented in parentheses below each percentage represent the number of cases on which the percentages are based. Thus, for example, the reader knows there are 383 women under 40 years of age in the sample, and 32 percent of them attend church weekly. We can calculate from this that 123 of those 383 women attend weekly and that the other 260 younger women (or 68 percent) attend less frequently. This new table is easier to read than the former one, and it does not sacrifice any detail.

For another simple example of multivariate analysis, let's return to the issue of gender and income. Many explanations have been advanced to account for the long-standing pattern of women in the labor force earning less than men. One explanation is that, because of traditional family patterns, women as a group have participated less in the labor force and many only began working outside the home after completing certain child-rearing tasks. Thus, women as a group will probably have less seniority at work than will men, and income increases with seniority. A 1984 study by the Census Bureau showed this reasoning to be *partly* true, as Table 14-14 shows.

Table 14-14 indicates, first of all, that job tenure does indeed affect income. Among both men and women, those with more years on the job earn more. This is seen by reading down the first two columns of the table.

The table also indicates that women earn less than men, regardless of job seniority. This can be seen by comparing average wages across the rows of the table, and the ratio of women-to-men wages is shown in the third column. Thus, years on the job is an important determinant of earnings, but seniority does not adequately explain the pattern of women earning less than men. In fact, we see that women with 10 or more years on the job earn substantially less ($7.91/hour) than men with less than two years ($8.46/hour).

While years on the job does not fully explain the difference between men's and women's pay, there are other possible explanations: level of education, child care responsibilities, and so forth. The researchers who calculated Table 14-14 also examined some of the other variables that might rea-

TABLE 14-13 **A Simplification of Table 14-12**

	Percent Who Attend about Weekly	
	Men	Women
Under 40	24 (325)	32 (383)
40 and older	33 (323)	48 (452)

TABLE 14-14 **Gender, Job Tenure, and Income (Full-time workers 21–64 years of age)**

Years working with current employer	Average hourly income		Women/Men ratio
	Men	Women	
Less than 2 years	$ 8.46	$6.03	.71
2 to 4 years	$ 9.38	$6.78	.72
5 to 9 years	$10.42	$7.56	.73
10 years or more	$12.38	$7.91	.64

Source: U.S. Bureau of the Census, Current Population Reports, Series P-70, No. 10, *Male-Female Differences in Work Experience, Occupation, and Earning, 1984* (Washington, DC: U.S. Government Printing Office, 1987): 4.

sonably explain the differences in pay without representing gender discrimination. In addition to the number of years with current employer, the variables they considered included these:

- Number of years in the current occupation
- Total years of work experience (any occupation)
- Whether they have usually worked full time
- Marital status
- Size of city or town they live in
- Whether covered by a union contract
- Type of occupation
- Number of employees in the firm
- Whether private or public employer
- Whether they left previous job involuntarily
- Time spent between current and previous job
- Race
- Whether they have a disability
- Health status
- Age of children
- Whether they took an academic curriculum in high school
- Number of math, science, and foreign language classes in high school
- Whether they attended private or public high school
- Educational level achieved
- Percentage of women in the occupation
- College major

Each of the variables listed here might reasonably affect earnings and, if women and men differ in these regards, could help to account for male/female income differences. When all these vari-

A QUANDARY **REVISITED**

This chapter began with a question about whether anything meaningful or useful could be learned from the analysis of data that have been stripped of many details in order to permit statistical manipulation. The answer, we've seen, is an unqualified "yes."

Quantitative analysis can be a tool for social change. For instance, calculating the average incomes of men and women, whites and minorities can demonstrate the inequalities that exist for people doing exactly the same job. Such quantitative analyses can overpower anecdotal evidence about particular women or minorities earning large salaries. We also seen that quantitative analyses of qualitative phenomena, such as voting intentions, can be done with precision and utility.

The key lesson is that both qualitative and quantitative research are legitimate and powerful approaches to understanding social life. They are particularly useful, moreover, when used together.

ables were taken into account, the researchers were able to account for 60 percent of the discrepancy between the incomes of men and women. The remaining 40 percent, then, is a function of other "reasonable" variables and/or prejudice. This kind of conclusion can be reached only by examining the effects of several variables at the same time—that is, through multivariate analysis.

I hope this example shows how the logic implicit in day-to-day conversations can be represented and tested in a quantitative data analysis like this. As another example, consider the common observation that minority group members are more likely to be denied bank loans than are white applicants. A counterexplanation might be that the minority applicants in question were more likely to have had a prior bankruptcy or that they had less collateral to guarantee the requested loan—both reasonable bases for granting or denying loans. However, the kind of multivariate analysis we've just examined could easily resolve the disagreement.

Let's say we look only at those who have not had a prior bankruptcy and who have a certain level of collateral. Are whites and minorities equally likely to get the requested loan? We could conduct the same analysis in subgroups determined by level of collateral. If whites and minorities were equally likely to get their loans in each of the subgroups, we would need to conclude that there was no ethnic discrimination. If minorities were still less likely to get their loans, however, that would indicate that bankruptcy and collateral differences were not the explanation—strengthening the case that discrimination was at work.

All this should make it clear that social research can play a powerful role in serving in the human community. It can help us determine the current state of affairs and can often point the way to where we want to go.

Welcome to the world of social research!

Main Points

- All data are initially qualitative: they must be quantified to permit statistical analysis.

- Some data, such as age and income, are intrinsically numerical.

- Often, quantification involves coding into categories that are then given numerical representations.

- Researchers may use existing coding schemes, such as the Census Bureau's categorization of occupations, or develop their own coding categories. In either case, the coding scheme must be appropriate to the nature and objectives of the study.

- A codebook is the document that describes the identifiers assigned to different variables and the codes assigned to represent the attributes of those variables.

- Univariate analysis is the analysis of a single variable. Because univariate analysis does not involve the relationships between two or more variables, its purpose is descriptive rather than explanatory.

- Several techniques allow researchers to summarize their original data to make them more manageable while maintaining as much of the original detail as possible. Frequency distributions, averages, grouped data, and measures of dispersion are all ways of summarizing data concerning a single variable.

- Subgroup comparisons can be used to describe similarities and differences among subgroups with respect to some variable.

- Bivariate analysis focuses on relationships between variables rather than comparisons of groups. Bivariate analysis explores the statistical association between the independent variable and the dependent variable. Its purpose is usually explanatory rather than merely descriptive.

- The results of bivariate analyses often are presented in the form of contingency tables, which are constructed to reveal the effects of the independent variable on the dependent variable.

❑ Multivariate analysis is a method of analyzing the simultaneous relationships among several variables. It may also be used to understand the relationship between two variables more fully.

❑ While the topics discussed in this chapter are primarily associated with quantitative research, the logic and techniques involved can also be valuable to qualitative researchers.

Key Terms

quantitative analysis	dispersion
codebook	standard deviation
univariate analysis	continuous variable
frequency distribution	discrete variable
average	bivariate analysis
mean	contingency tables
mode	multivariate analysis
median	

Review Questions

1. How might the various majors at your college be classified into categories? Create a coding system that would allow you to categorize them according to some meaningful variable. Then create a different coding system, using a different variable.

2. How many ways could you be described in numerical terms? What are some of your intrinsically numerical attributes? Could you express some of your qualitative attributes in quantitative terms?

3. How would you construct and interpret a contingency table from the following information: 150 Democrats favor raising the minimum wage, and 50 oppose it; 100 Republicans favor raising the minimum wage, and 300 oppose it.

4. Using the hypothetical data in the following table, how would you construct and interpret tables showing the following?

   a. The bivariate relationship between age and attitude toward abortion

   b. The bivariate relationship between political orientation and attitude toward abortion

   c. The multivariate relationship linking age, political orientation, and attitude toward abortion

Age	Political Orientation	Attitude toward Abortion	Frequency
Young	Liberal	Favor	90
Young	Liberal	Oppose	10
Young	Conservative	Favor	60
Young	Conservative	Oppose	40
Old	Liberal	Favor	60
Old	Liberal	Oppose	40
Old	Conservative	Favor	20
Old	Conservative	Oppose	80

Additional Readings

Babbie, Earl, Fred Halley, and Jeanne Zaino. 2000. *Adventures in Social Research.* Newbury Park, CA: Pine Forge Press. This book introduces you to the analysis of social research data through SPSS for Windows. Several of the basic statistical techniques used by social researchers are discussed and illustrated.

Bernstein, Ira H., and Paul Havig. 1999. *Computer Literacy: Getting the Most from Your PC.* Thousand Oaks, CA: Sage. Here's a quick overview of the various ways social scientists use computers, including many common applications programs.

Davis, James. 1971. *Elementary Survey Analysis.* Englewood Cliffs, NJ: Prentice-Hall. An extremely well-written and well-reasoned introduction to analysis. In addition to covering the materials just presented in Chapter 14, Davis's book is well worth reading in terms of measurement and statistics.

Ferrante, Joan, and Angela Vaughn. 1999. *Let's Go Sociology: Travels on the Internet.* Belmont, CA: Wadsworth. This accessible little book gives an excellent introduction to the Internet and suggests many Web sites of interest to social researchers.

Lewis-Beck, Michael. 1995. *Data Analysis: An Introduction.* Volume 103 in the Quantitative Application in the Social Sciences series. Thousand Oaks, CA: Sage. This is a very popular short book that makes statistical language accessible to the novice. You should enjoy the clarity of explanations and the thorough use of examples.

Newton, Rae R., and Kjell Erik Rudestam. 1999. *Your Statistical Consultant: Answers to Your Data Analysis Questions.* Thousand Oaks, CA: Sage. Excellent reader-friendly manual that will answer all sorts of questions you have or will have as soon as you begin to analyze quantitative data.

Ziesel, Hans. 1957. *Say It with Figures.* New York: Harper & Row. An excellent discussion of table construction and other elementary analyses. Though many years old, this is still perhaps the best available presentation of that specific topic. It is eminently readable and understandable and has many concrete examples.

Multimedia Resources

The Wadsworth Sociology Resource Center: Virtual Society
http://sociology.wadsworth.com/

Visit the companion Web site for the second edition of *The Basics of Social Research* to access a wide range of student resources. Begin by clicking on the Student Resources section of the book's Web site to access the following study tools:

- eBabbie Resource Center
- Planning a Research Project
- Doing Data Analysis
- Statistics Review

- Flash Cards
- Internet Links and Exercises
- InfoTrac College Edition: Exercises
- Quizzes

Visit the **eBabbie Resource Center** for an overview of each chapter and helpful online tutorials. Find information on budgeting and step-by-step examples of model research projects at **Planning a Research Project.** Learn how to use quantitative and qualitative data analysis programs at **Doing Data Analysis,** and brush up on your statistics at **Statistics Review.** You can also further your study by accessing **Internet Links and Exercises** related to chapter materials, **Flash Cards, Quizzes,** and many other learning tools.

InfoTrac College Edition
http://www.infotrac-college.com/ wadsworth/access.html

Access the latest news and journal articles with InfoTrac College Edition, an easy-to-use online database of reliable, full-length articles from hundreds of top academic journals. Conduct an electronic search using the following search terms:

Bivariate analysis
Central tendency
Continuous variable
Discrete variable
Multivariate analysis technique
Univariate analysis

APPENDIXES

APPENDIX A

Using the Library

INTRODUCTION

We live in a world filled with social science research reports. Our daily newspapers, magazines, professional journals, alumni bulletins, club newsletters—virtually everything we pick up to read may carry reports dealing with a particular topic. For formal explorations of a topic, of course, the best place to start is still a good college or university library. Today, there are two major approaches to finding library materials: the traditional paper system and the electronic route. Let's begin with the traditional method and then examine the electronic option.

GETTING HELP

When you want to find something in the library, your best friend is the reference librarian, who is specially trained to find things in the library. Some libraries have specialized reference librarians—for the social sciences, humanities, government documents, and so forth. Find the librarian who specializes in your field. Make an appointment. Tell the librarian what you're interested in. He or she will probably put you in touch with some of the many available reference sources.

REFERENCE SOURCES

You've probably heard the expression "information explosion." Your library is one of the main battlefields. Fortunately, a large number of reference volumes offer a guide to the information that's available.

Books in Print   This volume lists all the books currently in print in the United States—listed separately by author and by title. Out-of-print books can often be found in older editions of *Books in Print*.

Readers' Guide to Periodical Literature   This annual volume with monthly updates lists articles published in many journals and magazines. Because the entries are organized by subject matter, this is an excellent source for organizing your reading on a particular topic. Figure A-1 presents a sample page from the *Readers' Guide*.

In addition to these general reference volumes, you'll find a great variety of specialized references. Here are just a few:

- Sociological Abstracts
- Psychological Abstracts
- Social Science Index
- Social Science Citation Index
- Popular Guide to Government Publications
- New York Times Index
- Facts on File
- Editorial Research Reports
- Business Periodicals Index
- Monthly Catalog of Government Publications
- Public Affairs Information Service Bulletin
- Education Index
- Applied Science and Technology Index
- A Guide to Geographic Periodicals
- General Science Index
- Biological and Agricultural Index
- Nursing and Applied Health Index
- Nursing Studies Index
- Index to Little Magazines
- Popular Periodical Index
- Biography Index

MUSIC—*cont.*

Study and teaching
See also
Guitar—Study and teaching

Themes, motives, etc.
See also
Automobiles in music

Theory
See also
Atonality

Japan
The Japanese and Western music. L. Futoransky. il *The Courier (Unesco)* 40:38+ D '87

MUSIC, AMERICAN
See also
Jazz music

MUSIC, ELECTRONIC
See also
Computers—Musical use
Musical instruments, Electronic

MUSIC AND STATE
Viewpoint [government subsidies of opera] J. L. Poole. Opera News 52:4 F 13 '88

Soviet Union
Gorbachev sets the beat for Soviet rock. il *U.S. News & World Report* 104:8-9 F 8 '88

MUSIC AND THE BLIND
Call him Doc [D. Watson] F. L. Schultz. il pors *Country Journal* 15:44-53 F '88

MUSIC AND THE HANDICAPPED
See also
Guitarists, Handicapped

MUSIC CORPORATION OF AMERICA *See* MCA Inc.

MUSIC CRITICS AND CRITICISM
See also
Opera reviews

MUSIC FESTIVALS

Austria
Bregenz. H. Koegler. il *Opera News* 52:38 F 13 '88

Germany (West)
Bayreuth. J. H. Sutcliffe. il *Opera News* 52:36 Ja 30 '88

Great Britain
Buxton. E. Forbes. *Opera News* 52:40-1 F 13 '88

Italy
Torre del Lago (Puccini Festival) M. Hamlet-Mets. *Opera News* 52: 38-40 F 13 '88

Pennsylvania
Philadelphia [American Music Theater Festival] R. Baxter, *Opera News* 52:34 Ja 30 '88

MUSICAL COMEDIES, REVUES, ETC. *See* Musicals, revues, etc.

MUSICAL INSTRUMENTS, ELECTRONIC
It's alive with the sound of—well, just about everything (Synclavier synthesizer) L. Helm. il *Business Week* p75 F 8 '88

MUSICAL INSTRUMENTS INDUSTRY
See also
New England Digital Corporation

MUSICALS, REVUES, ETC.

Choreography
See Choreography

Reviews
Single works
Anything goes
Dance Magazine il 62:52-7 Ja '88. J. Gruen
Cabaret
Dance Magazine 62:73-4 Ja '88. H. M. Simpson
The chosen
The Nation 246:176 F 6 '88. T. M Disch
Into the woods
Dance Magazine 62:64 Ja '88. K. Grubb
Oil City Symphony
The Nation 246:175-6 F 6 '88. T. M. Disch
The phantom of the opera
Life il 11:88-92 F '88. M. Stasio

Maclean's il 101:51 F 8 '88. L. Black
New York il 21:89-90 F 8 '88. J. Simon
The New Yorker 63:97-8 F 8 '88. M. Kramer
Newsweek il por 111:68-70+ F 8 '88. J. Kroll
Rolling Stone il p26 F 25 '88. D. Handelman
Time il 131:83-4 F 8 '88. W. A. Henry

Stage setting and scenery
High-tech magic: follow that gondola [Phantom of the opera] J. Kroll. il *Newsweek* 111:70 F 8 '88

Writing
Changing the face of Broadway [A. Lloyd Webber] M. Stasio. il pors *Life* 11:88-92 F '88

MUSICIANS
See also
Drugs and musicians
Rock musicians

MUSKE, CAROL, 1945-
Skid [poem] *The New Yorker* 63:38 F 8 '88

MUSLIMS
See also
Islam

Afghanistan
Beyond the Afghan stalemate. L. Komisar. il *The New Leader* 71:5-6 Ja 11-25 '88

Middle East
The Islamic resurgence: a new phase? R. Wright. bibl f *Current History* 87:53-6+ F '88

MUTATION
See also
Transposons

MUTUAL FUNDS *See* Investment trusts
MUTUALISM (BIOLOGY) *See* Symbiosis
MUZIEKTHEATER (AMSTERDAM, NETHERLANDS)
See Opera houses

MYASTHENIA GRAVIS
Suzanne Rogers: "I looked at my face and thought, 'Who'd hire a freak?" A. W. Petrucelli. pors *Redbook* 170:104+F '88

MYCOBACTERIAL DISEASES
See also
Tuberculosis

MYCOTOXINS *See* Toxins and antitoxins

N

N. W. AYER & SON, INC.
Ayer to the throne [Burger King ad campaign] B. Kanner. il *New York* 21:24+ F 29 '88

NADIS, STEVEN J.
Robot observatories. il *Omni (New York, N.Y.)* 10:24+ Ja '88

NAEP *See* National Assessment of Educational Progress

NAKAGAMI, KENJI, 1946-
about
Two contemporary writers. D. Palmé. *The Courier (Unesco)* 40:44 D '87

NAKED SHORT SELLING *See* Securities—Short selling
NANDINA
Nandina does the unexpected. il *Southern Living* 23:50 Ja '88

NAPLES (ITALY)

Music
See also
Opera—Italy

NARCOTIC ADDICTS *See* Drug abuse

NARCOTICS LAWS AND REGULATIONS
See also
Boats in narcotics regulation
Robots in narcotics regulation

Austria
A five-year penalty call [Czech hockey legend J. Bubla serving prison sentence for smuggling heroin] J. Holland. il por *Maclean's* 101:6 F 8 '88

Colombia
Battling the drug lords [Attorney General C. Hoyos murdered] E. Tolmie. il *Maclean's* 101:26 F 8 '88

FIGURE A-1   **A Page from the *Reader's Guide to Periodical Literature***

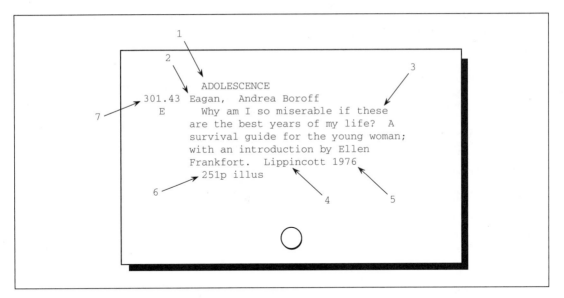

FIGURE A-2 **Sample Subject Catalog Card**

Source: Lilian L. Shapiro, *Teaching Yourself in Libraries* (New York: H. W. Wilson, 1978), 3–4. Used by permission.

- Congressional Quarterly Weekly Report
- Library Literature
- Bibliographic Index

USING THE STACKS

Serious research usually involves using the stacks, where most of the library's books are stored. This section provides information about finding books in the stacks.

The Card Catalog

In the traditional paper system, the card catalog is the main reference system for finding out where books are stored. Each book is described on three separate 3-by-5 cards. The cards are then filed in three alphabetical sets. One set is arranged by author, another by title, and the third by subject matter.

If you want to find a particular book, you can look it up in either the author file or the title file. If you only have a general subject area of interest, you should thumb through the subject catalog.

Figure A-2 presents a sample card in the card catalog. Notice the following elements:

1. Subject heading (always in capital letters)
2. Author's name (last name, first name)
3. Title of the book
4. Publisher
5. Date of publication
6. Number of pages in the book plus other information (such as whether the book contains illustrations)
7. Call number needed to find a nonfiction book on the library shelves; fiction is generally found in alphabetical order by the author's name

Library of Congress Classification

Here's a useful strategy to use when you're researching a topic. Once you've identified the call number for a particular book in your subject area, go to the stacks, find that book, and look over the other books on the shelves near it. Because the books are arranged by subject matter, this method will help you locate relevant books you didn't know about.

Alternatively, you may want to go directly to the stacks and look at books in your subject area. In most libraries, books are arranged and numbered according to a subject matter classification developed by the Library of Congress. (Some follow the Dewey decimal system.) The following is a selected list of Library of Congress categories.

Library of Congress Classifications (partial)

A GENERAL WORKS
B PHILOSOPHY, PSYCHOLOGY, RELIGION

	B-BD	Philosophy
	BF	Psychology
	BL-BX	Religion

C HISTORY-AUXILIARY SCIENCES
D HISTORY (except America)

	DA-DR	Europe
	DS	Asia
	DT	Africa

E-F HISTORY (America)

	E	United States
	E51–99	Indians of North America
	E185	Negroes in the United States
	F101–1140	Canada
	F1201–3799	Latin America

G GEOGRAPHY-ANTHROPOLOGY

	G-GF	Geography
	GC	Oceanology and oceanography
	GN	Anthropology
	GV	Sports, amusements, games

H SOCIAL SCIENCES

	H62.B2	*The Basics of Social Research*
	HB-HJ	Economics and business
	HM-HX	Sociology

J POLITICAL SCIENCE

	JK	United States
	JN	Europe
	JQ	Asia, Africa
	JX	International relations

K LAW
L EDUCATION
M MUSIC
N FINE ARTS

	NA	Architecture
	NB	Sculpture
	NC	Graphic arts
	ND	Painting
	NE	Engraving
	NK	Ceramics, textiles

P LANGUAGE AND LITERATURE

	RE	English language
	PG	Slavic language
	PJ-PM	Oriental language
	PN	Drama, oratory, journalism
	PQ	Romance literature
	PR	English literature
	PS	American literature
	PT	Germanic literature

Q SCIENCE

	QA	Mathematics
	QB	Astronomy
	QC	Physics
	QD	Chemistry
	QE	Geology
	QH-QR	Biology

R MEDICINE

	RK	Dentistry
	RT	Nursing

S AGRICULTURE—PLANT AND ANIMAL INDUSTRY
T TECHNOLOGY

	TA-TL	Engineering
	TR	Photography

U MILITARY SCIENCE
V NAVAL SCIENCE
Z BIBLIOGRAPHY AND LIBRARY SCIENCE

COMPUTERIZED LIBRARY FILES

Increasingly, library materials are catalogued electronically. While there are different computerized library systems, here's a typical example of how they work.

Sitting at a computer terminal—in the library, at a computer lab, or at home—you can type the title of a book and in seconds see a video display of a catalog card. If you want to explore the book

AU Kinloch-Graham-C.
TI The Changing Definition and Content of Sociology in Introductory Textbooks, 1894–1981.
SO International Review of Modern Sociology. 1984, 14, 1, spring, 89–103.
DE Sociology-Education; (D810300). Textbooks; (D863400).
AB An analysis of 105 introductory sociology textbooks published between 1894 & 1981 reveals historical changes in definitions of the discipline & major topics in relation to professional factors & changing societal contexts. Predominant views of sociology in each decade are discussed, with the prevailing view being that of a "scientific study of social structure in order to decrease conflict & deviance, thereby increasing social control." Consistencies in this orientation over time, coupled with the textbooks' generally low sensitivity to social issues, are explored in terms of their authors' relative homogeneity in age & educational backgrounds. 1 Table, 23 References. Modified HA.

FIGURE A-3   **A Research Summary from *Sociological Abstracts***

further, you can type an instruction at the terminal and see an abstract of the book.

Alternatively, you might type a subject name and see a listing of all the books and articles written on that topic. You could skim through the list and indicate which ones you want to see.

Many libraries today provide access to periodicals and books via the World Wide Web. Your library's computerized system should allow you to see which materials are available online. Sometimes whole dissertations or books can be downloaded. Most likely, your largest local library provides document delivery services to its members.

Many college libraries now have access to the Educational Resources Information Center (ERIC). This computer-based system allows you to search through hundreds of major educational journals to find articles published in the subject area of your interest (within the field of education). As a rule, each library Web site should have a list of the databases by discipline that you can visit, which may help you limit the number of titles related to a specific keyword. Make sure you narrow your search by limiting, for instance, language or period of the publication. Once you identify the articles you're interested in, the computer will print out their abstracts.

Of particular value to social science researchers, the publications *Sociological Abstracts* and *Psychological Abstracts* present summaries of books and articles—often prepared by the original authors—so that you can locate a great many relevant references easily and effectively. As you find

relevant references, you can track down the original works and see the full details. The summaries are available in both written and computerized forms.

Figure A-3 contains the abstract of an article obtained in a computer search of *Sociological Abstracts*. I began by asking for a list of articles dealing with sociology textbooks. After reviewing the list, I asked to see the abstracts of each of the listed articles. Here's an example of what I received seconds later: an article by the sociologist Graham C. Kinloch, published in the *International Review of Modern Sociology*.

In case the meaning of the abbreviations in Figure A-3 isn't immediately obvious, I should explain that AU is author; TI is title; SO is the source or location of the original publication; DE indicates classification codes under which the abstract is referenced; and AB is the abstract. The computerized availability of resources such as *Sociological Abstracts* provides a powerful research tool for modern social scientists. You'll have the option to download or print, with or without the abstract, any title you find through the library's browsers.

If a document is not available in the library itself or via the Web, you always have the resource of interlibrary loans, which often are free. Libraries don't own every document or multimedia material (CD-ROMs, videocassettes, laser disks, films), but many have loan agreements that can serve your needs. You need to be aware of the time you can expect between your request and actually receiving the book or article. In the case of a book that is located in an other library close by, for example, it may be faster for you to get it directly yourself. The key to a good library search is to become well in-formed. So start networking with librarians, faculty, and peers!

Additional Readings

Bart, Pauline, and Linda Frankel. 1986. *The Student Sociologist's Handbook.* New York: Random House. A survival kit for doing sociological research. Contains a step-by-step guide for writing research papers; chapters on periodicals, abstract and indexing services, bibliographies, bibliographical aids, and other secondary sources; and a complete guide to governmental and nongovernment sources of data. Special section on sex roles and women's studies.

Li, Tze-chung. 1990. *Social Science Reference Sources: A Practical Guide.* Westport, CT: Greenwood Press. Lists and describes all types of reference materials, including databases and archives as well as published sources. Organized into two parts: social sciences in general and by discipline.

Richlin-Klonsky, Judith, and Ellen Strenski, eds. 1998. *A Guide to Writing Sociology Papers.* New York: St. Martin's Press. This is a great little book with good advice on doing research. It's particularly useful for those who are new to sociology or other social science disciplines and have to learn about the most rudimentary aspects of research.

APPENDIX B

The Research Report

INTRODUCTION

This book has considered the variety of activities that compose the *doing* of social research. In this appendix, we'll turn to an often neglected subject: reporting the research to others. Unless research is properly communicated, all the efforts devoted to previously discussed procedures will go for naught. This means, first and foremost, that good social reporting requires good English (or Spanish or whatever language you use). Whenever we ask the figures "to speak for themselves," they tend to remain mute. Whenever we use unduly complex terminology or construction, communication is reduced. Every researcher should read and reread (at approximately three-month intervals) an excellent small book by William Strunk, Jr., and E. B. White, *The Elements of Style.** If you do this faithfully, and if even 10 percent of the contents rub off, you stand a good chance of making yourself understood and your findings appreciated.

Scientific reporting has several functions. First, the report communicates a body of specific data and ideas. The report should provide those specifics clearly and with sufficient detail to permit an informed evaluation by others. Second, the scientific report should be viewed as a contribution to the general body of scientific knowledge. While remaining appropriately humble, you should always regard your research report as an addition to what we know about social behavior. Finally, the report should stimulate and direct further inquiry.

---

*Fourth ed. (New York: Macmillan, 1999). Here's another useful reference on writing: R. W. Birchfield, *The New Fowler's Modern English Usage,* 3rd ed. (New York: Oxford University Press, 1998).

SOME BASIC CONSIDERATIONS

Despite these general guidelines, different reports serve different purposes. A report appropriate for one purpose might be wholly inappropriate for another. This section deals with some of the basic considerations in this regard.

Audience

Before drafting your report, ask yourself who you hope will read it. Normally you should make a distinction between scientists and general readers. If the report is written for the former, you may make certain assumptions about their existing knowledge and therefore summarize certain points rather than explain them in detail. Similarly, you may use more technical language than would be appropriate for a general audience.

At the same time, remain aware that any science has its factions and cults. Terms, assumptions, and special techniques familiar to your immediate colleagues may only confuse other scientists. The sociologist of religion writing for a general sociology audience, for example, should explain previous findings in more detail than he or she would if addressing an audience of sociologists of religion.

Form and Length of Report

My comments here apply to both written and oral reports. Each form, however, affects the nature of the report.

It's useful to think about the variety of reports that might result from a research project. To begin, you may wish to prepare a short *research note* for publication in an academic or technical journal. Such reports are approximately one to five pages long (typed, double-spaced) and should be concise and direct. In a small amount of space, you can't present the state of the field in any detail, so your methodological notes must be abbreviated. Basically, you should tell the reader why you feel a brief note is justified by your findings, then tell what those findings are.

Often researchers must prepare reports for the sponsors of their research. These reports may vary greatly in length. In preparing such a report, you should bear in mind its audience—scientific or lay—and their reasons for sponsoring the project in the first place. It is both bad politics and bad manners to bore the sponsors with research findings that have no interest or value to them. At the same time, it may be useful to summarize how the research has advanced basic scientific knowledge (if it has).

Working papers or *monographs* are another form of research reporting. In a large and complex project especially, you'll find comments on your analysis and the interpretation of your data useful. A working paper comprises a tentative presentation with an implicit request for comments. Working papers can also vary in length, and they may present all of the research findings of the project or only a portion of them. Because your professional reputation is not at stake in a working paper, feel free to present tentative interpretations that you can't altogether justify—identifying them as such and asking for evaluations.

Many research projects result in papers delivered at professional meetings. Often, these serve the same purpose that working papers do. You can present findings and ideas of possible interest to your colleagues and ask for their comments. Although the length of *professional papers* may vary depending on the organization of the meetings, it's best to say too little rather than too much. Although a working paper may ramble somewhat through tentative conclusions, conference participants should not be forced to sit through an oral unveiling of the same. Interested listeners can always ask for more details later, and uninterested ones can gratefully escape.

Probably the most popular research report is the *article* published in an academic journal. Again, lengths vary, and you should examine the lengths of articles previously published by the journal in question. As a rough guide, however, 25 typed pages is a good length. A subsequent section on the organization of the report is primarily based on the structure of a journal article, so I shall say no more at this point except to indicate that student term papers should follow this model. As a general rule, a term paper that would make a good journal article also makes a good term paper.

A *book,* of course, represents the most prestigious form of research report. It has the length and detail of the working paper, but it's more polished. Because publishing research findings as a book lends them greater substance and worth, you have a special obligation to your audience. Although some colleagues may provide comments, possibly leading you to revise your ideas, other readers may be led to accept your findings uncritically.

Aim of the Report

Earlier in this book, we considered the different purposes of social research projects. In preparing your report, you should keep these different purposes in mind.

Some reports focus primarily on the *exploration* of a topic. As such, their conclusions are tentative and incomplete. You should clearly indicate to your audience the exploratory aim of the study and point to the shortcomings of the particular project. An exploratory report points the way to more-refined research on the topic.

Most research reports have a descriptive element reflecting the descriptive purpose of the studies they document. Carefully distinguish those descriptions that apply only to the sample and those that apply to the population. Give your audience some indication of the probable range of error in any inferential descriptions you make.

Many reports have an explanatory aim: pointing to causal relationships among variables. Depending on your probable audience, carefully delineate the rules of explanation that lie behind your computations and conclusions. Also, as in the case of description, give your readers some guide to the relative certainty of your conclusions.

Finally, some research reports propose action. For example, the researcher of prejudice may suggest how prejudice can be reduced on the basis of the research findings. This aim often presents knotty problems, however, because the researcher's values and orientations may interfere with his or her proposals. Although it's perfectly legitimate for such proposals to be motivated by personal values, researchers must ensure that the specific actions proposed are warranted by the data. Thus, researchers should be especially careful to spell out the logic by which they move from empirical data to proposed action.

ORGANIZATION OF THE REPORT

Although the organization of reports differs somewhat in terms of form and purpose, a general format for presenting research data can be helpful. The following comments apply most directly to a journal article, but with some modification they apply to most forms of research reports as well.

Purpose and Overview

It's always helpful if you begin with a brief statement of the purpose of the study and the main findings of the analysis. In a journal article, this overview sometimes takes the form of an *abstract* or *synopsis.*

Some researchers find this difficult to do. For example, your analysis may have involved considerable detective work, with important findings revealing themselves only as a result of imaginative deduction and data manipulation. You may wish, therefore, to lead the reader through the same exciting process, chronicling the discovery process with a degree of suspense and surprise. To the extent that this form of reporting gives an accurate picture of the research process, it has considerable instructional value. Nevertheless, many readers may not be interested in following your entire research account, and not knowing the purpose and

general conclusions in advance may make it difficult for them to understand the significance of the study.

An old forensic dictum says, "Tell them what you're going to tell them; tell them; and tell them what you told them." You would do well to follow this dictum.

Review of the Literature

Next, you must indicate where your report fits in the context of the general body of scientific knowledge. After presenting the general purpose of your study, you should bring the reader up to date on the previous research in the area, pointing to general agreements and disagreements among the previous researchers.

In some cases, you may wish to challenge previously accepted ideas. Carefully review the studies that have led to the acceptance of those ideas, then indicate the factors that have not been previously considered or the logical fallacies present in the previous research.

When you're concerned with resolving a disagreement among previous researchers, you should summarize the research supporting one view, then summarize the research supporting the other, and finally suggest the reasons for the disagreement.

Your review of the literature serves a bibliographical function for readers, indexing the previous research on a given topic. This can be overdone, however, and you should avoid an opening paragraph that runs three pages and mentions every previous study in the field. The comprehensive bibliographical function can best be served by a bibliography at the end of the report, and the review of the literature should focus only on those studies that have direct relevance to the present one.

Avoiding Plagiarism

Whenever you're reporting on the work of others, you must be clear about who said what. That is, you must avoid *plagiarism:* the theft of another's words and/or ideas—whether intentional or accidental—and the presentation of those words and ideas as your own. Because this is a common and sometimes unclear problem for college students, let's examine it. Here are the ground rules regarding plagiarism:

- You cannot use another writer's exact words without using quotation marks and giving a complete citation, which indicates the source of the quotation such that your reader could locate that quotation in its original context. As a general rule, taking a passage of eight or more words without citation is a violation of federal copyright laws.
- It's also not acceptable to edit or paraphrase another's words and present the revised version as your own work.
- Finally, it's not even acceptable to present another's ideas as your own—even if you use totally different words to express those ideas.

The following examples should clarify what is or is not acceptable in the use of another's work.

The Original Work

Laws of Growth

Systems are like babies: once you get one, you have it. They don't go away. On the contrary, they display the most remarkable persistence. They not only persist; they grow. And as they grow, they encroach. The growth potential of systems was explored in a tentative, preliminary way by Parkinson, who concluded that administrative systems maintain an average growth of 5 to 6 percent per annum regardless of the work to be done. Parkinson was right so far as he goes, and we must give him full honors for initiating the serious study of this important topic. But what Parkinson failed to perceive, we now enunciate—the general systems analogue of Parkinson's Law.

The System Itself Tends To Grow
At 5 To 6 Percent Per Annum

Again, this Law is but the preliminary to the most general possible formulation, the Big-Bang Theorem of Systems Cosmology.

Systems Tend To Expand To Fill
The Known Universe
(GALL 1975:12–14)

Now let's look at some of the *acceptable* ways you might make use of Gall's work in a term paper.

- **Acceptable:** John Gall, in his work *Systemantics,* draws a humorous parallel between systems and infants: "Systems are like babies: once you get one, you have it. They don't go away. On the contrary, they display the most remarkable persistence. They not only persist; they grow." *
- **Acceptable:** John Gall warns that systems are like babies. Create a system and it sticks around. Worse yet, Gall notes, systems keep growing larger and larger.**
- **Acceptable:** It has also been suggested that systems have a natural tendency to persist, even grow and encroach (Gall 1975:12).

Note that the last format requires that you give a complete citation in your bibliography, as I do in this book. Complete footnotes or endnotes work as well. See the publication manuals of various organizations such as the APA or the ASA, as well as the *Chicago Manual of Style,* for appropriate citation formats.

Here now are some *unacceptable* uses of the same material, reflecting some common errors.

- **Unacceptable:** In this paper, I want to look at some of the characteristics of the social systems we create in our organizations. First, systems are like babies: once you get one, you have it. They don't go away. On the contrary, they display the most remarkable persistence. They not only persist; they grow. [It is unacceptable to directly quote someone else's materials without using quotation marks and giving a full citation.]

---

*John Gall, *Systemantics: How Systems Work and Especially How They Fail* (New York: Quadrangle, 1975), 12–14.
**John Gall, *Systemantics: How Systems Work and Especially How They Fail* (New York: Quadrangle, 1975), 12.

- **Unacceptable:** In this paper, I want to look at some of the characteristics of the social systems we create in our organizations. First, systems are a lot like children: once you get one, it's yours. They don't go away; they persist. They not only persist, in fact: they grow. [It is unacceptable to edit another's work and present it as your own.]
- **Unacceptable:** In this paper, I want to look at some of the characteristics of the social systems we create in our organizations. One thing I've noticed is that once you create a system, it never seems to go away. Just the opposite, in fact: they have a tendency to grow. You might say systems are a lot like children in that respect. [It is unacceptable to paraphrase someone else's ideas and present them as your own.]

Each of the preceding unacceptable examples is an example of plagiarism and represents a serious offense. Admittedly, there are some "gray areas." Some ideas are more or less in the public domain, not "belonging" to any one person. Or you may reach an idea on your own that someone else has already put in writing. If you have a question about a specific situation, discuss it with your instructor in advance.

I've discussed this topic in some detail because, although you must place your research in the context of what others have done and said, the improper use of their materials is a serious offense. Learning to avoid plagiarism is a part of your "coming of age" as a scholar.

Study Design and Execution

A research report containing interesting findings and conclusions can be very frustrating when the reader can't determine the methodological design and execution of the study. The worth of all scientific findings depends heavily on the manner in which the data were collected and analyzed.

In reporting the design and execution of a survey, for example, always include the following: the population, the sampling frame, the sampling method, the sample size, the data-collection method, the completion rate, and the methods of data processing and analysis. Comparable details should be given if other methods are used. The experienced researcher can report these details in a rather short space, without omitting anything required for the reader's evaluation of the study.

Analysis and Interpretation

Having set the study in the perspective of previous research and having described the design and execution of it, you should then present your data. This appendix momentarily will provide further guidelines in this regard. For now, a few general comments are in order.

The presentation of data, the manipulation of those data, and your interpretations should be integrated into a logical whole. It frustrates the reader to discover a collection of seemingly unrelated analyses and findings with a promise that all the loose ends will be tied together later in the report. Every step in the analysis should make sense at the time it is taken. You should present your rationale for a particular analysis, present the data relevant to it, interpret the results, and then indicate where that result leads next.

Summary and Conclusions

Following the forensic dictum mentioned earlier, summarizing the research report is essential. Avoid reviewing every specific finding, but review all the significant ones, pointing once more to their general significance.

The report should conclude with a statement of what you have discovered about your subject matter and where future research might be directed. Many journal articles end with the statement, "It is clear that much more research is needed." This conclusion is probably always true, but it has little value unless you can offer pertinent suggestions about the nature of that future research. You should review the particular shortcomings of your own study and suggest ways those shortcomings might be avoided.

GUIDELINES FOR REPORTING ANALYSES

The presentation of data analyses should provide a maximum of detail without being cluttered. You can accomplish this best by continually examining your report to see whether it achieves the following aims.

If you're using quantitative data, present them so the reader can recompute them. In the case of percentage tables, for example, the reader should be able to collapse categories and recompute the percentages. Readers should receive sufficient information to permit them to compute percentages in the table in the opposite direction from that of your own presentation.

Describe all aspects of a quantitative analysis in sufficient detail to permit a secondary analyst to replicate the analysis from the same body of data. This means that he or she should be able to create the same indexes and scales, produce the same tables, arrive at the same regression equations, obtain the same factors and factor loadings, and so forth. This will seldom be done, of course, but if the report allows for it, the reader will be far better equipped to evaluate the report than if it does not.

Provide details. If you're doing a qualitative analysis, you must provide enough detail that your reader has a sense of having made the observations with you. Presenting only those data that support your interpretations is not sufficient; you must also share those data that conflict with the way you've made sense of things. Ultimately, you should provide enough information that the reader might reach a different conclusion than you did—though you can hope your interpretation will make the most sense. The reader, in fact, should be in position to replicate the entire study independently, whether it involves participant observation among heavy metal groupies, an experiment regarding jury deliberation, or any other kind of study. Recall that replicability is an essential norm of science. A single study does not prove a point; only a series of studies can begin to do so. And unless studies can be replicated, there can be no meaningful series of studies.

Integrate supporting materials. I have previously mentioned the importance of integrating data and interpretations in the report. Here is a more specific guideline for doing this. Tables, charts, and figures, if any, should be integrated into the text of the report—appearing near that portion of the text discussing them. Sometimes students describe their analyses in the body of the report and place all the tables in an appendix. This procedure greatly impedes the reader, however. As a general rule, it is best to (1) describe the purpose for presenting the table, (2) present it, and (3) review and interpret it.

Draw explicit conclusions. Although research is typically conducted for the purpose of drawing general conclusions, you should carefully note the specific basis for such conclusions. Otherwise you may lead your reader into accepting unwarranted conclusions.

Point to any qualifications or conditions warranted in the evaluation of conclusions. Typically, you know best the shortcomings and tentativeness of your conclusions, and you should give the reader the advantage of that knowledge. Failure to do so can misdirect future research and result in a waste of research funds.

As I said at the outset of this appendix, research reports should be written in the best possible literary style. Writing lucidly is easier for some people than for others, and it is always harder than writing poorly. You are again referred to the Strunk and White book. Every researcher would do well to follow this procedure: Write. Read Strunk and White. Revise. Reread Strunk and White. Revise again. This will be a difficult and time-consuming endeavor, but so is science.

A perfectly designed, carefully executed, and brilliantly analyzed study will be altogether worthless unless you can communicate your findings to others. This appendix has attempted to provide some guidelines toward that end. The best guides are logic, clarity, and honesty. Ultimately, there is no substitute for practice.

APPENDIX C

Random Numbers

10480	15011	01536	02011	81647	91646	69179	14194	62590	36207	20969	99570	91291	90700
22368	46573	25595	85393	30995	89198	27982	53402	93965	34095	52666	19174	39615	99505
24130	48360	22527	97265	76393	64809	15179	24830	49340	32081	30680	19655	63348	58629
42167	93093	06243	61680	07856	16376	39440	53537	71341	57004	00849	74917	97758	16379
37570	39975	81837	16656	06121	91782	60468	81305	49684	60672	14110	06927	01263	54613
77921	06907	11008	42751	27756	53498	18602	70659	90655	15053	21916	81825	44394	42880
99562	72905	56420	69994	98872	31016	71194	18738	44013	48840	63213	21069	10634	12952
96301	91977	05463	07972	18876	20922	94595	56869	69014	60045	18425	84903	42508	32307
89579	14342	63661	10281	17453	18103	57740	84378	25331	12566	58678	44947	05585	56941
85475	36857	53342	53988	53060	59533	38867	62300	08158	17983	16439	11458	18593	64952
28918	69578	88231	33276	70997	79936	56865	05859	90106	31595	01547	85590	91610	78188
63553	40961	48235	03427	49626	69445	18663	72695	52180	20847	12234	90511	33703	90322
09429	93969	52636	92737	88974	33488	36320	17617	30015	08272	84115	27156	30613	74952
10365	61129	87529	85689	48237	52267	67689	93394	01511	26358	85104	20285	29975	89868
07119	97336	71048	08178	77233	13916	47564	81056	97735	85977	29372	74461	28551	90707
51085	12765	51821	51259	77452	16308	60756	92144	49442	53900	70960	63990	75601	40719
02368	21382	52404	60268	89368	19885	55322	44819	01188	65255	64835	44919	05944	55157
01011	54092	33362	94904	31273	04146	18594	29852	71585	85030	51132	01915	92747	64951
52162	53916	46369	58586	23216	14513	83149	98736	23495	64350	94738	17752	35156	35749
07056	97628	33787	09998	42698	06691	76988	13602	51851	46104	88916	19509	25625	58104
48663	91245	85828	14346	09172	30168	90229	04734	59193	22178	30421	61666	99904	32812
54164	58492	22421	74103	47070	25306	76468	26384	58151	06646	21524	15227	96909	44592
32639	32363	05597	24200	13363	38005	94342	28728	35806	06912	17012	64161	18296	22851
29334	27001	87637	87308	58731	00256	45834	15398	46557	41135	10367	07684	36188	18510
02488	33062	28834	07351	19731	92420	60952	61280	50001	67658	32586	86679	50720	94953
81525	72295	04839	96423	24878	82651	66566	14778	76797	14780	13300	87074	79666	95725
29676	20591	68086	26432	46901	20849	89768	81536	86645	12659	92259	57102	80428	25280
00742	57392	39064	66432	84673	40027	32832	61362	98947	96067	64760	64584	96096	98253
05366	04213	25669	26422	44407	44048	37397	63904	45766	66134	75470	66520	34693	90449
91921	26418	64117	94305	26766	25940	39972	22209	71500	64568	91402	42416	07844	69618
00582	04711	87917	77341	42206	35126	74087	99547	81817	42607	43808	76655	62028	76630
00725	69884	62797	56170	86324	88072	76222	36086	84637	93161	76038	65855	77919	88006
69011	65795	95876	55293	18988	27354	26575	08625	40801	59920	29841	80150	12777	48501
25976	57948	29888	88604	67917	48708	18912	82271	65424	69774	33611	54262	85963	03547
09763	83473	73577	12908	30883	18317	28290	35797	05998	41688	34952	37888	38917	88050
91567	42595	27958	30134	04024	86385	29880	99730	55536	84855	29080	09250	79656	73211
17955	56349	90999	49127	20044	59931	06115	20542	18059	02008	73708	83517	36103	42791
46503	18584	18845	49618	02304	51038	20655	58727	28168	15475	56942	53389	20562	87338
92157	89634	94824	78171	84610	82834	09922	25417	44137	48413	25555	21246	35509	20468
14577	62765	35605	81263	39667	47358	56873	56307	61607	49518	89656	20103	77490	18062
98427	07523	33362	64270	01638	92477	66969	98420	04880	45585	46565	04102	46880	45709
34914	63976	88720	82765	34476	17032	87589	40836	32427	70002	70663	88863	77775	69348
70060	28277	39475	46473	23219	53416	94970	25832	69975	94884	19661	72828	00102	66794
53976	54914	06990	67245	68350	82948	11398	42878	80287	88267	47363	46634	06541	97809
76072	29515	40980	07391	58745	25774	22987	80059	39911	96189	41151	14222	60697	59583
90725	52210	83974	29992	65831	38857	50490	83765	55657	14361	31720	57375	56228	41546
64364	67412	33339	31926	14883	24413	59744	92351	97473	89286	35931	04110	23726	51900
08962	00358	31662	25388	61642	34072	81249	35648	56891	69352	48373	45578	78547	81788
95012	68379	93526	70765	10592	04542	76463	54328	02349	17247	28865	14777	62730	92277
15664	10493	20492	38391	91132	21999	59516	81652	27195	48223	46751	22923	32261	85653
16408	81899	04153	53381	79401	21438	83035	92350	36693	31238	59649	91754	72772	02338
18629	81953	05520	91962	04739	13092	97662	24822	94730	06496	35090	04822	86774	98289
73115	35101	47498	87637	99016	71060	88824	71013	18735	20286	23153	72924	35165	43040
57491	16703	23167	49323	45021	33132	12544	41035	80780	45393	44812	12515	98931	91202
30405	83946	23792	14422	15059	45799	22716	19792	09983	74353	68668	30429	70735	25499
16631	35006	85900	98275	32388	52390	16815	69298	82732	38480	73817	32523	41961	44437
96773	20206	42559	78985	05300	22164	24369	54224	35083	19687	11052	91491	60383	19746
38935	64202	14349	82674	66523	44133	00697	35552	35970	19124	63318	29686	03387	59846
31624	76384	17403	53363	44167	64486	64758	75366	76554	31601	12614	33072	60332	92325
78919	19474	23632	27889	47914	02584	37680	20801	72152	39339	34806	08930	85001	87820

03931	33309	57047	74211	63445	17361	62825	39908	05607	91284	68833	25570	38818	46920
74426	33278	43972	10119	89917	15665	52872	73823	73144	88662	88970	74492	51805	99378
09066	00903	20795	95452	92648	45454	09552	88815	16553	51125	79375	97596	16296	66092
42238	12426	87025	14267	20979	04508	64535	31355	86064	29472	47689	05974	52468	16834
16153	08002	26504	41744	81959	65642	74240	56302	00033	67107	77510	70625	28725	34191
21457	40742	29820	96783	29400	21840	15035	34537	33310	06116	95240	15957	16572	06004
21581	57802	02050	89728	17937	37621	47075	42080	97403	48626	68995	43805	33386	21597
55612	78095	83197	33732	05810	24813	86902	60397	16489	03264	88525	42786	05269	92532
44657	66999	99324	51281	84463	60563	79312	93454	68876	25471	93911	25650	12682	73572
91340	84979	46949	81973	37949	61023	43997	15263	80644	43942	89203	71795	99533	50501
91227	21199	31935	27022	84067	05462	35216	14486	29891	68607	41867	14951	91696	85065
50001	38140	66321	19924	72163	09538	12151	06878	91903	18749	34405	56087	82790	70925
65390	05224	72958	28609	81406	39147	25549	48542	42627	45233	57202	94617	23772	07896
27504	96131	83944	10573	08619	64482	73923	36152	05184	94142	25299	84387	34925
37169	94851	39117	89632	00959	16487	65536	49071	39782	17095	02330	74301	00275	48280
11508	70225	51111	38351	19444	66499	71945	05422	13442	78675	84081	66938	93654	59894
37449	30362	06694	54690	04052	53115	62757	95348	78662	11163	81651	50245	34971	52924
46515	70331	85922	38329	57015	15765	97161	17869	45349	61796	66345	81073	49106	79860
30986	81223	42416	58353	21532	30502	32305	86482	05174	07901	54339	58861	74818	46942
63798	64995	46583	09785	44160	78128	83991	42865	92520	83531	80377	35909	81250	54238
82486	84846	99254	67632	43218	50076	21361	64816	51202	88124	41870	52689	51275	83556
21885	32906	92431	09060	64297	51674	64126	62570	26123	05155	59194	52799	28225	85762
60336	98782	07408	53458	13564	59089	26445	29789	85205	41001	12535	12133	14645	23541
43937	46891	24010	86355	33941	25786	54990	71899	15475	95434	98227	21824	19585
97656	63175	89303	16275	07100	92063	21942	18611	47348	20203	18534	03862	78095	50136
03299	01221	05418	38982	55758	92237	26759	86367	21216	98442	08303	56613	91511	75928
79626	06486	03574	17668	07785	76020	79924	25651	83325	88428	85076	72811	22717	50585
85636	68335	47539	03129	65651	11977	02510	26113	99447	68645	34327	15152	55230	93448
18039	14367	61337	06177	12143	46609	32989	74014	64708	00533	35398	58408	13261	47908
08362	15656	60627	36478	65648	16764	53412	09013	07832	41574	17639	82163	60859	75567
79556	29068	04142	16268	15387	12856	66227	38358	22478	73373	88732	09443	82558	05250
92608	82674	27072	32534	17075	27698	98204	63863	11951	34648	88022	56148	34925	57031
23982	25835	40055	67006	12293	02753	14827	23235	35071	99704	37543	11601	35503	85171
09915	96306	05908	97901	28395	14186	00821	80703	70426	75647	76310	88717	37890	40129
59037	33300	26695	62247	69927	76123	50842	43834	86654	70959	79725	93872	28117	19233
42488	78077	69882	61657	34136	79180	97526	43092	04098	73571	80799	76536	71255	64239
46764	86273	63003	93017	31204	36692	40202	35275	57306	55543	53203	18098	47625	88684
03237	45430	55417	63282	90816	17349	88298	90183	36600	78406	06216	95787	42579	90730
86591	81482	52667	61582	14972	90053	89534	76036	49199	43716	97548	04379	46370	28672
38534	01715	94964	87288	65680	43772	39560	12918	86537	62738	19636	51132	25739	56947

Abridged from *Handbook of Tables for Probability and Statistics,* 2nd ed., edited by William H. Beyer (Cleveland: The Chemical Rubber Company, 1968). Used by permission of The Chemical Rubber Company.

APPENDIX D

Distribution of Chi Square

				Probability			
df	.99	.98	.95	.90	.80	.70	.50
1	.0³157	.0³628	.00393	.0158	.0642	.148	.455
2	.0201	.0404	.103	.211	.446	.713	1.386
3	.115	.185	.352	.584	1.005	1.424	2.366
4	.297	.429	.711	1.064	1.649	2.195	3.357
5	.554	.752	1.145	1.610	2.343	3.000	4.351
6	.872	1.134	1.635	2.204	3.070	3.828	5.348
7	1.239	1.564	2.167	2.833	3.822	4.671	6.346
8	1.646	2.032	2.733	3.490	4.594	5.528	7.344
9	2.088	2.532	3.325	4.168	5.380	6.393	8.343
10	2.558	3.059	3.940	4.865	6.179	7.267	9.342
11	3.053	3.609	4.575	5.578	6.989	8.148	10.341
12	3.571	4.178	5.226	6.304	7.807	9.034	11.340
13	4.107	4.765	5.892	7.042	8.634	9.926	12.340
14	4.660	5.368	6.571	7.790	9.467	10.821	13.339
15	5.229	5.985	7.261	8.547	10.307	11.721	14.339
16	5.812	6.614	7.962	9.312	11.152	12.624	15.338
17	6.408	7.255	8.672	10.085	12.002	13.531	16.338
18	7.015	7.906	9.390	10.865	12.857	14.440	17.338
19	7.633	8.567	10.117	11.651	13.716	15.352	18.338
20	8.260	9.237	10.851	12.443	14.578	16.266	19.337
21	8.897	9.915	11.591	13.240	15.445	17.182	20.337
22	9.542	10.600	12.338	14.041	16.314	18.101	21.337
23	10.196	11.293	13.091	14.848	17.187	19.021	22.337
24	10.856	11.992	13.848	15.659	18.062	19.943	23.337
25	11.524	12.697	14.611	16.473	18.940	20.867	24.337
26	12.198	13.409	15.379	17.292	19.820	21.792	25.336
27	12.879	14.125	16.151	18.114	20.703	22.719	26.336
28	13.565	14.847	16.928	18.939	21.588	23.647	27.336
29	14.256	15.574	17.708	19.768	22.475	24.577	28.336
30	14.953	16.306	18.493	20.599	23.364	25.508	29.336

continued

For larger values of df, the expression $\sqrt{2\chi^2} - \sqrt{2df - 1}$ may be used as a normal deviate with unit variance, remembering that the probability of χ^2 corresponds with that of a single tail of the normal curve.

Source: I am grateful to the Literary Executor of the late Sir Ronald A. Fisher, F.R.S., to Dr. Frank Yates, F.R.S., and to Longman Group Ltd., London, for permission to reprint Table IV from their book *Statistical Tables for Biological, Agricultural, and Medical Research* (6th Edition, 1974).

Probability

df	.30	.20	.10	.05	.02	.01	.001
1	1.074	1.642	2.706	3.841	5.412	6.635	10.827
2	2.408	3.219	4.605	5.991	7.824	9.210	13.815
3	3.665	4.642	6.251	7.815	9.837	11.341	16.268
4	4.878	5.989	7.779	9.488	11.668	13.277	18.465
5	6.064	7.289	9.236	11.070	13.388	15.086	20.517
6	7.231	8.558	10.645	12.592	15.033	16.812	22.457
7	8.383	9.803	12.017	14.067	16.622	18.475	24.322
8	9.524	11.030	13.362	15.507	18.168	20.090	29.125
9	10.656	12.242	14.684	16.919	19.679	21.666	27.877
10	11.781	13.442	15.987	18.307	21.161	23.209	29.588
11	12.899	14.631	17.275	19.675	22.618	24.725	31.264
12	14.011	15.812	18.549	21.026	24.054	26.217	32.909
13	15.119	16.985	19.812	22.362	25.472	27.688	34.528
14	16.222	18.151	21.064	23.685	26.873	29.141	36.123
15	17.322	19.311	22.307	24.996	28.259	30.578	37.697
16	18.841	20.465	23.542	26.296	29.633	32.000	39.252
17	15.511	21.615	24.769	27.587	30.995	33.409	40.790
18	20.601	22.760	25.989	28.869	32.346	34.805	42.312
19	21.689	23.900	27.204	30.144	33.687	36.191	43.820
20	22.775	25.038	28.412	31.410	35.020	37.566	45.315
21	23.858	26.171	29.615	32.671	36.343	38.932	46.797
22	24.939	27.301	30.813	33.924	37.659	40.289	48.268
23	26.018	28.429	32.007	35.172	38.968	41.638	49.728
24	27.096	29.553	33.196	36.415	40.270	42.980	51.179
25	28.172	30.675	34.382	37.652	41.566	44.314	52.620
26	29.246	31.795	35.563	38.885	42.856	45.642	54.052
27	30.319	32.912	36.741	40.113	44.140	46.963	55.476
28	31.391	34.027	37.916	41.337	45.419	48.278	56.893
29	32.461	35.139	39.087	42.557	46.693	49.588	58.302
30	35.530	36.250	40.256	43.773	47.962	50.892	59.703

APPENDIX E

Normal Curve Areas

z	.00	.01	.02	.03	.04	.05	.06	.07	.08	.09
0.0	.0000	.0040	.0080	.0120	.0160	.0199	.0239	.0279	.0319	.0359
0.1	.0398	.0438	.0478	.0517	.0557	.0596	.0636	.0675	.0714	.0753
0.2	.0793	.0832	.0871	.0910	.0948	.0987	.1026	.1064	.1103	.1141
0.3	.1179	.1217	.1255	.1293	.1331	.1368	.1406	.1443	.1480	.1517
0.4	.1554	.1591	.1628	.1664	.1700	.1736	.1772	.1808	.1844	.1879
0.5	.1915	.1950	.1985	.2019	.2054	.2088	.2123	.2157	.2190	.2224
0.6	.2257	.2291	.2324	.2357	.2389	.2422	.2454	.2486	.2517	.2549
0.7	.2580	.2611	.2642	.2673	.2704	.2734	.2764	.2794	.2823	.2852
0.8	.2881	.2910	.2939	.2967	.2995	.3023	.3051	.3078	.3106	.3133
0.9	.3159	.3186	.3212	.3238	.3264	.3289	.3315	.3340	.3365	.3389
1.0	.3413	.3438	.3461	.3485	.3508	.3531	.3554	.3577	.3599	.3621
1.1	.3643	.3665	.3686	.3708	.3729	.3749	.3770	.3790	.3810	.3830
1.2	.3849	.3869	.3888	.3907	.3925	.3944	.3962	.3980	.3997	.4015
1.3	.4032	.4049	.4066	.4082	.4099	.4115	.4131	.4147	.4162	.4177
1.4	.4192	.4207	.4222	.4236	.4251	.4265	.4279	.4292	.4306	.4319
1.5	.4332	.4345	.4357	.4370	.4382	.4394	.4406	.4418	.4429	.4441
1.6	.4452	.4463	.4474	.4484	.4495	.4505	.4515	.4525	.4535	.4545
1.7	.4554	.4564	.4573	.4582	.4591	.4599	.4608	.4616	.4625	.4633
1.8	.4641	.4649	.4656	.4664	.4671	.4678	.4686	.4693	.4699	.4706
1.9	.4713	.4719	.4726	.4732	.4738	.4744	.4750	.4756	.4761	.4767
2.0	.4772	.4778	.4783	.4788	.4793	.4798	.4803	.4808	.4812	.4817
2.1	.4821	.4826	.4830	.4834	.4838	.4842	.4846	.4850	.4854	.4857
2.2	.4861	.4864	.4868	.4871	.4875	.4878	.4881	.4884	.4887	.4890
2.3	.4893	.4896	.4898	.4901	.4904	.4906	.4909	.4911	.4913	.4916
2.4	.4918	.4920	.4922	.4925	.4927	.4929	.4931	.4932	.4934	.4936
2.5	.4938	.4940	.4941	.4943	.4945	.4946	.4948	.4949	.4951	.4952
2.6	.4953	.4955	.4956	.4957	.4959	.4960	.4961	.4962	.4963	.4964
2.7	.4965	.4966	.4967	.4968	.4969	.4970	.4971	.4972	.4973	.4974
2.8	.4974	.4975	.4976	.4977	.4977	.4978	.4979	.4979	.4980	.4981
2.9	.4981	.4982	.4982	.4983	.4984	.4984	.4985	.4985	.4986	.4986
3.0	.4987	.4987	.4987	.4988	.4988	.4989	.4989	.4989	.4990	.4990

Abridged from Table I of *Statistical Tables and Formulas,* by A. Hald (New York: John Wiley & Sons, Inc., 1952). Used by permission of John Wiley & Sons, Inc.

APPENDIX F

Estimated Sampling Error

How to use this table: Find the intersection between the sample size and the approximate percentage distribution of the binomial in the sample. The number appearing at this intersection represents the estimated sampling error, at the 95 percent confidence level, expressed in percentage points (plus or minus).

Example: In the sample of 400 respondents, 60 percent answer yes and 40 percent answer no. The sampling error is estimated at plus or minus 4.9 percentage points. The confidence interval, then, is between 55.1 percent and 64.9 percent. We would estimate (95 percent confidence) that the proportion of the total population who would say yes is somewhere within that interval.

Sample Size	Binomial Percentage Distribution				
	50/50	60/40	70/30	80/20	90/10
100	10	9.8	9.2	8	6
200	7.1	6.9	6.5	5.7	4.2
300	5.8	5.7	5.3	4.6	3.5
400	5	4.9	4.6	4	3
500	4.5	4.4	4.1	3.6	2.7
600	4.1	4	3.7	3.3	2.4
700	3.8	3.7	3.5	3	2.3
800	3.5	3.5	3.2	2.8	2.1
900	3.3	3.3	3.1	2.7	2
1000	3.2	3.1	2.9	2.5	1.9
1100	3	3	2.8	2.4	1.8
1200	2.9	2.8	2.6	2.3	1.7
1300	2.8	2.7	2.5	2.2	1.7
1400	2.7	2.6	2.4	2.1	1.6
1500	2.6	2.5	2.4	2.1	1.5
1600	2.5	2.4	2.3	2	1.5
1700	2.4	2.4	2.2	1.9	1.5
1800	2.4	2.3	2.2	1.9	1.4
1900	2.3	2.2	2.1	1.8	1.4
2000	2.2	2.2	2	1.8	1.3

GLOSSARY

anonymity   Anonymity is guaranteed in a research project when neither the researchers nor the readers of the findings can identify a given response with a given respondent. See Chapter 3.

attributes   Characteristics of persons or things. See *variables* and Chapter 1.

average   An ambiguous term generally suggesting typical or normal—a central tendency. The *mean, median,* and *mode* are specific examples of mathematical averages. See Chapter 14.

bias   (1) That quality of a measurement device that tends to result in a misrepresentation of what is being measured in a particular direction. For example, the questionnaire item "Don't you agree that the president is doing a good job?" would be biased in that it would generally encourage favorable responses. See Chapter 9. (2) The thing inside you that makes other people or groups seem consistently better or worse than they really are. (3) What a nail looks like after you hit it crooked. (If you drink, don't drive.)

bivariate analysis   The analysis of two variables simultaneously, for the purpose of determining the empirical relationship between them. The construction of a simple percentage table or the computation of a simple correlation coefficient are examples of bivariate analyses. See Chapter 14.

Bogardus social distance scale   A measurement technique for determining the willingness of people to participate in social relations—of varying degrees of closeness—with other kinds of people. It is an especially efficient technique in that several discrete answers may be summarized without losing any of the original details of the data. See Chapter 6.

case-oriented analysis   (1) An analysis that aims to understand a particular case or several cases by looking closely at the details of each. See Chapter 13. (2) A private investigator's billing system.

case study   The in-depth examination of a single instance of some social phenomenon, such as a village, a family, or a juvenile gang. See Chapter 10.

closed-ended questions   Survey questions in which the respondent is asked to select an answer from among a list provided by the researcher. Popular in survey research because they provide a greater uniformity of responses and are more easily processed than are *open-ended questions.* See Chapter 9.

cluster sampling   (1) A multistage sampling in which natural groups (clusters) are sampled initially, with the members of each selected group being subsampled afterward. For example, you might select a sample of U.S. colleges and universities from a directory, get lists of the students at all the selected schools, then draw samples of students from each. This procedure is discussed in Chapter 7. (2) Pawing around in a box of macadamia nut clusters to take all the big ones for yourself.

codebook   (1) The document used in data processing and analysis that tells the location of different data items in a data file. Typically, the codebook identifies the locations of data items and the meaning of the codes used to represent different attributes of variables. See Chapter 14. (2) The document that cost you 38 box tops just to learn that Captain Marvelous wanted you to brush your teeth and always tell the truth. (3) The document that allows CIA agents to learn that Captain Marvelous wants them to brush their teeth.

coding (1) The process whereby raw data are transformed into standardized form suitable for machine processing and analysis. See Chapter 11. (2) A strong drug you may take when you hab a bad code.

cohort study A study in which some specific subpopulation, or cohort, is studied over time, although data may be collected from different members in each set of observations. A study of the occupational history of the class of 1970, in which questionnaires were sent every five years, for example, would be a cohort study. See Chapter 4 for more on this topic (if you want more). See also *longitudinal study, panel study,* and *trend study.*

concept mapping (1) The graphical display of concepts and their interrelations, useful in the formulation of theory. See Chapter 13. (2) A masculine technique for finding locations by logic and will, without asking for directions.

conceptualization (1) The mental process whereby fuzzy and imprecise notions (concepts) are made more specific and precise. So you want to study prejudice. What do you mean by prejudice? Are there different kinds of prejudice? What are they? See Chapter 5, which is all about conceptualization and its pal, operationalization. (2) Sexual reproduction among intellectuals.

confidence interval (1) The range of values within which a population parameter is estimated to lie. A survey, for example, may show 40 percent of a sample favoring Candidate A (poor devil). Although the best estimate of the support existing among all voters would also be 40 percent, we would not expect it to be exactly that. We might, therefore, compute a confidence interval (such as from 35 to 45 percent) within which the actual percentage of the population probably lies. Note that we must specify a confidence level in connection with every confidence interval. See Chapter 7. (2) How close you dare to get to an alligator.

confidence level (1) The estimated probability that a population parameter lies within a given confidence interval. Thus, we might be 95 percent confident that between 35 and 45 percent of all voters favor Candidate A. See Chapter 7. (2) How sure you are that the ring you bought from a street vendor for $10 is really a three-carat diamond.

confidentiality A research project guarantees confidentiality when the researcher can identify a given person's responses but promises not to do so publicly. See Chapter 3.

constant comparative method (1) A component of the Grounded Theory Method in which observations are compared with one another and with the evolving inductive theory. See Chapter 13. (2) A blind-dating technique.

construct validity The degree to which a measure relates to other variables as expected within a system of theoretical relationships. See Chapter 5.

content analysis The study of recorded human communications, such as books, Web sites, paintings, and laws. See Chapter 11.

content validity The degree to which a measure covers the range of meanings included within a concept. See Chapter 5.

contingency question A survey question intended for only some respondents, determined by their responses to some other question. For example, all respondents might be asked whether they belong to the Cosa Nostra, and only those who said yes would be asked how often they go to company meetings and picnics. The latter would be a contingency question. See Chapter 9.

contingency table (1) A format for presenting the relationships among variables as percentage distributions. See Chapter 14 for several illustrations and guides to making such tables. (2) The card table you keep around in case your guests bring their seven kids with them to dinner.

continuous variable A variable whose attributes form a steady progression, such as age or income. Thus, the ages of a group of people might include 21, 22, 23, 24, and so forth and could even be broken down into fractions of years. Contrast this with discrete variables, such as *gender* or *religious affiliation,* whose attributes form discontinuous chunks. See Chapter 14.

control group (1) In experimentation, a group of subjects to whom no experimental stimulus is administered and who resemble the experimental group in all other respects. The comparison of the control group and the *experimental group* at the end of the experiment points to the effect of the experimental stimulus. See Chapter 8. (2) American Association of Managers.

control variable See *test variable.*

conversation analysis A meticulous analysis of the details of conversation, based on a complete transcript that includes pauses, hems, and also haws. See Chapter 13.

correlation (1) An empirical relationship between two variables such that (a) changes in one are associated with changes in the other or (b) particular attributes of one variable are associated with particular attributes of the other. Thus, for example, we say that *education* and *income* are correlated in that higher levels of education are associated with higher levels of income. Correlation in and of itself does not constitute a causal relationship between the two variables, but it is one criterion of causality. (2) Someone you and your friend are both related to.

criterion-related validity The degree to which a measure relates to some external criterion. For example, the validity of the college board is shown in their ability to predict the college success of students. Also called predictive validity. See Chapter 5.

cross-case analysis An analysis that involves an examination of more than one case, either a variable-oriented or case-oriented analysis. See Chapter 13.

cross-sectional study A study based on observations representing a single point in time. Contrasted with a *longitudinal study.*

debriefing (1) Interviewing subjects to learn about their experience of participation in the project. Especially important if there's a possibility that they have been damaged by that participation. See Chapter 3. (2) Pulling someone's shorts down. Don't do that. It's not nice.

deduction (1) The logical model in which specific expectations of hypotheses are developed on the basis of general principles. Starting from the general principle that all deans are meanies, you might anticipate that this one won't let you change courses. This anticipation would be the result of deduction. See also *induction.* (2) What the Internal Revenue Service said your good-for-nothing moocher of a brother-in-law technically isn't. (3) Of a duck.

dependent variable (1) A variable assumed to depend on or be caused by another (called the independent variable). If you find that *income* is partly a function of *amount of formal education, income* is being treated as a dependent variable. (2) A wimpy variable.

dimension A specifiable aspect of a concept. See Chapter 5.

discrete variable A variable whose attributes are separate from one another, or discontinuous, as in the case of *gender* or *religious affiliation.* Contrast this with *continuous variables,* in which one attribute shades off into the next. Thus, in age (a continuous variable), the attributes progress steadily from 21 to 22 to 23, and so forth, whereas there is no progression from male to female in the case of gender. See Chapter 14.

dispersion The distribution of values around some central value, such as an average. The range is a simple example of a measure of dispersion. Thus, we may report that the mean age of a group is 37.9, and the range is from 12 to 89. See Chapter 14.

double-blind experiment An experimental design in which neither the subjects nor the experimenters know which is the experimental group and which is the control. See Chapter 8.

ecological fallacy Erroneously drawing conclusions about individuals based solely on the observation of groups. See Chapter 4.

element That unit of which a population is comprised and which is selected in a sample. Distinguished from *units of analysis,* which are used in data analysis. See Chapter 7.

EPSEM (equal probability of selection method) A sample design in which each member of a population has the same chance of being selected into the sample. See Chapter 7.

ethnography A report on social life that focuses on detailed and accurate description rather than explanation. See Chapter 10.

ethnomethodology An approach to the study of social life that focuses on the discovery of implicit, usually unspoken assumptions and agreements; this method often involves the intentional breaking of agreements as a way of revealing their existence. See Chapter 10.

evaluation research Research undertaken for the purpose of determining the impact of some social intervention, such as a program aimed at solving a social problem. See Chapter 12.

experimental group In experimentation, a group of subjects to whom an experimental stimulus is administered. Compare with *control group*. See Chapter 8.

extended case method A technique developed by Michael Burawoy in which case study observations are used to discover flaws in and to improve existing social theories. See Chapter 10.

external invalidity Refers to the possibility that conclusions drawn from experimental results may not be generalizable to the "real" world. See Chapter 8 and also *internal invalidity*.

external validation The process of testing the validity of a measure, such as an index or scale, by examining its relationship to other, presumed indicators of the same variable. If the index really measures prejudice, for example, it should correlate with other indicators of prejudice. See Chapter 6 for a fuller discussion and illustrations.

face validity (1) That quality of an indicator that makes it seem a reasonable measure of some variable. That the frequency of church attendance is some indication of a person's religiosity seems to make sense without a lot of explanation. It has face validity. See Chapter 5. (2) When your face looks like your driver's license photo (rare and perhaps unfortunate).

focus group A group of subjects interviewed together, prompting a discussion. The technique is frequently used by market researchers, who ask a group of consumers to evaluate a product or discuss a type of commodity, for example. See Chapter 10.

frequency distribution (1) A description of the number of times the various attributes of a variable are observed in a sample. The report that 53 percent of a sample were men and 47 percent were women would be a simple example of a frequency distribution. Another example would be the report that 15 of the cities studied had populations under 10,000, 23 had populations between 10,000 and 25,000, and so forth. See Chapter 14. (2) A radio dial.

grounded theory (1) An inductive approach to the study of social life that attempts to generate a theory from the constant comparing of unfolding observations. This is very different from hypothesis testing, in which theory is used to generate hypotheses to be tested through observations. See Chapter 10. (2) A theory that is not allowed to fly.

Grounded Theory Method (GTM) An inductive approach to research introduced by Barney Glaser and Anselm Strauss in which theories are generated solely from an examination of data rather than being derived deductively. See Chapter 13.

Guttman scale (1) A type of composite measure used to summarize several discrete observations and to represent some more-general variable. See Chapter 6. (2) The device Louis Guttman weighs himself on.

historical/comparative research The examination of societies (or other social units) over time and in comparison with one another. See Chapter 11.

hypothesis A specified testable expectation about empirical reality that follows from a more general proposition; more generally, an expectation about the nature of things derived from a theory. It is a statement of something that ought to be observed in the real world if the theory is correct. See *deduction* and also Chapter 2.

idiographic An approach to explanation in which we seek to exhaust the idiosyncratic causes of a particular condition or event. Imagine trying to list all the reasons why you chose to attend your particular college. Given all those reasons, it's difficult to imagine your making any other choice. By contrast, see *nomothetic*.

independent variable　(1) A variable with values that are not problematical in an analysis but are taken as simply given. An independent variable is presumed to cause or determine a dependent variable. If we discover that religiosity is partly a function of gender—women are more religious than are men—*gender* is the independent variable and *religiosity* is the dependent variable. Note that any given variable might be treated as independent in one part of an analysis and dependent in another part of it. *Religiosity* might become an independent variable in the explanation of crime. See *dependent variable.* (2) A variable that refuses to take advice.

index　A type of composite measure that summarizes and rank-orders several specific observations and represents some more general dimension. Contrasted with *scale.* See Chapter 6.

indicator　An observation that we choose to consider as a reflection of a variable we wish to study. Thus, for example, attending church might be considered an indicator of religiosity. See Chapter 5.

induction　(1) The logical model in which general principles are developed from specific observations. Having noted that Jews and Catholics are more likely to vote Democratic than are Protestants, you might conclude that religious minorities in the United States are more affiliated with the Democratic party and explain why. This would be an example of induction. See also *deduction.* (2) The culinary art of stuffing ducks.

informant　Someone well versed in the social phenomenon that you wish to study and who is willing to tell you what he or she knows about it. If you were planning participant observation among the members of a religious sect, you would do well to make friends with someone who already knows about them—possibly a member of the sect—who could give you some background information about them. Not to be confused with a *respondent.*

informed consent　A norm in which subjects base their voluntary participation in research projects on a full understanding of the possible risks involved. See Chapter 3.

institutional ethnography　A research technique in which the personal experiences of individuals are used to reveal power relationships and other characteristics of the institutions within which they operate. See Chapter 10.

internal invalidity　Refers to the possibility that the conclusions drawn from experimental results may not accurately reflect what went on in the experiment itself. See Chapter 8 and also *external invalidity.*

interval measure　A level of measurement describing a variable whose attributes are rank-ordered and have equal distances between adjacent attributes. The Fahrenheit temperature scale is an example of this, because the distance between 17 and 18 is the same as that between 89 and 90. See also Chapter 5 and *nominal measure, ordinal measure,* and *ratio measure.*

interview　A data-collection encounter in which one person (an interviewer) asks questions of another (a respondent). Interviews may be conducted face-to-face or by telephone. See Chapter 9 for more information on interviewing as a method of survey research.

item analysis　An assessment of whether each of the items included in a composite measure makes an independent contribution or merely duplicates the contribution of other items in the measure. See Chapter 6.

judgmental sampling　(1) See *purposive sampling* and Chapter 7. (2) A sampling of opinionated people.

latent content　(1) As used in connection with content analysis, the underlying meaning of communications as distinguished from their *manifest content.* See Chapter 11. (2) What you need to make a latent.

Likert scale　A type of composite measure developed by Rensis Likert in an attempt to improve the levels of measurement in social research through the use of standardized response categories in survey questionnaires to determine the relative intensity of different items. Likert items are those using such response categories as strongly agree, agree, disagree, and strongly disagree. Such items may be used in the construction of true Likert scales as well as other types of composite measures. See Chapter 6.

longitudinal study A study design involving the collection of data at different points in time, as contrasted with a *cross-sectional study.* See also Chapter 4 and *cohort study, panel study,* and *trend study.*

macrotheory A theory aimed at understanding the "big picture" of institutions, whole societies, and the interactions among societies. Karl Marx's examination of the class struggle is an example of macrotheory. By contrast, see *microtheory.*

manifest content (1) In connection with content analysis, the concrete terms contained in a communication, as distinguished from *latent content.* See Chapter 11. (2) What you have after a manifest bursts.

matching In connection with experiments, the procedure whereby pairs of subjects are matched on the basis of their similarities on one or more variables, and one member of the pair is assigned to the experimental group and the other to the control group. See Chapter 8.

mean (1) An average computed by summing the values of several observations and dividing by the number of observations. If you now have a grade point average of 4.0 based on 10 courses, and you get an F in this course, your new grade point (mean) average will be 3.6. (2) The quality of the thoughts you might have if your instructor did that to you.

median (1) An average representing the value of the "middle" case in a rank-ordered set of observations. If the ages of five men are 16, 17, 20, 54, and 88, the median would be 20. (The mean would be 39.) (2) The dividing line between safe driving and exciting driving.

memoing Writing memos that become part of the data for analysis in qualitative research such as grounded theory. Memos can describe and define concepts, deal with methodological issues, or offer initial theoretical formulations. See Chapter 13.

microtheory A theory aimed at understanding social life at the intimate level of individuals and their interactions. Examining how the play behavior of girls differs from that of boys would be an example of microtheory. By contrast, see *macrotheory.*

mode (1) An average representing the most frequently observed value or attribute. If a sample contains 1,000 Protestants, 275 Catholics, and 33 Jews, "Protestant" is the modal category. See Chapter 14 for more thrilling disclosures about averages. (2) Better than apple pie à la median.

multiple time-series designs The use of more than one set of data that were collected over time, as in accident rates over time in several states or cities, so that comparison can be made. See Chapter 12.

multivariate analysis The analysis of the simultaneous relationships among several variables. Examining simultaneously the effects of *age, gender,* and *social class* on *religiosity* would be an example of multivariate analysis. See Chapter 14.

naturalism An approach to field research based on the assumption that an objective social reality exists and can be observed and reported accurately. See Chapter 10.

nominal measure A variable whose attributes have only the characteristics of exhaustiveness and mutual exclusiveness. In other words, a level of measurement describing a variable that has attributes that are merely different, as distinguished from *ordinal, interval,* or *ratio measures.* Gender would be an example of a nominal measure. See Chapter 5.

nomothetic An approach to explanation in which we seek to identify a few causal factors that generally impact a class of conditions or events. Imagine the two or three key factors that determine which colleges students choose, such as proximity, reputation, and so forth. By contrast, see *idiographic.*

nonequivalent control group A control group that is similar to the experimental group but is not created by the random assignment of subjects. This sort of control group does differ significantly from the experimental group in terms of the dependent variable or variables related to it. See Chapter 12.

nonprobability sampling Any technique in which samples are selected in some way not suggested by probability theory. Examples include reliance on available subjects as well as *purposive (judgmental), quota,* and *snowball sampling.* See Chapter 7.

null hypothesis (1) In connection with hypothesis testing and tests of statistical significance, that hypothesis that suggests there is no relationship among the variables under study. You may conclude that the variables are related after having statistically rejected the null hypothesis. (2) An expectation about nulls.

open coding The initial classification and labeling of concepts in qualitative data analysis. In open coding, the codes are suggested by the researchers' examination and questioning of the data. See Chapter 13.

open-ended questions Questions for which the respondent is asked to provide his or her own answers. In-depth, qualitative interviewing relies almost exclusively on open-ended questions. See Chapters 9 and 10.

operational definition The concrete and specific definition of something in terms of the operations by which observations are to be categorized. The operational definition of "earning an A in this course" might be "correctly answering at least 90 percent of the final exam questions." See Chapter 2.

operationalization (1) One step beyond conceptualization. Operationalization is the process of developing operational definitions, or specifying the exact operations involved in measuring a variable. (2) Surgery on intellectuals.

ordinal measure A level of measurement describing a variable with attributes we can rank-order along some dimension. An ex-

ample would be *socioeconomic status* as composed of the attributes high, medium, low. See also Chapter 5 and *nominal measure, interval measure,* and *ratio measure.*

panel study A type of longitudinal study, in which data are collected from the same set of people (the sample or panel) at several points in time. See Chapter 4 and *cohort, longitudinal,* and *trend study.*

paradigm (1) A model or framework for observation and understanding, which shapes both what we see and how we understand it. The conflict paradigm causes us to see social behavior one way, the interactionist paradigm causes us to see it differently. (2) $0.20.

parameter The summary description of a given variable in a population. See Chapter 7.

participatory action research An approach to social research in which the people being studied are given control over the purpose and procedures of the research; intended as a counter to the implicit view that researchers are superior to those they study. See Chapter 10.

population The theoretically specified aggregation of the elements in a study. See Chapter 7.

posttesting The remeasurement of a dependent variable among subjects after they've been exposed to a stimulus representing an independent variable. See *pretesting* and Chapter 8.

PPS (probability proportionate to size) (1) This refers to a type of multistage cluster sample in which clusters are selected, not with equal probabilities (see *EPSEM*) but with probabilities proportionate to their sizes—as measured by the number of units to be subsampled. See Chapter 7. (2) The odds on who gets to go first: you or the 275-pound fullback.

pretesting The measurement of a dependent variable among subjects before they are exposed to a stimulus representing an independent variable. See *posttesting* and Chapter 8.

probability sampling The general term for samples selected in accord with probability theory, typically involving some random-selection mechanism. Specific types of probability sampling include *EPSEM, PPS, simple* *random sampling,* and *systematic sampling.* See Chapter 7.

probe   A technique employed in interviewing to solicit a more complete answer to a question. It is a nondirective phrase or question used to encourage a respondent to elaborate on an answer. Examples include "Anything more?" and "How is that?" See Chapter 9 for a discussion of interviewing.

purposive sampling   A type of nonprobability sampling in which you select the units to be observed on the basis of your own judgment about which ones will be the most useful or representative. Another name for this is *judgmental sampling.* See Chapter 7.

qualitative analysis   (1) The nonnumerical examination and interpretation of observations, for the purpose of discovering underlying meanings and patterns of relationships. This is most typical of field research and historical research. See Chapter 13. (2) A classy analysis.

qualitative interview   Contrasted with survey interviewing, the qualitative interview is based on a set of topics to be discussed in depth rather than the use of standardized questions. See Chapter 10.

quantitative analysis   (1) The numerical representation and manipulation of observations for the purpose of describing and explaining the phenomena that those observations reflect. See Chapter 14. (2) A BIG analysis.

quasi experiments   Nonrigorous inquiries somewhat resembling controlled experiments but lacking key elements such as pre- and posttesting and/or control groups. See Chapter 12.

questionnaire   A document containing questions and other types of items designed to solicit information appropriate for analysis. Questionnaires are used primarily in survey research but also in experiments, field research, and other modes of observation. See Chapter 9.

quota sampling   A type of nonprobability sampling in which units are selected into a sample on the basis of prespecified characteristics, so that the total sample will have the same distribution of characteristics assumed to exist in the population being studied. See Chapter 7.

random selection   A sampling method in which each element has an equal chance of selection independent of any other event in the selection process. See Chapter 7.

randomization   A technique for assigning experimental subjects to experimental and control groups randomly. See Chapter 8.

ratio measure   A level of measurement describing a variable with attributes that have all the qualities of nominal, ordinal, and interval measures and in addition are based on a "true zero" point. Age would be an example of a ratio measure. See also Chapter 5 and *nominal measure, interval measure,* and *ordinal measure.*

reductionism   (1) A fault of some researchers: a strict limitation (reduction) of the kinds of concepts to be considered relevant to the phenomenon under study. (2) The cloning of ducks.

reliability   (1) That quality of measurement method that suggests that the same data would have been collected each time in repeated observations of the same phenomenon. In the context of a survey, we would expect that the question "Did you attend church last week?" would have higher reliability than the question "About how many times have you attended church in your life?" This is not to be confused with *validity.* See Chapter 5. (2) Quality of repeatability in untruths.

replication   The duplication of an experiment to expose or reduce error.

representativeness   (1) That quality of a sample of having the same distribution of characteristics as the population from which it was selected. By implication, descriptions and explanations derived from an analysis of the sample may be assumed to represent similar ones in the population. Representativeness is enhanced by probability sampling and provides for generalizability and the use of inferential statistics. See Chapter 7. (2) A noticeable quality in the presentation-of-self of some members of the U.S. Congress.

respondent A person who provides data for analysis by responding to a survey questionnaire. See Chapter 9.

response rate The number of people participating in a survey divided by the number selected in the sample, in the form of a percentage. This is also called the completion rate or, in self-administered surveys, the return rate: the percentage of questionnaires sent out that are returned. See Chapter 9.

sampling error The degree of error to be expected in probability sampling. The formula for determining sampling error contains three factors: the parameter, the sample size, and the standard error. See Chapter 7.

sampling frame That list or quasi list of units composing a population from which a sample is selected. If the sample is to be representative of the population, it is essential that the sampling frame include all (or nearly all) members of the population. See Chapter 7.

sampling interval The standard distance between elements selected from a population for a sample. See Chapter 7.

sampling ratio The proportion of elements in the population that are selected to be in a sample. See Chapter 7.

sampling unit That element or set of elements considered for selection in some stage of sampling. See Chapter 7.

scale (1) A type of composite measure composed of several items that have a logical or empirical structure among them. Examples of scales include *Bogardus social distance, Guttman, Likert,* and *Thurstone scales.* Contrasted with *index.* See also Chapter 6. (2) One of the less appetizing parts of a fish.

secondary analysis (1) A form of research in which the data collected and processed by one researcher are reanalyzed—often for a different purpose—by another. This is especially appropriate in the case of survey data. Data archives are repositories or libraries for the storage and distribution of data for secondary analysis. (2) Estimating the weight and speed of an opposing team's linebackers.

semantic differential A questionnaire format in which the respondent is asked to rate something in terms of two, opposite adjectives (e.g., rate textbooks as "boring" or "exciting"), using qualifiers such as "very," "somewhat," "neither," "somewhat," and "very" to bridge the distance between the two opposites. See Chapter 6.

semiotics The study of signs and the meanings associated with them. This is commonly associated with content analysis. See Chapter 13.

simple random sampling (1) A type of probability sampling in which the units composing a population are assigned numbers. A set of random numbers is then generated, and the units having those numbers are included in the sample. Although probability theory and the calculations it provides assume this basic sampling method, it's seldom used, for practical reasons. An equivalent alternative is the systematic sample (with a random start). See Chapter 7. (2) A random sample with a low IQ.

snowball sampling (1) A nonprobability sampling method often employed in field research whereby each person interviewed may be asked to suggest additional people for interviewing. See Chapter 7. (2) Picking the icy ones to throw at your methods instructor.

social artifact Any product of social beings or their behavior. Can be a unit of analysis. See Chapter 4.

social indicators Measurements that reflect the quality or nature of social life, such as crime rates, infant mortality rates, number of physicians per 100,000 population, and so forth. Social indicators are often monitored to determine the nature of social change in a society. See Chapter 12.

spurious relationship A coincidental statistical correlation between two variables, shown to be caused by some third variable. For example, there is a positive relationship between the number of fire trucks responding to a fire and the amount of damage done: the more trucks, the more damage. The third variable is the size of the fire. They send lots of fire trucks to a large fire and a lot of damage is done because of the size of the fire. For a little fire, they just send a little fire truck, and not much damage is done because it's a small fire. Sending more fire trucks does not cause more damage. For a given size of fire, in fact, sending more trucks would reduce the amount of damage.

standard deviation   (1) A measure of dispersion around the mean, calculated so that approximately 68 percent of the cases will lie within plus or minus one standard deviation from the mean, 95 percent will lie within plus or minus two standard deviations, and 99.9 percent will lie within three standard deviations. Thus, for example, if the mean age in a group is 30 and the standard deviation is 10, then 68 percent have ages between 20 and 40. The smaller the standard deviation, the more tightly the values are clustered around the mean; if the standard deviation is high, the values are widely spread out. See Chapter 14. (2) Routine rule-breaking.

statistic   The summary description of a variable in a sample, used to estimate a population parameter. See Chapter 7.

stratification   The grouping of the units composing a population into homogeneous groups (or strata) before sampling. This procedure, which may be used in conjunction with *simple random, systematic,* or *cluster sampling,* improves the representativeness of a sample, at least in terms of the variables used for stratification. See Chapter 7.

study population   That aggregation of elements from which a sample is actually selected. See Chapter 7.

systematic sampling   (1) A type of probability sampling in which every *k*th unit in a list is selected for inclusion in the sample—for example, every 25th student in the college directory of students. You compute *k* by dividing the size of the population by the desired sample size; *k* is called the *sampling interval.* Within certain constraints, systematic sampling is a functional equivalent of *simple random sampling* and usually easier to do. Typically, the first unit is selected at random. See Chapter 7. (2) Picking every third one whether it's icy or not. See *snowball sample* (2).

theory   A systematic explanation for the observations that relate to a particular aspect of life: juvenile delinquency, for example, or perhaps social stratification or political revolution.

Thurstone scale   A type of composite measure, constructed in accord with the weights assigned by "judges" to various indicators of some variables. See Chapter 6.

time-series design   A research design that involves measurements made over some period, such as the study of traffic accident rates before and after lowering the speed limit. See Chapter 12.

trend study   A type of longitudinal study in which a given characteristic of some population is monitored over time. An example would be the series of Gallup Polls showing the electorate's preferences for political candidates over the course of a campaign, even though different samples were interviewed at each point. See Chapter 4 and *cohort, longitudinal,* and *panel study.*

typology   (1) The classification (typically nominal) of observations in terms of their attributes on two or more variables. The classification of newspapers as liberal-urban, liberal-rural, conservative-urban, or conservative-rural would be an example. See Chapter 6. (2) Apologizing for your neckwear.

units of analysis   The what or whom being studied. In social science research, the most typical units of analysis are individual people. See Chapter 4.

univariate analysis   The analysis of a single variable, for purposes of description. Frequency distributions, averages, and measures of dispersion would be examples of univariate analysis, as distinguished from *bivariate* and *multivariate analysis.* See Chapter 14.

unobtrusive research   Methods of studying social behavior without affecting it. This includes *content analysis,* analysis of existing statistics, and *historical/comparative research.* See Chapter 11.

validity A term describing a measure that accurately reflects the concept it is intended to measure. For example, your IQ would seem a more valid measure of your intelligence than would the number of hours you spend in the library. Though the ultimate validity of a measure can never be proven, we may agree to its relative validity on the basis of *face validity, criterion-related validity, content validity, construct validity, internal validation,* and *external validation.* This must not be confused with *reliability.* See Chapter 5.

variable-oriented analysis An analysis that describes and/or explains a particular variable. See Chapter 13.

variables Logical groupings of attributes. The variable *gender* is made up of the attributes male and female.

weighting Assigning different weights to cases that were selected into a sample with different probabilities of selection. In the simplest scenario, each case is given a weight equal to the inverse of its probability of selection. When all cases have the same chance of selection, no weighting is necessary. See Chapter 7.

BIBLIOGRAPHY

Abdulhadi, Rabab. 1998. "The Palestinian Women's Autonomous Movement: Emergence, Dynamics, and Challenges." *Gender and Society* 12 (6): 649.

Anderson, Andy B., Alexander Basilevsky, and Derek P. J. Hum. 1983. "Measurement: Theory and Techniques." Pp. 231–87 in *Handbook of Survey Research,* edited by Peter H. Rossi, James D. Wright, and Andy B. Anderson. New York: Academic Press.

Anderson, Walt. 1990. *Reality Isn't What It Used to Be: Theatrical Politics, Ready-to-Wear Religion, Global Myths, Primitive Chic, and Other Wonders of the Postmodern World.* San Francisco: Harper & Row.

Andorka, Rudolf. 1990. "The Importance and the Role of the Second Economy for the Hungarian Economy and Society." *Quarterly Journal of Budapest University of Economic Sciences* 12 (2): 95–113.

Aneshensel, Carol S., Rosina Becerra, Eve Fielder, and Roberleigh Schuler. 1989. "Participation of Mexican American Female Adolescents in a Longitudinal Panel Survey." *Public Opinion Quarterly* 53:548–62.

Asch, Solomon. 1958. "Effects of Group Pressure upon the Modification and Distortion of Judgments." Pp. 174–83 in *Readings in Social Psychology,* 3rd ed., edited by Eleanor E. Maccoby et al. New York: Holt, Rinehart & Winston.

Asher, Ramona M., and Gary Alan Fine. 1991. "Fragile Ties: Sharing Research Relationships with Women Married to Alcoholics." Pp. 196–205 in *Experiencing Fieldwork: An Inside View of Qualitative Research,* edited by William B. Shaffir and Roberta A. Stebbins. Thousand Oaks, CA: Sage.

Auster, Carol J. 1985. "Manuals for Socialization: Examples from Girl Scout Handbooks 1913–1984." *Qualitative Sociology* 8 (4): 359–67.

Babbie, Earl. 1966. "The Third Civilization." *Review of Religious Research,* Winter, pp. 101–21.

——— 1967. "A Religious Profile of Episcopal Churchwomen," *Pacific Churchman,* January, pp. 6–8, 12.

——— 1970. *Science and Morality in Medicine.* Berkeley: University of California Press.

——— 1982. *Social Research for Consumers.* Belmont, CA: Wadsworth.

——— 1985. *You Can Make a Difference.* New York: St. Martin's Press.

——— 1990. *Survey Research Methods.* Belmont, CA: Wadsworth.

——— 1994. *The Sociological Spirit.* Belmont, CA: Wadsworth.

——— 1998. *Observing Ourselves: Essays in Social Research.* Prospect Heights, IL: Waveland Press.

Bailey, William C. 1975. "Murder and Capital Punishment." In *Criminal Law in Action,* edited by William J. Chambliss. New York: Wiley.

Bailey, William C., and Ruth D. Peterson. 1994. "Murder, Capital Punishment, and Deterrence: A Review of the Evidence and an Examination on Police Killings." *Journal of Social Issues* 50:53–74.

Baker, Vern, and Charles Lambert. 1990. "The National Collegiate Athletic Association and the Governance of Higher Education." *Sociological Quarterly* 31 (3): 403–21.

Ball-Rokeach, Sandra J., Joel W. Grube, and Milton Rokeach. 1981. "Roots: The Next Generation—Who Watched and with What Effect." *Public Opinion Quarterly* 45:58–68.

Bart, Pauline, and Linda Frankel. 1986. *The Student Sociologist's Handbook.* Morristown, NJ: General Learning Press.

Bart, Pauline, and Patricia O'Brien. 1985. *Stopping Rape: Successful Survival Strategies.* New York: Pergamon.

Becker, Howard S. 1997. *Tricks of the Trade: How to Think about Your Research While You're Doing It.* Chicago: University of Chicago.

Bednarz, Marlene. 1996. "Push Polls Statement." Report to the AAPORnet listserv, April 5 [Online at http://www.aapor.org/ethics/pushpoll.html].

Belenky, Mary Field, Blythe McVicker Clinchy, Nancy Rule Goldberger, and Jill Mattuck Tarule. 1986. *Women's Ways of Knowing: The Development of Self, Voice, and Mind.* New York: Basic Books.

Bellah, Robert N. 1957. *Tokugawa Religion*. Glencoe, IL: Free Press.

———. 1967. "Research Chronicle: Tokugawa Religion." Pp. 164–85 in *Sociologists at Work*, edited by Phillip E. Hammond. Garden City, NY: Anchor Books.

———. 1970. "Christianity and Symbolic Realism." *Journal for the Scientific Study of Religion* 9:89–96.

———. 1974. "Comment on the Limits of Symbolic Realism." *Journal for the Scientific Study of Religion* 13:487–89.

Benton, J. Edwin, and John L. Daly. 1991. "A Question Order Effect in a Local Government Survey." *Public Opinion Quarterly* 55:640–42.

Berbrier, Mitch. 1998. "'Half the Battle': Cultural Resonance, Framing Processes, and Ethnic Affectations in Contemporary White Separatist Rhetoric." *Social Problems* 45 (4): 431–50.

Berg, Bruce L. 1989. *Qualitative Research Methods for the Social Sciences*. Boston: Allyn & Bacon.

———. 1998. *Qualitative Research Methods for the Social Sciences*. 3rd ed. Boston: Allyn & Bacon.

Berg, Richard, and Peter H. Rossi. 1998. *Thinking about Program Evaluation*. Thousand Oaks, CA: Sage.

Berger, Joseph, Morris Zelditch, Jr., and Bo Anderson, eds. 1989. *Sociological Theories in Progress*. Thousand Oaks, CA: Sage.

Beveridge, W. I. B. 1950. *The Art of Scientific Investigation*. New York: Vintage Books.

Beyer, William H. 1968. *Handbook of Tables for Probability and Statistics*. 2nd ed. Cleveland, OH: Chemical Rubber Company.

Bian, Yanjie. 1994. *Work and Inequality in Urban China*. Albany: State University of New York Press.

Bickman, Leonard, and Debra J. Rog, eds. 1998. *Handbook of Applied Social Research Methods*. Thousand Oaks, CA: Sage.

Bielby, William T., and Denise Bielby. 1999. "Organizational Mediation of Project-Based Labor Markets: Talent Agencies and the Careers of Screenwriters." *American Sociological Review* 64:64–85.

Birchfield, R. W. 1998. *The New Fowler's Modern English Usage*. 3rd ed. New York: Oxford University Press.

Black, Donald. 1970. "Production of Crime Rates." *American Sociological Review* 35 (August): 733–48.

Blair, Johnny, Shanyang Zhao, Barbara Bickart, and Ralph Kuhn. 1995. *Sample Design for Household Telephone Surveys: A Bibliography 1949–1995*. College Park: Survey Research Center, University of Maryland.

Blaunstein, Albert, and Robert Zangrando, eds. 1970. *Civil Rights and the Black American*. New York: Washington Square Press.

Bobo, Lawrence, and Frederick C. Licari. 1989. "Education and Political Tolerance: Testing the Effects of Cognitive Sophistication and Target Group Effect." *Public Opinion Quarterly* 53:285–308.

Bohrnstedt, George W. 1983. "Measurement." Pp. 70–121 in *Handbook of Survey Research*, edited by Peter H. Rossi, James D. Wright, and Andy B. Anderson. New York: Academic Press.

Bolstein, Richard. 1991. "Comparison of the Likelihood to Vote among Preelection Poll Respondents and Nonrespondents." *Public Opinion Quarterly* 55:648–50.

Bottomore, T. B., and Maximilien Rubel, eds. [1843] 1956. *Karl Marx: Selected Writings in Sociology and Social Philosophy*. Translated by T. B. Bottomore. New York: McGraw-Hill.

Bradburn, Norman M., and Seymour Sudman. 1988. *Polls and Surveys: Understanding What They Tell Us*. San Francisco: Jossey-Bass.

Brownlee, K. A. 1975. "A Note on the Effects of Nonresponse on Surveys." *Journal of the American Statistical Association* 52 (227): 29–32.

Burawoy, M., A. Burton, A. A. Ferguson, K. J. Fox, J. Gamson, N. Gartrell, L. Hurst, C. Kurzman, L. Salzinger, J. Schiffman, and S. Ui, eds. 1991. *Ethnography Unbound: Power and Resistance in the Modern Metropolis*. Berkeley: University of California Press.

Campbell, Donald, and Julian Stanley. 1963. *Experimental and Quasi-Experimental Designs for Research*. Chicago: Rand McNally.

Campbell, M. L. 1998. "Institutional Ethnography and Experience as Data." *Qualitative Sociology* 21 (1): 55–73.

Carmines, Edward G., and Richard A. Zeller. 1979. *Reliability and Validity Assessment*. Thousand Oaks, CA: Sage.

Carpini, Michael X. Delli, and Scott Keeter. 1991. "Stability and Change in the U.S. Public's Knowledge of Politics." *Public Opinion Quarterly* 55:583–612.

Carr, C. Lynn. 1998. "Tomboy Resistance and Conformity: Agency in Social Psychological Gender Theory." *Gender and Society* 12 (5): 528–53.

Casley, D. J., and D. A. Lury. 1987. *Data Collection in Developing Countries*. Oxford: Clarendon Press.

Census Bureau *See* U.S. Bureau of the Census

Centers for Disease Control and Prevention, National Center for Health Statistics. 1998. *Monthly Vital Statistics Report* 46 (11, Suppl.): 29.

———. 2000. *Births, Marriages, Divorces, and Deaths: Provisional Data for October 1999*. Washington, DC: U.S. Government Printing Office.

Chen, Huey-Tsyh. 1990. *Theory-Driven Evaluations*. Thousand Oaks, CA: Sage.

Chossudovsky, Michel. 1997. *The Globalization of Poverty: Impacts of IMF and World Bank Reforms.* London: Zed Books.

Clark, Roger, Rachel Lennon, and Leana Morris. 1993. "Of Caldecotts and Kings: Gendered Images in Recent American Children's Books by Black and Non-Black Illustrators." *Gender and Society* 7 (2): 227–45.

Cole, Stephen. 1992. *Making Science: Between Nature and Society.* Cambridge, MA: Harvard University Press.

Collins, G. C., and Timothy B. Blodgett. 1981. "Sexual Harassment . . . Some See It . . . Some Won't." *Harvard Business Review,* March–April, pp. 76–95.

Comstock, Donald. 1980. "Dimensions of Influence in Organizations." *Pacific Sociological Review,* January, pp. 67–84.

Conrad, Clifton F. 1978. "A Grounded Theory of Academic Change." *Sociology of Education* 51:101–12.

Cook, Thomas D., and Donald T. Campbell. 1979. *Quasi-Experimentation: Design and Analysis Issues for Field Settings.* Chicago: Rand McNally.

Cooper, Harris M. 1989. *Integrating Research: A Guide for Literature Reviews.* Thousand Oaks, CA: Sage.

Cooper-Stephenson, Cynthia, and Athanasios Theologides. 1981. "Nutrition in Cancer: Physicians' Knowledge, Opinions, and Educational Needs." *Journal of the American Dietetic Association,* May, pp. 472–76.

Craig, R. Stephen. 1992. "The Effect of Television Day Part on Gender Portrayals in Television Commercials: A Content Analysis." *Sex Roles* 26 (5/6): 197–211.

Crawford, Kent S., Edmund D. Thomas, and Jeffrey J. Fink. 1980. "Pygmalion at Sea: Improving the Work Effectiveness of Low Performers." *Journal of Applied Behavioral Science,* October–December, pp. 482–505.

Cunningham, J. Barton. 1993. *Action Research and Organizational Development.* Westport, CT: Praeger.

Davis, Fred. 1973. "The Martian and the Convert: Ontological Polarities in Social Research." *Urban Life* 2 (3): 333–43.

Davis, James A. 1992. "Changeable Weather in a Cooling Climate atop the Liberal Plateau: Conversion and Replacement in Forty-Two General Social Survey Items, 1972–1989." *Public Opinion Quarterly* 56:261–306.

Death Penalty Information Center. 1999. [Online]. Available: http://www.deathpenaltyinfo.org/deter.html

DeFleur, Lois. 1975. "Biasing Influences on Drug Arrest Records: Implications for Deviance Research." *American Sociological Review* 40:88–103.

Denzin, Norman K., and Yvonna S. Lincoln. 1994. *Handbook of Qualitative Research.* Thousand Oaks, CA: Sage.

DeVault, Marjorie L. 1999. *Liberating Method: Feminism and Social Research.* Philadelphia: Temple University Press.

Dillman, Don A. 1978. *Mail and Telephone Surveys: The Total Design Method.* New York: Wiley.

Donald, Marjorie N. 1960. "Implications of Nonresponse for the Interpretation of Mail Questionnaire Data." *Public Opinion Quarterly* 24:99–114.

Doyle, Sir Arthur Conan. [1891] 1892. "A Scandal in Bohemia." First published in *The Strand,* July 1891. Reprinted in *The Original Illustrated Sherlock Holmes,* pp. 11–25. Secaucus, NJ: Castle.

Durkheim, Emile. [1893] 1964. *The Division of Labor in Society.* Translated by George Simpson. New York: Free Press.

———. [1897] 1951. *Suicide.* Glencoe, IL: Free Press.

Elder, Glen H., Jr., Eliza K. Pavalko, and Elizabeth C. Clipp. 1993. *Working with Archival Data: Studying Lives.* Thousand Oaks, CA: Sage.

Ellison, Christopher G., and Darren E. Sherkat. 1990. "Patterns of Religious Mobility among Black Americans." *Sociological Quarterly* 31 (4): 551–68.

Emerson, Robert M., ed. 1988. *Contemporary Field Research.* Boston: Little, Brown.

Emerson, Robert M., Kerry O. Ferris, and Carol Brooks Gardner. 1998. "On Being Stalked." *Social Problems* 45 (3): 289–314.

Evans, William. 1996. "Computer-Supported Content Analysis: Trends, Tools, and Techniques." *Social Science Computer Review* 14 (3): 269–79.

Fairness and Accuracy in Reporting. 1999. "Action Alert: ABC News Gives up on Accuracy?" September 28 [Online]. Available: http://www.fair.org/activism/stossel-america.html.

Feick, Lawrence F. 1989. "Latent Class Analysis of Survey Questions That Include Don't Know Responses." *Public Opinion Quarterly* 53:525–47.

Festinger, L., H. W. Reicker, and S. Schachter. 1956. *When Prophecy Fails.* Minneapolis: University of Minnesota Press.

Fisher, Ronald A., and Frank Yates. 1974. *Statistical Tables for Biological, Agricultural, and Medical Research.* 6th ed. London: Longman.

"5-Nation Survey Finds Hope for U.N." 1985. *New York Times,* June 26, p. 6.

Ford, David A. 1989. "Preventing and Provoking Wife Battery through Criminal Sanctioning: A Look at the Risks." September, unpublished manuscript.

Ford, David A., and Mary Jean Regoli. 1992. "The Preventive Impacts of Policies for Prosecuting Wife Batterers." Pp. 181–208 in *Domestic Violence: The Changing*

Criminal Justice Response, edited by E. S. Buzawa and C. G. Buzawa. New York: Auburn.

Foschi, Martha, G. Keith Warriner, and Stephen D. Hart. 1985. "Standards, Expectations, and Interpersonal Influence." *Social Psychology Quarterly* 48 (2): 108–17.

Fowler, Floyd J., Jr. 1995. *Improving Survey Questions: Design and Evaluation.* Thousand Oaks, CA: Sage.

Fox, Katherine J. 1991. "The Politics of Prevention: Ethnographers Combat AIDS among Drug Users." Pp. 227–49 in *Ethnography Unbound: Power and Resistance in the Modern Metropolis,* edited by M. Burawoy, A. Burton, A. A. Ferguson, K. J. Fox, J. Gamson, N. Gartrell, L. Hurst, C. Kurzman, L. Salzinger, J. Schiffman, and S. Ui. Berkeley: University of California Press.

Frankfort-Nachmias, Chava, and Anna Leon-Guerrero. 1997. *Social Statistics for a Diverse Society.* Newbury Park, CA: Pine Forge Press.

Gall, John. 1975. *Systemantics: How Systems Work and Especially How They Fail.* New York: Quadrangle.

Gallup, George. 1984. "Where Parents Go Wrong." *San Francisco Chronicle,* December 13, p. 7.

Gallup, George, Jr., Burns Roper, Daniel Yankelovich et al. 1990. "Polls That Made a Difference." *Public Perspective,* May–June, pp. 17–21.

Gamson, William A. 1992. *Talking Politics.* New York: Cambridge University Press.

Gans, Herbert. 1971. The Uses of Poverty: The Poor Pay All. *Social Policy,* July–August, pp. 20–24.

Garant, Carol. 1980. "Stalls in the Therapeutic Process." *American Journal of Nursing,* December, pp. 2166–67.

Garfinkel, H. 1967. *Studies in Ethnomethodology.* Englewood Cliffs, NJ: Prentice-Hall.

Gaventa, J. 1991. "Towards a Knowledge Democracy: Viewpoints on Participatory Research in North America." Pp. 121–31 in *Action and Knowledge: Breaking the Monopoly with Participatory Action-Research,* edited by O. Fals-Borda and M. A. Rahman. New York: Apex Press.

Glaser, Barney, and Anselm Strauss. 1967. *The Discovery of Grounded Theory.* Chicago: Aldine.

Glock, Charles Y., Benjamin B. Ringer, and Earl R. Babbie. 1967. *To Comfort and to Challenge.* Berkeley: University of California Press.

Goffman, Erving. 1961. *Asylums: Essays on the Social Situation of Mental Patients and Other Inmates.* Chicago: Aldine.

———. 1963. *Stigma: Notes on the Management of a Spoiled Identity.* Englewood Cliffs, NJ: Prentice-Hall.

———. 1974. *Frame Analysis.* Cambridge, MA: Harvard University Press.

———. 1979. *Gender Advertisements.* New York: Harper & Row.

Gottlieb, Bruce. 1999. "Cooking the School Books: How U.S. News Cheats in Picking Its 'Best American Colleges.'" Article posted by Slate [Online], August 31. Available: http://www.slate.com/crapshoot/99-08-31/crapshoot.asp

Gould, Julius, and William Kolb. 1964. *A Dictionary of the Social Sciences.* New York: Free Press.

Graham, Laurie, and Richard Hogan. 1990. "Social Class and Tactics: Neighborhood Opposition to Group Homes." *Sociological Quarterly* 31 (4): 513–29.

Greenwood, Peter W., C. Peter Rydell, and Karyn Model. 1996. *Diverting Children from a Life of Crime: Measuring Costs and Benefits.* Santa Monica, CA: Rand Corporation.

Greenwood, Peter W., et al. 1994. *Three Strikes and You're Out: Estimated Benefits and Costs of California's New Mandatory-Sentencing Law.* Santa Monica, CA: Rand Corporation.

Griffith, Alison I. 1995. "Mothering, Schooling, and Children's Development." Pp. 108–21 in *Knowledge, Experience, and Ruling Relations: Studies in the Social Organization of Knowledge,* edited by M. Campbell and A. Manicom. Toronto, Canada: University of Toronto Press.

Grimes, Michael D. 1991. *Class in Twentieth-Century American Sociology: An Analysis of Theories and Measurement Strategies.* New York: Praeger.

Groves, Robert M. 1990. "Theories and Methods of Telephone Surveys." Pp. 221–40 in *Annual Review of Sociology* (vol. 16), edited by W. Richard Scott and Judith Blake. Palo Alto, CA: Annual Reviews.

Gubrium, Jaber F., and James A. Holstein. 1997. *The New Language of Qualitative Method.* New York: Oxford University Press.

Hald, A. 1952. *Statistical Tables and Formulas.* New York: Wiley.

Hamnett, Michael P., Douglas J. Porter, Amarjit Singh, and Krishna Kumar. 1984. *Ethics, Politics, and International Social Science Research.* Honolulu: University of Hawaii Press.

Hedrick, Terry E., Leonard Bickman, and Debra J. Rog. 1993. *Applied Research Design: A Practical Guide.* Thousand Oaks, CA: Sage.

Hempel, Carl G. 1952. "Fundamentals of Concept Formation in Empirical Science." *International Encyclopedia of United Science II,* no. 7.

Heritage, J. 1984. *Garfinkel and Ethnomethodology.* Cambridge: Polity Press.

Heritage, John, and David Greatbatch. 1992. "On the Institutional Character of Institutional Talk." In *Talk at Work,* edited by P. Drew and J. Heritage. Cambridge, England: Cambridge University Press.

Higginbotham, A. Leon, Jr. 1978. *In the Matter of Color: Race and the American Legal Process.* New York: Oxford University Press.

Hilts, Philip J. 1981. "Values of Driving Classes Disputed." *San Francisco Chronicle,* June 25, p. 4.

Hogan, Richard, and Carolyn C. Perrucci. 1998. "Producing and Reproducing Class and Status Differences: Racial and Gender Gaps in U.S. Employment and Retirement Income." *Social Problems* 45 (4): 528–49.

Homan, Roger. 1991. *The Ethics of Social Research.* London: Longman.

Hoover, Kenneth R. 1992. *The Elements of Social Scientific Thinking.* New York: St. Martin's Press.

Horowitz, Irving Louis. 1967. *The Rise and Fall of Project Camelot.* Cambridge, MA: MIT Press.

"How the Poll Was Conducted." 1995. *New York Times,* October 1, p. 15.

Howard, Edward N., and Darlene M. Norman. 1981. "Measuring Public Library Performance." *Library Journal,* February, pp. 305–8.

Howell, Joseph T. 1973. *Hard Living on Clay Street.* Garden City, NY: Doubleday Anchor.

Huberman, A. Michael, and Matthew B. Miles. 1994. "Data Management and Analysis Methods." Pp. 428–44 in *Handbook of Qualitative Research,* edited by Norman K. Denzin and Yvonna S. Lincoln. Thousand Oaks, CA: Sage.

Hughes, Michael. 1980. "The Fruits of Cultivation Analysis: A Reexamination of Some Effects of Television Watching." *Public Opinion Quarterly* 44:287–302.

Humphreys, Laud. 1970. *Tearoom Trade: Impersonal Sex in Public Places.* Chicago: Aldine.

Hunt, Morton. 1985. *Profiles of Social Research: The Scientific Study of Human Interactions.* New York: Basic Books.

Hurst, Leslie. 1991. "Mr. Henry Makes a Deal." Pp. 183–202 in

Ethnography Unbound: Power and Resistance in the Modern Metropolis, edited by M. Burawoy, A. Burton, A. A. Ferguson, K. J. Fox, J. Gamson, N. Gartrell, L. Hurst, C. Kurzman, L. Salzinger, J. Schiffman, and S. Ui. Berkeley: University of California Press.

Hutchby, Ian, and Robin Wooffitt. 1998. *Conversation Analysis: Principles, Practices and Applications.* Cambridge, England: Polity Press.

Indrayan, A., M. J. Wysocki, A. Chawla, R. Kumar, and N. Singh. 1999. "Three-Decade Trend in Human Development Index in India and Its Major States." *Social Indicators Research* 46 (1): 91–120.

Irwin, John, and James Austin. 1997. *It's About Time: America's Imprisonment Binge.* Belmont, CA: Wadsworth.

Iversen, Gudmund R. 1991. *Contextual Analysis.* Thousand Oaks, CA: Sage.

Isaac, Larry W., and Larry J. Griffin. 1989. "A Historicism in Time-Series Analyses of Historical Process: Critique, Redirection, and Illustrations from U.S. Labor History." *American Sociological Association* 54:873–90.

Jackman, Mary R., and Mary Scheuer Senter. 1980. "Images of Social Groups: Categorical or Qualified?" *Public Opinion Quarterly* 44:340–61.

Jacobs, Bruce A., and Jody Miller. 1998. "Crack Dealing, Gender, and Arrest Avoidance." *Social Problems* 45 (4): 550–69.

Jasso, Guillermina. 1988. "Principles of Theoretical Analysis." *Sociological Theory* 6:1–20.

Jensen, Arthur. 1969. "How Much Can We Boost IQ and Scholastic Achievement?" *Harvard Educational Review* 39:273–74.

Jobes, Patrick C., Andra Aldea, Consttantin Cernat, Ioana-Minerva Icolisan, Gabriel Iordache, Sabastian Lazeru, Catalin Stoica, Gheorghe Tibil, and Eugenia Udangiu. 1997. "Shopping as a Social Problem: A Grounded Theoretical Analysis of Experiences among Romanian Shoppers." *Journal of Applied Sociology* 14 (1): 124–46.

Johnson, Jeffrey C. 1990. *Selecting Ethnographic Informants.* Thousand Oaks, CA: Sage.

Johnston, Hank, and David A. Snow. 1998. "Subcultures and the Emergence of the Estonian Nationalist Opposition 1945–1990." *Sociological Perspectives* 41 (3): 473–97.

Jones, James H. 1981. *Bad Blood: The Tuskegee Syphilis Experiments.* New York: Free Press.

Jones, Stephen R. G. 1990. "Worker Independence and Output: The Hawthorne Studies Reevaluated." *American Sociological Review* 55:176–90.

Kalton, Graham. 1983. *Introduction to Survey Sampling.* Thousand Oaks, CA: Sage.

Kaplan, Abraham. 1964. *The Conduct of Inquiry.* San Francisco: Chandler.

Kasl, Stanislav V., Rupert F. Chisolm, and Brenda Eskenazi. 1981. "The Impact of the Accident at Three Mile Island on the Behavior and Well-Being of Nuclear Workers." *American Journal of Public Health,* May, pp. 472–95.

Kasof, Joseph. 1993. "Sex Bias in the Naming of Stimulus Persons." *Psychological Bulletin* 113 (1): 140–63.

Kebede, Alemseghed, and J. David Knottnerus. 1998. "Beyond the Pales of Babylon: The Ideational Components and Social Psychological Foundations of Rastafari." *Sociological Perspectives* 42 (3): 499–517.

Kelle, Udo, ed. 1995. *Computer-Aided Qualitative Data Analysis: Theory, Methods, and Practice.* Thousand Oaks, CA: Sage.

Kendall, Diana. 2001. *Sociology in Our Times.* 3rd ed. Belmont, CA: Wadsworth.

Khayatt, Didi. 1995. "Compulsory Heterosexuality: Schools and Lesbian Students." Pp. 149–63 in *Knowledge, Experience, and Ruling Relations: Studies in the Social Organization of Knowledge,* edited by M. Campbell and A. Manicom. Toronto, Canada: University of Toronto Press.

Kilburn, John C., Jr. 1998. "It's a Matter of Definition: Dilemmas of Service Delivery and Organizational Structure in a Growing Voluntary Organization." *Journal of Applied Sociology* 15 (1): 89–103.

Kinnell, Ann Marie, and Douglas W. Maynard. 1996. "The Delivery and Receipt of Safer Sex Advice in Pretest Counseling Sessions for HIV and AIDS." *Journal of Contemporary Ethnography* 24:405–37.

Kish, Leslie. 1965. *Survey Sampling.* New York: Wiley.

Krueger, Richard A. 1988. *Focus Groups.* Thousand Oaks, CA: Sage.

Kuhn, Thomas. 1970. *The Structure of Scientific Revolutions.* Chicago: University of Chicago Press.

Kvale, Steinar. 1996. *InterViews: An Introduction to Qualitative Research Interviewing.* Thousand Oaks, CA: Sage.

Lazarsfeld, Paul, Ann Pasanella, and Morris Rosenberg, eds. 1972. *Continuities in the Language of Social Research.* New York: Free Press.

Lazarsfeld, Paul F., and Morris Rosenberg, eds. 1955. *The Language of Social Research.* New York: Free Press.

Lee, Raymond. 1993. *Doing Research on Sensitive Topics.* Thousand Oaks, CA: Sage.

Lever, Janet. 1986. "Sex Differences in the Complexity of Children's Play and Games." Pp. 74–89 in *Structure and Process,* edited by Richard J. Peterson and Charlotte A. Vaughan. Belmont, CA: Wadsworth.

Li, Tze-chung. 1990. *Social Science Reference Sources: A Practical Guide.* Westport, CT: Greenwood Press.

Linton, Ralph. 1937. *The Study of Man.* New York: D. Appleton-Century.

Literary Digest. 1936a. "Landon, 1,293,669: Roosevelt, 972,897." October 31, pp. 5–6.

———. 1936b. "What Went Wrong with the Polls?" November 14, pp. 7–8.

Lofland, John, and Lyn H. Lofland. 1995. *Analyzing Social Settings: A Guide to Qualitative Observation and Analysis.* 3rd. ed. Belmont, CA: Wadsworth.

Lynd, Robert S., and Helen M. Lynd. 1929. *Middletown.* New York: Harcourt, Brace.

———. 1937. *Middletown in Transition.* New York: Harcourt, Brace.

Madison, Anna-Marie. 1992. "Primary Inclusion of Culturally Diverse Minority Program Participants in the Evaluation Process." *New Directions for Program Evaluation,* no. 53, pp. 35–43.

Maltz, Michael D. 1998. "Visualizing Homicide: A Research Note." *Journal of Quantitative Criminology* 15 (4): 397–410.

Manning, Peter K., and Betsy Cullum-Swan. 1994. "Narrative, Content, and Semiotic Analysis." Pp. 463–77 in *Handbook of Qualitative Research,* edited by Norman K. Denzin and Yvonna S. Lincoln. Thousand Oaks, CA: Sage.

Marshall, Catherine, and Gretchen B. Rossman. 1995. *Designing Qualitative Research.* Thousand Oaks, CA: Sage.

Martin, David W. 1996. *Doing Psychology Experiments.* 4th ed. Monterey, CA: Brooks/Cole.

Marx, Karl. [1867] 1967. *Capital.* New York: International Publishers.

———. [1880] 1956. *Revue Socialist,* July 5. Reprinted in *Karl Marx: Selected Writings in Sociology and Social Philosophy,* edited by T. B. Bottomore and Maximilien Rubel. New York: McGraw-Hill.

Maxwell, Joseph A. 1996. *Qualitative Research Design.* Thousand Oaks, CA: Sage.

McAlister, Alfred, Cheryl Perry, Joel Killen, Lee Ann Slinkard, and Nathan Maccoby. 1980. "Pilot Study of Smoking, Alcohol, and Drug Abuse Prevention." *American Journal of Public Health,* July, pp. 719–21.

McGrane, Bernard. 1994. *The Un-TV and the 10 mph Car: Experiments in Personal Freedom and Everyday Life.* Fort Bragg, CA: Small Press.

McIver, John P., and Edward G. Carmines. 1981. *Unidimensional Scaling.* Thousand Oaks, CA: Sage.

Menard, Scott. 1991. *Longitudinal Research.* Thousand Oaks, CA: Sage.

Merton, Robert K. 1938. "Social Structure and Anomie." *American Sociological Review* 3:672–82.

Milgram, Stanley. 1963. "Behavioral Study of Obedience." *Journal of Abnormal Social Psychology* 67:371–78.

———. 1965. "Some Conditions of Obedience and Disobedience to Authority." *Human Relations* 18:57–76.

Miller, Delbert. 1991. *Handbook of Research Design and Social Measurement.* Thousand Oaks, CA: Sage.

Mitchell, Richard G., Jr. 1991. "Secrecy and Disclosure in Field Work." Pp. 97–108 in *Experiencing Fieldwork: An Inside View of Qualitative Research,* edited by William B. Shaffir and Robert A. Stebbins. Thousand Oaks, CA: Sage.

Mitofsky, Warren J. 1999. "Miscalls Likely in 2000." *Public Perspective* 10 (5): 42–43.

Morgan, David L., ed. 1993. *Successful Focus Groups: Advancing the State of the Art.* Thousand Oaks, CA: Sage.

Morgan, Lewis H. 1870. *Systems of Consanguinity and Affinity.* Washington, DC: Smithsonian Institution.

Moskowitz, Milt. 1981. "The Drugs That Doctors Order." *San Francisco Chronicle,* May 23, p. 33.

Moynihan, Daniel. 1965. *The Negro Family: The Case for National Action.* Washington, DC: U.S. Government Printing Office.

Myrdal, Gunnar. 1944. *An American Dilemma.* New York: Harper & Row.

Naisbitt, John, and Patricia Aburdene. 1990. *Megatrends 2000: Ten New Directions for the 1990's.* 1st ed. New York: Morrow.

Neuman, W. Lawrence. 1998. "Negotiated Meanings and State Transformation: The Trust Issue in the Progressive Era." *Social Problems* 45 (3): 315–35.

Nicholls, William L., II, Reginald P. Baker, and Jean Martin. 1996. "The Effect of New Data Collection Technology on Survey Data Quality." In *Survey Measurement and Process Quality,* edited by L. Lyberg, P. Biemer, M. Collins, C. Dippo, N. Schwarz, and D. Trewin. New York: Wiley.

O'Neill, Harry W. 1992. "They Can't Subpoena What You Ain't Got." *AAPOR News* 19 (2): 4, 7.

Øyen, Else, ed. 1990. *Comparative Methodology: Theory and Practice in International Social Research.* Thousand Oaks, CA: Sage.

Perinelli, Phillip J. 1986. "Nonsuspecting Public in TV Call-in Polls." *New York Times,* February 14, letter to the editor.

Petersen, Larry R., and Judy L. Maynard. 1981. "Income, Equity, and Wives' Housekeeping Role Expectations." *Pacific Sociological Review,* January, pp. 87–105.

Picou, J. Steven. 1996a. "Compelled Disclosure of Scholarly Research: Some Comments on High Stakes Litigation." *Law and Contemporary Problems* 59 (3): 149–57.

———. 1996b. "Sociology and Compelled Disclosure: Protecting Respondent Confidentiality." *Sociological Spectrum* 16 (3): 207–38.

Picou, J. Steven, Duane A. Gill, and Maurie J. Cohen, eds. 1999. *The Exxon Valdez Disaster: Readings on a Modern Social Problem.* Dubuque, IA: Kendall-Hunt.

Polivka, Anne E., and Jennifer M. Rothgeb. 1993. "Redesigning the CPS Questionnaire." *Monthly Labor Review* 116 (9): 10–28.

Population Communications International. 1996. *International Dateline* [February]. New York: Population Communications International.

Population Reference Bureau. 2000. *2000 World Data Sheet.* Washington, DC: Population Reference Bureau.

Powell, Elwin H. 1958. "Occupation, Status, and Suicide: Toward a Redefinition of Anomie." *American Sociological Review* 23:131–39.

Presser, Stanley, and Johnny Blair. 1994. "Survey Pretesting: Do Different Methods Produce Different Results?" Pp. 73–104 in *Sociological Methodology 1994,* edited by Peter Marsden. San Francisco, CA: Jossey-Bass.

Ragin, Charles C., and Howard S. Becker. 1992. *What Is a Case? Exploring the Foundations of Social Inquiry.* Cambridge, England: Cambridge University Press.

Rasinski, Kenneth A. 1989. "The Effect of Question Wording on Public Support for Government Spending." *Public Opinion Quarterly* 53:388–94.

Ray, William J. 2000. *Methods toward a Science of Behavior and Experience.* 6th ed. Belmont, CA: Wadsworth.

Redfield, Robert. 1941. *The Folk Culture of Yucatan.* Chicago: University of Chicago Press.

Reinharz, Shulamit. 1992. *Feminist Methods in Social Research.* New York: Oxford University Press.

Richlin-Klonsky, Judith, and Ellen Strenski, eds. 1998. *A Guide to Writing Sociology Papers.* New York: St. Martin's Press.

Riecken, Henry W., and Robert F. Boruch. 1974. *Social Experimentation: A Method for Planning and Evaluating Social Intervention.* New York: Academic Press.

Rimer, Sara. 1985. "Poll Sees Landslide for Santa: Of U.S. Children, 87% Believe." *New York Times,* December 24.

Ritzer, George. 1988. *Sociological Theory.* New York: Knopf.

Roethlisberger, F. J., and W. J. Dickson. 1939. *Management and the Worker.* Cambridge, MA: Harvard University Press.

Rogers, Everett M., Peter W. Vaughan, Ramadhan M. A. Swalehe, Nagesh Rao, and Suruchi Sood. 1996. "Effects of an Entertainment-Education Radio Soap Opera on Family Planning and HIV/AIDS Prevention Behavior in Tanzania." Report presented at a technical briefing on the Tanzania Entertainment-Education Project, Rockefeller Foundation, New York, March 27.

Roper, Burns. 1992. ". . . But Will They Give the Poll Its Due?" *AAPOR News* 19 (2): 5–6.

Rosenberg, Morris. 1968. *The Logic of Survey Analysis.* New York: Basic Books.

Rosenthal, Robert, and Lenore Jacobson. 1968. *Pygmalion in the Classroom.* New York: Holt, Rinehart & Winston.

Rossi, Peter H., and Howard E. Freeman. 1996. *Evaluation: A Systematic Approach.* Thousand Oaks, CA: Sage.

Rothman, Ellen K. 1981. "The Written Record." *Journal of Family History,* Spring, pp. 47–56.

Rubin, Herbert J., and Riene S. Rubin. 1995. *Qualitative Interviewing: The Art of Hearing Data.* Thousand Oaks, CA: Sage.

Sacks, Jeffrey J., W. Mark Krushat, and Jeffrey Newman. 1980. "Reliability of the Health Hazard Appraisal." *American Journal of Public Health*, July, pp. 730–32.

Sanders, William B. 1994. *Gangbangs and Drive-bys: Grounded Culture and Juvenile Gang Violence.* New York: Aldine De Gruyter.

Schiflett, Kathy L., and Mary Zey. 1990. "Comparison of Characteristics of Private Product Producing Organizations and Public Service Organizations." *Sociological Quarterly* 31 (4): 569–83.

Schmitt, Frederika E., and Patricia Yancey Martin. 1999. "Unobtrusive Mobilization by an Institutionalized Rape Crisis Center: 'All We Do Comes from Victims.'" *Gender and Society* 13 (3): 364–84.

Schutz, Alfred. 1967. *The Phenomenology of the Social World.* Evanston, IL: Northwestern University Press. 1970. *On Phenomenology and Social Relations.* Chicago: University of Chicago Press.

Shaffir, William B., and Robert A. Stebbins, eds. 1991. *Experiencing Fieldwork: An Inside View of Qualitative Research.* Thousand Oaks, CA: Sage.

Shapiro, Lilian L. 1978. *Teaching Yourself in Libraries.* New York: H. W. Wilson.

Shea, Christopher. 2000. "Don't Talk to the Humans: The Crackdown on Social Science Research." *Lingua Franca,* 10, (6). Available online at http://www.lingua franca.com/print/0009/humans.html

Sheatsley, Paul F. 1983. "Questionnaire Construction and Item Writing." Pp. 195–230 in *Handbook of Survey Research,* edited by Peter H. Rossi, James D. Wright, and Andy B. Anderson. New York: Academic Press.

Shostak, Arthur, ed. 1977. *Our Sociological Eye: Personal Essays on Society and Culture.* Port Washington, NY: Alfred.

Silverman, David. 1993. *Interpreting Qualitative Data.* Thousand Oaks, CA: Sage. 1999. *Doing Qualitative Research: A Practical Handbook.* Thousand Oaks, CA: Sage.

Smith, Andrew E., and G. F. Bishop. 1992. *The Gallup Secret Ballot Experiments: 1944–1988.* Paper presented at the annual conference of the American Association for Public Opinion Research, St. Petersburg, FL, May.

Smith, Dorothy E. 1978. *The Everyday World as Problematic.* Boston: Northeastern University Press.

Smith, Eric R. A. N., and Peverill Squire. 1990. "The Effects of Prestige Names in Question Wording." *Public Opinion Quarterly* 54:97–116.

Smith, Vicki. 1998. "The Fractured World of the Temporary Worker: Power, Participation, and Fragmentation in the Contemporary Workplace." *Social Problems* 45 (4): 411–30.

Snow, David A., and Anderson, Leon. 1987. "Identity Work among the Homeless: The Verbal Construction and Avowal of Personal Identities." *American Journal of Sociology* 92:1336–71.

Spohn, Cassie, and Julie Horney. 1990. "A Case of Unrealistic Expectations: The Impact of Rape Reform Legislation in Illinois." *The Criminal Justice Policy Review* 4 (1): 1–18.

Sprague, Joey. 1997. "Holy Men and Big Guns: The Can[n]on in Social Theory." *Gender and Society* 11 (1): 88–107.

Srole, Leo. 1956. "Social Integration and Certain Corollaries: An Exploratory Study." *American Sociological Review* 21:709–16.

"State Saved $21.7 Billion with Five-Year-Old 'Three Strikes' Law." 1999. *BayInsider*, March 1 [Online]: http://www.bayinsider.com/news/1999/03/01/three_strikes.html

Stearns, Cindy A. 1999. "Breastfeeding and the Good Maternal Body." *Gender and Society* 13 (3): 308.

Steele, Stephen F., and Joyce Miller Iutcovich, eds. 1997. *Directions in Applied Sociology.* Arnold, MD: Society for Applied Sociology.

Strang, David, and James N. Baron. 1990. "Categorical Imperatives: The Structure of Job Titles in California State Agencies." *American Sociological Review* 55:479–95.

Strauss, Anselm, and Juliet Corbin. 1990. *Basics of Qualitative Research: Grounded Theory Procedures and Techniques.* Thousand Oaks, CA: Sage. 1994. "Grounded Theory Methodology: An Overview." Pp. 273–85 in *Handbook of Qualitative Research,* edited by Norman K. Denzin and Yvonna S. Lincoln. Thousand Oaks, CA: Sage.

Strunk, William, Jr., and E. B. White. 1999. *The Elements of Style.* 4th ed. Boston: Allyn & Bacon.

Sudman, Seymour. 1983. "Applied Sampling." Pp. 145–94 in *Handbook of Survey Research,* edited by Peter H. Rossi, James D. Wright, and Andy B. Anderson. New York: Academic Press.

Survey Sampling, Inc. 2000. "Increase Response Rates of Online Sampling." November [Online]. Available:

http://www.worldopinion.com/the_frame/frame4.html

Swafford, Michael. 1992. "Soviet Survey Research: The 1970's vs. the 1990's." *AAPOR News* 19 (3): 3–4.

Swalehe, Ramadhan, Everett M. Rogers, Mark J. Gilboard, Krista Alford, and Rima Montoya. 1995. *A Content Analysis of the Entertainment-Education Radio Soap Opera 'Twende na Wakati' (Let's Go with the Times) in Tanzania.* Arusha, Tanzania: Population Family Life and Education Programme (POFLEP), Ministry of Community Development, Women Affairs, and Children, November 15.

Sweet, Stephen. 1999. "Using a Mock Institutional Review Board to Teach Ethics in Sociological Research." *Teaching Sociology* 27 (January): 55–59.

Takeuchi, David. 1974. "Grass in Hawaii: A Structural Constraints Approach." M.A. thesis, University of Hawaii.

Tan, Alexis S. 1980. "Mass Media Use, Issue Knowledge and Political Involvement." *Public Opinion Quarterly* 44:241–48.

Tandon, Rajesh, and L. Dave Brown. 1981. "Organization-Building for Rural Development: An Experiment in India." *Journal of Applied Behavioral Science,* April–June, pp. 172–89.

Taylor, Humphrey, and George Terhanian. 1999. "Heady Days Are Here Again: Online Polling Is Rapidly Coming of Age." *Public Perspective* 10 (4): 20–23.

Thomas, W. I., and Florian Znaniecki. 1918. *The Polish Peasant in Europe and America.* Chicago: University of Chicago Press.

Tiano, Susan. 1994. *Patriarchy on the Line: Labor, Gender, and Ideology in the Mexican Maquila Industry.* Philadelphia, PA: Temple University Press.

Tourangeau, Roger, Kenneth A. Rasinski, Norman Bradburn, and Roy D'Andrade. 1989. "Carryover Effects in Attitude Surveys." *Public Opinion Quarterly* 53: 495–524.

Trials of War Criminals before the Nuremberg Military Tribunals under Control Council Law No. 10. Nuremberg, October 1946–April 1949. 1949–1953. Washington, DC: U.S. Government Printing Office. Also available online at http://www.ushmm.org/research/doctors/indiptx.htm

Tuckel, Peter S., and Barry M. Feinberg. 1991. "The Answering Machine Poses Many Questions for Telephone Survey Researchers. *Public Opinion Quarterly* 55:200–217.

Turk, Theresa Guminski. 1980. "Hospital Support: Urban Correlates of Allocation Based on Organizational Prestige." *Pacific Sociological Review,* July, pp. 315–32.

Turner, Jonathan H., ed. 1989. *Theory Building in Sociology: Assessing Theoretical Cumulation.* Thousand Oaks, CA: Sage.

Turner, Stephen Park, and Jonathan H. Turner. 1990. *The Impossible Science: An Institutional Analysis of American Sociology.* Thousand Oaks, CA: Sage.

United Nations. 1995. "Human Development Report 1995." New York: United Nations Development Program. [Summarized in Population Communication International. 1996. *International Dateline,* February, pp. 1–4.]

U.S. Bureau of the Census. 1978. Current Population Reports, Series P-23, No. 100. *Statistical Portrait of Women in the United States: 1978.* Washington, DC: U.S. Government Printing Office.

———. 1987. Current Population Reports, Series P-70, No. 10, *Male-Female Differences in Work Experience, Occupation, and Earning, 1984.* Washington, DC: U.S. Government Printing Office.

———. 1996a. *Statistical Abstract of the United States, 1995.* CD-ROM CD-SA-95, April.

———. 1996b. *Statistical Abstract of the United States, 1996, National Data Book and Guide to Sources.* Washington, DC: U.S. Government Printing Office.

———. 1999. *Statistical Abstract of the United States.* Washington, DC: U.S. Government Printing Office.

U.S. Department of Health and Human Services. 1992. *Survey Measurement of Drug Use.* Washington, DC: U.S. Government Printing Office.

U.S. News and World Report. 1999. "America's Best Colleges." August 30.

Uwe, Flick. 1998. *An Introduction to Qualitative Research.* Thousand Oaks, CA: Sage.

Veroff, Joseph, Shirley Hatchett, and Elizabeth Douvan. 1992. "Consequences of Participating in a Longitudinal Study of Marriage." *Public Opinion Quarterly* 56:325–27.

Votaw, Carmen Delgado. 1979. *Women's Rights in the United States.* United States Commission on Civil Rights, Inter-American Commission on Women. Washington, DC: Clearinghouse Publications.

Walker, Jeffery T. 1994. "Fax Machines and Social Surveys: Teaching an Old Dog New Tricks." *Journal of Quantitative Criminology* 10 (2): 181–88.

Walker Research. 1988. *Industry Image Study.* 8th ed. Indianapolis, IN: Walker Research.

Wallace, Walter. 1971. *The Logic of Science in Sociology.* Chicago: Aldine.

Ward, Lester. 1906. *Applied Sociology.* Boston: Ginn.

Warner, W. Lloyd. 1949. *Democracy in Jonesville.* New York: Harper.

Webb, Eugene J., Donald T. Campbell, Richard D. Schwartz, Lee Sechrest, and Janet Belew Grove. 1981. *Nonreactive Measures in the Social Sciences.* Boston: Houghton Mifflin.

Weber, Max. [1905] 1958. *The Protestant Ethic and the Spirit of Capitalism.* Translated by Talcott Parsons. New York: Scribner.

　　 [1925] 1946. "Science as a Vocation." Pp. 129–56 in *From Max Weber: Essays in Sociology,* edited and translated by Hans Gerth and C. Wright Mills. New York: Oxford University Press.

　　 [1934] 1951. *The Religion of China.* Translated by Hans H. Gerth. New York: Free Press.

　　 [1934] 1952. *Ancient Judaism.* Translated by Hans H. Gerth and Don Martindale. New York: Free Press.

　　 [1934] 1958. *The Religion of India.* Translated by Hans H. Gerth and Don Martindale. New York: Free Press.

Weber, Robert Philip. 1990. *Basic Content Analysis.* Thousand Oaks, CA: Sage.

Weiss, Carol. 1972. *Evaluation Research.* Englewood Cliffs, NJ: Prentice-Hall.

Weitzman, Lenore J., Deborah Eifler, Elizabeth Hokada, and Catherine Ross. 1972. "Sex-Role Socialization in Picture Books for Preschool Children." *American Journal of Sociology* 77:1125–50.

Wellman, David. 1995. *The Union Makes Us Stronger: Radical Unionism on the San Francisco Waterfront.* Cambridge, MA: Cambridge University Press.

Wharton, Amy S., and James N. Baron. 1987. "So Happy Together? The Impact of Gender Segregation on Men at Work." *American Sociological Review* 52:574–87.

White, William S. 1997. Communication to APP-SOC listserv (appsoc@miagra.ucs.edu) from wwite @jaguar1.usouthal.edu, October 11 [Online].

Whyte, W. F. 1943. *Street Corner Society.* Chicago: University of Chicago Press.

Whyte, W. F., D. J. Greenwood, and P. Lazes. 1991. "Participatory Action Research: Through Practice to Science in Social Research." Pp. 19–55 in *Participatory Action Research,* edited by W. F. Whyte. New York: Sage.

Wieder, D. L. 1988. *Language and Social Reality: The Case of Telling the Convict Code.* Landman, MD: University Press of America.

Williams, Robin M., Jr. 1989. "The American Soldier: An Assessment, Several Wars Later." *Public Opinion Quarterly* 53:155–74.

Wilson, Camilo. 1999. Private email, September 8.

Worcester, Robert. In press. *WAPOR Newsletter,* Winter 2001.

Yammarino, Francis J., Steven J. Skinner, and Terry L. Childers. 1991. "Understanding Mail Survey Response Behavior: A Meta-Analysis." *Public Opinion Quarterly* 55:613–39.

Yinger, J. Milton, et al. 1977. *Middle Start: An Experiment in the Educational Enrichment of Young Adolescents.* London: Cambridge University Press.

INDEX

461